Lecture Notes in Computer Science

Lecture Notes in Bioinformatics   **13920**

The series Lecture Notes in Bioinformatics (LNBI) was established in 2003 as a topical subseries of LNCS devoted to bioinformatics and computational biology.

The series publishes state-of-the-art research results at a high level. As with the LNCS mother series, the mission of the series is to serve the international R & D community by providing an invaluable service, mainly focused on the publication of conference and workshop proceedings and postproceedings.

Ignacio Rojas · Olga Valenzuela ·
Fernando Rojas Ruiz · Luis Javier Herrera ·
Francisco Ortuño
Editors

Bioinformatics and Biomedical Engineering

10th International Work-Conference, IWBBIO 2023
Meloneras, Gran Canaria, Spain, July 12–14, 2023
Proceedings, Part II

Editors
Ignacio Rojas 🆔
University of Granada
Granada, Spain

Olga Valenzuela 🆔
University of Granada
Granada, Spain

Fernando Rojas Ruiz 🆔
University of Granada
Granada, Spain

Luis Javier Herrera 🆔
University of Granada
Granada, Spain

Francisco Ortuño 🆔
University of Granada
Granada, Spain

ISSN 0302-9743   ISSN 1611-3349 (electronic)
Lecture Notes in Bioinformatics
ISBN 978-3-031-34959-1   ISBN 978-3-031-34960-7 (eBook)
https://doi.org/10.1007/978-3-031-34960-7

LNCS Sublibrary: SL8 – Bioinformatics

This Springer imprint is published by the registered company Springer Nature Switzerland AG
The registered company address is: Gewerbestrasse 11, 6330 Cham, Switzerland

Preface

We are proud to present the final set of accepted full papers for the 10th "International Work-Conference on Bioinformatics and Biomedical Engineering" (IWBBIO 2023), held in Gran Canaria, Spain, during July 12–14, 2023.

IWBBIO 2023 provided a discussion forum for scientists, engineers, educators, and students about the latest ideas and realizations in the foundations, theory, models, and applications for interdisciplinary and multidisciplinary research encompassing disciplines of computer science, mathematics, statistics, biology, bioinformatics, and biomedicine.

The aim of IWBBIO 2023 was to create a friendly environment that could lead to the establishment or strengthening of scientific collaborations and exchanges among attendees, and therefore IWBBIO 2023 solicited high-quality original research papers (including significant work in progress) on any aspect of bioinformatics, biomedicine, and biomedical engineering.

New computational techniques and methods in machine learning; data mining; text analysis; pattern recognition; data integration; genomics and evolution; next-generation sequencing data; protein and RNA structure; protein function and proteomics; medical informatics and translational bioinformatics; computational systems biology; modelling and simulation; and their application in the life science domain, biomedicine, and biomedical engineering were especially encouraged. The list of topics in the call for papers also evolved, resulting in the following list for the present edition:

1. **Computational proteomics**. Analysis of protein-protein interactions. Protein structure modelling. Analysis of protein functionality. Quantitative proteomics and PTMs. Clinical proteomics. Protein annotation. Data mining in proteomics.
2. **Next-generation sequencing and sequence analysis**. De novo sequencing, resequencing, and assembly. Expression estimation. Alternative splicing discovery. Pathway analysis. Chip-seq and RNA-Seq analysis. Metagenomics. SNPs prediction.
3. **High performance in bioinformatics**. Parallelization for biomedical analysis. Biomedical and biological databases. Data mining and biological text processing. Large-scale biomedical data integration. Biological and medical ontologies. Novel architecture and technologies (GPU, P2P, Grid, etc.) for bioinformatics.
4. **Biomedicine**. Biomedical computing. Personalized medicine. Nanomedicine. Medical education. Collaborative medicine. Biomedical signal analysis. Biomedicine in industry and society. Electrotherapy and radiotherapy.
5. **Biomedical engineering**. E-Computer-assisted surgery. Therapeutic engineering. Interactive 3D modelling. Clinical engineering. Telemedicine. Biosensors and data acquisition. Intelligent instrumentation. Patient monitoring. Biomedical robotics. Bio-nanotechnology. Genetic engineering.
6. **Computational systems for modelling biological processes**. Inference of biological networks. Machine learning in bioinformatics. Classification for biomedical

data. Microarray data analysis. Simulation and visualization of biological systems. Molecular evolution and phylogenetic modelling.
7. **Healthcare and diseases**. Computational support for clinical decisions. Image visualization and signal analysis. Disease control and diagnosis. Genome-phenome analysis. Biomarker identification. Drug design. Computational immunology.
8. **E-Health**. E-Health technology and devices. E-Health information processing. Telemedicine/E-Health application and services. Medical image processing. Video techniques for medical images. Integration of classical medicine and E-Health.
9. **COVID-19**. A special session was organized in which different aspects, fields of application, and technologies that have been applied against COVID-19 were analyzed.

After a careful peer review and evaluation process (each submission was reviewed by at least 2, and on the average 3.2, Program Committee members or additional reviewers), 79 papers were accepted, according to the recommendations of reviewers and the authors' preferences, to be included in the LNBI proceedings.

During IWBBIO 2023 several Special Sessions were carried out. Special Sessions are a very useful tool in order to complement the regular program with new and emerging topics of particular interest for the participating community. Special Sessions that emphasized multidisciplinary and transversal aspects as well as cutting-edge topics were especially encouraged and welcomed, and in this edition of IWBBIO 2023 the following were received:

– **SS1. High-Throughput Genomics: Bioinformatic Tools and Medical Applications.**
Genomics is concerned with the sequencing and analysis of an organism's genome. It is involved in the understanding of how every single gene can affect the entire genome. This goal is mainly afforded using the current, cost-effective, high-throughput sequencing technologies. These technologies produce a huge amount of data that usually require high-performance computing solutions and opens new ways for the study of genomics, but also transcriptomics, gene expression, and systems biology, among others. The continuous improvements and broader applications of sequencing technologies is producing a continuous new demand for improved high-throughput bioinformatics tools.

In this context, the generation, integration, and interpretation of genetic and genomic data is driving a new era of healthcare and patient management. Medical genomics (or genomic medicine) is an emerging discipline that involves the use of genomic information about a patient as part of clinical care with diagnostic or therapeutic purposes to improve the health outcomes. Moreover, it can be considered a subset of precision medicine that has an impact in the fields of oncology, pharmacology, rare and undiagnosed diseases, and infectious diseases. The aim of this Special Session was to bring together researchers in medicine, genomics, and bioinformatics to translate medical genomics research into new diagnostic, therapeutic, and preventive medical approaches. Therefore, we invited authors to submit original research, new tools or pipelines, as well as update and review articles on relevant topics, such as (but not limited to):

- Tools for data pre-processing (quality control and filtering)
- Tools for sequence mapping
- Tools for the comparison of two read libraries without an external reference
- Tools for genomic variants (such as variant calling or variant annotation)
- Tools for functional annotation: identification of domains, orthologues, genetic markers, and controlled vocabulary (GO, KEGG, InterPro,etc.)
- Tools for gene expression studies and tools for Chip-Seq data
- Integrative workflows and pipelines

Organizers: **M. Gonzalo Claros**, *University of Málaga, Spain.*
Javier Pérez Florido, *Fundación Progreso y Salud, Spain.*
Francisco M. Ortuño, *University of Granada, Spain.*

- **SS2. Feature Selection, Extraction, and Data Mining in Bioinformatics: Approaches, Methods, and Adaptations.**
Various applications of bioinformatics, system biology, and biophysics measurement data mining require proper, accurate, and precise preprocessing or data transformation before the analysis itself. Here, the most important issues are covered by the feature selection and extraction techniques to translate the raw data into the inputs for the machine learning and multi-variate statistic algorithms. This is a complex task; it requires reducing the problem dimensionality, removal of redundant of irrelevant data, without affecting significantly the present information. The methods and approaches are often conditioned by the physical properties of the measurement process, mathematically congruent description and parameterization, as well as biological aspects of specific tasks. With the current increase of artificial intelligence methods adoption into bioinformatics problem solutions, it is necessary to understand the conditionality of such algorithms, to choose and use the correct approach and avoid misinterpretations, artefacts, and aliasing affects. The adoption often uses already existing knowledge from different fields, and direct application might underestimate the required conditions and corrupt the analysis results. This special session saw discussion on the multidisciplinary overlaps, development, implementation, and adoption of feature and selection methods for datasets of biological origin in order to set up a pipeline from the measurement design through signal processing to knowledge obtaining. The topic covered theoretical questions, practical examples, and results verifications.

Organizer: **Jan Urban**, *University of South Bohemia, Czech Republic*

- **SS3. Sensor-Based Ambient Assisted Living Systems and Medical Applications.**
Many advancements in medical technology are possible, including developing intelligent systems for the treatment, diagnosis, and prevention of many healthcare issues. The field of surgery is now experiencing an increase in intelligent systems. Medical rules should be in place during the creation of these technologies to ensure market acceptance. Hospital systems can be connected to mobile devices or other specialized equipment, boosting patient monitoring.

Technology-based tools may monitor, treat, and reduce several health-related issues. The system and concepts discussed here can use sensors found in mobile devices and other sensors found in intelligent environments, and sensors utilized with other equipment. The advancements in this area right now will be incredibly beneficial for treating various ailments.

The main points of this topic are the presentation of cutting-edge, active projects, conceptual definitions of devices, systems, services, and sensor-based advanced healthcare efforts.

The special session covered, but was not limited to:

- Assistive technology and adaptive sensing systems
- Diagnosis and treatment with mobile sensing systems
- Healthcare self-management systems
- M-Heath, eHealth, and telemedicine systems
- Body-wearer/implemented sensing devices
- Sensing vital medical metrics
- Sensing for persons with limited capabilities
- Motion and path-tracking medical systems
- Artificial intelligence with sensing data
- Patient empowerment with technological equipment
- Virtual and augmented reality in medical systems
- Mobile systems usability and accessibility
- Medical regulations in mobile systems
- Medical regulations and privacy

Organizer: **Ivan Miguel Pires**, *Universidade da Beira Interior, Portugal.*
Norberto Jorge Gonçalves, *Universidade de Trás-os-Montes e Alto Douro, Portugal.*
Paulo Jorge Coelho, *Polytechnic Institute of Leiria, Portugal.*

– **SS4. Analysis of Molecular Dynamics Data in Proteomics.**
Molecular dynamics (MD) simulations have become a key method for exploring the dynamic behavior of macromolecules and studying their structure-to-function relationships. In proteomics, they are crucial for extending the understanding of several processes related to protein function, e.g., protein conformational diversity, binding pocket analysis, protein folding, ligand binding, and its influence on signaling, to name a few. Nevertheless, the investigation of the large amounts of information generated by MD simulations is a far from trivial challenge.

This special session addressed new research on computational techniques and machine learning (ML) algorithms that can provide efficient solutions for the diverse problems that entail the analysis of MD simulation data in their different areas of application.

Topics of interest included, but were not limited to:

- Sampling techniques in MD simulations
- Potential energy surfaces
- Detection of rare events
- Transition pathways analytics
- Visualization techniques for MD
- Feature representations for molecular structures

- Deep learning architectures for MD simulations
- Generative models for MD

Organizer: **Caroline König**, *Universitat Politècnica de Catalunya, Spain.*
Alfredo Vellido, *Universitat Politècnica de Catalunya.*

- **SS5. Image Visualization and Signal Analysis.**
Signal processing focuses on analysing, modifying and synthesizing signals such as sound, images, and biological measurements. Signal processing techniques can be used to improve transmission, storage efficiency, and subjective quality and also to emphasize or detect components of interest in a measured signal and are a very relevant topic in medicine.

Any signal transduced from a biological or medical source could be called a biosignal. The signal source could be at the molecular level, cell level, or a systemic or organ level. A wide variety of such signals are commonly encountered in the clinic, research laboratory, and sometimes even at home. Examples include the electrocardiogram (ECG), or electrical activity from the heart; speech signals; the electroencephalogram (EEG), or electrical activity from the brain; evoked potentials (EPs, i.e., auditory, visual, somatosensory, etc.), or electrical responses of the brain to specific peripheral stimulation; the electroneurogram, or field potentials from local regions in the brain; action potential signals from individual neurons or heart cells; the electromyogram (EMG), or electrical activity from the muscle; the electroretinogram from the eye; and so on.

From the other side, medical imaging is the technique and process of creating visual representations of the interior of a body for clinical analysis and medical intervention, as well as visual representation of the function of some organs or tissues (physiology). Medical imaging seeks to reveal internal structures hidden by the skin and bones, as well as to diagnose and treat disease. Medical imaging also establishes a database of normal anatomy and physiology to make it possible to identify abnormalities. Although imaging of removed organs and tissues can be performed for medical reasons, such procedures are usually considered part of pathology instead of medical imaging.

Organizer: **L. Wang**, *University of California, San Diego, USA.*

- **SS6. Computational Approaches for Drug Design and Personalized Medicine.**
With continuous advancements of biomedical instruments and the associated ability to collect diverse types of valuable biological data, numerous recent research studies have focused on how to best extract useful information from the big biomedical data currently available. While drug design has been one of the most essential areas of biomedical research, the drug design process for the most part has not fully benefited from the recent explosive growth of biological data and bioinformatics algorithms. With the incredible overhead associated with the traditional drug design process in terms of time and cost, new alternative methods, possibly based on computational approaches, are very much needed to propose innovative ways to propose effective

drugs and new treatment options. Employing advanced computational tools for drug design and precision treatments has been the focus of many research studies in recent years. For example, drug repurposing has gained significant attention from biomedical researchers and pharmaceutical companies as an exciting new alternative for drug discovery that benefits from computational approaches. This new development also promises to transform healthcare to focus more on individualized treatments, precision medicine, and lower risks of harmful side effects. Other alternative drug design approaches that are based on analytical tools include the use of medicinal natural plants and herbs as well as using genetic data for developing multi-target drugs.

Organizer: **Hesham H. Ali**, *University of Nebraska at Omaha.*

It is important to note that for the sake of consistency and readability of the book, the presented papers are classified under 15 chapters. The organization of the papers is in two volumes arranged basically following the topics list included in the call for papers. The first volume (LNBI 13919), entitled "Bioinformatics and Biomedical Engineering. Part I", is divided into nine main parts and includes the contributions on:

1. Analysis Of Molecular Dynamics Data In Proteomics
2. Bioinformatics
3. Biomarker Identification
4. Biomedical Computing
5. Biomedical Engineering
6. Biomedical Signal Analysis
7. Computational Support for Clinical Decisions
8. COVID-19 Advances in Bioinformatics and Biomedicine

The second volume (LNBI 13920), entitled "Bioinformatics and Biomedical Engineering. Part II", is divided into seven main parts and includes the contributions on:

1. Feature Selection, Extraction, and Data Mining in Bioinformatics
2. Genome-Phenome Analysis
3. Healthcare and Diseases
4. High-Throughput Genomics: Bioinformatic Tools and Medical Applications
5. Image Visualization And Signal Analysis
6. Machine Learning in Bioinformatics and Biomedicine
7. Medical Image Processing
8. Next-Generation Sequencing and Sequence Analysis
9. Sensor-Based Ambient Assisted Living Systems And Medical Applications

This 10th edition of IWBBIO was organized by the University of Granada. We wish to thank our main sponsor as well as the following institutions: Department of Computer Engineering, Automation and Robotics, CITIC-UGR from the University of Granada, International Society for Computational Biology (ISCB) for their support and grants. We also wish to thank the editors in charge of various international journals for their interest in editing special issues from a selection of the best papers of IWBBIO 2023. At IWBBIO 2023, there were two awards (best contribution award and best contribution from student participant) sponsored by the editorial office of Genes, an MDPI journal.

We would also like to express our gratitude to the members of the different committees for their support, collaboration, and good work. We especially thank the Program Committee, the reviewers, and Special Session organizers. We also want to express our gratitude to the EasyChair platform. Finally, we wish to thank the staff of Springer, for their continuous support and cooperation.

April 2023

Ignacio Rojas
Olga Valenzuela
Fernando Rojas
Luis Javier Herrera
Francisco Ortuño

Organization

Conference Chairs

Ignacio Rojas	University of Granada, Spain
Olga Valenzuela	University of Granada, Spain
Fernando Rojas	University of Granada, Spain
Luis Javier Herrera	University of Granada, Spain
Francisco Ortuño	University of Granada, Spain

Steering Committee

Miguel A. Andrade	University of Mainz, Germany
Hesham H. Ali	University of Nebraska at Omaha, USA
Oresti Baños	University of Granada, Spain
Alfredo Benso	Politecnico di Torino, Italy
Larbi Boubchir	University of Paris 8, France
Giorgio Buttazzo	Superior School Sant'Anna, Italy
Gabriel Caffarena	University CEU San Pablo, Spain
Mario Cannataro	University Magna Graecia of Catanzaro, Italy
Jose María Carazo	Spanish National Center for Biotechnology (CNB), Spain
Jose M. Cecilia	Universidad Católica San Antonio de Murcia (UCAM), Spain
M. Gonzalo Claros	University of Malaga, Spain
Joaquin Dopazo	Fundacion Progreso y Salud, Spain
Werner Dubitzky	University of Ulster, UK
Afshin Fassihi	Universidad Católica San Antonio de Murcia (UCAM), Spain
Jean-Fred Fontaine	University of Mainz, Germany
Humberto Gonzalez	University of the Basque Country (UPV/EHU), Spain
Concettina Guerra	Georgia Tech, USA
Roderic Guigo	Pompeu Fabra University, Spain
Andy Jenkinson	Karolinska Institute, Sweden
Craig E. Kapfer	Reutlingen University, Germany
Narsis Aftab Kiani	European Bioinformatics Institute (EBI), UK
Natividad Martinez	Reutlingen University, Germany
Marco Masseroli	Politecnico di Milano, Italy

Federico Moran	Complutense University of Madrid, Spain
Cristian R. Munteanu	University of A Coruña, Spain
Jorge A. Naranjo	New York University Abu Dhabi, UAE
Michael Ng	Hong Kong Baptist University, China
Jose L. Oliver	University of Granada, Spain
Juan Antonio Ortega	University of Seville, Spain
Fernando Rojas	University of Granada, Spain
Alejandro Pazos	University of A Coruña, Spain
Javier Perez Florido	Fundación Progreso y Salud, Spain
Violeta I. Pérez Nueno	Inria Nancy Grand Est, LORIA, France
Horacio Pérez-Sánchez	Universidad Católica San Antonio de Murcia (UCAM), Spain
Alberto Policriti	Università di Udine, Italy
Omer F. Rana	Cardiff University, UK
M. Francesca Romano	Superior School Sant'Anna, Italy
Yvan Saeys	Ghent University, Belgium
Vicky Schneider	The Genome Analysis Centre (TGAC), UK
Ralf Seepold	HTWG Konstanz, Germany
Mohammad Soruri	University of Birjand, Iran
Yoshiyuki Suzuki	Tokyo Metropolitan Institute of Medical Science, Japan
Shusaku Tsumoto	Shimane University, Japan
Renato Umeton	Dana-Farber Cancer Institute and Massachusetts Institute of Technology, USA
Jan Urban	University of South Bohemia, Czech Republic
Alfredo Vellido	Polytechnic University of Catalunya, Spain
Wolfgang Wurst	GSF National Research Center of Environment and Health, Germany

Program Committee and Additional Reviewers

Heba Afify	MTI University, Egypt
Fares Al-Shargie	American University of Sharjah, United Arab Emirates
Jesus Alcala-Fdez	University of Granada, Spain
Hesham Ali	University of Nebraska at Omaha, USA
Georgios Anagnostopoulos	Florida Institute of Technology, USA
Cecilio Angulo	Universitat Politècnica de Catalunya, Spain
Masanori Arita	National Institute of Genetics, Japan
Gajendra Kumar Azad	Patna University, India
Hazem Bahig	Ain Sham University, Egypt
Oresti Banos	University of Granada, Spain

Ugo Bastolla — Centro de Biologia Molecular "Severo Ochoa", Spain

Payam Behzadi — Islamic Azad University, Iran

Sid Ahmed Benabderrahmane — University of Edinburgh, UK

Alfredo Benso — Politecnico di Torino, Italy

Anna Bernasconi — Politecnico di Milano, Italy

Mahua Bhattacharya — Indian Institute of Information Technology and Management, Gwalior, India

Paola Bonizzoni — Università di Milano-Bicocca, Italy

Larbi Boubchir — University of Paris 8, France

Gabriel Caffarena — University CEU San Pablo, Spain

Mario Cannataro — University Magna Graecia of Catanzaro, Italy

Jose Maria Carazo — National Center for Biotechnology, CNB-CSIC, Spain

Francisco Carrillo Pérez — Universidad de Granada, Spain

Claudia Cava — IBFM-CNR, Italy

Francisco Cavas-Martínez — Technical University of Cartagena, Spain

Ting-Fung Chan — Chinese University of Hong Kong, China

Kun-Mao Chao — National Taiwan University, Taiwan

Chuming Chen — University of Delaware, USA

Javier Cifuentes Faura — University of Murcia, Spain

M. Gonzalo Claros — Universidad de Málaga, Spain

Darrell Conklin — University of the Basque Country, Spain

Alexandre G. De Brevern — Université Denis Diderot Paris 7 and INSERM, France

Javier De Las Rivas — CiC-IBMCC, CSIC/USAL/IBSAL, Spain

Paolo Di Giamberardino — Sapienza University of Rome, Italy

Maria Natalia Dias Soeiro Cordeiro — University of Porto, Portugal

Marko Djordjevic — University of Belgrade, Serbia

Joaquin Dopazo — Fundación Progreso y Salud, Spain

Mohammed Elmogy — Mansoura University, Egypt

Gionata Fragomeni — Magna Graecia University, Italy

Pugalenthi Ganesan — Bharathidasan University, India

Hassan Ghazal — Mohammed I University, Morocco

Razvan Ghinea — University of Granada, Spain

Luis Gonzalez-Abril — University of Seville, Spain

Humberto Gonzalez-Diaz — UPV/EHU, IKERBASQUE, Spain

Morihiro Hayashida — National Institute of Technology, Matsue College, Japan

Luis Herrera — University of Granada, Spain

Ralf Hofestaedt — Bielefeld University, Germany

Jingshan Huang	University of South Alabama, USA
Xingpeng Jiang	Central China Normal University, China
Hamed Khodadadi	Khomeinishahr Branch, Islamic Azad University, Iran
Narsis Kiani	Karolinska Institute, Sweden
Dongchul Kim	University of Texas Rio Grande Valley, USA
Tomas Koutny	University of West Bohemia, Czech Republic
Konstantin Krutovsky	Georg-August-University of Göttingen, Germany
José L. Lavín	Neiker Tecnalia, Spain
Chen Li	Monash University, Australia
Hua Li	Bio-Rad, USA
Li Liao	University of Delaware, USA
Zhi-Ping Liu	Shandong University, China
Francisco Martínez-Álvarez	Universidad Pablo de Olavide, Spain
Marco Masseroli	Politecnico di Milano, Italy
Roderick Melnik	Wilfrid Laurier University, Canada
Enrique Muro	Johannes Gutenberg University, Germany
Kenta Nakai	Institute of Medical Science, University of Tokyo, Japan
Isabel Nepomuceno	University of Seville, Spain
Dang Ngoc Hoang Thanh	University of Economics Ho Chi Minh City, Vietnam
Anja Nohe	University of Delaware, USA
José Luis Oliveira	University of Aveiro, Portugal
Yuriy Orlov	Institute of Cytology and Genetics, Russia
Juan Antonio Ortega	University of Seville, Spain
Andres Ortiz	University of Malaga, Spain
Francisco Manuel Ortuño	Fundación Progreso y Salud, Spain
Motonori Ota	Nagoya University, Japan
Mehmet Akif Ozdemir	Izmir Katip Celebi University, Turkey
Joel P. Arrais	University of Coimbra, Portugal
Paolo Paradisi	ISTI-CNR, Italy
Taesung Park	Seoul National University, South Korea
Alejandro Pazos	University of A Coruña, Spain
Antonio Pinti	I3MTO Orléans, France
Yuri Pirola	Univ. degli Studi di Milano-Bicocca, Italy
Joanna Polanska	Silesian University of Technology, Poland
Alberto Policriti	University of Udine, Italy
Hector Pomares	University of Granada, Spain
María M Pérez	University of Granada, Spain
Julietta V. Rau	Istituto di Struttura della Materia, Italy
Khalid Raza	Jamia Millia Islamia, India

Contents – Part II

High-Throughput Genomics: Bioinformatic Tools and Medical Applications

Image Visualization and Signal Analysis

Machine Learning in Bioinformatics and Biomedicine

Medical Image Processing

Next Generation Sequencing and Sequence Analysis

Sensor-Based Ambient Assisted Living Systems and Medical Applications

Contents – Part I

Biomedical Computing

Biomedical Engineering

Biomedical Signal Analysis

Computational Proteomics

Computational Support for Clinical Decisions

Feature Selection, Extraction, and Data Mining in Bioinformatics

Agent Based Modeling of Fish Shoal Behavior

Pavla Urbanova[1,2(✉)], Ievgen Koliada[2], Petr Císař[2], and Miloš Železný[1]

[1] Department of Cybernetics, Faculty of Applied Sciences, University of West Bohemia in Pilsen, Univerzitní 8, 306 14 Plzeň, Czech Republic
[2] Laboratory of Signal and Image Processing, Institute of Complex Systems, Faculty of Fisheries and Protection of Waters, South Bohemian Research Center of Aquaculture and Biodiversity of Hydrocenoses, University of South Bohemia in Ceske Budejovice, Zamek 136, 37 333 Nove Hrady, Czech Republic
urbanovp@kky.zcu.cz

Abstract. Fish require a sufficient amount of dissolved oxygen in the water to breathe and maintain their metabolic functions. Insufficient levels of dissolved oxygen can lead to stress, illness, and even death among the fish population. Therefore, it is crucial to model and simulate the relationship between dissolved oxygen levels and fish behavior in order to optimize aquarium design and management.

One approach to studying this relationship is through multiagent based modeling. This method involves creating a virtual environment in which multiple agents, representing individual fish, interact with each other and the environment based on a set of predefined rules. In the context of aquarium simulation, the agents would represent individual fish, and the environment would represent the aquarium water and its parameters.

Keywords: Agent based modeling · Fish shoal behavior · Simulation

1 Introduction

With the increasing population on the planet, the consumption of food follows the similar trend. Nutrition demands are responsible for the soil exhaustion and sea replenish. The recent green and blue revolutions offered greenhouses, hydroponic, extensive aquaculture, and biotechnology [48]. Aquaculture belongs to the list of key human resources and the traditional way of maintenance prevails [7,23,41]. Aside from the classical aquaculture, the current trends in blue revolution rediscover and reincorporate the smart technology towards the Industry 4.0 [4,5,11,32,47]. The main point is, that work which could be done by machines, actually should be done by machines.

Pavla Urbanova: The study was financially supported by the Ministry of Education, Youth and Sports of the Czech Republic - project CENAKVA (LM2018099), the CENAKVA Centre Development ($No.CZ.$1.05/2.1.00/19.0380), and SGS of the University of West Bohemia, project No. $SGS - 2022 - 017$. Authors thank to J.Urban for discussion and consultation.

I. Rojas et al. (Eds.): IWBBIO 2023, LNBI 13920, pp. 3–13, 2023.
https://doi.org/10.1007/978-3-031-34960-7_1

Here we can see the attempts to intensive aquaculture, smart cages (or tanks, or ponds), IoT (Internet of Things) monitoring and control, remote measurements, and use of the software applications for data handling, processing, and analysis. Since the technology for data acquisition become available and computational power increased, the demands for data mining, processing and analysis are on the rise. The basic element in aquaculture is obviously the fish and major concern on the fish production. The final amount is dependent (conditioned) by many attributes. Not all of them are well defined.

However, from the biological point of view, there are questions of the food optimization, fish metabolism, disease, water quality/pollution, and welfare. Each topic is again conditioned by species, sex or age. Fish welfare (general well-being) is becoming a very popular topic, where the exact definitions of welfare indicators are currently under development. Lot of the indicators are related to the fish behavior. Fish experts are able from the observation of the fish school estimate the conditions. In generally, fish behavior is very well described. Unfortunately, the terms are often vague, and it is difficult to parametrize the used description.

From the technical point of view, the welfare indicators (healthy/disease, saturated/hungry, oxygen deprivation/normal, ...) could be outputs. The inputs are coming from environment (food, pathogen, temperature, dissolved oxygen, ...). Water quality or pollution are also input variables (ammonia level). What remain is the behavior. But, the behavior is defined as the response of the system to inputs. Moreover, from information point of view [12], behavior consists of

- actor,
- operation,
- interactions,
- and their properties.

Lot of the welfare indicators are just various classifications of the responses (sick fish interacts differently). Therefore, fish behavior is the output and environmental variables are the inputs. Logically, the individual fish is the actor, operating with what environment gives, and interact with other fish. From that point, it is reasonable to consider to average the fish behavior in the shoal to smooth or blur the nonrelevant movements of single fish. Thus we can apply the central limit theorem, consider Gaussian type distributions and distinguish different behavior distributions by central moments. This helps also in modeling, that we do not need model representing exactly each individual fish, but to have only shoal with the same distribution of the behavior.

Why is modeling of fish behavior necessary? In the first place, the aquacalture research needs models to investigate shoal formation, energetics, stability, changes, and hierarchy to understand the relations between environment and shoal behavior [18,35,40,50]. The models help to simulate the specific conditions and investigate behavior patterns. Shoals are highly dynamic and varying in response to changes in their physiological stage and the external environment. Modeling of fish shoals provide an excellent opportunity to investigate the functions and mechanisms of group living. Fish decision system, which links observation and response, is of critical importance in resolving it. Fish shoal are an archetypal example of how local interactions lead to complex global decisions and motions [9, 24, 25, 30, 35, 37].

The cybernetics is offering tools and methods to create adequate fish shoaling model, based on the experimental observation. More over, the extraction of the itself could by automatized and self-parametrized. The model will serve for different biological purposes from aquaponics, intensive aquaculture, food optimization, disease, polution, welfare analysis or early warning systems.

Changes in swimming behaviour could reflect how a fish is sensing and responding to its environment. Every swimming activity can be related to a particular swimming behaviour. A number of common aquacultural water quality parameters can have an effect upon swimming behaviour. For example, reduced dissolved oxygen levels (hypoxia) can reduce the swimming speeds and activity of a number of species including white sturgeon Acipenser transmontanus and Atlantic cod. This reduced activity may enable fish to survive prolonged and widespread exposure to hypoxic conditions. In contrast, Tang and Boisclair (1995) have reported that the brook charr Salvelinus fontinalis increase swimming speeds in response to hypoxic conditions, and this is suggested to be an escape response from an area of potentially localised hypoxia. Thus, any deviations from normal, be it reduced or elevated, swimming speeds can be used as a potential indicator of hypoxia, although other water quality factors can have an effect upon swimming speeds. Hyperoxic conditions can also affect the swimming behaviour of fish. Atlantic salmon exhibit reduced swimming speeds (measured by reduced tail beat frequency) when held in super-oxygenated water for prolonged periods. Other water quality parameters, in addition to oxygen levels, have been shown to affect swimming behaviour. These include low water pH levels, increased ammonia levels, carbon dioxide exposure and trace element exposure.

Just for the complementary sake, as a shoal is considered any group of fish, even quite detached, while the fish school is reserved for the highly organised and aligned groups of fish. In other words, fish school is a shoal, which moves together synchronically.

A school is described as a synchronized, polarized aggregation of fish. Approximately 25% of species show schooling behaviour at some point throughout their life. Animals living in groups make movement decisions depending on social interaction between group members. In a similar way, schooling enables individuals to maximize the flow of information about swimming behaviour between neighbours from either visual or lateral line cues, usually to rapidly transfer threat information to other fish, such as an oncoming predator. Density and internal organization of a fish school affects the extent to which information can transfer through the school, consequently affecting the strength of these collective behavioural responses.

School of fish is a phenomenon where a group of fish swim together in a coordinated manner. This behavior is often observed in the wild, where fish swim in schools to reduce the risk of predation or to find food. However, the behavior of schooling fish can also be leveraged in aquaculture or aquaponics to create optimal conditions for fish growth.

On an individual level, the behavior of schooling fish can also be used to blur out insignificant behavior. By swimming in a school, individual fish are less likely to be targeted by predators, as their movements are less noticeable. Additionally, by swimming in a coordinated manner, individual fish can conserve energy and improve their chances of finding food or mates.

In summary, the behavior of schooling fish is a complex phenomenon that can be leveraged in a variety of contexts, from improving fish growth and health in aquaculture to optimizing nutrient cycling in aquaponics.

Monitoring the behavior of a school of fish can provide valuable insights that can be used for a variety of purposes. One such application is the optimization of feeding, particularly in the case of salmon farming. By closely monitoring the behavior of salmon, farmers can determine the ideal feeding times and quantities, leading to improved growth rates and reduced waste.

Another important use of behavior monitoring is in disease prevention. By observing changes in the behavior of fish, such as reduced activity levels or swimming near the surface, early signs of illness can be detected, allowing for prompt intervention and treatment.

Behavior monitoring can also be used as an early warning system, as demonstrated in the case of a brewery that used a school of fish to detect water pollution. When the fish began exhibiting abnormal behavior, such as swimming erratically, it signaled the presence of toxic substances in the water.

In addition, monitoring the behavior of fish can provide valuable data on environmental changes. For example, changes in water temperature, pH levels, or salinity can affect the behavior of fish, providing a useful indicator of environmental conditions.

Behavior monitoring can also shed light on sexual behavior, particularly in the case of invasive species. By observing the behavior of these species, researchers can gain insights into their reproductive strategies and identify potential methods for control and eradication.

Furthermore, monitoring the behavior of fish can help identify pollution sources, such as the presence of algae preparations that release toxins into the water. By detecting changes in fish behavior, researchers can identify potential sources of pollution and take steps to address the issue.

Finally, monitoring the behavior of fish can help prevent oxygen deprivation, which can lead to the death of the entire pond. By observing changes in fish behavior, such as gasping at the surface, oxygen levels can be monitored and steps taken to prevent a potentially catastrophic loss of fish.

In conclusion, monitoring the behavior of fish has a wide range of applications, from improving the efficiency of salmon farming to detecting water pollution and preventing the death of fish due to oxygen deprivation. By leveraging the insights provided by behavior monitoring, we can better understand and manage the delicate balance of aquatic ecosystems.

2   Modeling

Modeling is the construction of the abstract system from the real object according to the reason. Such a construction contains simplified characteristics, which are exactly defined. It could be modeled a chain of subsystems which leads to the key subsystem of interest reason. It is necessary to find the desirable amount of attributes.

Model and real object are homomorphic if the behaviors of output variables are similar.

Model and real object are isomorphic if also the structures are similar.

The main reasons for modelling of a system are to understand and control it. Therefore, the model helps to predict the system dynamics or behaviour. The software applications for modeling should allow us to do three consequent tasks:

- the modelling itself,
- the simulations of the model(s) and
- optimization of the model and/or simulation.

Simulations are the experiments with the model. It is the simplified imitation of the behavior of the real object (usually too complex). Model works in time (reference attribute). During the simulations are observed the changes and behavior in time. Tuning of the attributes evaluated the functionality and the robustness of the model. Extreme simplification may leads to the absurd results.

Models were developed for the prediction of dissolved oxygen [6,13,34], enviromental impacts [43], and mostly modeling of the population dynamic growth [2,8,21,43,49] or predator-prey interaction [22,43]. The dynamic of the fish shoaling is considered mainly from the point of self organization patterns [14,20,33,38] and stability [8,29,36], based on change of turbidity [15,16] or water quality [1,8].

3   Methods

Three basic behavioural rules in schooling

- Cohesion − the attraction rule This enables fish to group with conspecifics in order to produce aggregation. Vision drives this rule.
- Directional orientation − the alignment rule. Fish match the behaviour of their neighbours in allelomimetric behaviour.
- Collision − avoidance − the repulsion rule. Fish maintain a certain distance from their nearest neighbour. Lateral line drives this rule.

This takes the highest priority.

The compressing-stretching-tearing hypothesis suggests that **inter-fish distances and polarization level depends on state of environment**. In a low stress environment, fish show individualist and exploratory behaviour, increased inter-fish distances and lower polarization.

One of the most relevant work about fish schooling and behavior was published by Kubo et al. in 2016 [31], where alignment, cohesion, and separation are properly defined. The model is in relation with attempts to behavior description [19,26,28]. Let x_i be the position vector of the $i-th$ fish and U_i the neighborhood individuals around $i-th$ fish. Then

$$u_j = \|x_j - x_i\|; \tag{1}$$

$$a_i = \sum_{j \in U_i} u_j; \tag{2}$$

$$c_i = \sum_{j \in U_i} (x_j - x_i); \tag{3}$$

$$s_i = \sum_{j \in U_i} \frac{(x_j - x_i)}{\|x_j - x_i\|}; \tag{4}$$

where a_i represents $i-th$ fish alignment, c_i fish cohesion, and s_i separation tendency. Kubo concluded, that the patterns in behavior could be distinguished by the ratios of these parameters.

3.1   Multi agents Based Modeling and Simulation

The agent based modeling represents another stage in the modeling. While in the cellular automat is the neighborhood fixed, in the case of agents, the nearest neighbours vary with time, as the agents are free to move and interact with each other and their environment.

Thus the agent based modeling is focused on the investigation of the agents behavior. Each of the agent has to fulfill following atributes:

– autonomy,
– local views,
– decentralization.

On the other hand, the multi agent modeling is focused on the structure of the agents. Generally, there could be three types of agents:

– passive, like obstacles;
– active;
– learning or cognitive.

To model the behavior of agents, in is necessary, their interaction with the environment. In the fish study, the environment is well defined. The water crucial properties are volume, temperature, oxygen level, pH, and ammonia. The dissolved gas follows the diffusion rules, physics and chemistry describable by Ordinary Differential Equations. The diffusion is multidirectional, depends on pressure and temperature, capilar effects, evaporation, supersaturation, and turbulences could be taken in account.

The environment is therefore represented by freshwater volume and surface level. Oxygen is absorbed from the surface, the level is saturated with atmospheric oxygen (8.1 mg/l). Dissolved oxygen of tanks and aquaria varies from suffocating values (below 1.35 mg/l) to optimal saturation (above 6.75 mg/l). The standard diffusion rate is $(1.7\,m^2/s^-9)$ and the density of one mole $(26\,g/cm^3)$ of dissolved oxygen in water.

To model the fish as agents, it is necessary to setup the structure and parameters of the agent. A school exhibits many contrasts. It is made up of discrete fish yet overall motion seems fluid. It is simple in concept yet is so visually complex. Schools are an example of self-organized behaviour in a group.

The modeling and simulation accepted very well the concept of functional programming, like Lisp [3,42,51], then object oriented programming [17,27,39, 51], and currently NetLogo [44,46,51].

NetLogo is a multi-agent programmable modeling environment. It is environment enables exploration of emergent phenomena. The NetLogo is designed as multi agent simulation environment, however most of the use is in agent based point of view cases.

The modeling of the fish shoal emergent behavior could be achieved via the multi agent modeling, using software applications like NetLogo [51]. The alignment, cohesion, and separation are relevant parameter for the description of the fish behavior [26,31]. The ratios between the parameters distinguish from different behavior patterns [31]. Fish cohesion and alignment depend on state of the environment. For the structure of the agent could be adopted the concept of BDI strategy for agent decision modeling [10], identification of decision parameters in fish, and environmental conditionality [45].

The inductive approach to determining water parameters from behavior involves using individual agents that each follow a set of simple rules to create complex behavior for the entire group. In this approach, mobile agents such as fish and fixed agents such as the environment are divided into a grid of 0.0625 l.

4 Results and Discussion

This article describes the approach to study, model and simulate the relationship between dissolved oxygen levels and fish behavior via the multi agent modeling, using software applications like NetLogo. Three parameters (alignment, cohesion, and separation) described by Kubo were used for modeling of agents.

This model successfully models the oxygen saturation for an aquarium of the given size, in agreement with the time of the real experiment. The model aquarium will hold the corresponding number of fish without additional aeration. It is possible to influence the behavior of the shoal by choosing a parameter and, for example, predict the development of a simulation run on the model based on the estimate of a real experiment. The fish in the school have a distribution of the probability of these parameters, so none of the fish behaves exactly the same and the model better describes the randomness of a real school (Figs. 1, 2 and 3).

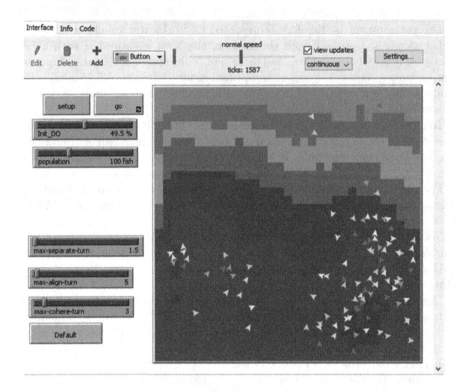

Fig. 1. NetLogo model for oxygen consumption by fish school.

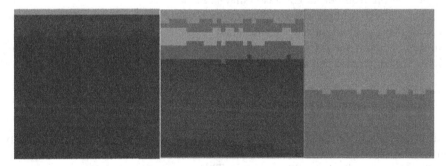

Fig. 2. NetLogo model for oxygen saturation after 1 min, 10 min and 40 min.

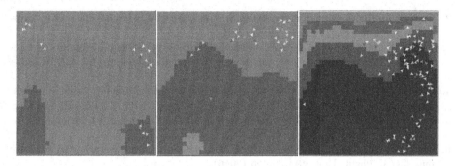

Fig. 3. NetLogo model for oxygen consumption by fish school: 20 fish after 40 min, 30 fish after 4 h and 100 fish after 40 min.

In conclusion, multiagent based modeling is a powerful tool for predicting and simulating the saturation and consumption of dissolved oxygen by fish in aquariums. By understanding the relationship between dissolved oxygen levels and fish behavior, researchers can optimize aquarium design and management to ensure the welfare of the fish population.

References

1. Abba, S., et al.: Evolutionary computational intelligence algorithm coupled with self-tuning predictive model for water quality index determination. J. Hydrol. **587**, 124974 (2020)
2. Anderson, J.: A stochastic model for the size of fish schools. Fish. Bull. **79**(2), 315–323 (1981)
3. Anderson, J.R., Farrell, R., Sauers, R.: Learning to program in LISP. Cogn. Sci. **8**(2), 87–129 (1984)
4. Antonucci, F., Costa, C.: Precision aquaculture: a short review on engineering innovations. Aquac. Int. **28**(1), 41–57 (2020)
5. Asche, F., Guttormsen, A.G., Nielsen, R.: Future challenges for the maturing Norwegian salmon aquaculture industry: an analysis of total factor productivity change from 1996 to 2008. Aquaculture **396**, 43–50 (2013)
6. Barbu, M., Ceangă, E., Caraman, S.: Water quality modeling and control in recirculating aquaculture systems. Urban Agric **2**, 64 (2018)
7. Beveridge, M.C., Little, D.C.: The history of aquaculture in traditional societies. Ecol. Aquac. Evol. Blue Revolut. 3–29 (2002)
8. Bjørkvoll, E., Grøtan, V., Aanes, S., Sæther, B.E., Engen, S., Aanes, R.: Stochastic population dynamics and life-history variation in marine fish species. Am. Nat. **180**(3), 372–387 (2012)
9. de Boer, L.: What makes fish school? (2010)
10. Bratman, M., et al.: Intention, Plans, and Practical Reason, vol. 10. Harvard University Press, Cambridge (1987)
11. Campbell, I., et al.: Biosecurity policy and legislation for the global seaweed aquaculture industry. J. Appl. Phycol. **32**(4), 2133–2146 (2019). https://doi.org/10.1007/s10811-019-02010-5

12. Cao, L.: In-depth behavior understanding and use: the behavior informatics approach. Inf. Sci. **180**(17), 3067–3085 (2010)
13. Cheng, X., Xie, Y., Zhu, D., Xie, J.: Modeling re-oxygenation performance of fine-bubble-diffusing aeration system in aquaculture ponds. Aquacult. Int. **27**(5), 1353–1368 (2019)
14. Crosato, E., et al.: Informative and misinformative interactions in a school of fish. Swarm Intell. **12**(4), 283–305 (2018). https://doi.org/10.1007/s11721-018-0157-x
15. Curatolo, M., Teresi, L.: The virtual aquarium: simulations of fish swimming. In: Proceedings of the European COMSOL Conference (2015)
16. Curatolo, M., Teresi, L.: Modeling and simulation of fish swimming with active muscles. J. Theor. Biol. **409**, 18–26 (2016)
17. Dahl, O.J., Nygaard, K.: SIMULA: an ALGOL-based simulation language. Commun. ACM **9**(9), 671–678 (1966)
18. Dumont, H., Protsch, P., Jansen, M., Becker, M.: Fish swimming into the ocean: how tracking relates to students' self-beliefs and school disengagement at the end of schooling. J. Educ. Psychol. **109**(6), 855 (2017)
19. Faucher, K., Parmentier, E., Becco, C., Vandewalle, N., Vandewalle, P.: Fish lateral system is required for accurate control of shoaling behaviour. Anim. Behav. **79**(3), 679–687 (2010)
20. Filella, A., Nadal, F., Sire, C., Kanso, E., Eloy, C.: Model of collective fish behavior with hydrodynamic interactions. Phys. Rev. Lett. **120**(19), 198101 (2018)
21. Føre, M., et al.: Modelling growth performance and feeding behaviour of Atlantic salmon (Salmo salar L.) in commercial-size aquaculture net pens: model details and validation through full-scale experiments. Aquaculture **464**, 268–278 (2016)
22. Free, B.A., McHenry, M.J., Paley, D.A.: Probabilistic analytical modelling of predator-prey interactions in fishes. J. R. Soc. Interface **16**(150), 20180873 (2019)
23. Gambelli, D., Naspetti, S., Zander, K., Zanoli, R.: Organic aquaculture: economic, market and consumer aspects. In: Lembo, G., Mente, E. (eds.) Org. Aquac., pp. 41–63. Springer, Cham (2019). https://doi.org/10.1007/978-3-030-05603-2_3
24. Hensor, E., Couzin, I.D., James, R., Krause, J.: Modelling density-dependent fish shoal distributions in the laboratory and field. Oikos **110**(2), 344–352 (2005)
25. Hoare, D., Krause, J., Peuhkuri, N., Godin, J.: Body size and shoaling in fish. J. Fish Biol. **57**(6), 1351–1366 (2000)
26. Huth, A., Wissel, C.: The simulation of the movement of fish schools. J. Theor. Biol. **156**(3), 365–385 (1992)
27. Keesman, K.J., et al.: Aquaponics systems modelling. In: Goddek, S., Joyce, A., Kotzen, B., Burnell, G.M. (eds.) Aquaponics Food Production Systems, pp. 267–299. Springer, Cham (2019). https://doi.org/10.1007/978-3-030-15943-6_11
28. Kelley, J.L., Phillips, B., Cummins, G.H., Shand, J.: Changes in the visual environment affect colour signal brightness and shoaling behaviour in a freshwater fish. Anim. Behav. **83**(3), 783–791 (2012)
29. Kolm, N., Hoffman, E.A., Olsson, J., Berglund, A., Jones, A.: Group stability and homing behavior but no kin group structures in a coral reef fish. Behav. Ecol. **16**(3), 521–527 (2005)
30. Krause, J., et al.: Fish shoal composition: mechanisms and constraints. Proc. Royal Soc. London. Ser. B: Biol. Sci. **267**(1456), 2011–2017 (2000)
31. Kubo, Y., Iwasa, Y.: Phase diagram of a multiple forces model for animal group formation: marches versus circles determined by the relative strength of alignment and cohesion. Popul. Ecol. **58**(3), 357–370 (2016). https://doi.org/10.1007/s10144-016-0544-3

32. Leow, B.T., Tan, H.K.: Technology-driven sustainable aquaculture for eco-tourism. In: Wang, C.M., Lim, S.H., Tay, Z.Y. (eds.) WCFS2019. LNCE, vol. 41, pp. 209–218. Springer, Singapore (2020). https://doi.org/10.1007/978-981-13-8743-2_11
33. Lewy, P., Nielsen, A.: Modelling stochastic fish stock dynamics using Markov chain monte Carlo. ICES J. Mar. Sci. **60**(4), 743–752 (2003)
34. Li, C., Li, Z., Wu, J., Zhu, L., Yue, J.: A hybrid model for dissolved oxygen prediction in aquaculture based on multi-scale features. Inf. Process. Agric. **5**(1), 11–20 (2018)
35. Li, G., Kolomenskiy, D., Liu, H., Thiria, B., Godoy-Diana, R.: On the energetics and stability of a minimal fish school. PLoS ONE **14**(8), e0215265 (2019)
36. Li, G., Kolomenskiy, D., Liu, H., Thiria, B., Godoy-Diana, R.: A pair of swimming fish: energetics and stability. In: ECCOMAS MSF 2019 THEMATIC CONFERENCE (2019)
37. Lopez, U., Gautrais, J., Couzin, I.D., Theraulaz, G.: From behavioural analyses to models of collective motion in fish schools. Interface Focus **2**(6), 693–707 (2012)
38. Niwa, H.S.: Self-organizing dynamic model of fish schooling. J. Theor. Biol. **171**(2), 123–136 (1994)
39. Nygaard, K., Dahl, O.J.: The development of the SIMULA languages. In: History of Programming Languages, pp. 439–480 (1978)
40. Rahman, S.R., Sajjad, I., Mansoor, M.M., Belden, J., Murphy, C., Truscott, T.T.: School formation characteristics and stimuli based modeling of tetra fish. Bioinspiration Biomimetics **15**(6), 065002 (2020)
41. Raychaudhuri, S., Mishra, M., Salodkar, S., Sudarshan, M., Thakur, A.: Traditional aquaculture practice at East Calcutta Wetland: the safety assessment (2008)
42. Rhodes, C., Costanza, P., D'Hondt, T., Lemmens, A., Hübner, H.: LISP. In: Cebulla, M. (ed.) ECOOP 2007. LNCS, vol. 4906, pp. 1–6. Springer, Heidelberg (2008). https://doi.org/10.1007/978-3-540-78195-0_1
43. Ruth, M., Hannon, B.: Modeling dynamic biological systems. In: Modeling dynamic biological systems. Modeling Dynamic Systems, pp. 3–27. Springer, New York (1997). https://doi.org/10.1007/978-1-4612-0651-4_1
44. Sklar, E.: NetLogo, a multi-agent simulation environment. Artif. Life **13**(3), 303–311 (2007)
45. Štys, D., et al.: 5iD viewer-observation of fish school behaviour in labyrinths and use of semantic and syntactic entropy for school structure definition. World Acad. Sci. Eng. Technol. Int. J. Comput. Electr. Autom. Control Inf. Eng. **9**(1), 281–285 (2015)
46. Tisue, S., Wilensky, U.: NetLogo: a simple environment for modeling complexity. In: International Conference on Complex Systems, vol. 21, pp. 16–21. Boston, MA (2004)
47. Trivelli, L., Apicella, A., Chiarello, F., Rana, R., Fantoni, G., Tarabella, A.: From precision agriculture to industry 4.0. Br. Food J. **121**, 1730–1743 (2019)
48. Urbanová, P.: Modeling in aquaponics system (2018)
49. Varga, M., Berzi-Nagy, L., Csukas, B., Gyalog, G.: Long-term dynamic simulation of environmental impacts on ecosystem-based pond aquaculture. Environ. Model. Softw. **134**, 104755 (2020)
50. Weihs, D.: Hydromechanics of fish schooling. Nature **241**(5387), 290–291 (1973)
51. Wilensky, U., Rand, W.: An Introduction to Agent-Based Modeling: Modeling Natural, Social, and Engineered Complex Systems with NetLogo. MIT Press, Cambridge (2015)

Entropy Approach of Processing for Fish Acoustic Telemetry Data to Detect Atypical Behavior During Welfare Evaluation

Jan Urban[✉]

Laboratory of Signal and Image Processing, Institute of Complex Systems, Faculty of Fisheries and Protection of Waters, South Bohemian Research Center of Aquaculture and Biodiversity of Hydrocenoses, University of South Bohemia in Ceske Budejovice, Zamek 136, 37 333 Nove Hrady, Czech Republic
urbanj@frov.jcu.cz

Abstract. Fish telemetry is an important tool for studying fish behavior, allowing to monitor fish movements in real-time. Analyzing telemetry data and translating it into meaningful indicators of fish welfare remains a challenge. This is where entropy approaches can provide valuable insights. Methods based on information theory can quantify the complexity and unpredictability of animal behavior distribution, providing a comprehensive understanding of the animal state. Entropy-based techniques can analyze telemetry data and detect changes in fish behavior, or irregularity. By analyzing the accelerometer data, using entropy approach, it is possible to identify atypical behavior that may be indicative of compromised welfare

Keywords: Shannon entropy · fish telemetry · biotelemetry · data processing · atypical behavior · typical behavior · fish welfare

1 Introduction

Since we are living in the digital era, the progress in technology, availability, size and capacity is affecting also the problems we are solving as well as the solutions in all aspects. Heaps of data values are acquired, sampled, or generated, stored, processed, analyzed, shared, and published. As the process expanded, we are now dealing with the big data approaches, cloud and online solutions accompanied by the artificial intelligence methods [1–3].

Acquisition of data is now faster and more accurate due to advancements in digital sensors and instrumentation. Data storage is more efficient due the development of high−capacity storage devices and cloud-based solutions. The analysis of data becomes more sophisticated with the use of machine learning algorithms and other advanced computational techniques. Sharing and publishing of data has become easier with the development of online platforms and

I. Rojas et al. (Eds.): IWBBIO 2023, LNBI 13920, pp. 14–26, 2023.
https://doi.org/10.1007/978-3-031-34960-7_2

social media. Online solutions have also emerged to allow users to access their data from anywhere, anytime [4,5].

The rise of artificial intelligence, neural networks, and deep learning enabled computers to process and analyze data more accurately and efficiently than ever before. Industry 4.0 and the Internet of Things have introduced smart sensors and cyber-physical systems, which can collect and analyze data in real-time, leading to improved decision-making and automation. These technologies have also facilitated the emergence of cognitive computing, where computers can learn and adapt from data, as well as automation, where machines can perform tasks without human intervention. The expansion of these processes has led to a more interconnected and efficient digital world [1,4,6–10].

Smart aquaculture involves the use of advanced technologies such as sensors, automation, and remote access to optimize fish farming practices in tanks or cages. This approach aims to enhance efficiency, reduce labor costs, minimize environmental impact, and improve fish welfare. Monitoring systems enable continuous real-time measurement of water quality, temperature, and other important variables. Automatic control systems can adjust water flow rates, oxygen levels, and feeding rates to maintain optimal conditions for fish growth and health. Early warning systems can alert farmers to potential problems such as disease outbreaks or water quality issues [4,11–17].

The concept of digital twins was adopted from industry, its application in aquaculture is in the early stages of development. Digital twin technology can help in creating virtual replicas of aquaculture systems and fish behavior, which can be used to optimize system design, predict or prevent events, and enhance decision-making processes. With advancements in technology and increased interest in sustainable food production, the use of digital twin technology in aquaculture is expected to grow in the coming years [18–21].

By integrating advanced technologies and analytics, fish farmers can improve their operations and maximize productivity while minimizing their impact on the environment and ensuring the welfare of their fish.

1.1   Biotelemetry

Fish telemetry is a method used to track the movement and behavior of fish in their habitat. It involves the use of various types of tags or transmitters attached to the fish, which then transmit data to receivers either on land or underwater. The data collected from the tags can provide information on fish behavior, such as migration patterns, feeding habits, and habitat use [22–24].

The question of tagging deals with the development and improvement of telemetry tags and transmission systems, like research on the design of tags suitable for different types of fish and environments, as well as the development of new encoding and compressing technologies to improve data transfer and reduce power consumption. Transmitting signals in water can be challenging due to several factors, since water absorbs and scatters signals, high-frequencies are absorbed more readily than low-frequency signals. Water is a noisy environment, with natural sounds of waves, wind, marine life, and human-generated noise.

Development of smaller, more lightweight tags that are less invasive and stressful for fish, is demanded for the fish welfare [25–29].

The analysis of telemetry data starts with classical statistical methods and data analysis techniques to select and extract fish movement or behavior patterns and relationships, regression models, environmental factors identification, spatial, density, or activity analysis, and estimation fish abundance and survival [30–34]. New methods of analysis adopts the machine learning and artificial intelligence to identify and classify fish parameters, and simulate fish behavior and movement patterns.

For the welfare purpose, there is a trend to replace tagging by non-invasive methods and other remote sensing technologies, research on the use of cameras and other sensors and their applicability [14,27,35]. On the other hand, telemetry could first help to study and improve fish welfare, e.g. assessing habitat use, determine which areas of environment fish are using most frequently or avoiding, measuring stress, changes in behavior patterns as indicators of stress, when and where stress may be occurring, improving aquaculture practices, handling, feeding, and stocking by providing valuable insights into the fish behavior, and improving their wellfare by managing informed decisions [22,25,27].

1.2 Fish Welfare

Fish welfare refers to the physical and mental well-being of fish, including their ability to engage natural behaviors, and to avoid negative experiences such as pain and stress. Welfare can be affected by a variety of factors, the conditions of keeping and harvesting, availability of food, presence of predators or other stressors. Fish behavior in the wild is generally characterized by a wide range of natural activities, in contrast, fish behavior in captivity, such as in a cages or tanks, may be more limited due to the confined space and reduced environmental complexity. Captive fish may exhibit repetitive or stereotypic behaviors, such as pacing or circling in the same area. Additionally, captives subject stressors not present in the wild. The reduced environmental complexity and exposure to additional stressors in captivity can lead to significant differences in behavior and well-being [36–39].

In order to step towards improving fish wellfare, the principles of $3R$ [40] could be adopted to fish telemetry as well:

– Replacement, finding alternatives to animal use, such as using non-animal models or computer simulations, exploring alternative approaches to telemetry research like remote sensing technologies without physically capturing or implanting tags in the fish.
– Reduction, minimizing the number of animals used in research or other activities, by improving experimental design, and analysis of the data, as well as re-analysis of the existing data and usage in other and novel studies than original measurement purpose.
– Refinement, minimizing stress, pain, and suffering, associated with tagging, and ensuring that they are housed in environments that meet their physical and behavioral needs.

Behavior analysis of telemetry data could help with parameterization of definition of typical behavior and development of tools for detection of atypical behavior. There are often used the estimations of behavior attributes probabilities and distribution functions [41,42]. In this article, we propose instead of using statistical tests like Smirnov-Kolmogorov [43,44] describe and distinguish the typicality by much simpler yet more powerfull informational entropy [45–47].

Entropy is a concept to measure complexity or variability of a system. In the context of fish welfare, by analyzing swimming patterns over time calculation of entropy values generates quantitative measures of the complexity and diversity of fish behavior. These measures can be used as indicators of fish welfare.

2   Dataset

The fish telemetry dataset was obtained during analysis of crowding in Atlantic salmon [23] in order to reuse existing data. The used tags cover RMS accaleration, maximal depth, minimal depth, average depth, and standard deviation of the depth within defined time interval, 30 s for acceleration and 150 s for depth measurements. The transmision interval varied from 220 to 380 s.

In order to demonstrate the applicability of entropy evaluation on telemetry data, the acceleration data were selected. The possible values range from 0 to 255 representing 0.014 m/s^2 per increment, covering the range 0 to 3.465 m/s^2 [23]. The duration of the experiment was 148 days with approximately 92 000 values for one fish (approximately one measurement every 2 min).

To showcase how entropy evaluation can be applied to telemetry data, the acceleration data was specifically chosen. The acceleration data comprises values ranging from 0 to 255, with each increment representing 0.014 m/s^2, encompassing the full range of 0 to 3.465 m/s^2 [23]. The experiment lasted for a duration of 148 days, with approximately 92,000 values obtained for a single fish, corresponding to a measurement taken every 2 min on average.

3   Methods

The concept of entropy was introduced into information theory by C. Shannon in 1948 [48]. It was adopted from thermodynamics as a measure of randomness in a system. Entropy is based on the idea that information could be considered as a reduction in uncertainty. The entropy concept revolutionized the information theory field, providing a powerful tool for analyzing and quantifying information.

Entropy has since become a fundamental concept and has found applications in a wide range of fields, including computer science, physics, and statistics. We can distinguish several basic types of entopy:

– Shannon entropy for a random discrete variable X, with probability $p(X)_i$:

$$S(X) = -\sum_{i=1}^{n} p(X)_i \log_2 p(X)_i, \tag{1}$$

where, $S(X)$ is the entropy of X, and n is the number of possible outcomes for X.

– Conditional entropy, introduced by Norbert Wiener, and extended by Shannon [48, 49]:

$$S(Y \mid X) = -\sum_{x \in X} \sum_{y \in Y} p(x, y) \log_2 p(y \mid x), \qquad (2)$$

for conditional probabilities $p(y \mid x)$.

– Rényi entropy citerenyi1961measures,baez2011renyi is a generalization of the Shannon concept for different probabilities distributions

$$S_\alpha(p) = \frac{1}{1 - \alpha} \log_2 \left(\sum_x p^\alpha(x) \right), \qquad (3)$$

where the Rényi entropy reduces to Shannon for $alpha = 1$.

– Tsalis entropy [50] is another generalization

$$S_q = \frac{1 - \sum_{i=1}^{W} p_i^q}{q - 1} \qquad (4)$$

Tsalis entropy could be also considered as a special case of a Rényi entropy, where α is replaced by $1 - q$.

– Havrda–Charvát entropy [51, 52] could be considered as equivalent as Tsalis entropy, applicable on pair variables X, Y:

$$S_H(X, Y) = -\sum_i \sum_j p_{ij}^H \log_2(p_{ij}^H), \qquad (5)$$

where for $H = 1$ it is again the Shannon entropy.

– Kullback-Leibler divergence [53] is the measure of difference between two probability distributions, with various types of methods and applications [54, 55].

In all cases, we need the variable enough observations to be able to estimate probability distribution function of the variable values. Such distribution estimation is the input for entropy calculation.

4   Results

The whole fish telemetry dataset from the crowding experiment [23] was restricted to time and acceleration values. For individual fish independently, it was divided into 24 h intervals and normalized before probability estimation. Every day was evaluated independently for the amount of Shannon entropy in bits. The Fig. 1 represents the entropy values for each day.

Basicaly, it is difficult to define typical fish behavior with exact mathematical attributes and their accepted range of values, conditionalities, and effects.

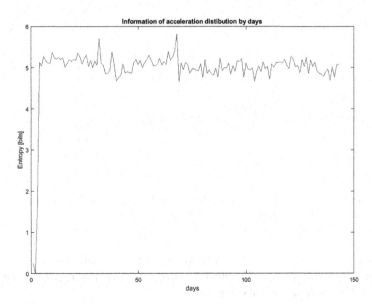

Fig. 1. Information entropy values, avaluated from acceleration telemetry, for individual days of measurement.

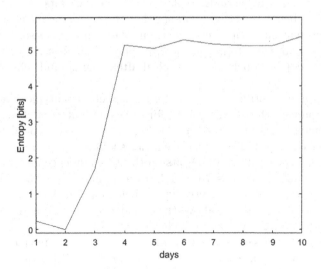

Fig. 2. Begining of the experiment, information entropy values for days. First three days are distinctively different.

However, it is possible to accept behavior distributions of the same level of information as a behavior which is not surprising. On the other hand, dramatical change in the behavior distribution information is a surprise.

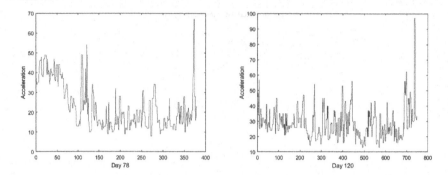

Fig. 3. Examples of typical behavior with teh similar level of information entropy, left day 78, entropy 4.93 bits, right day 120, entropy 4.97 bits.

In the context of data analysis and anomaly detection, surprise refers to unexpected or atypical behavior that deviates from the norm or expected patterns. Surprise is therefore always atypical behavior.

The information entropy time evolution of 1 illustrates the typical behavior within the long measurement period as the behavior with the informative value around 5 bits (4.96) with actually minor fluctuation 0.67 bits.

The atypical behavior is clearly observed at the beginning of the experiment, where the information of the days is extremely low in comparison with typical entropy 2. There, directly from the entropy approach analysis, the fish required three days to adopt the behavior to typical after tagging and reintroducing to the shoal.

Since the value of entropy cold serve for the detection of atypical behavior, we can directly compare the original values for typical and atypical days. Examples of typical behavior are illustrated at the Fig. 3, while the atypical behavior is represented at the Fig. 4. The typical behavior should be expected on the range from $(4.96 - 0.67)$ to $(4.96 + 0.67)$ bits on the 1σ confidence interval.

The two examples of days 78 and 120 on 3 represent 4.93 and 4.97 bits respectively, therefore almost exactly typical behavior. On the other hand, value 5.71 bits of the day 32 is just above the confidence interval, and pinpoint the potential candidate for atypical day behavior 4.

The estimated values of Shannon entropy are aditive by principle, therefore it is possible to compute basic statistic on them, and deal with them as with usual variables, for example binning, thresholding, normalization, standardization, or central moments estimation.

All computation were performed in Matlab environment.

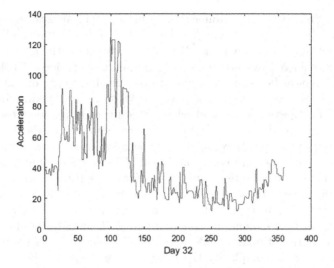

Fig. 4. Example of atypical behavior by information entropy, day 32, entropy 5.71 bits.

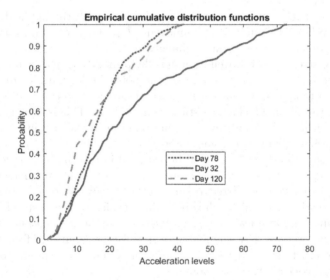

Fig. 5. Comparison of acceleration day cumulative distribution functions to illustrate the differences between typical days (d78 and d120) and atypical day (d32).

5   Conclusion

The main problem of the definition of the typical behavior is in the question of qualitative analysis, not the quantitative analysis. First of all, the normality of data is not to be expected and all tests have to be nonparametric. Second, due to the common fluctuations of circadian rhytmes, the median of the distribution could be a moving variable, therefore even nonparametric test (Wilcoxon rank sum, Mann-Whitney U-test, Kolmogorov-Smirnov) will always reject the null hypothesis of the same population distribution (the normalization and standardization do not affect the situation). Third, the extremes with small probability will not be considered by such tests, even if they have periodical behavior. In other words, typical behavior does have to be one single type of distributions, but the whole set. But this is not a priori what we are looking for, we are looking for an a posteriori method which unequivocally pinpoint the atypical behavior. It was already proven, that entropy concept overpass the nonparametric test for such task [46]. Here, in this article we evaluated and confirmed the approach power on $1/sigma$ confidence level. This could be simply illustrated by the plotting of the empirical cumulative distribution functions (cdf) of the days in Fig. 5. None of the cdfs will pass the similarity test, however the day 32 is clearly less similar to typical days. So far, the entropy is clearly distinctive method with much lower computational burden (1.75 times faster in comparison of only cdf empiric estimation). Moreover, single calculation of nonparametric test or entrop) are of complexity $O(n)$, but for comparisons, the n in entropy grows linearly, while in tests binomicaly (since we need to compare each distribution to each).

Thus, information entropy can potentially help with the application of the 3Rs in fish welfare, by analyzing the movement patterns of fish, improving the efficiency of data analysis, identify changes in behavior or activity levels that may indicate stress or other welfare issues, help modify experimental protocols or improve living conditions to minimize the potential for harm or distress.

The entropy approach, simple in principle, offers additional evaluation of telemetry datasets to the classical statistical analysis. This is becoming important for the fish welfare indicators estimation, since the definition of typical behavior is yet under investigation [36,56–58].

The entropy values analysis could be variated in principally simple subtasks like continuous analysis of the behavior, division of the sets to smaller subsets, indicating when the fish started to behave atypically. Moreover, the analysis could hel to distinguish which fish is behaving atypically, which speed, depth, or even pair values are atypical or typical. Other potential applications of entropy analysis in fish welfare may include analyzing the variability of feeding behavior, social interactions, changes in behavior or welfare in response to different environmental conditions or stressors.

Entropy approach has a potential to be a valuable tool in fish welfare assessment, as it allows for quantifying the complexity and variability of fish behavior, which can be used as an indicator of their welfare. By analyzing the entropy of fish movement patterns and behavior, it may be possible to identify abnormal behavior and detect potential welfare issues in aquaculture systems. Additionally, by combining entropy analysis with other telemetry and sensor technologies, it may be possible to develop early warning systems for detecting and addressing welfare issues before they become serious. Overall, while entropy analysis is not a comprehensive solution for assessing fish welfare, it can provide valuable insights into the complexity and variability of fish behavior and may be a useful tool for developing more effective management practices and promoting the welfare of captive fish.

Furthermore, entropy-based techniques can also help to standardize telemetry data analysis and provide objective indicators of fish welfare, which can be useful for management and regulatory purposes. In conclusion, entropy approaches hold significant potential for improving fish welfare assessment and management in the context of fish telemetry.

Acknowledgments. The study was financially supported by the Ministry of Education, Youth and Sports of the Czech Republic - project CENAKVA (*LM*2018099) and the European Union's research and innovation program under grant agreement No. 652831 (AQUAEXCEL3.0). Author thanks to M.Føre for data access and discussion.

References

1. Prapti, D.R., Mohamed Shariff, A.R., Che Man, H., Ramli, N.M., Perumal, T., Shariff, M.: Internet of things (IoT)-based aquaculture: an overview of IoT application on water quality monitoring. Rev. Aquac. **14**(2), 979–992 (2022)
2. Rowan, N.J.: The role of digital technologies in supporting and improving fishery and aquaculture across the supply chain-Quo Vadis? Aquaculture and Fisheries (2022)
3. Zhang, H., Gui, F.: The application and research of new digital technology in marine aquaculture. J. Mar. Sci. Eng. **11**(2), 401 (2023)
4. O'Donncha, F., Grant, J.: Precision aquaculture. IEEE Internet Things Mag. **2**(4), 26–30 (2019)
5. Bárta, A., Souček, P., Bozhynov, V., Urbanová, P., Bekkozhayeova, D.: Trends in online biomonitoring. In: Rojas, I., Ortuño, F. (eds.) IWBBIO 2018. LNCS, vol. 10813, pp. 3–14. Springer, Cham (2018). https://doi.org/10.1007/978-3-319-78723-7_1
6. Mustapha, U.F., Alhassan, A.W., Jiang, D.N., Li, G.L.: Sustainable aquaculture development: a review on the roles of cloud computing, internet of things and artificial intelligence (CIA). Rev. Aquac. **13**(4), 2076–2091 (2021)
7. Yadav, A., Noori, M.T., Biswas, A., Min, B.: A concise review on the recent developments in the internet of things (IoT)-based smart aquaculture practices. Rev. Fish. Sci. Aquac. **31**(1), 103–118 (2023)
8. Abinaya, T., Ishwarya, J., Maheswari, M.: A novel methodology for monitoring and controlling of water quality in aquaculture using internet of things (IoT). In: 2019 International Conference on Computer Communication and Informatics (ICCCI), pp. 1–4. IEEE (2019)
9. Rastegari, H., et al.: Internet of things in aquaculture: a review of the challenges and potential solutions based on current and future trends. Smart Agric. Technol. **4**, 100187 (2023)
10. Sun, M., Yang, X., Xie, Y.: Deep learning in aquaculture: a review. J. Comput. **31**(1), 294–319 (2020)
11. Gladju, J., Kamalam, B.S., Kanagaraj, A.: Applications of data mining and machine learning framework in aquaculture and fisheries: a review. Smart Agric. Technol. **4**, 100061 (2022)
12. Antonucci, F., Costa, C.: Precision aquaculture: a short review on engineering innovations. Aquac. Int. **28**(1), 41–57 (2020)

13. Hu, Z., Li, R., Xia, X., Yu, C., Fan, X., Zhao, Y.: A method overview in smart aquaculture. Environ. Monit. Assess. **192**(8), 1–25 (2020). https://doi.org/10.1007/s10661-020-08409-9

14. Saberioon, M., Gholizadeh, A., Cisar, P., Pautsina, A., Urban, J.: Application of machine vision systems in aquaculture with emphasis on fish: state-of-the-art and key issues. Rev. Aquac. **9**(4), 369–387 (2017)

15. Brijs, J., Føre, M., Gräns, A., Clark, T., Axelsson, M., Johansen, J.: Bio-sensing technologies in aquaculture: how remote monitoring can bring us closer to our farm animals. Philos. Trans. R. Soc. B **376**(1830), 20200218 (2021)

16. Pramana, R., Suprapto, B.Y., Nawawi, Z.: Remote water quality monitoring with early-warning system for marine aquaculture. In: E3S Web of Conferences. vol. 324, p. 05007. EDP Sciences (2021)

17. Davidson, K., et al.: HABreports: online early warning of harmful algal and biotoxin risk for the Scottish shellfish and finfish aquaculture industries. Front. Mar. Sci. **8**, 631732 (2021)

18. Zhabitskii, M., Andryenko, Y., Malyshev, V., Chuykova, S., Zhosanov, A.: Digital transformation model based on the digital twin concept for intensive aquaculture production using closed water circulation technology. In: IOP Conference Series: Earth and Environmental Science, vol. 723, p. 032064. IOP Publishing (2021)

19. Lima, A.C., Royer, E., Bolzonella, M., Pastres, R.: Digital twin prototypes in flow-through systems for finfish. Aquaculture **2021** (2021)

20. Lan, H.Y., Ubina, N.A., Cheng, S.C., Lin, S.S., Huang, C.T.: Digital twin architecture evaluation for intelligent fish farm management using modified analytic hierarchy process. Appl. Sci. **13**(1), 141 (2022)

21. Ahmed, A., Zulfiqar, S., Ghandar, A., Chen, Y., Hanai, M., Theodoropoulos, G.: Digital twin technology for aquaponics: towards optimizing food production with dynamic data driven application systems. In: Tan, G., Lehmann, A., Teo, Y.M., Cai, W. (eds.) AsiaSim 2019. CCIS, vol. 1094, pp. 3–14. Springer, Singapore (2019). https://doi.org/10.1007/978-981-15-1078-6_1

22. Muñoz, L., Aspillaga, E., Palmer, M., Saraiva, J.L., Arechavala-Lopez, P.: Acoustic telemetry: a tool to monitor fish swimming behavior in sea-cage aquaculture. Front. Mar. Sci. **7**, 645 (2020)

23. Føre, M.: Using acoustic telemetry to monitor the effects of crowding and delousing procedures on farmed Atlantic salmon (Salmo salar). Aquaculture **495**, 757–765 (2018)

24. Brownscombe, J.W., Griffin, L.P., Brooks, J.L., Danylchuk, A.J., Cooke, S.J., Midwood, J.D.: Applications of telemetry to fish habitat science and management. Can. J. Fish. Aquat. Sci. **79**(8), 1347–1359 (2022)

25. Brownscombe, J.W., et al.: Conducting and interpreting fish telemetry studies: considerations for researchers and resource managers. Rev. Fish Biol. Fish. **29**, 369–400 (2019)

26. Gesto, M., Zupa, W., Alfonso, S., Spedicato, M.T., Lembo, G., Carbonara, P.: Using acoustic telemetry to assess behavioral responses to acute hypoxia and ammonia exposure in farmed rainbow trout of different competitive ability. Appl. Anim. Behav. Sci. **230**, 105084 (2020)

27. Hassan, W., Føre, M., Urke, H.A., Ulvund, J.B., Bendiksen, E., Alfredsen, J.A.: New concept for measuring swimming speed of free-ranging fish using acoustic telemetry and doppler analysis. Biosys. Eng. **220**, 103–113 (2022)

28. Alfonso, S., Zupa, W., Spedicato, M.T., Lembo, G., Carbonara, P.: Use of telemetry sensors as a tool for health/welfare monitoring of European sea bass (Dicentrarchus

labrax) in aquaculture. In: 2021 International Workshop on Metrology for the Sea; Learning to Measure Sea Health Parameters (MetroSea), pp. 262–267. IEEE (2021)

29. Azevedo, J., Bartolomeu, T., Teixeira, S., Teixeira, J.: Design concept of a non-invasive tagging device for blue sharks. In: Innovations in Mechanical Engineering II. icieng 2022. Lecture Notes in Mechanical Engineering, pp. 80–90. Springer, Cham (2022). https://doi.org/10.1007/978-3-031-09382-1_8

30. Nguyen, V.M., Young, N., Brownscombe, J.W., Cooke, S.J.: Collaboration and engagement produce more actionable science: quantitatively analyzing uptake of fish tracking studies. Ecol. Appl. **29**(6), e01943 (2019)

31. Williamson, M.J.: Analysing detection gaps in acoustic telemetry data to infer differential movement patterns in fish. Ecol. Evol. **11**(6), 2717–2730 (2021)

32. Bohaboy, E.C., Guttridge, T.L., Hammerschlag, N., Van Zinnicq Bergmann, M.P., Patterson III, W.F.: Application of three-dimensional acoustic telemetry to assess the effects of rapid recompression on reef fish discard mortality. ICES J. Mar. Sci. **77**(1), 83–96 (2020)

33. Matley, J.K., et al.: Global trends in aquatic animal tracking with acoustic telemetry. Trends Ecol. Evol. **37**(1), 79–94 (2022)

34. Lees, K.J., MacNeil, M.A., Hedges, K.J., Hussey, N.E.: Estimating survival in a remote community-based fishery using acoustic telemetry. Can. J. Fish. Aquat. Sci. **79**(11), 1830–1842 (2022)

35. Bassing, S.B., et al.: Are we telling the same story? comparing inferences made from camera trap and telemetry data for wildlife monitoring. Ecol. Appl. **33**(1), e2745 (2023)

36. Hvas, M., Folkedal, O., Oppedal, F.: Fish welfare in offshore salmon aquaculture. Rev. Aquac. **13**(2), 836–852 (2021)

37. Arechavala-Lopez, P., Cabrera-Álvarez, M.J., Maia, C.M., Saraiva, J.L.: Environmental enrichment in fish aquaculture: a review of fundamental and practical aspects. Rev. Aquac. **14**(2), 704–728 (2022)

38. Jones, N.A., Webster, M.M., Salvanes, A.G.V.: Physical enrichment research for captive fish: time to focus on the details. J. Fish Biol. **99**(3), 704–725 (2021)

39. Macaulay, G., Bui, S., Oppedal, F., Dempster, T.: Challenges and benefits of applying fish behaviour to improve production and welfare in industrial aquaculture. Rev. Aquac. **13**(2), 934–948 (2021)

40. Sloman, K.A., Bouyoucos, I.A., Brooks, E.J., Sneddon, L.U.: Ethical considerations in fish research. J. Fish Biol. **94**(4), 556–577 (2019)

41. Runde, B.J., Michelot, T., Bacheler, N.M., Shertzer, K.W., Buckel, J.A.: Assigning fates in telemetry studies using hidden Markov models: an application to deepwater groupers released with descender devices. North Am. J. Fish. Manag. **40**(6), 1417–1434 (2020)

42. Elliott, C.W., Ridgway, M.S., Blanchfield, P.J., Tufts, B.L.: Novel insights gained from tagging walleye (Sander vitreus) with pop-off data storage tags and acoustic transmitters in Lake Ontario. J. Great Lakes Res. **49**, 51–530 (2023)

43. Smirnov, N.: Ob uklonenijah empiriceskoi krivoi raspredelenija. Recl. Math.(Matematiceskii Sb.) NS **6**(48), 3–26 (1939)

44. Kolmogorov, A.: On determination of empirical low of distribution. J. Ital. Inst. Actuaries **4**, 83–91 (1933)

45. Lee, S., Kim, M.: On entropy-based goodness-of-fit test for asymmetric student-t and exponential power distributions. J. Stat. Comput. Simul. **87**(1), 187–197 (2017)

46. Evren, A., Tuna, E.: On some properties of goodness of fit measures based on statistical entropy. Int. J. Res. Rev. Appl. Sci. **13**, 192–205 (2012)

47. Shoaib, M., Siddiqui, I., Rehman, S., ur Rehman, S., Khan, S.: Speed distribution analysis based on maximum entropy principle and Weibull distribution function. Environ. Prog. Sustain. Energy **36**(5), 1480–1489 (2017)
48. Shannon, C.E.: A mathematical theory of communication. Bell Syst. Tech. J. **27**(3), 379–423 (1948)
49. Wiener, N.: Cybernetics or Control and Communication in the Animal and the Machine. MIT Press, Cambridge (2019)
50. Tsallis, C.: Possible generalization of Boltzmann-Gibbs statistics. J. Stat. Phys. **52**, 479–487 (1988)
51. Havrda, J., Charvát, F.: Quantification method of classification processes. concept of structural a-entropy. Kybernetika **3**(1), 30–35 (1967)
52. Jizba, P., Korbel, J., Zatloukal, V.: Tsallis thermostatics as a statistical physics of random chains. Phys. Rev. E **95**(2), 022103 (2017)
53. Kullback, S., Leibler, R.A.: On information and sufficiency. Ann. Math. Stat. **22**(1), 79–86 (1951)
54. Urban, J., Vanek, J., Stys, D.: Preprocessing of microscopy images via Shannon's entropy (2009)
55. Urban, J.: Information Entropy. Applications from Engineering with MATLAB Concepts, p. 43 (2016)
56. Martins, C.I., et al.: Behavioural indicators of welfare in farmed fish. Fish Physiol. Biochem. **38**, 17–41 (2012)
57. Sánchez-Suárez, W., Franks, B., Torgerson-White, L.: From land to water: taking fish welfare seriously. Animals **10**(9), 1585 (2020)
58. Fife-Cook, I., Franks, B.: Positive welfare for fishes: rationale and areas for future study. Fishes **4**(2), 31 (2019)

Determining HPV Status in Patients with Oropharyngeal Cancer from 3D CT Images Using Radiomics: Effect of Sampling Methods

Kubra Sarac$^{(\boxtimes)}$ (iD) and Albert Guvenis (iD)

Bogazici University, Istanbul 34684, Turkey
kubra.sarac@boun.edu.tr

Abstract. Non-invasive detection of human papillomavirus (HPV) status is important for the treatment planning of patients with oropharyngeal cancer (OPC). In this work, three-dimensional (3D) head and neck computed tomography (CT) scans are utilized to identify HPV infection status in patients with OPC by applying radiomics and several resampling methods to handle highly imbalanced data. 1142 radiomic features were obtained from the segmented CT images of 238 patients. The features used were selected through correlation coefficient analysis, feature importance analysis, and backward elimination. The fifty most important features were chosen. Six different sampling methods, which are Synthetic Minority Oversampling Technique (SMOTE), Support Vector Machine Synthetic Minority Oversampling Technique (SVMSMOTE), Adaptive Synthetic Sampling Method (ADASYN), NearMiss, Condensed Nearest Neighbors (CNN), and Tomek's Link, were performed on the training set for each of the positive and negative HPV classes. Two different machine learning (ML) algorithms, a Light Gradient Boosting Machine (LightGBM) and Extreme Gradient Boosting (XGBoost), were applied as predictive classification models. Model performances were assessed separately on 20% of the data. Oversampling methods displayed better performance than undersampling methods. The best performance was seen in the combination of SMOTE and XGBoost algorithms, which had an area under the curve (AUC) of 0.93 (95% CI: 82–99) and an accuracy of 90% (95% CI: 78–96). Our work demonstrated a reasonable accuracy in the forecast of HPV status using 3D imbalanced and small datasets. Further work is needed to test the algorithms on larger, balanced, and multi-institutional data.

Keywords: HPV Status · Radiomics · Resampling · Machine Learning

1 Introduction

With over 630,000 new cases of diagnosis each year and more than 350,000 deaths, head and neck cancers (HNC) rank as the sixth most prevalent malignancy in the world [1]. HNC refers to cancers arising from very different anatomical locations, encompassing the oral cavity, oropharynx, nasopharynx, larynx, and hypopharynx. Each anatomical site's clinical behavior differs to varying degrees, along with the therapeutic options

© The Author(s), under exclusive license to Springer Nature Switzerland AG 2023
I. Rojas et al. (Eds.): IWBBIO 2023, LNBI 13920, pp. 27–41, 2023.
https://doi.org/10.1007/978-3-031-34960-7_3

[2]. Oropharyngeal cancers (OPC), subtypes of head and neck carcinomas, have been linked to human papillomavirus (HPV) infection over the past 15 years [3] and have lately increased strikingly among males younger than 50 [4]. Even though more than 100 different subtypes of HPV are recognized, research demonstrates that HPV-16 is the root reason for 70% of instances of OPC [5]. For this reason, in addition to typical risk factors including tobacco, alcohol, nutrition, radiation, ethnicity, immunosuppression, and familial and genetic predisposition, HPV infection has been identified as the primary cause of OPC [6].

Knowing the HPV status can be important for the treatment planning of OPC patients. The HPV status determines the prognosis and can be a determining factor in selecting the proper dose for radiotherapy [7].

The status of HPV may be detected through the use of invasive laboratory methods such as target amplification techniques (Polymerase Chain Reaction [PCR]) and signal amplification techniques (hybridization methods in the liquid phase) [8]. The use of image analysis in quantitative terms, such as radiomics, which have a dependable power to identify HPV status, is currently being studied as an alternative way to determine HPV status [9–20].

Radiomics, which is a highly effective method of extracting quantitative medical image features from diagnostic imaging, is becoming more remarkable in the cancer industry since it allows data to be gathered information and performed in the decision phase of clinical practice to obtain prognostic, diagnostic, and predictive information [15]. Radiomic analysis can be applied to medical images from a variety of modalities, including magnetic resonance imaging (MRI), positron emission tomography (PET), and computed tomography (CT). The authors' choices and the particular study conditions continue to have a significant impact on the findings, and even the most recent research suggests that there is still lacking stability [19].

In radiomic studies [9–20], the datasets often consist of imbalanced data. The imbalanced dataset is one of the situations that negatively affect the classification model's performance. There is still more work to be done in this area due to the uneven distribution of the classes, which frequently leads to the misclassification of predictions [21]. Before using classifiers for training, resampling techniques change the imbalanced distribution of classes to be predicted [22]. There are many algorithms used in two different resampling methods, undersampling and oversampling, in machine learning. Undersampling techniques can lose potentially important features that are inextricably linked to their selection method, whereas oversampling techniques that merely repeat members of the minority class are vulnerable to overfitting [23]. Nevertheless, there has not been any research that compares the efficacy of various sampling approaches when designing a machine learning classifier for a CT radiomics-based model in predicting HPV status.

In this work, we propose to address the difficulties of imbalanced and small data for HPV status determination and enhance prediction performance by using different resampling methods and three-dimensional (3D) features as a novel contribution to the literature on the subject. As oversampling methods, Synthetic Minority Oversampling Technique (SMOTE) [24], Adaptive Synthetic Sampling Method (ADASYN) [24], and Support Vector Machine-Based Synthetic Minority Oversampling Technique

(SVMSMOTE) [20] were applied; as undersampling methods, NearMiss [25], Tomek Link (TL) [26], and Condensed Nearest Neighbors (CNN) [27] were implemented on imbalanced data. The general goal was to determine an accurate HPV infection status for patients with OPC noninvasively. Two advanced machine learning (ML) methods, Light Gradient Boosting Machine (LightGBM) and Extreme Gradient Boosting (XGBoost), were implemented. The trained models were interpreted using an independent test dataset. The results of our study were then compared to similar studies found in the literature.

2   Material and Methods

2.1   Data Set

In the dataset from The Cancer Imaging Archive (TCIA) [28], 495 patients were treated between 2005 and 2012 at the University of Texas MD Anderson Cancer Center. 257 patients were excluded since their HPV status was not reported (see Fig. 1). Contrast-enhanced CT images (segmentation labeled) of 238 patients, 204 of whom were HPV positive and diagnosed with OPC, were used in the study. All subjects in the dataset used have HPV status information in the form of p16 protein positivity. Table 1 shows a summary of the clinical characteristics of the patients selected for the study. In radiomic studies [13, 20], the gross primary tumor volume (GTVp), which is manually segmented by specialists, is considered. Since our main interest is to determine HPV status, we take into account segments that qualify as GTVp in the used dataset.

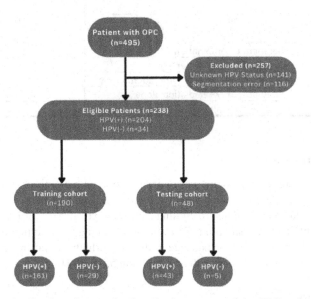

Fig. 1. The flow diagram for the selection process of the OPC patients.

Table 1. Clinical characteristics of patients whose data were used [29].

Characteristics	Statistics (n=238)
Age (mean) (yr)	58.02 ± 9.28
Sex (No.) (%)	
Female	42 (17.65%)
Male	196 (82.35%)
Smoking (No.) (%)	
Current	59 (24.78%)
Former	90 (37.82%)
Never	89 (37.40%)
HPV Status (No.) (%)	
Negative	34 (14.28%)
Positive	204 (85.72%)
T Category (No.) (%)	
T1	59 (24.78%)
T2	94 (39.50%)
T3	48 (20.16%)
T4	37 (15.56%)
N Category (No.) (%)	
N0	22 (%9.24)
N1	30 (%12.60)
N2	182 (%76.48)
N3	4 (%1.68)
Overall stage (AJCC seventh edition)	
I	3 (1.27%)
II	8 (3.36%)
III	39 (16.38%)
IV	188 (78.99%)
Therapeutic Combination	
Concurrent chemoradiotherapy	121 (50.85%)
Induction chemotherapy + Concurrent chemoradiotherapy	67 (28.15%)
Induction chemotherapy + Radiation alone	13 (5.46%)
Radiation alone	37 (15.54%)

2.2 Image Pre-processing

All CT images were resampled and normalized before the radiomics feature extraction process (see Fig. 2). Some patients' data were not used since there were no regions of interest (ROI) related to the data. To eliminate different voxel sizes [30], 1 mm × 1 mm × 1 mm was obtained by resampling CT scans. Resampling and interpolation processes were performed using the Python package Radiomics, which was taken from the library Pyradiomics [31] in the 3D slicer software (version 5.0.3) [14].

2.3 Feature Extraction

Quantitative features from medical images can be obtained through a process known as feature extraction [32]. After image preprocessing, 3D Slicer software (version 5.0.3)

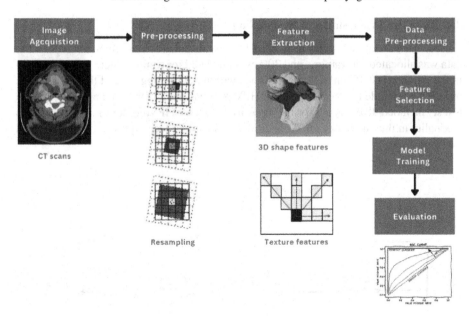

Fig. 2. The workflow of the study.

[20] was performed for the feature extraction process. Gray-level discretization was implemented by making the standard bin width 10 to decrease heterogeneity [33]. Features of the original images, wavelet-transformed images, and Laplacian of Gaussian (Log) filtered images were provided (see Fig. 3). Extracted features define the spatial relationships between pixels and the signal intensity distribution within a region of interest [34]. Feature extraction quantifies various ROI/volume of interest (VOI) components and provides a numerical representation of phenotypic traits. The created data are used to train and test ML models after the images have been transformed into features. For instance, the gray-level co-occurrence matrix (GLCM), which is one of the radiomics features frequently utilized as an indicator of heterogeneity and may offer details about the tumor microenvironment, Quantized intensities within a VOI are used to create GLCMs. The spatial correlation of the pixels is quantified by GLCM by merging neighboring pixel values to form a matrix [35]. These features are used to compare images quantitatively [36]. Full definitions of the radiomics features may be found in [31].

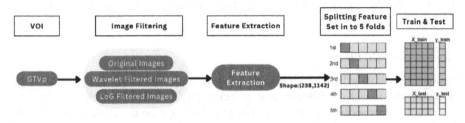

Fig. 3. Process of feature extraction and data allocation for five-fold cross-validation.

2.4 Data Pre-processing and Resampling

Z-score normalization was performed on large-scale data for 1142 features. 80% of the data was allocated for training, and 20% was allocated for testing. There was a disparity in the number of HPV infection conditions within the training set and the test set. The training set included 190 cases (161 for HPV positive and 29 for HPV negative), and the test set included 48 cases (43 for HPV positive and 5 for HPV negative). Because of this inequality in the number of HPV conditions, resampling techniques were implemented (see Fig. 4).

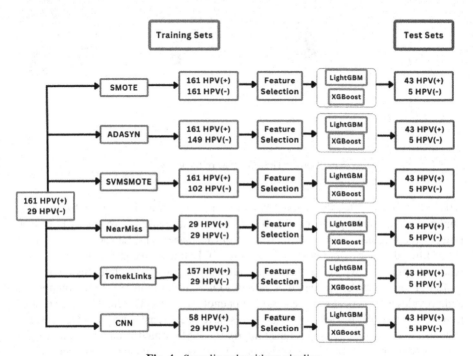

Fig. 4. Sampling algorithms pipeline.

As oversampling methods, SMOTE [24], ADASYN [24], and SVMSMOTE [20]; as undersampling methods, NearMiss [25], TL [26], and CNN [27] were performed in Python (version 3.9) using the Imblearn library [37] on the training data to create artificial data. Only training data was subject to resampling algorithms in order to avoid overly optimistic results.

2.5 Feature Selection

Radiomics methods usually result in high dimensionality, which can lead to over-fitting, escalating the model's confusion, and reducing prediction performance. Therefore, choosing features seeks to address this issue by choosing the most important features while eliminating unnecessary and redundant ones [38]. There are many functional feature selection methods [39]. In this study, correlation coefficient analysis, random forest

(RF) feature importance analysis, and backward elimination techniques were applied sequentially (see Fig. 5). As a filter-based method, correlation coefficient analysis (CCA) was initially used to remove unnecessary features that were very strongly connected (absolute correlation coefficient > 0.9). The random forest model was performed with the Gini impurity metric, which provides a superior means for measuring feature relevance [40]. The sequential backward selection algorithm was utilized to choose the fifty most important features (k-nearest neighbor was used as an estimator). The MLXtend library and Scikit-learn library were employed in Python (version 3.9) to perform the feature selection algorithms [41].

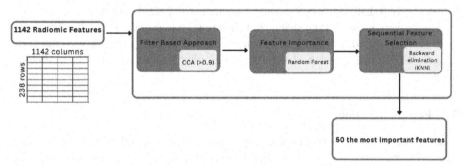

Fig. 5. Workflow of the feature selection operation.

2.6 Model Training and Evaluation

LightGBM and XGBoost machine learning classification algorithms were applied for predicting HPV status in Python software using the Scikit-learn library. LightGBM and XGBoost are gradient-boosting frameworks that are powerful ML algorithms. Using training data, 29 HPV-negative cases were increased by SMOTE, ADASYN, and SVMSMOTE, and 161 HPV-positive cases were decreased by NearMiss, TL, and CNN model training assessed. LightGBM and XGBoost classifiers were created and optimized using five-fold cross-validation, a very powerful method to avoid overfitting [37]. To determine the optimum hyperparameters with five-fold nested cross-validation, more than 500 hyperparameter combinations were tested on LightGBM and XGBoost models using the random search approach on training feature sets. Model performance was assessed separately based on the 20% of the initial imbalanced test data that was not used. The area under the curve (AUC) was found to evaluate the validation and test results of the models.

3   Results

3.1 Data Pre-processing and Resampling

All CT images used were resampled to a spatial resolution of 1 mm × 1 mm × 1 mm. In the data, there were 161 HPV-positive and 29 HPV-negative cases. In the implementation of resampling algorithms, for 29 HPV-negative cases, 161 synthetic data were created by the SMOTE oversampling algorithm, 149 synthetic data were created by the ADASYN oversampling algorithm, and 102 synthetic data were created by the SVMSMOTE oversampling algorithm. 161 HPV-positive cases were reduced to 29 cases by the NearMiss undersampling algorithm, to 157 cases by the TomekLinks undersampling algorithm, and to 58 cases by the CNN undersampling algorithm.

3.2 Feature Extraction

The radiomic features were extracted in eight different categories: first order, gray level co-occurrence matrix (GLCM), gray level difference matrix (GLDM), gray level run length matrix (GLRLM), gray level size zone matrix (GLSZM), neighboring gray-tone difference matrix (NGTDM), shape, and shape two-dimensional (2D). The full range of features extracted was 1142, of which 735 were extracted from wavelet-transform images, 270 from Laplacian of Gaussian (log) filter images, and 137 from original images.

3.3 Feature Selection

Since 831 features were found to be highly correlated with each other, 311 features were selected in the first round of elimination by correlation coefficient analysis. 105 of the 311 features were chosen in the second round of implementation by RF. Finally, using the sequential backward selection algorithm, the most crucial 50 features were picked.

3.4 Performance Evaluation

The models were interpreted by considering only the independent test dataset. HPV status was predicted in the combination of SMOTE and LightGBM models with an accuracy of 92% (95% CI: 83–99) and an AUC of 0.89 on the test data, and in the SMOTE and XGBoost models combined with an accuracy of 90% (95% CI: 78–96) and an AUC of 0.93 on the test data. The accuracy was found to be 0.90 (95% CI: 80–98) and the AUC was 0.82 in the ADASYN and LightGBM combination, while in the ADASYN and XGBoost combination, the accuracy was found to be 0.87 (95% CI: 78–96) and the AUC was 0.83. In the SVMSMOTE and LightGBM combination, the accuracy was found to be 0.88 (95% CI: 75–95) and the AUC was 0.84, and in the SVMSMOTE and XGBoost combination, the accuracy was found to be 0.88 (95% CI: 80–98) and the AUC was found to be 0.91. Table 2 gives the summary of the results obtained by using oversampling algorithms by demonstrating the AUC, accuracy (ACC), macro and weighted average (AVG), recall, precision (Prec), and F1 score results.

Considering the results of the models applied using undersampling algorithms, in the NearMiss and LightGBM combination, accuracy was found to be 0.65 (95% CI: 33–62) and AUC was 0.78; in the combination of NearMiss and XGBoost, accuracy was found to be 0.82 (95% CI: 33–62) and AUC was 0.74. Using the combination of TomekLinks and LightGBM, the accuracy was found to be 0.82 (95% CI: 83–99) and the AUC was 0.81, and using the combination of TomekLinks and XGBoost, the accuracy was found to be 0.80 (95% CI: 83–99) and the AUC was 0.67. The accuracy was found to be 0.58

(95% CI: 70–92) and the AUC was 0.62 for the combination of CNN and LightGBM, and for the combination of CNN and XGBoost, the accuracy was found to be 0.54 (95% CI: 67–90) and the AUC was 0.61. Table 3 gives a summary of the results obtained by using undersampling algorithms.

The best performance was provided by the combination of SMOTE oversampling and the XGBoost ML classification algorithm. The ROC curve for the most effective performance result is demonstrated in Fig. 6, and the confusion matrix for the best-performing algorithm on the test data is demonstrated in Fig. 7.

Table 2. Classification results for oversampling algorithms.

OVERSAMPLING	ML	AUC	ACC	MACRO AVG			WEIGHTED AVG		
				RECALL	PREC	F1	RECALL	PREC	F1
SMOTE	LightGBM	0.89	0.92	0.87	0.77	0.81	0.92	0.93	0.92
SMOTE	XGBoost	0.93	0.90	0.84	0.71	0.75	0.88	0.92	0.89
ADASYN	LightGBM	0.82	0.90	0.77	0.73	0.74	0.90	0.91	0.90
ADASYN	XGBoost	0.83	0.87	0.75	0.69	0.71	0.88	0.90	0.88
SVMSMOTE	LightGBM	0.84	0.88	0.74	0.66	0.69	0.85	0.89	0.87
SVMSMOTE	XGBoost	0.91	0.88	0.77	0.73	0.74	0.90	0.91	0.90

Table 3. Classification results for undersampling algorithms.

UNDERSAMPLING	ML	AUC	ACC	MACRO AVG			WEIGHTED AVG		
				RECALL	PREC	F1	RECALL	PREC	F1
NearMiss	LightGBM	0.78	0.65	0.62	0.55	0.42	0.48	0.87	0.57
NearMiss	XGBoost	0.74	0.82	0.62	0.55	0.42	0.48	0.87	0.57
TomekLinks	LightGBM	0.81	0.82	0.60	0.96	0.64	0.92	0.92	0.89
TomekLinks	XGBoost	0.67	0.80	0.60	0.96	0.64	0.92	0.92	0.89
CNN	LightGBM	0.77	0.51	0.68	0.72	0.69	0.90	0.89	0.89
CNN	XGBoost	0.66	0.55	0.68	0.72	0.69	0.90	0.89	0.89

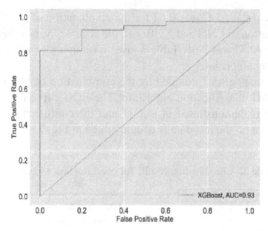

Fig. 6. ROC Curve of XGBoost model with the SMOTE oversampling algorithm.

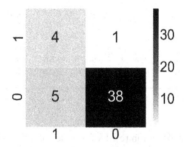

Fig. 7. Confusion matrix of XGBoost model with the SMOTE oversampling algorithm.

4 Discussion

HPV status is crucial at the decision-making stages of patients' treatment planning. In this work, we investigated if the resampling techniques in radiomic analysis of 3D CT images can be performed with the purpose of predicting the HPV status of OPC patients using an imbalanced dataset. To address the issue of an unbalanced dataset, many resampling techniques were employed. The oversampling algorithms performed better than the undersampling algorithms. The results of the study have shown promising performance. Hence, this research can have a clinical effect, as these HPV status predictions are acquired preoperatively and noninvasively.

In certain ways, this study is unique from other reported works in the literature. Similar studies to identify HPV status were shown in Table 4 with their algorithms and results.

In one of these studies [12], 60 OPSCC patients were included, and 1618 features were obtained from the patients' MR scans. LR, RF, and XGboost models were used to predict HPV status, resulting in AUCs of 0.77, 0.76, and 0.70, respectively. In this study, resampling algorithms were not used for the data. Thus, the models were evaluated with an imbalanced dataset. In another study [13], 651 OPC patients' CT images were

Table 4. Summary of similar studies and the present study.

Author	# of patients	Imaging Modality	Sampling Algorithm	Prediction Algorithm	AUC
Leijenaar et al. (2018)	778	CT	Dataset balanced	Multi Linear Regression	0.76
Boot et al. (2023)	249	MR	Dataset balanced	LR and RF	0.79 (for both)
Zhinan et al. (2022)	651	CT	Not applied	LR, SVM, RF, XGB, LGB	0.54, 0.64, 0.63, 0.60, and 0.61 (respectively)
Suh et al. (2020)	60	MR	Not applied	LR, RF, XGB	0.77, 0.76, and 0.71 (respectively)
Bogowicz et al. (2017)	149	CT	Not applied	Multi Linear Regression	0.78
Bagher-Ebadian et al. (2020)	187	CT	Not applied	General Linear Model (GLM)	0.86
Yu et al. (2017)	315	CT	Not applied	LR	0.91
Bagher-Ebadian et al. (2022)	128	CT	Not applied	Lasso-GLM	0.78
Altinok et al. (2022)	246	CT	ROSE	Bayesian Networks	0.72
Reiazi et al. (2021)	1294	CT	Under-sampling	RF	0.79
Our proposed model	238	CT	SMOTE	XGB	0.93

Notes: LR: Logistic Regression. RF: Random Forest. SVM: Support Vector Machine. LGB: Light Gradient Boosting Machine.

used. Dataset resampling methods have not been applied. Ten different machine-learning classification models were applied with 1316 features to identify HPV infection status. The best performance was obtained using a random forest classification algorithm with an AUC of 0.63 and 0.78 for the two testing cohorts used. In another study carried out [14], 246 OPC patients were included, and after image pre-processing applications, 851 characteristics were retrieved. In the data pre-processing step, there was an unequal distribution of HPV cases for training and test sets. They preferred to apply the ROSE oversampling algorithm to make datasets balanced. After feature selection, the Bayesian network model yielded 0.72 AUC results on the test set as a predictor for HPV infection status. In another study [15], the data of 149 HNSCC patients were used, 100 of whom had oropharyngeal cancer. 317 features were extracted from CT images by applying four feature extraction algorithms. Resampling algorithms have not been performed on imbalanced data. Multivariable logistic regression was applied as a classifier algorithm, with the four most important features chosen. The AUC score for the test dataset was 0.78 in this study. In [16], 187 OPC patients were included in the study, and 172 features were gathered from the CT images. Data that is unbalanced has not been subjected to resampling methods. A general linear model (GLM) classification algorithm was performed with the 12 most important features to predict HPV status. The AUC was found to be 0.86 with the GLM classifier model. Another study conducted on 315 OPC patients was also published [17]. In the training set, there was an inequality for HPV cases (128 positive and 22 negative cases). The dataset was not balanced using any sampling methods. The logistic regression model was performed with an AUC of 0.91

in the study. In [20], 1294 patients were deemed appropriate for the study. The authors of the study applied undersampling algorithms to HPV-positive cases in their data. A random forest (RF) classifier with an AUC of 0.79 was obtained.

In the present study, the dataset used was small and highly imbalanced. The data were balanced with six different resampling algorithms, three of which were oversampling and three of which were undersampling. The models were evaluated with two different classification models. The best performance AUC was obtained at 0.93 (95% CI: 82–99) with an accuracy of 90% (95% CI: 78–96) on the test data with the combination of SMOTE oversampling and XGBoost classification algorithms. The high AUC obtained in our study is probably due to our specific combination of algorithms, the use of 3D features, and optimal sampling to correct the imbalanced data. In spite of the fact that our models can predict HPV status with high accuracy and AUC, further investigations should be conducted using larger clinical datasets to confirm the performance of the developed ML model.

5 Conclusion

We proposed an HPV status prediction system using different resampling algorithms applied to the radiomics features extracted from 3D CT images of patients with OPC. The outcome of the study indicates that accurate HPV status can be implemented non-invasively by applying resampling and ML classification algorithms to small and imbalanced datasets. It is shown that the appropriate resampling algorithm can help the classification model perform better. Further work is needed to test the algorithms on larger, balanced, and multi-institutional data.

Acknowledgement. This study was supported by Boğaziçi University Research Fund Grant Number 19703P.

References

1. Vigneswaran, N., Williams, M.D.: Epidemiologic trends in head and neck cancer and aids in diagnosis. Oral Maxillofac. Surg. Clin. North Am. **26**, 123–141 (2014). https://doi.org/10.1016/j.coms.2014.01.001
2. Howard, J.D., Chung, C.H.: Biology of human papillomavirus-related oropharyngeal cancer. Semin. Radiat. Oncol. **22**, 187–193 (2012). https://doi.org/10.1016/j.semradonc.2012.03.002
3. Kreimer, A.R., Clifford, G.M., Boyle, P., Franceschi, S.: Human papillomavirus types in head and neck squamous cell carcinomas worldwide: a systemic review. Cancer Epidemiol. Biomark. Prev. **14**, 467–475 (2005). https://doi.org/10.1158/1055-9965.EPI-04-0551
4. Sathish, N., Wang, X., Yuan, Y.: Human Papillomavirus (HPV)-associated oral cancers and treatment strategies. J. Dent. Res. **93**, 29S-36S (2014). https://doi.org/10.1177/0022034514527969
5. HPV and Throat/Oral Cancer FAQs | Mount Sinai - New York. https://www.mountsinai.org/locations/head-neck-institute/cancer/oral/hpv-faqs. Accessed 14 Mar 2023
6. Kumar, M., Nanavati, R., Modi, T., Dobariya, C.: Oral cancer: etiology and risk factors: a review. J. Cancer Res. Ther. **12**, 458–463 (2016). https://doi.org/10.4103/09731482.186696

7. Eide, M.L., Debaque, H.: HPV detection methods and genotyping techniques in screening for cervical cancer. Ann. Pathol. **32**, e15–e23 (2012). https://doi.org/10.1016/J.ANNPAT.2012.09.231

8. Göttgens, E.L., Ostheimer, C., Span, P.N., Bussink, J., Hammond, E.M.: HPV, hypoxia and radiation response in head and neck cancer. Br. J. Radiol. **92**, 20180047 (2019). https://doi.org/10.1259/BJR.20180047

9. Leijenaar, R.T.H., et al.: Development and validation of a radiomic signature to predict HPV (p16) status from standard CT imaging: a multicenter study. Br. J. Radiol. **91**, 1–8 (2018). https://doi.org/10.1259/bjr.20170498

10. Song, B., et al.: Radiomic features associated with HPV status on pretreatment computed tomography in oropharyngeal squamous cell carcinoma inform clinical prognosis. Front. Oncol. **11**, 744250 (2021). https://doi.org/10.3389/FONC.2021.744250

11. Boot, P.A., et al.: Magnetic resonance imaging based radiomics prediction of human papillomavirus infection status and overall survival in oropharyngeal squamous cell carcinoma. Oral Oncol. **137**, 106307 (2023). https://doi.org/10.1016/j.oraloncology.2023.106307

12. Suh, C.H., et al.: Oropharyngeal squamous cell carcinoma: radiomic machine-learning classifiers from multiparametric MR images for determination of HPV infection status. Sci. Rep. **10**(1), 17525 (2020). https://doi.org/10.1038/s41598-020-74479-x

13. Zhinan, L.: Prediction of HPV status in oropharyngeal squamous cell carcinoma based on radiomics and machine learning algorithms: a multi-cohort study, pp. 1–16 (2022)

14. Altinok, O., Guvenis, A.: Interpretable Radiomics Method for Predicting Human Papillomavirus Statusin Oropharyngeal Cancer using Bayesian Networks (2022). https://doi.org/10.1101/2022.06.29.22276890

15. Bogowicz, M., et al.: Computed tomography radiomics predicts HPV status and local tumor control after definitive radiochemotherapy in head and neck squamous cell carcinoma. Int. J. Radiat. Oncol. Biol. Phys. **99**, 921–928 (2017). https://doi.org/10.1016/j.ijrobp.2017.06.002

16. Bagher-Ebadian, H., et al.: Application of radiomics for the prediction of HPV status for patients with head and neck cancers. Med. Phys. **47**, 563–575 (2020). https://doi.org/10.1002/MP.13977

17. Yu, K., et al.: Radiomic analysis in prediction of Human Papilloma Virus status. Clin. Transl. Radiat. Oncol. **7**, 49–54 (2017). https://doi.org/10.1016/J.CTRO.2017.10.001

18. Bagher-Ebadian, H., et al.: Radiomics outperforms clinical factors in characterizing human papilloma virus (HPV) for patients with oropharyngeal squamous cell carcinomas. Biomed. Phys. Eng. Express. **8**, 045010 (2022). https://doi.org/10.1088/2057-1976/AC39AB

19. van Timmeren, J.E., Cester, D., Tanadini-Lang, S., Alkadhi, H., Baessler, B.: Radiomics in medical imaging—"how-to" guide and critical reflection. Insights Imaging **11**, 1–16 (2020). https://doi.org/10.1186/S13244-020-00887-2/TABLES/3

20. Reiazi, R., et al.: Prediction of human papillomavirus (HPV) association of oropharyngeal cancer (OPC) using radiomics: the impact of the variation of CT scanner. Cancers (Basel) **13**, 2269 (2021). https://doi.org/10.3390/CANCERS13092269/S1

21. He, H., Garcia, E.A.: Learning from imbalanced data. IEEE Trans. Knowl. Data Eng. **21**, 1263–1284 (2009). https://doi.org/10.1109/TKDE.2008.239

22. Xie, C., et al.: Effect of machine learning re-sampling techniques for imbalanced datasets in 18F-FDG PET-based radiomics model on prognostication performance in cohorts of head and neck cancer patients. Eur. J. Nucl. Med. Mol. Imaging **47**(12), 2826–2835 (2020). https://doi.org/10.1007/s00259-020-04756-4

23. Chawla, N.V.: Data mining for imbalanced datasets: an overview. In: Maimon, O., Rokach, L. (eds.) Data Mining and Knowledge Discovery Handbook. Springer, Boston (2009). https://doi.org/10.1007/978-0-387-09823-4_45

24. Rich, B., et al.: Radiomics predicts for distant metastasis in locally advanced human papillomavirus-positive oropharyngeal squamous cell carcinoma. Cancers (Basel) **13**, 5689 (2021). https://doi.org/10.3390/CANCERS13225689

25. Yen, S.J., Lee, Y.S.: Under-sampling approaches for improving prediction of the minority class in an imbalanced dataset. In: Huang, D.S., Li, K., Irwin, G.W. (eds.) Intelligent Control and Automation. Lecture Notes in Control and Information Sciences, vol. 344. Springer, Heidelberg (2006). https://doi.org/10.1007/978-3-540-37256-1_89

26. Pereira, R.M., Costa, Y.M.G., Silla, C.N.: MLTL: a multi-label approach for the Tomek Link undersampling algorithm. Neurocomputing **383**, 95–105 (2020). https://doi.org/10.1016/J.NEUCOM.2019.11.076

27. Siddappa, N.G., Kampalappa, T.: Adaptive condensed nearest neighbor for imbalance data classification. Int. J. Intell. Eng. Syst. **12**, 104–113 (2019). https://doi.org/10.22266/IJIES2019.0430.11

28. Clark, K., et al.: The cancer imaging archive (TCIA): maintaining and operating a public information repository. J. Digit. Imaging **26**(6), 1045–1057 (2013). https://doi.org/10.1007/s10278-013-9622-7

29. Radiomics outcome prediction in Oropharyngeal cancer - TCIA DOIs - Cancer Imaging Archive Wiki. https://wiki.cancerimagingarchive.net/display/DOI/Radiomics+outcome+prediction+in+Oropharyngeal+cancer. Accessed 14 Mar 2023

30. Anderson, M.M.D., Quantitative, N., Working, I.: Matched computed tomography segmentation and demographic data for oropharyngeal cancer radiomics challenges. Sci. Data **4**, 170077 (2017). https://doi.org/10.1038/sdata.2017.77

31. Van Griethuysen, J.J.M., et al.: Computational radiomics system to decode the radiographic phenotype. Cancer Res. **77**, e104–e107 (2017). https://doi.org/10.1158/0008-5472.CAN-17-0339

32. Bharodiya, A.K.: Feature extraction methods for CT-scan images using image processing. Comput. Scan. 63 (2022). https://doi.org/10.5772/INTECHOPEN.102573

33. Larue, R.T.H.M., et al.: Influence of gray level discretization on radiomic feature stability for different CT scanners, tube currents and slice thicknesses: a comprehensive phantom study. Acta Oncol. **56**, 1544–1553 (2017). https://doi.org/10.1080/0284186X.2017.1351624

34. Shur, J.D., et al.: Radiomics in oncology: a practical guide. Radiographics **41**, 1717–1732 (2021). https://doi.org/10.1148/rg.2021210037

35. Kim, Y.J.: Machine learning model based on radiomic features for differentiation between COVID-19 and pneumonia on chest X-ray. Sensors **22**, 6709 (2022). https://doi.org/10.3390/s22176709

36. Tamal, M.: Grey Level Co-occurrence Matrix (GLCM) as a radiomics feature for Artificial Intelligence (AI) assisted Positron Emission Tomography (PET) images analysis. In: IOP Conference Series: Materials Science and Engineering, vol. 646 (2019). https://doi.org/10.1088/1757-899X/646/1/012047

37. Rahman, S., Mithila, S.K., Akther, A., Alans, K.M.: An empirical study of machine learning-based Bangla news classification methods. In: 2021 12th International Conference on Computing Communication and Networking Technologies, ICCCNT 2021 (2021). https://doi.org/10.1109/ICCCNT51525.2021.9579655

38. Laajili, R., Said, M., Tagina, M.: Application of radiomics features selection and classification algorithms for medical imaging decision: MRI radiomics breast cancer cases study. Inform. Med. Unlocked **27**, 100801 (2021). https://doi.org/10.1016/J.IMU.2021.100801

39. Chandrashekar, G., Sahin, F.: A survey on feature selection methods. Comput. Electr. Eng. **40**, 16–28 (2014). https://doi.org/10.1016/J.COMPELECENG.2013.11.024

40. Menze, B.H., et al.: A comparison of random forest and its Gini importance with standard chemometric methods for the feature selection and classification of spectral data. BMC Bioinform. **10**, 1–16 (2009). https://doi.org/10.1186/1471-2105-10-213/TABLES/4

41. Stancin, I., Jovic, A.: An overview and comparison of free Python libraries for data mining and big data analysis. In: 2019 42nd International Convention on Information and Communication Technology, Electronics and Microelectronics, MIPRO 2019 – Proceedings, pp. 977–982 (2019). https://doi.org/10.23919/MIPRO.2019.8757088

MetaLLM: Residue-Wise Metal Ion Prediction Using Deep Transformer Model

Fairuz Shadmani Shishir[1(✉)], Bishnu Sarker[2], Farzana Rahman[3], and Sumaiya Shomaji[1]

[1] Electrical Engineering and Computer Science, University of Kansas, Lawrence, USA
shishir@ku.edu
[2] Computer Science and Data Science, Meharry Medical College, Nashville, USA
[3] School of Computer Science and Mathematics, Kingston University London, London, UK

Abstract. Proteins bind to metals such as copper, zinc, magnesium, etc., serving various purposes such as importing, exporting, or transporting metal in other parts of the cell as ligands and maintaining stable protein structure to function properly. A metal binding site indicates the single amino acid position where a protein binds a metal ion. Manually identifying metal binding sites is expensive, laborious, and time-consuming. A tiny fraction of the millions of proteins in UniProtKB – the most comprehensive protein database – are annotated with metal binding sites, leaving many millions of proteins waiting for metal binding site annotation. Developing a computational pipeline is thus essential to keep pace with the growing number of proteins. A significant shortcoming of the existing computational methods is the consideration of the long-term dependency of the residues. Other weaknesses include low accuracy, absence of positional information, hand-engineered features, and a pre-determined set of residues and metal ions. In this paper, we propose MetaLLM, a metal binding site prediction technique, by leveraging the recent progress in self-supervised attention-based (e.g. Transformer) large language models (LLMs) and a considerable amount of protein sequences publicly available. LLMs are capable of modelling long residual dependency in a sequence. The proposed MetaLLM uses a transformer pre-trained on an extensive database of protein sequences and later fine-tuned on metal-binding proteins for multi-label metal ions prediction. A stratified 10-fold cross-validation shows more than 90% precision for the most prevalent metal ions. Moreover, the comparative performance analysis confirms the superiority of the proposed MetaLLM over classical machine-learning techniques.

Keywords: metal binding-site prediction · deep learning · attention · self-supervised learning language model · transformers · bio-transformers

© The Author(s), under exclusive license to Springer Nature Switzerland AG 2023
I. Rojas et al. (Eds.): IWBBIO 2023, LNBI 13920, pp. 42–55, 2023.
https://doi.org/10.1007/978-3-031-34960-7_4

1   Introduction

Proteins are biomolecules composed of amino acid chains that form the building blocks of life and play fundamental roles in the entire cell cycle. They perform multitudes of functions including catalyzing reactions as enzymes, participating in the body's defense mechanism as antibodies, forming structures, and transporting important chemicals. In addition, they interact with other molecules including proteins, DNAs, RNAs, and drug molecules to act on metabolic and signaling pathways, cellular processes, and organismal systems. Protein structures and interactions describe the molecular mechanism of diseases and can convey important insights about disease prevention, diagnosis, and treatments. Likewise, proteins bind to different metal ions, such as zinc, iron, copper etc. to play necessary roles in many biological processes, including enzyme catalysis, regulation of gene expression, and oxygen transport. Metal ions are often bound to specific sites on proteins, known as metal binding sites, which play a key role in determining protein's structure and function. Identifying metal binding sites manually using various experimental procedures such as mass spectrometry, electrophoretic mobility shift assay, metal ion affinity column chromatography, gel electrophoresis, nuclear magnetic resonance spectroscopy, absorbance spectroscopy, X-ray crystallography, and electron microscopy is an expensive, laborious, and time-consuming process [23]. A very small fraction of the millions of proteins stored in UniProtKB [1] – the most comprehensive protein database – are annotated with metal binding sites. Millions of other proteins are awaiting for metal binding site annotation. To keep pace with the exponential increase of protein sequences in the public databases, it is essential to develop computational approaches for predicting metal binding sites in proteins. Considering the benefits it can provide in understanding function and structure of proteins as well as having practical implications in drug design and biotechnology, automatic prediction of metal binding sites in proteins is considered to be an important problem in computational biology.

Metal Binding Site Prediction: Predicting the binding sites for metals is a challenging problem in computational biology. Decades of research has been dedicated to discovering computational approaches that can accurately predict the metal ions as well as the positions where they bind to the proteins [3,5,9,19]. A comprehensive review of recent advances in computational approaches for predicting metal binding sites can be found in [34]. Broadly, these approaches can be categorized into following three groups based on the type of attributes they take into account: 1) structure-based methods that use three dimensional secondary structure of proteins as primary data; 2) sequence-based methods that use amino acid sequence as primary data; and 3) combined methods that leverage both structure and sequence attributes.

Structured-Based: Structure-based approaches for predicting metal binding sites use a combination of geometric, chemical, and electrostatic criteria to explain the metal-protein interaction that eventually works to identifypossible binding sites in a protein structure. A relatively early work described in [29] uses electrostatic energy computation [10, 22] to find binding affinity of metal ion to a site in a protein structure. In [4], the proposed method learns geometric constraints to differentiate binding sites for different metals based on the statistical analysis of structures of metal binding proteins. Another structured-based method is proposed in [38] for predicting only the zinc binding sites. A template-based method is proposed in [21] where a database of pre-computed structural templates for metal binding sites is searched against each residue in a query protein to find which metal binds to it. mFASD [13] is a structure based model to predict metal binding sites. From the structures of metal binding proteins, mFASD computes the functional atom set (FAS) - the set of the atoms that are in contact with the metal - for each metal, and store it as reference. The distance between FASs of different metals is used to distinguish binding sites for different metals. Given a query protein structure, mFASD scans the database reference FASs against FAS of each sites, and computes the distance. The decision is made based on how many reference FASs matched for a given metal. One of the shortcomings of the structure based models is that they are dependent on structural databases such as Protein Data Bank (PDB) which is very limited in terms of amount of protein structures in the database.

Sequence-Based: On the other hand, sequence-based methods for predicting metal binding sites use sequence conservation, alignment, and similarity to identify metal binding sites. For example, [2] is a sequence-based method that find the patterns of binding sites from the metal binding proteins. Sliding window-based feature extraction and biological feature encoding techniques are proposed to predict the protein metal-binding amino acid residues from its sequence information using neural network in [17] and using support vector in machine [18]. MetalDetector [20] is a sequence-based technique that uses decision tree to classify histidine residues in proteins into one of two states as 1) free, or 2) metal bound; and cysteine residues into one of three states as 1) free, 2) metal bound, or 3) disulfide bridged. A two stage machine learning model is proposed in [24] that includes support vector machine as local classifier in the first stage, and a recurrent neural network (RNN) [14] in second stage to refine the classification based on dependencies among residues. A combined approach proposed in [30] where support vector machine, sequence homology, and position specific scoring matrix (PSSM) are put together to predict zinc-binding Cys, His, Asp and Glu residues.

Additionally, there are combined methods for predicting metal binding sites that use an ensemble of sequence, structural, and physicochemical features. For example, MetSite [32] is a method that uses both the sequence profile information and approximate structural data - PSSM scores together with secondary structure, site residue distances, and solvent accessibility - are fed into neural network machine learning technique.

Deep Learning and Transformers: Classical machine learning techniques such as decision tree, random forests, support vector machine etc. have been widely applied to the problem of metal binding site prediction. These model face several challenges, such as 1) they mostly depends on hand-engineered features e.g. K-mers, 2) they can not model variable length sequences data, and 3) they fail to comprehend long-distance residual dependency. Moreover, scaling these models to work on millions of variable length sequences is another big challenge.

Over the last decade, Deep Learning based techniques such as convolutional neural network (CNN) [16], long short term memory (LSTM) [14], transformers [33] etc. have grown to be extensively powerful. Generally, Deep Learning is a type of machine learning built using neural networks inspired by the structure, function and deep interconnection of the brain neurons [16]. In a typical deep neural network, many layers of interconnected nodes process vast amounts of complex data for learning from the experience. The strength of the deep learning lies in its ability to automatic feature engineering by learning low rank vector representation of data points. These advanced models are very efficient in handling large datasets. For example, [11,12] applied different deep learning architectures namely 2D CNN, LSTM, RNN coupled with various feature extraction techniques for predicting metal-binding sites of Histidines (HIS) and Cysteines (CYS) amino acids.

Very recently Transformer - a deep learning model for natural language processing tasks such as language translation, text summarization, question answering etc. - is introduced by historic paper "Attention is All You Need" [33]. Generally, a transformer comprises of an encoder and a decoder neural network. Encoder and decoder composed of multiple layers of self-attention layers. The input of the encoder consists of queries as well as keys of dimension d_k, and values of dimension d_v. There are dot products of the query with all keys, which will divide each by $\sqrt{d_k}$, and apply a softmax function to obtain the attention weights which is shown in Eq. 1. The attention function on a set of queries was done simultaneously, packed together into a matrix Q. The keys and values are also packed together into matrices K and V. Transformer is capable of looking into the data from multiple perspectives following a multi-head mechanism shown in Eq. 3. Each head computes an attention weights that provides a new perspective that is shown in Eq. 2.

$$Attention(Q, K, V) = softmax(\frac{QK^T}{\sqrt{d_k}})V \tag{1}$$

$$head_i = Attention(QW_i^Q, KW_i^K, VW_i^V) \tag{2}$$

$$MultiHead(Q, K, V) = Concat(head_1, ..., head_h)W^O \tag{3}$$

Transformer-based language models overcome the challenge of modeling long-distance dependency in natural text by introducing attention weights. This property ideally make it suitable for using in protein sequence modeling that long chain of amino acids.

While many of the existing methods are performing well, predicting metal binding sites is still a challenging problem as well as an open problem in computational biology. Partly because metals exhibit similar chemical properties making it hard for machine learning models to differentiate. A major shortcoming of the existing computational methods is in taking into account the long-distance dependency of the residues to distinguish the presence of distinct metal ions. There are other shortcomings such as low accuracy, absence of positional information, hand-engineered features, and pre-determined set of residues and metal ions. Building high performing prediction model would require a comprehensive understanding of the structural, chemical, and biological factors that influence metal binding. Therefore, ongoing research effort is important to improve our understanding of metal binding in proteins and to develop more accurate and efficient computational pipeline to keep pace with the growing number of proteins. Considering the challenges and the availability of protein data, in this paper, we propose MetaLLM, a metal and binding site prediction technique by leveraging the recent progress in self-supervised attention-based (e.g. Transformer) large language models (LLMs) and huge amount of protein sequences publicly available. LLMs are capable of modeling long residual dependency in a sequence. The proposed MetaLLM uses a transformer pre-trained on large database of protein sequences, and later fine-tuned on metal binding proteins for multi-label metal ions prediction. In the fine-tuning step, the low rank sequence embeddings are concatenated with positional one-hot embeddings and fed into the fully connected neural network layer for metal ions prediction. A stratified 10-fold cross-validation shows more than 90% precision for the most prevalent metal ions.

2   Methodology

In this section, the proposed methodology and overall workflow have been discussed thoroughly. The workflow starts by computing the sequence embeddings for training and testing datasets using pre-trained protein language model provided by the Bio-Transformers[1] python wrapper of ProtTrans [8]. After that, each sequence embedding is combined with the positional one-hot encoding of metal binding sites. Finally, this combined vector is fed to a fully connected multi-layer deep neural network to predict the metal ion that finds the one-hot encoded position. The performance is validated on a set of text examples following the same workflow. The working principle is depicted in Fig. 1. In the following subsections, we detail the workflow of model development and evaluation.

2.1   MetaLLM: Residue-Wise Metal Ion Prediction

MetalLLM - the proposed metal ion prediction model - is built on Large Language Model (LLM) pretrained on millions of protein sequences to provide sequence embeddings. Powerful LLMs are built using encoder-decoder based deep learning technique called Transformer [26,33]. Thanks to the capability of transformers in ingesting large amount of data and learning meaningful representation, it has already found application in various life science and biomedicine tasks such as protein folding, drug discovery, and gene expression prediction [35]

---

[1] https://github.com/DeepChainBio/bio-transformers.

in academia and industry alike. ProtTrans [8] is a such model using transformer architecture that includes self-attention layers in both encoder and decoder and is trained on large amount of sequence data. Pretrained ProtTrans provides fixed dimensional embeddings for given sequence. We used ProtTrans as the background model to transform the variable length protein sequence into a fixed length numerical vector for the purpose of predicting metal ion that binds to the protein. To include the binding site in the pipeline, we have combined the positional one-hot encoding with sequence embedding.

Problem Statement: Therefore, the problem that we solve in this paper is defined as predicting metal ion given an input protein sequence and the residue position we are interested to look at as possible binding site. We formulate this as a machine learning problem where the sequence is represented by latent features computed using ProtTrans combined with positional enconding vector constitute the features, and finally we train a fully connected deep neural network for predicting the probability of metal ions as a downstream task.

Using transformer-based LLM has many benefits; 1) pretrained model reduce the computation cost significantly, 2) can be easly fine-tuned towards transfer learning for new data as well as new tasks, 3) they are capable to capturing long distance dependancy among the residues and tokens, 4) easily scalable.

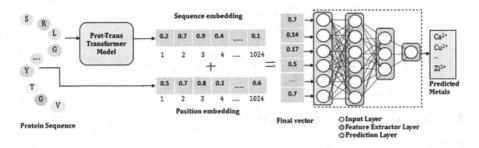

Fig. 1. The overall architecture of the proposed framework of MetaLLM: (i) the protein sequences are delivered as an input to the pre-trained language model to produce the sequence embedding, (ii) the embedding for identifying the binding positions of the amino acids are extracted (iii) both the embedding are concatenated to obtain a final vector, and (iv) the final vectors are fed into a deep neural network to predict the metal ions.

Model Description: The proposed model is shown in Fig. 1 initially employs a pre-trained language model called ProtTrans which effectively generate embedded and informative protein sequence representation from the raw protein sequence. ProtTrans, a large protein language model, was used as the back-end of our proposed architecture. Moreover, we have also tested our representation for the classification of various metals to justify the novelty of the language model. Our proposed architecture leverages ProtTrans model for the feature extraction, which was trained on computationally expensive hardwares [8]. Then, the embedded 1024-dimensional features were extracted from the last layer of Prot-Trans model and classification was done integrating the sites information with the protein sequence embedding. The fixed 1024 dimension was chosen through the investigating in our experiment and 1024 dimentional embedding has given most discriminant features set which performed good for our methods. To achieve better performance from our model, we also concatenated the positional information by one-hot encoding since the specific binding location of a the metal ion is very important in a protein sequence. Our novel contribution on this paper is to carefully design a neural network which is more robust and cost effective with limited number of parameters without dropping the accuracy of the prediction.

The metal prediction was done by the classification model on top of the language model. The input of the classification layer was the 1024 size embedding from the language model and then concatenated with the same dimensional sites information. The weights of the four layers of classification model was initialized randomly and for the non-linearity, relu activation function was used. In order to get the final prediction, sigmoid activation function was considered. We also used stratified 10-fold cross validation to evaluate the whole dataset better and maximize the performance of our model. The mathematical representation are given in equations respectively.

$$\sigma(z) = \frac{1}{1 + e^{-z}} \tag{4}$$

$$Relu(z) = max(0, z) \tag{5}$$

$$BCE = -(y \log(p) + (1 - y) \log(1 - p)) \tag{6}$$

$$multiclass\ classification = -\sum_{c=1}^{M} y_{o,c} \log(p_{o,c}) \tag{7}$$

3   Experiments and Results

3.1   Experimental Details

Dataset Description: To conduct the experiment, we have prepared a dataset with sequences retrieved from UniProt Knowledge Base (UniProtKB) [1]. UniProtKB is the largest protein database divided into two parts; 1) UniProtKB/SwissProt contains manually reviewed protein data and 2) UniProtKB/TrEMBL contains unreviewed protein data. We have accessed UniProtKB/SwissProt in March 2022 to retrieve protein sequences, binding sites and metal ions. Therefore, all of the protein data selected for this experiment were manually reviewed. To eliminate the possible fragmented protein sequences, the sequences smaller than 50 amino acids long are removed. We have also removed the sequences longer than 1000 amino acids long. Moreover, metals that does not have a sufficient number of instances are removed during further pre-processing. Finally, it contains 18,348 protein sequences holding 6 different metals in total. The distribution of metals is shown in the Fig. 2.

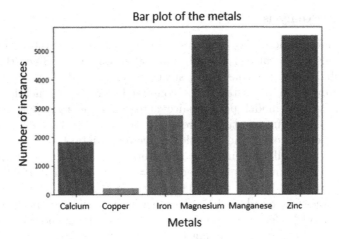

Fig. 2. This figure shows the number of samples of our dataset.

Implementation Details: MetaLLM is implemented using Bio-Transformers[2] python package that wraps the implementation of ProtTrans (Protein Transformer) [8] and ESM (Evolutionary Scale Modeling) [25–27] - two of the large language models trained on million of protein sequences to predict embeddings.

MetaLLM gets the sequence embeddings using Bio-Transformer with Prot-Trans backend. Sequence embeddings are added to multi-hot position encodings and the resultant vector is fed into the fully connected neural network. MetaLLM proposes a four-layer fully connected neural network with 500 hidden neurons in the first layer, 300 neurons in the second and third layer, and the final layer is with 100 hidden neurons before it goes to the soft-max layer for predicting the likelihood of different metals. To prevent the over-fitting of the network, we carefully used a dropout with a rate of 0.02. Batch size of 64 and learning rate in the range of 0.0001 to 0.1 was found to be optimal. The model is trained for 200 epochs with Adam optimizer to optimize binary cross-entropy loss function. The training takes 45 min on Nvidia GeForce RTX 3090 GPU-enabled computer. Once the model is trained, it takes 5 s to make a prediction for one batch of protein sequences.

Evaluation Metrics: To evaluate the performance of our proposed model, we have considered the following evaluation metrics; 1) $Accuracy = (\frac{TP+TN}{TP+TN+FP+FN})$, 2) $Precision = \frac{TP}{TP+FP}$, 3) $Recall = \frac{TP}{TP+FN}$, and 4) $F1\text{-}max = \frac{2*Precision*Recall}{Precision+Recall}$. F1-max score is also computed as $\frac{2*TP}{2*TP+FP+FN}$. Here, TP = true positives and TN = true negatives denote correctly predicted positive and negative examples respectively, whereas FP = false positives and FN = false negatives denote the number of incorrectly predicted metals respectively.

---

[2] https://github.com/DeepChainBio/bio-transformers.

3.2   Result Analysis

Model Performance: We evaluated MetaLLM by Precision, recall, and F1-score using the stratified 10-fold cross-validation (CV). Since the dataset is class imbalanced, we considered stratified cross-validation which ensures that the proportion of instances is conserved across the metals per fold. Table 1 shows the model performance across various metals for with and without positional information. Precision, recall, and F1-scores are computed as average over the 10-folds. Overall, MetaLLM has performed very well across all the metals. However, the superior performance is observed for ion like magnesium (**Precision = 0.97, Recall = 0.94**) and iron (**Precision = 0.96, Recall = 0.94**) considering positional information. Table 1 also lists precision, recall, and F1-scores without considering positional information - the model is fed with a protein sequence embedding and asked to predict the likelihood of metal ions binding to it. It is evident from the Table 1 that performance dropped significantly for copper, iron, and magnesium. MetaLLM achieved an average CV accuracy of 90% when positional encoding is used. The box plots shown in Fig. 3 depict the spread of performance scores along the folds and across the metals. Finally, Fig. 4 shows the training and validation accuracy of our proposed model for a single fold across the epochs.

Table 1. Performance metrics of the proposed model

Metal Name	With Position			Without Position		
	Precision	Recall	F1	Precision	Recall	F1
Calcium	0.83	0.84	0.84	0.94	0.71	0.81
Copper	0.94	0.67	0.78	0.67	0.93	0.78
Iron	0.96	0.94	0.95	0.89	0.90	0.90
Magnesium	0.97	0.94	0.95	0.72	0.94	0.82
Manganese	0.92	0.93	0.92	0.71	0.47	0.56
Zinc	0.91	0.89	0.90	0.84	0.95	0.89

Comparison with Classical Machine Learning Methods: To briefly discuss our proposed method, we first consider the Bio-Transformer model for generating the embedding from the protein sequences, next we add the positional information for metal binding to the embeddings to create new representations, and finally, we train a fully connected deep neural network considering the new representation to classify the metal ions. By leveraging the promising feature representation techniques from the individual steps, we maximize the prediction of unknown metal ions. However, unlike ours, there exist several studies which rely on the traditional classifiers for predicting the binding sites and metal ions [37]. To compare the performance of our work with the existing classical

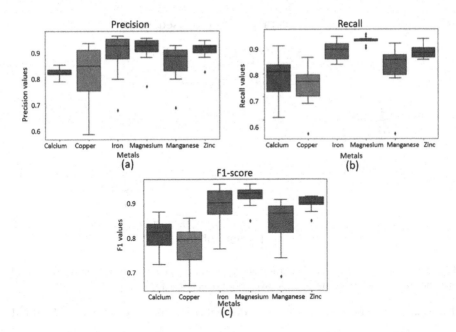

Fig. 3. The box-plot (a) shows the precision of 10-fold CV results for 6 metals. A comparison of the metals clearly indicates that zinc and magnesium have stable precision values of $93\pm3\%$ and $91\pm3\%$ respectively. The overall distribution of these two metals is not very scattered and also has low variance. On the contrary, copper, iron, and manganese have larger percentile values with some outliers also. In box-plot (b) zinc has outperformed compared to other metals with a recall value of $95\pm2\%$. However, there are some outliers in that distribution due to the limitation of the samples. The calcium and copper, although having limited number of samples, performed well with a recall value of $82\pm5\%$ and $78\pm4\%$. The box plot (c) shows the F1-score of CV. Iron, magnesium, and zinc have performed well in terms of F1-score and variation is also low compared to other metals.

ones, we determined five widely used methods: K-Nearest Neighbor [36], Logistic Regression [7], Xgb classifier [31], Gaussian Naïve Bayes [15], and Support Vector Machine [28]. To train the classical models with our dataset, we have considered the embeddings extracted from Bio-Transformer and added positional information to create the final feature set for training. Once the models are trained, we compared the findings from these five methods with that of our proposed study. As shown in Table 2, the classical methods offer $42\% \sim 76\%$ of F1- score for Calcium. Similarly, it offers $19\% \sim 62\%$ for Copper, $73\% \sim 90\%$ for Iron, $53\% \sim 90\%$ for Magnesium, $59\% \sim 85\%$ for Manganese, and $65\% \sim 86\%$ for Zinc respectively. Whereas our proposed model achieved F1- score of 84% for Calcium, 78% for Copper, 95% for Iron, 95% for Magnesium, 92% for Manganese, and 90% for Zinc. As deliberated earlier, our proposed method is established on state-of-the-art deep learning models which provide promising distinctive features, therefore it can offer better prediction F1-score than the classical methods.

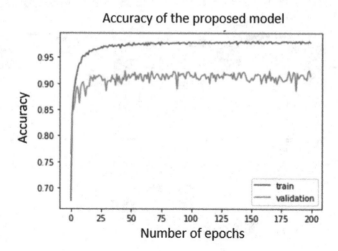

Fig. 4. Train and validation accuracy score of the proposed model. The overall trend of validation curve remained stable over the epochs during training.

Table 2. Metal wise F1-scores of the traditional classifiers compare against MetaLLM

Classifier	Calcium	Copper	Iron	Magnesium	Manganese	Zinc
Logistic Regression	0.72	0.59	0.88	0.89	0.81	0.84
Naive Bayes	0.42	0.19	0.73	0.53	0.59	0.65
SVM	0.75	0.51	0.89	0.90	0.84	0.86
KNN	0.76	0.38	0.90	0.84	0.85	0.86
XGB	0.63	0.62	0.84	0.80	0.70	0.74
MetaLLM	**0.84**	**0.78**	**0.95**	**0.95**	**0.92**	**0.90**

The Impact of Positional Information in Proposed Model: It has been estimated that more than one-third of the entire proteomes are metal-binding proteins [6]. Therefore, metal position in a protein plays a significant role in protein binding. Metal ions, such as zinc and copper, can act as co-factors in enzyme reactions and can also play a role in stabilizing protein-protein interactions. In some cases, metal ions can act as a "switch" in protein activity, allowing or preventing binding to other molecules. Considering the aforementioned reasons, our experiment was designed with and without concatenating sites information with the sequences. We also investigated further to understand whether site of a metal can contribute to increase accuracy while making prediction or not. The network architecture was developed based-on that workflow. It was found in the experiment that the model performed better with the positional information than without providing it while making the prediction of the metals, which is indeed one of the major findings in our work.

4   Conclusion

In conclusion, predicting metal binding sites in proteins is a complex and multifaceted problem that requires integrating multiple types of information and developing sophisticated computational methods. While significant progress has been made in this area, there is still much to be learned about the roles of metal ions in biological systems and the computational approaches appropriate to predict where they bind to proteins. To this end, in this paper, we have described a transformer-based model called MetaLLM to incorporate the benefits of highly sophisticated self-supervised deep learning technique to propose an end-to-end computational pipeline that is capable of predicting the presence of metal ions as well as the binding sites with state-of-art performance in terms of precision, recall and F1-score. The performance of MetaLLM is validated on a dataset of protein sequences extracted from UniPortKB/SwissProt. Being a transformer-based model, MetaLLM leverage the attention mechanism for capturing long-distance residual dependency which is crucial for identifying and distinguishing the binding sites. MetaLLM is limited to sequences in the range of 50 to 1000 amino acids. In the future, we envision to extend the model for longer sequences and compare our results with some of the existing tools for metal binding site prediction. Furthermore, we aim to include structural pipeline as a separate stream to validate the prediction from structural context.

References

1. Uniprot: the universal protein knowledgebase in 2023. Nucleic Acids Res. **51**(D1), D523–D531 (2023)
2. Andreini, C., Bertini, I., Rosato, A.: A hint to search for metalloproteins in gene banks. Bioinformatics **20**(9), 1373–1380 (2004)
3. Aptekmann, A.A., Buongiorno, J., Giovannelli, D., Glamoclija, M., Ferreiro, D.U., Bromberg, Y.: mebipred: identifying metal-binding potential in protein sequence. Bioinformatics **38**(14), 3532–3540 (2022). https://doi.org/10.1093/bioinformatics/btac358
4. Babor, M., Gerzon, S., Raveh, B., Sobolev, V., Edelman, M.: Prediction of transition metal-binding sites from apo protein structures. Proteins: Struct. Funct. Bioinf. **70**(1), 208–217 (2008)
5. Bromberg, Y., et al.: Quantifying structural relationships of metal-binding sites suggests origins of biological electron transfer. Sci. Adv. **8**(2), eabj3984 (2022)
6. Cheng, Y., et al.: Co-evolution-based prediction of metal-binding sites in proteomes by machine learning. Nat. Chem. Biol., 1–8 (2023)
7. Dreiseitl, S., Ohno-Machado, L.: Logistic regression and artificial neural network classification models: a methodology review. J. Biomed. Inf. **35**(5–6), 352–359 (2002)
8. Elnaggar, A., et al.: Prottrans: towards cracking the language of life's code through self-supervised deep learning and high performance computing. arXiv preprint arXiv:2007.06225 (2020)
9. Gucwa, M., et al.: CMM-An enhanced platform for interactive validation of metal binding sites. Protein Sci. **32**, e4525 (2022)

10. Guerois, R., Serrano, L.: The sh3-fold family: experimental evidence and prediction of variations in the folding pathways. J. Mol. Biol. **304**(5), 967–982 (2000)
11. Haberal, İ., Oğul, H.: DeepMBS: prediction of protein metal binding-site using deep learning networks. In: 2017 Fourth International Conference on Mathematics and Computers in Sciences and in Industry (MCSI), pp. 21–25. IEEE (2017)
12. Haberal, İ, Oğul, H.: Prediction of protein metal binding sites using deep neural networks. Mol. Inf. **38**(7), 1800169 (2019)
13. He, W., Liang, Z., Teng, M., Niu, L.: mFASD: a structure-based algorithm for discriminating different types of metal-binding sites. Bioinformatics **31**(12), 1938–1944 (2015)
14. Hochreiter, S., Schmidhuber, J.: Long short-term memory. Neural Comput. **9**(8), 1735–1780 (1997)
15. Jahromi, A.H., Taheri, M.: A non-parametric mixture of gaussian naive bayes classifiers based on local independent features. In: 2017 Artificial Intelligence and Signal Processing Conference (AISP). pp. 209–212 (2017). https://doi.org/10.1109/AISP.2017.8324083
16. LeCun, Y., Bengio, Y., Hinton, G.: Deep learning. Nature **521**(7553), 436–444 (2015)
17. Lin, C.T., Lin, K.L., Yang, C.H., Chung, I.F., Huang, C.D., Yang, Y.S.: Protein metal binding residue prediction based on neural networks. Int. J. Neural Syst. **15**(01n02), 71–84 (2005)
18. Lin, H., et al.: Prediction of the functional class of metal-binding proteins from sequence derived physicochemical properties by support vector machine approach. In: BMC Bioinformatics, vol. 7, pp. 1–10. BioMed Central (2006)
19. Lin, Y.F., Cheng, C.W., Shih, C.S., Hwang, J.K., Yu, C.S., Lu, C.H.: MIB: metal ion-binding site prediction and docking server. J. Chem. Inf. Model. **56**(12), 2287–2291 (2016)
20. Lippi, M., Passerini, A., Punta, M., Rost, B., Frasconi, P.: Metaldetector: a web server for predicting metal-binding sites and disulfide bridges in proteins from sequence. Bioinformatics **24**(18), 2094–2095 (2008)
21. Lu, C.H., Lin, Y.F., Lin, J.J., Yu, C.S.: Prediction of metal ion-binding sites in proteins using the fragment transformation method. PLoS ONE **7**(6), e39252 (2012)
22. Mendes, J., Guerois, R., Serrano, L.: Energy estimation in protein design. Curr. Opin. Struct. Biol. **12**(4), 441–446 (2002)
23. Mohamadi, A., Cheng, T., Jin, L., Wang, J., Sun, H., Koohi-Moghadam, M.: An ensemble 3d deep-learning model to predict protein metal-binding site. Cell Rep. Phys. Sci. **3**(9), 101046 (2022)
24. Passerini, A., Punta, M., Ceroni, A., Rost, B., Frasconi, P.: Identifying cysteines and histidines in transition-metal-binding sites using support vector machines and neural networks. Proteins: Struct. Funct. Bioinf. **65**(2), 305–316 (2006)
25. Rao, R., et al.: MSA transformer. bioRxiv (2021). https://doi.org/10.1101/2021.02.12.430858, https://www.biorxiv.org/content/10.1101/2021.02.12.430858v1
26. Rao, R.M., Meier, J., Sercu, T., Ovchinnikov, S., Rives, A.: Transformer protein language models are unsupervised structure learners. bioRxiv (2020). https://doi.org/10.1101/2020.12.15.422761, https://www.biorxiv.org/content/10.1101/2020.12.15.422761v1
27. Rives, A., et al.: Biological structure and function emerge from scaling unsupervised learning to 250 million protein sequences. bioRxiv (2019). https://doi.org/10.1101/622803, https://www.biorxiv.org/content/10.1101/622803v4
28. Rossi, F., Villa, N.: Support vector machine for functional data classification. Neurocomputing **69**(7–9), 730–742 (2006)

29. Schymkowitz, J.W., Rousseau, F., Martins, I.C., Ferkinghoff-Borg, J., Stricher, F., Serrano, L.: Prediction of water and metal binding sites and their affinities by using the fold-x force field. Proc. Natl. Acad. Sci. **102**(29), 10147–10152 (2005)

30. Shu, N., Zhou, T., Hovmöller, S.: Prediction of zinc-binding sites in proteins from sequence. Bioinformatics **24**(6), 775–782 (2008)

31. Shwartz-Ziv, R., Armon, A.: Tabular data: deep learning is not all you need. Inf. Fusion **81**, 84–90 (2022)

32. Sodhi, J.S., Bryson, K., McGuffin, L.J., Ward, J.J., Wernisch, L., Jones, D.T.: Predicting metal-binding site residues in low-resolution structural models. J. Mol. Biol. **342**(1), 307–320 (2004)

33. Vaswani, A., et al.: Attention is all you need. In: Advances in Neural Information Processing Systems, vol. 30 (2017)

34. Ye, N., et al.: A comprehensive review of computation-based metal-binding prediction approaches at the residue level. BioMed Res. Int. 2022 (2022)

35. Yuan, Q., Chen, S., Wang, Y., Zhao, H., Yang, Y.: Alignment-free metal ion-binding site prediction from protein sequence through pretrained language model and multi-task learning. bioRxiv (2022)

36. Zhang, S., Li, X., Zong, M., Zhu, X., Wang, R.: Efficient kNN classification with different numbers of nearest neighbors. IEEE Trans. Neural Netw. Learn. Syst. **29**(5), 1774–1785 (2018). https://doi.org/10.1109/TNNLS.2017.2673241

37. Zhao, J., Cao, Y., Zhang, L.: Exploring the computational methods for protein-ligand binding site prediction. Comput. Struct. Biotechnol. J. **18**, 417–426 (2020)

38. Zhao, W., et al.: Structure-based de novo prediction of zinc-binding sites in proteins of unknown function. Bioinformatics **27**(9), 1262–1268 (2011)

Genome-Phenome Analysis

Prediction of Functional Effects of Protein Amino Acid Mutations

Óscar Álvarez-Machancoses[1] , Eshel Faraggi[2] , Enrique J. de Andrés-Galiana[1] ,
Juan Luis Fernández-Martínez[1] , and Andrzej Kloczkowski[3,4(✉)]

[1] Department of Mathematics, University of Oviedo, 33007 Oviedo, Spain
[2] Department of Physics, Indiana University Purdue University, Indianapolis, IN 46202, USA
[3] The Steve and Cindy Rasmussen Institute for Genomic Medicine, Nationwide Children Hospital, Columbus, OH 43205, USA
Andrzej.Kloczkowski@nationwidechildrens.org
[4] Department of Pediatrics, The Ohio State University, Columbus, OH 43205, USA

Abstract. Human Single Amino Acid Polymorphisms (SAPs) or Single Amino Acid Variants (SAVs) usually named as nonsynonymous Single Nucleotide Variants nsSNVs) represent the most frequent type of genetic variation among the population. They originate from non-synonymous single nucleotide variations (missense variants) where a single base pair substitution alters the genetic code in such a way that it produces a different amino acid at a given position. Since mutations are commonly associated with the development of various genetic diseases, it is of utmost importance to understand and predict which variations are deleterious and which are neutral. Computational tools based on machine learning are becoming promising alternatives to tedious and highly costly mutagenic experiments. Generally, varying quality, incompleteness and inconsistencies of nsSNVs datasets degrade the usefulness of machine learning approaches. Consequently, robust and more accurate approaches are essential to address these issues. In this paper, we present the application of a consensus classifier based on the holdout sampling, which shows robust and accurate results, outperforming currently available tools. We generated 100 holdouts to sample different classifiers' architectures and different classification variables during the training stage. The best performing holdouts were selected to construct a consensus classifier and tested by blindly utilizing a k-fold ($1 \leq k \leq 5$) cross-validation approach. We also performed an analysis of the best protein attributes for predicting the effects of nsSNVs by calculating their discriminatory power. Our results show that our method outperforms other currently available tools, and provides robust results, with small standard deviations among folds and high accuracy. The superiority of our algorithm is based on the utilization of a tree of holdouts, where different machine learning algorithms are sampled with different boundary conditions or different predictive attributes.

Keywords: Protein Mutations · Single Amino Acid Mutations · Deleterious Mutations · Neutral Mutations

I. Rojas et al. (Eds.): IWBBIO 2023, LNBI 13920, pp. 59–71, 2023.
https://doi.org/10.1007/978-3-031-34960-7_5

1 Introduction

Identifying amino acid substitutions that have impact on protein function and their involvement in disease is one of the forefront challenges in proteomics, metabolomics and medical genomics [1]. It is estimated that each individual could experience from 24,000 to 40,000 amino acid variations [2], with nonsynonymous Single Nucleotide Variants (nsSNVs) being the most frequent type of genetic variation for humans, accounting for at least 90% of sequence dissimilarities [3, 4]. While some amino acid variations are neutral [5, 6], some have an effect on gene expression, on the transcriptome, and/or on translated protein function. These effects can lead to the disease development [7]. Roughly a half of known disease-related mutations are due to non-synonymous variants [2, 8], expressed as amino-acid mutations. Therefore, it is important to unravel the links between nonsynonymous Single Nucleotide Variants and associated diseases to discriminate between pathogenic and neutral substitutions [9]. It has been found that these substitutions could be directly related to pathological effects such as Parkinson's or Alzheimer's diseases, or to the involvement in complex diseases, such as cancer development [10]. However, despite the massive amount of data collected in the last few years and enormous efforts invested in unravelling these links, a comprehensive understanding of these disorders and interpretation of the metabolomics of various diseases remains unknown, due to the high complexity of the underlying problem. Consequently, computational and machine learning methodologies have emerged as an important tool in the screening of nsSNVs data.

A wide range of computational tools are available to predict the effects of amino acid substitutions for a given phenotype. Of special interest are methods based on machine learning and classification that integrate different types of biological information and attributes to discriminate between deleterious or neutral amino acid substitutions [11]. The use of machine learning and AI approaches is necessary because the forward model that links the amino acids substitutions with their effect is a priori unknown.

Some of these tools utilize evolutionary information to discriminate whether a particular amino acid mutation is deleterious or neutral, e.g., Provean [12, 13], SIFT [14], PANTHER [15] or the Evolutionary Action method [16, 17]. The major drawback is that they utilize only evolutionary conservation information for a given position in multiply aligned homologous sequences. Although this information is being characterized by a single score, it represents several molecular effects such as protein stability, function and solubility, among others. Aiming to improve evolutionary data-based predictions, other methods have been developed with physicochemical attributes on top of evolutionary and sequence information. Mutation Taster [18], CADD [19], Polyphen-2 [20], SNP&GO [21], PhD-SNP [22], PredictSNP [23] or MAPP [24] are few examples of available tools. These methodologies employ several machine learning techniques to predict the effect of missense mutation based on a set of features. However, the accuracy of these tools remains limited due to high false-positive rate, i.e., predicted deleterious mutations are often experimentally observed to be neutral [25]. Another important issue is that prediction tools tend to focus on achieving high accuracies, without understanding the links and connections between different classification attributes, the overall performance and the uncertainty analysis of the prediction. Even though, protein structure, sequence and evolutionary information as a whole play an important factor on the protein function, it is seldom exploited by classification methods [26–28]. Finally, another major drawback of these supervised machine learning methods is that decision rules are derived from overlapping training datasets, leading to overestimates of the performance [29, 30]. Consequently, it is very important to train, test and validate the performance of these tools on completely independent datasets [31]. This problem is further enhanced by the lack of variability in training datasets, leading to biased predictions [32].

To avoid the limitations listed above, we aim here to predict the effect of mutations by enhanced sampling of mutant protein attributes to comprehend the relationship between protein structural stability and variant deleteriousness. Furthermore, in order to avoid biased predictions, the dataset has been randomly divided into different holdouts in which training and testing is performed by randomly selecting the nsSNVs attributes. We then select those holdouts that provide the best predictions and combine them into a consensus classifier. It is generally well known that individual classifiers can be combined into a consensus form [33]; recent examples of consensus-based tools are Meta-SNP [34], CONDEL [35] or PredictSNP [23], among others.

Our consensus classifier outperforms the individual holdouts' predictions and offers a robust and insightful prediction scheme, since it is possible to detect and analyze protein attributes that better classify the amino acid substitutions according to their deleteriousness.

2 Methods

2.1 nsSNV Datasets

A dataset of disease-causing and neutral protein mutations has been extracted from the UniProt/SwissVar database [36] in order to train, test and validate the algorithm. We have applied k-fold ($1 \leq k \leq 5$) cross validation, in which the 5th fold was utilized for blind validations and the rest has been used to train the model by generating 100 random holdouts, split such that 75% of the holdout data was used for training and 25% of the data was used for testing. The whole set is composed of 38,460 single point variations from 9,067 proteins.

2.2 Protein Mutation Prediction Methodology: The Holdout- nsSNV Algorithm

Predicting the effect of a nsSNVs is a problem of high uncertainty, therefore; there is a need for a robust method to sample correctly the uncertainty space, while performing efficient classification. In essence, any classification methods could be understood as a simple linear regression. In this sense, the least square fitting of a linear model consists of finding a set of parameters $m = (a_0, a_1)$, so that the distance between the observed data vector $y^{obs} = \begin{pmatrix} y_1 \\ \vdots \\ y_s \end{pmatrix}$ and the predicted values $y^{pre}(m) = \begin{pmatrix} y_1^{pre}(m) \\ \vdots \\ y_s^{pre}(m) \end{pmatrix}$ is minimized according to the Euclidean distance in R^s. If we express the problem in a matrix form, then the problem consists of finding the solution of a linear system $Fm = y^{obs}$, where the matrix $F = [1_{R^s}, x]$ depends on the abscissas of the data points $x = \begin{pmatrix} x_1 \\ \vdots \\ x_s \end{pmatrix}$.

With respect to the uncertainty, its analysis is straightforward and is based on sampling the set of equivalent models $m = (a_0, a_1)$, that predict the observed data y^{obs} with a specified error tolerance, M_{tol}:

$$M_{tol} = \left\{ m = (a_0, a_1) : \frac{\left\| y^{obs} - y^{pre}(m) \right\|_2}{\left\| y^{obs} \right\|_2} < tol \right\} \tag{1}$$

Fernández-Martínez et al. [37–39] proved that the topography of the cost function corresponds to a straight flat valley in case of linear problems, while in the nonlinear case, the cost functions has one or more curvilinear valleys of low misfits eventually connected by saddle points. In this simple linear regression problem, the equivalent models belong to an ellipse whose axes and orientations are related to the eigenvalues and eigenvectors of the matrix [37–39].

It could be observed that these model parameters belong to the region of uncertainty and the equivalent model parameters that fit the observed data within a given error bound are preferably located throughout the longer axis of the ellipse of uncertainty corresponding to the lower singular value of F. This is the simplest case where the forward operator is linear. In the case of nonlinear problems, the uncertainty region is a curvilinear flat valley composed of one or several disconnected basins [37]. The main conclusion of this analysis is twofold: (1) the uncertainty space in any classification or decision-making problem has a deterministic nature and, (2) a simple and robust way of sampling these equivalent model parameters in a linear regression model consists of dividing the training dataset into different random data bags and finding the least-squares solution for them.

The nsSNV prediction problem can be considered as a generalized regression problem between the sets of discriminatory nsSNV attributes that characterize the effect of mutation and the set of classes that are part of the training dataset.

This methodology has been recently successfully used by us to predict stability changes due to single and multiple point protein mutations [40], for robust sampling of defective pathways and phenotype prediction in Alzheimer's disease [41], in Multiple Myeloma [42], and in Inclusion Body Myositis [43, 44]. We named it in earlier publications the Holdout sampler. The terminology "the Holdout sampler" is an abbreviation used by us for the 5-fold cross validation algorithm based on pre-processing of our data in holdouts.

A key step in developing machine learning classifiers based on consensus methodology is defining and implementing a computational framework that optimally selects the partial results that are combined to produce the overall prediction. In this research paper, we propose a consensus prediction based on holdouts [45] and a combination of two classifiers: an Extreme Learning Machine (ELM) and a Random Forest (RF) algorithm [46]. After the training and testing, only the best holdouts (data bags) are utilized for blind validation.

A k-fold ($1 \leq k \leq 5$) blind cross-validation has been used in such a way that 75% of the Protein Dataset has been utilized to generate 100 holdouts. Each holdout is a data bag where 75% of the data is used for training and 25% of the data is used for testing of the algorithm. For each holdout a random selection of attributes and of classification activation function was applied to overcome two major issues in classification problems:

(i) To avoid overfitting or underfitting problems each holdout is trained using a random selection of attributes. The holdouts selected by consensus to perform the blind-validation in each fold reflect only those attributes that better predict the targeted class.

(ii) Robustness of the method is ensured due to the k-fold ($1 \leq k \leq 5$) blind cross-validation and a consensus holdout selection with random training features.

The Holdout- nsSNV algorithm predicts the impact of protein amino-acid substitutions by classifying them into deleterious or neutral by considering a combination of sequence, evolutionary and structural information. The nature of the algorithm that combines random holdout selection and classification avoids the redundancy information problem. That is, the algorithm is sufficiently robust to avoid underfitting and overfitting problems that may arise due to an excess or deficiencies of classification attributes, since during the training the classifier network architecture and the attributes utilized in the training are not defined, they vary at each holdout and only the best holdouts predictors are utilized for blind validation.

The Holdout- nsSNV algorithm attributes are generated for all protein sequences. For a given protein amino acid sequence, the sequence profile is computed by using the Protein-Protein Basic Local Alignment Search Tool (BLASTP) algorithm [47] by finding homologous sequences in the UniRef90 database in UniProt [36] with default parameters and a threshold e-value of 10^{-9}. From each BLASTP run, we obtain (i) the sequence alignment score, (ii) the alignment ratio and (iii) the number of aligned sequences. The algorithm also computes (iv) the sequence profile, which consists of a 20 values array, corresponding to the frequency of occurrence of each of 20 amino acids at each position in the sequence. Finally, (v) the mutation vector is computed, defined as a vector of 20 values corresponding to each amino acid, where the wild-type residue is assigned the value -1 and the mutated residue a value of $+1$, while the rest of the non-mutated amino acids are assigned the value of 0; and (vi) the residue position.

The evolutionary information is obtained via PANTHER which estimates the length of time that a given amino acid was preserved in the lineage for a given protein, so that the longer is the preservation time, the greater is the likelihood of functional impact [48, 49]. Finally, the structural information is obtained from the Solvent Accessible Area computed with ASAquick [50]. All this information is combined in a matrix form and utilized to train, test and validate the Holdout-SAV algorithm.

2.3 Consensus Holdout Training and Selection

Holdout methodologies have been exclusively utilized for model validation (1), however, in the present study, we have developed a different scheme in which holdouts are generated not only to validate a model, but also to optimally perform a consensus classification by selecting the holdouts that better predict the targeted classes at the learning stage.

This methodology, inspired by the idea of bootstrapping (2) has been earlier successfully applied to phenotype prediction problems by Fernández-Martínez et al. [52].

Likewise, in the linear regression problem, the major drawback in the prediction of effects of nsSNVs is the lack of a robust model that is capable of relating the different attributes of nsSNVs to the phenotype prediction (prediction of deleterious or neutral effect of the variant). In this sense, a classifier L^* must be constructed, as a link between the nsSNVs attributes and the set of classes {Deleterious, Neutral} into which the effect is divided:

$$L^*(m) : m \in \mathbb{R}^s \to C = \{Deleterious, Neutral\}, \tag{2}$$

with s being the length of the attribute that has been selected for the classifier L^*.

To robustly predict the class of a nsSNV, we have to find discriminatory attributes of the nsSNV corresponding to $L^*(m)$ that better fit the observed class vector c^{obs}. Consequently, modeling the effect of nsSNVs consists of two different steps: learning and validation. The learning step involves the selection of a subset of samples $\mathbf{T}$ (training dataset) whose class vector is known c^{obs}; i.e., finding the minimal subset of attributes that maximizes the learning accuracy:

$$Acc(m) = 100 - \left\| L^*(m) - c^{obs} \right\|_1 \tag{3}$$

where $\left\| L^*(m) - c^{obs} \right\|_1$ stands for the prediction error.

The effect of the holdout procedure in the sampling is comparable, according to Bayes rule, to modifying the evidence of c^{obs} with respect to the classifier $L^*(m)$, since part of the samples used for blind testing/validation have not been used (observed in training).

$$P\left(m/c^{obs}\right) = P(m)P\left(c^{obs}/m\right) \Big/ P\left(c^{obs}\right) \tag{4}$$

where $P(m)$ is called the prior probability, $P\left(c^{obs}/m\right)$ the likelihood, and $P\left(c^{obs}\right)$ the evidence.

This method grounds on the statistical technique of bootstrapping, or arbitrary sampling with replacement [45], which is used to build the confidence intervals in sample estimates and to estimate the sampling distribution of any statistics via a random sampler. In prior works and other fields, this methodology was utilized to optimally sample model parameters posterior distribution via the least squares fitting of different data bags [53, 54]. The Holdout algorithm works as follows:

(i) Data bagging: We randomly divide the data set into a 75/25 dataset holdouts, where 75% of the data is used for learning and 25% for testing/validation. In our paper, 1000 different bags were generated. For each holdout, a random selection of attributes, m_i, is carried out, before performing classification of the proteins p_k, according to the observations, c^{obs}.

(ii) Data Testing: After completing the learning stage, where classifier was trained to classify protein mutations according to a set of attributes, m_i, and the observations, c^{obs}, a testing is carried out. The testing procedure is performed to compute the predictive accuracy of each holdout. Consequently, it is possible to obtain a distribution of holdouts according to their predictive capability and select those that perform best.

(iii) Holdout Selection: Once the mutated proteins were classified in the testing dataset T_k, the holdout accuracy is computed and the best holdout predictors are chosen, afterwards, for the blind validation. The holdouts that fulfill the following condition: $Acc_{HD,i} > 0.90 \cdot Acc_{HD,max}$ are selected to perform the blind validation.

2.4 Extreme Learning Machine

In this paper, the Extreme Learning Machines (ELM) classifier was also utilized to categorize protein single amino acid polymorphisms as neutral or deleterious. Extreme Learning Machines are a new generation of feedforward neural networks that could be used for classification, regression, clustering, compression or feature learning. They consist of a single or of multiple layers of hidden nodes, where the hidden nodes hyper-parameters must be tuned. The hidden nodes could be either randomly assigned and updated or transferred from the ancestors without modifications. In many cases, the weights of each hidden neuron are learned. Extreme Learning Machine models are capable of creating good generalization performance and can be trained faster than networks utilizing back propagation methodologies since ELMs use the Moore-Penrose pseudoinverse. ELM input weights are randomly selected, and the output weights are analytically calculated. Since an analytic learning process is used in updating the output weights, the success rate increases because the resolution time and the error value is reduced, minimizing the possibility of getting trapped in a local minimum. In our research paper, the ELM architecture is not predefined, consequently, at each holdout, it is randomly sampled, trained and tested.

2.5 Random Forests

In addition to ELM, Random Forests (RF) were also utilized in the prediction. Random forests or random decision forests are an ensemble learning method for classification, regression and other tasks that operates by constructing a multitude of decision trees at training time and outputting the class that is the mode of the classes (classification) or mean prediction (regression) of the individual trees.

The Random Forest applies a bootstrap technique for each holdout, consequently, a data bagging technique (Random decision tree) is applied for each data bag (holdout).

Since a bootstrap is performed over a bootstrapped dataset, this produces a better model performance, since the model variance is decreased dramatically without practically increasing the bias. This is due to the fact that, while the prediction of each tree is highly sensitive and noisy, the global average over all the trees is not, since they are not correlated. Moreover, since each RF takes place in a holdout (data bag), the bias is further reduced. As a result, the combination of Holdouts and Random Forest decision trees is a robust way to dramatically decorrelate the trees by ensuring that each tree is trained with different datasets.

3 Results

The Holdout- nsSNV algorithm combines a set of features that range from sequence to structural, evolutionary and stability attributes. The algorithm reads the sequence of a given mutant protein, determines all the attributes and performs the classification by using the Holdout Sampler. The Holdout Sampler has been used in this work as a consensus classifier. The utilization of consensus classifiers generally leads to a significant improvement in the prediction performance, by reducing the bias while increasing the robustness and accuracy. The idea of consensus classifier is inspired by the principle of Condorcet; for given a set of independent decision-makers, the most voted decision tends to be right when the number of decision-makers increases. In this sense, it could be expected that consensus classifiers are accurate and robust alternatives to traditional and individual machine learning tools [23].

In our case, the Holdout Sampler employs a consensus strategy connecting different decision boundaries. An important feature of this algorithm is its explorative character, as observed in Table 1:

Table 1. Training Performance of Holdout- nsSNV prediction tool in each individual K-Fold.

K-Fold	Mean Accuracy	Median Accuracy	Accuracy Std	Accuracy Uncertainty	Minimum Accuracy	Maximum Accuracy
1	**75.40**	**80.99**	15.56	28.82	48.63	**90.86**
2	**74.31**	**68.58**	16.36	32.55	50.11	**91.04**
3	**74.46**	**70.08**	15.98	30.98	49.22	**91.03**
4	**72.93**	**67.52**	16.08	31.11	49.63	**90.83**
5	**74.97**	**68.83**	15.78	28.72	50.16	**90.83**
Overall	**74.41**	**71.20**	15.95	30.44	49.55	**90.92**

For each K-Fold, different holdouts or data bags where trained and tested varying the choice of the classifier (ELM or RF) and the attributes selected to perform the classification. Due to this, it is possible to observe a high dispersion and uncertainty in the testing accuracy for each holdout, that leads to a distribution of holdouts containing low performing, intermediate and high performing data bags. A further analysis reveals that the sharp increase around the 50th percentile of holdouts depends on the choice of the classifier; more specifically, the Random Forests offer accuracies that clearly outperform ELM. In summary, the algorithm gives an overall accuracy of 90.2% and an average value of Mathews Correlation Coefficient (MCC) of 0.80 with the Areas Under the Curve (AUCs) that range between 0.9005 and 0.9041.

The analysis of the discriminatory power of different attributes provides a deeper insight regarding the performance of other popular bioinformatics tools published in the literature in comparison to Holdout- nsSNV. Table 2 summarizes the results of such comparison with other publicly available algorithms, such as: PROVEAN, Mutation Taster, CADD, PPH-2, SNP&GO, PhD-SNP, PredictSNP, MAPP, Meta-SNP and CONDEL. Our Holdout-SAV algorithm ranks top among the methods reported in Table 2, mostly based on a similar philosophy.

The Holdout- nsSNV method is followed by PhD-SNP, Meta-SNP and SNP&GO tools in terms of performance. PhD-SNP and SNP&GO algorithms utilize a combination of mutation and sequence information to train the classifiers. Based on the discriminatory

Table 2. Comparison of performance of Holdout- nsSNV sampler with other most popular prediction tools

Performance Metrics	PROVEAN	CADD	PPH-2	SNP&GO	PhD-SNP	Predict SNP	MAPP	MetaSNP	CONDEL	Holdout-SAV
Dataset	Swiss-Prot	UniProt HumVar	Swiss-Prot	Swiss-Prot	Swiss-Prot	Swiss-Prot	Swiss-Prot	Swiss-Prot	Swiss-Prot	Swiss-Prot
Accuracy	79%	76%	70%	83%	88%	75%	71%	**87%**	75%	**90%**
MC	0.74	*Not reported	0.41	0.67	0.72	0.49	0.41	0.74	0.51	**0.80**
AUC	0.85	0.86	0.78	**0.91**	**0.91**	0.81	0.77	0.91	0.82	0.90
Ref.	[12]	[19]	[20]	[21]	[22]	[23]	[24]	[34]	[35]	-

power analysis, both algorithms utilize the variables with the highest predictive capability. In our opinion, the Holdout- nsSNV algorithm surpasses PhD-SNP and performance due to the utilization of consensus, which reduces biases while increasing accuracy. On the other hand, SNP&GO algorithm, despite utilizing mutation and sequence information, includes also other features, such as evolutionary, stability or structural information that induce some bias due to noise or uncertainty. As consequence, the performance of this method is slightly lower than the one presented here. The fact that Meta-SNP uses consensus could explain its better performance. It would be interesting to further explore Meta-SNP by including a Fisher's ratio analysis to improve the prediction accuracy.

PROVEAN, CADD MAPP and PolyPhen-2 utilize the variables with the highest discriminatory power. However, their accuracy is lower than ours due to the utilization of classifiers that are relatively sensitive to noise. Finally, it is worth mentioning Predict-SNP and CONDEL tool, which despite utilizing consensus lack a way of discriminating between high predictive and low predictive variables; therefore, the overall accuracy of this method is reduced due to the introduction of noise in the classification. An interesting feature, which is further supported by the training performance analysis of the Holdout Sampler is the fact that least-square methods for classification, such as ELM or the ones included in CADD, PROVEAN, PPH-2, experience a lower accuracy rate with respect to other RF-like methodologies. The comparison of the performance of various methods has been based on results published in corresponding references. An ideal comparison should be done on the same data set applied to each method. In practice this is very difficult, and very rarely done in literature, because of various reasons (unavailability of the codes, unavailability of the data that was used for training by other methods to eliminate bias, etc.). The perfect blind test should be performed on a completely new data set, that hasn't been used in the past by the existing machine learning-based methods. The best possibility would be to apply our method in the Critical Assessment of Genome Interpretation (CAGI) - a community experiment to objectively assess computational methods for predicting the phenotypic impacts of genomic variation. CAGI participants are provided experimentally studied (but yet unpublished) genetic variants and make blind predictions of resulting phenotype. We plan to participate in the CAGI experiment to blindly test the performance of our method and compare it with other leading methodologies. This will be a perfect possibility to blindly test our methodology.

4 Conclusions and Future Directions

In this work, we present a new methodology to predict deleteriousness of SAVs by utilizing a deep sampling algorithm. Since this is a problem of high uncertainty, the utilization of a deep sampling technique to model the SAV attributes that are most discriminatory is of utmost importance to reduce bias and improve accuracy and robustness of predictions. The methodology consists of sampling a set of classifiers architectures and boundary conditions for different holdouts or data bags during the training stage. The best performing holdouts with the highest predictive capability are selected to construct a consensus-based classifier. Our algorithm has been trained and blindly validated utilizing a k-fold ($1 \leq k \leq 5$) cross validation procedure. The results show that our method outperforms other currently available tools, including other consensus-based algorithms, and provides robust results, with small standard deviations among folds and high accuracy. The superiority of our algorithm is based on the utilization of a tree of holdouts, where different machine learning algorithms are sampled with different boundary conditions or different predictive attributes. This pre-parametrization allows us to construct a consensus classifier based on the best holdouts. In future work we will try to improve our methodology by using not only human data, but also additional data from other species. We will try also to add to the sequential information the structural information.

Acknowledgment. AK acknowledges the financial support from NSF grant DBI 1661391, and NIH grants R01GM127701, and R01HG012117.

References

1. Sunyaev, S., Ramensky, V., Bork, P.: Towards a structural basis of human non-synonymous single nucleotide polymorphisms. Trends Genet. **16**, 198–200 (2000)
2. Cargill, M., et al.: Characterization of single-nucleotide polymorphisms in coding regions of human genes. Nat. Genet. **22**, 231–238 (1999)
3. Collins, F.S., Brooks, L.D., Chakravarti, A.: A DNA polymorphism discovery resource for research on human genetic variation. Genome Res. **8**, 1229–1231 (1998)
4. Abecasis, G.R., et al.: A map of human genome variation from population-scale sequencing. Nature **467**, 1061–1073 (2010)
5. Collins, F.S., Guyer, M.S., Charkravarti, A.: Variations on a theme: cataloging human DNA sequence variation. Science **278**, 1580–1581 (1997)
6. Risch, N., Merikangas, K.: The future of genetic studies of complex human diseases. Science **273**, 1516–1517 (1996)
7. Studer, R.A., Dessailly, B.H., Orengo, C.A.: Residue mutations and their impact on protein structure and function: detecting beneficial and pathogenic changes. Biochem. J. **449**, 581–594 (2013)
8. Halushka, M.K., et al.: Patterns of single-nucleotide polymorphisms in candidate genes for blood-pressure homeostasis. Nat. Genet. **22**, 239–247 (1999)
9. Capriotti, E., Nehrt, N.L., Kann, M.G., Bromberg, Y.: Bioinformatics for personal genome interpretation. Brief. Bioinform. **13**, 495–512 (2012)
10. Niu, B.: Protein-structure-guided discovery of functional mutations across 19 cancer types. Nat. Genet. **2016**(48), 827–837 (2016)

11. Goode, D.L., et al.: A simple consensus approach improves somatic mutation prediction accuracy. Genome Med. **5**, 90 (2013)
12. Choi, Y., Sims, G.E., Murphy, S., Miller, J.R., Chan, A.P.: Predicting the functional effect of amino acid substitutions and indels. PLoS ONE **7**, e46688 (2012)
13. Choi, Y., Chan, A.P.: PROVEAN web server: a tool to predict the functional effect of amino acid substitutions and indels. Bioinformatics **31**, 2745–2747 (2015)
14. Kumar, P., Henikoff, S., Ng, P.C.: Predicting the effects of coding non-synonymous variants on protein function using the SIFT algorithm. Nat. Protoc. **4**, 1073–1081 (2009)
15. Tang, H., Thomas, P.D.: PANTHER-PSEP: predicting disease-causing genetic variants using position-specific evolutionary preservation. Bioinformatics **32**, 2230–2232 (2016)
16. Katsonis, P., Lichtarge, O.: A formal perturbation equation between genotype and phenotype determines the evolutionary action of protein-coding variations on fitness. Genome Res. **24**, 2050–2058 (2014)
17. Gallion, J., et al.: Predicting phenotype from genotype: improving accuracy through more robust experimental and computational modeling. Hum. Mutat. **38**, 569–580 (2017)
18. Schwarz, J.M., Rödelsperger, C., Schuelke, M., Seelow, D.: MutationTaster evaluates disease-causing potential of sequence alterations. Nat. Methods **7**, 575–576 (2010)
19. Reva, B., Antipin, Y., Sander, C.: Predicting the functional impact of protein mutations: application to cancer genomics. Nucleic Acids Res. **39**, e118 (2011)
20. Adzhubei, I.A., et al.: A method and server for predicting damaging missense mutations. Nat. Methods **7**, 248–249 (2010)
21. Capriotti, E., et al.: WS-SNPs&GO: a web server for predicting the deleterious effect of human protein variants using functional annotation. BMC Genomics **14**, S6 (2013)
22. Capriotti, E., Calabrese, R., Casadio, R.: Predicting the insurgence of human genetic diseases associated to single point protein mutations with support vector machines and evolutionary information. Bioinformatics **22**, 2729–2734 (2006)
23. Bendl, J., et al.: PredictSNP: robust and accurate consensus classifier for prediction of disease-related mutations. PLoS Comput. Biol. **10**, e1003440 (2014)
24. Stone, E.A., Sidow, A.: Physicochemical constraint violation by missense substitutions mediates impairment of protein function and disease severity. Genome Res. **15**, 978–986 (2005)
25. Miosge, L.A.: Comparison of predicted and actual consequences of missense mutations. Proc. Natl. Acad. Sci. USA **112**, 189–198 (2015)
26. Saunders, C.T., Baker, D.: Evaluation of structural and evolutionary contributions to deleterious mutation prediction. J. Mol. Biol. **322**, 891–901 (2002)
27. Stefl, S., Nishi, H., Petukh, M., Panchenko, A.R., Alexov, E.: Molecular mechanisms of disease-causing missense mutations. J. Mol. Biol. **425**, 3919–3936 (2013)
28. Pires, D.E.V., Chen, J., Blundell, T.L., Ascher, D.B.: In silico functional dissection of saturation mutagenesis: interpreting the relationship between phenotypes and changes in protein stability, interactions and activity. Sci. Rep. **6**, 19848 (2016)
29. Castaldi, P.J., Dahabreh, I.J., Ioannidis, J.P.A.: An empirical assessment of validation practices for molecular classifiers. Brief. Bioinform. **12**, 189–202 (2011)
30. Baldi, P., Brunak, S.: Bioinformatics: The Machine Learning Approach. MIT Press, Cambridge (2001)
31. Thusberg, J., Olatubosun, A., Vihinen, M.: Performance of mutation pathogenicity prediction methods on missense variants. Hum. Mutat. **32**, 358–368 (2011)
32. Ng, P.C., Henikoff, S.: Predicting the effects of amino acid substitutions on protein function. Annu. Rev. Genomics Hum. Genet. **7**, 61–80 (2006)
33. Polikar, R.: Ensemble based systems in decision making. IEEE Circuits Syst. Mag. **6**, 21–45 (2006)

34. Capriotti, E., Altman, R.B., Bromberg, Y.: Collective judgment predicts disease-associated single nucleotide variants. BMC Genomics **14**, S2 (2013)
35. González-Pérez, A., López-Bigas, N.: Improving the assessment of the outcome of nonsynonymous SNVs with a consensus deleteriousness score. Condel. Am. J. Hum. Genet. **88**, 440–449 (2011)
36. The UniProt Consortium: The universal protein resource (UniProt). Nucleic Acids Res. **36**, D190–D195 (2008)
37. Fernández-Martínez, J.L., Fernández-Muñiz, Z., Tompkins, M.J.: On the topography of the cost functional in linear and nonlinear inverse problems. Geophysics **77**, W1–W5 (2012)
38. Fernández-Martínez, J.L., Pallero, J.L.G., Fernández-Muñiz, Z., Pedruelo-González, L.M.: From Bayes to Tarantola: new insights to understand uncertainty in inverse problems. J. App. Geophys. **98**, 62–72 (2013)
39. Fernández-Martínez, J.L., Fernández-Muñiz, Z.: The curse of dimensionality in inverse problems. J. Comput. Appl. Math. **369**, 112571 (2020)
40. Álvarez-Machancoses, Ó., deAndrés-Galiana, J.E., Fernández-Martínez, J.L., Kloczkowski, A.: Robust prediction of single and multiple point protein mutations stability changes. Biomolecules **10**, 67 (2020)
41. Fernández-Martínez, J.L., Álvarez-Machancoses, Ó., deAndrés-Galiana, E.J., Bea, G., Kloczkowski, A.: Robust sampling of defective pathways in Alzheimer's disease. Implications in drug repositioning. Int. J. Mol. Sci. **10**, 3594 (2020)
42. Fernández-Martínez, J.L., deAndrés-Galiana, E.J., Fernández-Ovies, F.J., Cernea, A., Kloczkowski, A.: Robust sampling of defective pathways in multiple myeloma. Int. J. Mol. Sci. **20**, 4681 (2019)
43. deAndrés-Galiana, E.J., Fernández-Ovies, F.J., Cernea, A., Fernández-Martínez, J.L., Kloczkowski, A.: Deep neural networks for phenotype prediction in rare disease inclusion body myositis: a case study. In: Artificial Intelligence in Precision Health. From Concept to Applications (Debmalya Barth, Editor), pp. 189–202. Elsevier, Amsterdam (2020)
44. Álvarez-Machancoses, Ó., deAndrés-Galiana, E., Fernández-Martínez, J.L., Kloczkowski, A.: The utilization of different classifiers to perform drug repositioning in inclusion body myositis supports the concept of biological invariance. In: Rutkowski, L., Scherer, R., Korytkowski, M., Pedrycz, W., Tadeusiewicz, R., Zurada, J.M. (eds.) ICAISC 2020. LNCS (LNAI), vol. 12415, pp. 589–598. Springer, Cham (2020). https://doi.org/10.1007/978-3-030-61401-0_55
45. Efron, B., Tibshirani, R.: An Introduction to Bootstrap. Chapman & Hall, Boca Raton (1993)
46. Breiman, L.: Random forests. Mach. Learn. **45**, 5–32 (2001)
47. Altschul, S.F., Gish, W., Miller, W., Myers, E.W., Lipman, D.J.: Basic local alignment search tool. J. Mol. Biol. **215**, 403–410 (1990)
48. Thomas, P.D., et al.: PANTHER: a library of protein families and subfamilies indexed by function. Genome Res. **13**, 2129–2141 (2003)
49. Thomas, P.D., et al.: Applications for protein sequence-function evolution data: mRNA/protein expression analysis and coding SNP scoring tools. Nucleic Acids Res. **34**, W645–W650 (2006)
50. Faraggi, E., Zhou, Y., Kloczkowski, A.: Accurate single-sequence prediction of solvent accessible surface area using local and global features. Proteins: Struct. Funct. Bioinform. **82**, 3170–3176 (2014)
51. Kohavi, R.: A study of cross-validation and bootstrap for accuracy estimation and model selection. In: Proceedings of the 14th International Joint Conference on Artificial Intelligence 2 (Montreal 20–25 August), pp. 1137–1145 (1995)
52. Fernández-Martínez, J.L., et al.: Sampling defective pathways in phenotype prediction problems via the holdout sampler. In: Rojas, I., Ortuño, F. (eds.) IWBBIO 2018. LNCS, vol. 10814, pp. 24–32. Springer, Cham (2018). https://doi.org/10.1007/978-3-319-78759-6_3

53. Fernández-Muñiz, Z., Hassan, K., Fernández-Martínez, J.L.: Data kit inversion and uncertainty analysis. J. Appl. Geophys. **161**, 228 (2019)
54. Fernández-Martínez, J.L., Fernández-Muñiz, Z., Breysse, D.: The uncertainty analysis in linear and nonlinear regression revisited: application to concrete strength estimation. Inverse Probl. Sci. Eng. **27**, 1740–1764 (2018)
55. Huang, G.B., Zhu, Q.Y., Siew, C.K.: Extreme learning machine: theory and applications. Neurocomputing **70**, 489–501 (2006)
56. Huang, G.B.: An insight into extreme learning machines: random neurons, random features and kernels. Cogn. Comput. **6**, 376–390 (2014)
57. Huang, G.B., Lei, C., Chee-Kheong, S.: Universal approximation using incremental constructive feedforward networks with random hidden nodes. IEEE Trans. Neural Netw. **17**, 879–892 (2006)
58. Huang, G.B.: What are extreme learning machines? Filling the gap between Frank Rosenblatt's Dream and John von Neumann's Puzzle. Cogn. Comput. **7**, 263–278 (2015)
59. Huang, G.B., Hongming, Z., Xiaojian, D., Rui, Z.: Extreme learning machine for regression and multiclass classification. IEEE Trans. Syst. Man Cybern. - Part B: Cybern. **42**, 513–529 (2012)
60. Ertugrul, O.F., Tagluk, M.E., Kaya, Y., Tekin, R.: EMG signal classification by extreme learning machine. In: 21st 2013 Signal Processing and Communications Applications Conference (SIU), April 24, pp. 1–4 (2013)
61. Huang, G.B., Zhu, Q.Y., Siew, C.K.: Extreme learning machine: a new learning scheme of feedforward neural networks. In: Neural Networks. Proceedings of the 2004 IEEE International Joint Conference on 2004 July 25, vol. 2, pp. 985–990 (2004)
62. Ho, T.K.: Random decision forest. In: Proceedings of the 3rd International Conference on Document Analysis and Recognition (Montreal) 14–16, pp. 278–282 (1995)

Optimizing Variant Calling for Human Genome Analysis: A Comprehensive Pipeline Approach

Miguel Pinheiro[1], Jorge Miguel Silva[2]([✉]), and José Luis Oliveira[2]

[1] iBiMED, Department of Medical Sciences, University of Aveiro, Aveiro, Portugal
[2] IEETA, DETI, LASI, University of Aveiro, Aveiro, Portugal
`jorge.miguel.ferreira.silva@ua.pt`

Abstract. The identification of genetic variations in large cohorts is a critical issue to identify patient cohorts, disease risks, and to develop more effective treatments. To help this analysis, we improved a variant calling pipeline for the human genome using state-of-the-art tools, including GATK (Hard Filter/VQSR) and DeepVariant. The pipeline was tested in a computing cluster where it was possible to compare Illumina Platinum genomes using different approaches. Moreover, by using a secure data space we provide a solution to privacy and security concerns in genomics research. Overall, this variant calling pipeline has the potential to advance the field of genomics research significantly, improve healthcare outcomes, and simplify the analysis process. Therefore, it is critical to rigorously evaluate these pipelines' performance before implementing them in clinical settings.

Keywords: Variant Calling · Genomics · Cohorts · Pipeline

1 Introduction

The sequencing of the human genome has revolutionised our understanding of human health and disease by enabling the identification of genetic variations between individuals [1]. These variations can significantly affect susceptibility to disease, drug response, and other traits. As a result, analysing genetic variations in large cohorts is crucial for identifying disease risks and developing effective treatments.

One significant challenge associated with creating and analysing genetic variations in a population is the limited availability of raw genomic data [2]. Without access to raw genomic data, researchers cannot reproduce results or conduct further analysis, hindering scientific progress. The lack of raw data is due to ethical considerations, and data privacy regulations differ from country to country and limit the sharing and distribution of genomic data [3]. Furthermore, data storage and management is costly, usually privately owned, and isolated [4], making it difficult for researchers to access and use these silos of raw data. As a result, performing these studies requires collaborative efforts and data-sharing initiatives to ensure sufficient data is available for analysis [5].

I. Rojas et al. (Eds.): IWBBIO 2023, LNBI 13920, pp. 72–85, 2023.
https://doi.org/10.1007/978-3-031-34960-7_6

One such effort is the European Genomic Data Infrastructure (GDI) project [6], a European project that builds upon the Beyond 1 Million Genomes (B1MG) initiative, to sequence over one million genomes and to provide vast genomic data for research and clinical reference [7]. However, sharing and accessing these data remains a challenging issue. The GDI project aims to unlock this network of over one million genome sequences by creating a secure data space, which will provide clinicians, scientists, and healthcare policymakers secure and authorised access to genomic and clinical data across Europe [6]. In addition, the project seeks to create a federated query system that supports genotypic and phenotypic queries in natural language, allowing for easier and more effective analysis of large genomic datasets. Creating a controlled access program with such a large amount of genomic data will provide unprecedented opportunities for transnational and multi-stakeholder actions in personalised medicine for common, rare, and infectious diseases [6]. In addition, the project will make non-sensitive and aggregated data openly discoverable through the European Genome Dashboard [6].

An essential format to represent and to analyse genetic variations is the variant calling format (VCF) [8]. A VCF file contains textual variant information and quality metrics from an individual or multiple individuals relative to a reference genome. In the case of having multiple individuals (cohort VCF), its primary purpose is to enable the comparison of genetic variations between them. The researchers can then use these files to identify candidate genes or mutations associated with specific diseases or traits by comparing variations and identifying shared patterns. Cohort VCFs can also be used to study population-level variations, essential in understanding a population's genetic diversity and evolution.

However, creating cohort VCFs can be challenging due to the massive amounts of raw genomic data generated from sequencing large cohorts [9]. This is especially true in the case of large-scale projects such as the GDI project, where sophisticated bioinformatics pipelines are required.

Selecting fast and accurate bioinformatics tools and techniques is vital when creating these pipelines [10,11]. Besides accommodating large amounts of data and accurately identifying genetic variations across a large cohort of individuals, they must be capable of identifying sequencing errors and filtering out non-relevant variations. Access to high-performance computing clusters is also essential for the processing and analysing raw genomic data. These clusters provide the computational resources needed to handle the large amounts of data generated from sequencing large cohorts. Additionally, these clusters allow for the parallelization of computational tasks, allowing for faster processing and analysis of the data.

In this paper, we present a variant call pipeline that is a fork of [12] with improved and validated components in the context of the GDI project. The pipeline comprises several critical components, each essential in generating high-quality cohort Variant Call Format (VCF) files. The first step involves quality control, followed by alignment, variant calling, variant filtration, and variant annotation. The final output is a cohort VCF.

To determine the optimal design for our pipeline, we assessed its performance using Illumina Platinum genomes specifically designed for developing and evaluating bioinformatics tools and pipelines. Using this dataset, we measured the accuracy and speed of the different tools, allowing us to identify the most effective design. In addition, our pipeline has been optimised to run on high-performance computing clusters, reflecting the increasing need for efficient data processing capabilities in genomic research. This will allow the handling of large amounts of raw data, quickly and accurately. The final result is a standardised and streamlined pipeline, which can generate high-quality cohort VCFs and be used for downstream analysis and content discovery.

2   Background

Variant calling pipelines are essential tools in analysing human genomic data [13]. These pipelines are used to identify single nucleotide variants (SNVs), insertions and deletions (indels), and structural variations in genomic data obtained from next-generation sequencing (NGS) platforms.

The pipeline typically involves a series of steps, including read alignment to a reference genome, duplicate marking, base quality score recalibration, local realignment around indels, and variant calling [14]. The resulting VCF file contains variant information and quality metrics from multiple individuals, which can be used for downstream analyses such as identifying disease-associated variants and predicting drug response.

One of the most widely used variant calling pipelines is the Genome Analysis Toolkit (GATK), developed by the Broad Institute [15]. The GATK pipeline has been used in several large-scale genomic studies, including the 1000 Genomes Project and the UK Biobank. The pipeline includes modules for quality control, read alignment, local realignment, and variant calling, among others. The GATK pipeline is available as open-source software and is constantly updated.

Another popular variant calling pipeline is DeepVariant, developed by Google Brain in collaboration with the Broad Institute [16]. This tool uses deep learning techniques to make highly accurate variant calls, especially for complex and rare variants, with a lower false positive rate compared to other pipelines. DeepVariant's accuracy is attributed to its training on millions of high-quality human genomes and the use of a convolutional neural network to predict the likelihood of a variant call. Several large-scale genomic studies, including the PrecisionFDA Truth Challenge, have successfully utilized DeepVariant due to its superior performance in detecting SNVs and indels compared to other pipelines [17].

Other popular variant calling pipelines include the SAMtools suite [18], FreeBayes [19], and VarScan [20]. SAMtools is a suite of tools for processing and analysing high-throughput sequencing data, including variant calling, based on the SAM/BAM format. FreeBayes is a Bayesian genetic variant detector designed for NGS data, which can call SNVs, indels, and structural variants. It uses a haplotype-based model to provide accurate and sensitive variant calls. VarScan is a Java-based tool for identifying SNVs and indels in NGS data. It detects variants by comparing the frequency of alleles between normal and abnormal samples.

Several cloud-based variant calling pipelines have been developed in recent years to enable easy access to high-performance computing resources. For example, the Broad Institute offers the FireCloud platform, which provides a user-friendly interface for running GATK and other bioinformatics pipelines on the Google Cloud Platform [21]. The Seven Bridges Platform is another cloud-based solution offering a wide range of bioinformatics pipelines, including several calling pipelines [22]. These platforms provide a more accessible and scalable approach to genomic analysis, making it easier for researchers to analyse large datasets and collaborate.

In conclusion, variant calling pipelines are essential for analysing human genomic data. They allow for the identification of genetic variations between individuals, which can be used to understand disease risk and develop targeted treatments. The GATK pipeline, SAMtools, FreeBayes, and VarScan are among the most widely used variant calling pipelines, and cloud-based solutions such as FireCloud and Seven Bridges Platform are becoming increasingly popular for easy access to high-performance computing resources.

3 Methods

3.1 Reference

As a reference, we chose GRCh38/hg38[1], with no ALT contigs and with the human herpesvirus [23,24]. GRCh38/hg38 serves as the current version of the human reference genome, a crucial standard against which genomic sequencing data is mapped and analysed to uncover genetic variations such as single nucleotide polymorphisms (SNPs) and small insertions or deletions (indels) [25]. One of the critical benefits of this GRCh38/hg38 reference is its exclusion of ALT contigs - large variations with flanking sequences almost identical to the primary human assembly [26], including unplaced and un-localized contigs to not force reads originating from these contigs to be mapped to the chromosomal assembly, leading to false variant calls. This reference genome helps mitigate the risk of false positive variant calls, which can be especially problematic in large-scale genomic analyses.

3.2 Dataset

The Platinum Genomes dataset [27,28], generated using the HiSeq 2000 system at 50× depth, has become a widely used resource in human genetics research. This dataset comprises high-quality whole genome sequencing data from a 17-member CEPH pedigree (1463). It also includes two samples (NA12877 and

---

[1] ftp://ftp.ncbi.nlm.nih.gov/genomes/all/GCA/000/001/405/GCA_000001405.15_GR Ch38/seqs_for_alignment_pipelines.ucsc_ids/GCA_000001405.15_GRCh38_no_alt_ana lysis_set.fna.gz.

NA12878) sequenced at an even higher depth of 200× (PRJEB3246). One of the most significant benefits of the Platinum Genomes dataset is the provision of a truthset of high-confidence variants for the samples (NA12877 and NA12878), which have been extensively characterized and validated using multiple sequencing technologies and independent analysis methods (PRJEB3381). Moreover, a BED file contains the confident regions for the truthset variants, enabling researchers to focus on high-confidence regions in their analysis. The Platinum Genomes dataset has become a gold standard in developing and evaluating bioinformatics tools and pipelines for variant calling and genotyping. In addition, its well-characterized nature and high-quality sequencing data make it an indispensable resource for benchmarking and improving the accuracy of genetic analysis methods.

To test and validate the results of this pipeline, we tested the samples shown in Table 1.

Table 1. Data used to validate pipeline. Samples, coverage, and run names.

Sample	Coverage	Run Names
NA12877	50	ERR194146
NA12878	50	ERR194147
NA12889	50	ERR194158
NA12890	50	ERR194159
NA12877	100	ERR1743[10-23]
NA12878	100	ERR1743[24-41]

The table provides information about the samples used to test and validate the pipeline results. The samples include NA12877, NA12878, NA12889, and NA12890, each sequenced at a coverage depth of 50× using the ERR194146, ERR194147, ERR194158, and ERR194159 runs, respectively. In addition, NA12877 and NA12878 were also sequenced at a coverage depth of 100× using the ERR1743[10-23] and ERR1743[24-41] runs, respectively. The coverages were obtained after removing duplicated reads of mapped files.

3.3   Quality and Control

Evaluating the quality and control of our variant calls were not the only measures we implemented to ensure data quality throughout the analysis. We also utilized several software tools, such as Trimmomatic [29] and FastQC [30], for read cleaning and quality assessment, as well as Picard Tools [31] for duplicate removal. Additionally, we generated reports on mapping, duplicates, and MultiQC [32], to aggregate all information about quality and trimming in one single file. By doing so, we were able to monitor the quality of our data and identify potential issues early on, thereby improving our analysis's overall accuracy and reliability.

3.4 Pipeline

This work aims to determine the best workflow for analysing GDI genomic data. The pipeline should be robust to generate high-quality variant calls and annotations. The pipeline should be adaptable and easily modified to include additional tools or steps as needed for specific queries. More importantly, the pipeline must efficiently and accurately analyze genomic data, providing accurate cohort VCFs.

To best comprehensive, we devised a four-part pipeline, each playing a crucial role in ensuring the accuracy and reliability of the analysis. Then, using a golden standard dataset, we used different configurations to assess the accuracy of the tools in generating VCFs. The pipeline structure with possible variations in parts 2 and 3 is shown in Fig. 1.

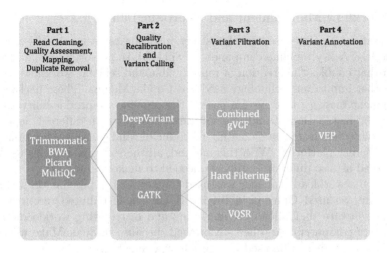

Fig. 1. Pipeline Structure with Variations

The first part of the pipeline involves read cleaning, quality assessment, mapping, and duplicate removal. First, Trimmomatic is used to impose quality on the reads, and FastQC is used for quality assessment. Next, BWA maps and sorts the reads against the GRCh38 reference [33], identifies duplicate reads, and removes them using Picard Tools. Quality reports are then generated using MultiQC, providing valuable information about read mapping, duplicates, and quality.

The second part of the pipeline involves base quality recalibration and variant calling. This step is only valid if the user chooses the GATK variant call. The GATK-HaplotypeCaller in gVCF mode creates a gVCF file containing information about all sites, not just those with a variation. GenotypeGVCFs is then used to call SNPs and INDELs [33], using information across all samples to call variants that cannot be called with information from one sample alone creating a cohort VCF. After this, variant filtration occurs (part 3) to determine which detected variants are true and false positives. The filtration step can be performed using a Hard filter with predefined values or be done using Variant Quality Score Recalibration (VQSR), a sophisticated filtering technique that uses machine learning to filter out probable artefacts from the call set. With DeepVariant, phase 2 goes for variant calling and filtration. Phase 3 is responsable for creation of unique cohort VCF from each gVCF file produced in Part 2. GLnexus is the software that performs this task.

The fourth and final part of the pipeline involves variant annotation. The Variant Effect Predictor (VEP) is utilized for VCF annotation, providing information about genes and transcripts affected by variants, the location of variants, consequences of variants on the protein sequence, known variants that match the variants being analyzed, and associated minor allele frequencies. SIFT and PolyPhen-2 scores are also provided for changes to the protein sequence.

3.5   Workflow Management and Reproducibility

The workflow pipeline used in this study incorporates several software tools and models. A key pipeline component is SnakeMake, a workflow management system that enables the creation of reproducible and scalable workflows.

Another important technology used was Conda/Mamba. These package and environment management system ensure the pipeline's reproducibility since it enables the installation of multiple software packages with different dependencies in isolated environments, creating reproducible and consistent environments across different platforms. With this method, all the software versions used will be recorded as also different databases needed to make the analysis.

Finally, a workload manager was used to execute the pipeline on a cluster. Specifically, we used Open Grid Engine (OGE), a distributed resource management software that allows for managing and distributing workloads across a cluster of computers. Another advantaged, because of SnakeMake workflow, different workloads can be used in a simple way, like Slurm.

3.6   Benchmarking

To ensure the accuracy and precision of our pipeline, we followed best practices for benchmarking germline small-variant calls in human genomes, as recommended by the Global Alliance for Genomics and Health (GA4GH) [34].

To evaluate the quality and control of our variant calls, we utilized hap.py, a tool recommended by GA4GH for stratified variant evaluations. By comparing our variant calls against the truthset variants provided by the Genome in a Bottle Consortium (GIAB)[2], we were able to calculate true positives (TP), false positives (FP), false negatives (FN), and other statistics derived from these values. This allowed us to identify and address potential errors or biases in our pipeline and improve the quality of our variant calls.

In this study, we aimed to optimize the performance of our pipeline by implementing several key improvements. We began by evaluating the initial performance of the pipeline, which required a significant amount of time to run. Then,

---

[2] https://ftp-trace.ncbi.nlm.nih.gov/giab/ftp/release/.

we implemented several vital improvements to address this, including parallelizing GATK, adding coverage analysis, and setting the Variant Quality Score Recalibration (VQSR)/Hard Filter filter in VCF files.

By parallelizing GATK we used the computational resources of our cluster more efficiently, resulting in a considerable reduction in the time required for the pipeline to complete. After implementing the modifications, we significantly reduced the time required for the pipeline to be complete, achieving a time of 64 h for four samples on a cluster with four nodes.

Additionally, adding coverage analysis to our pipeline enabled us to evaluate the depth of coverage across all samples and ensure the generation of reliable results. Furthermore, by setting the VQSR filter in VCF files, we reduced the number of false positives and improved the quality of our variant calls. This improvement allowed us to refine our results further, making them even more precise and reliable.

3.7 Computational Resources

The variant calling pipeline was executed on a high-performance computing cluster with the specifications shown in Table 2. The cluster contains one head node with two processors of Intel(R) Xeon(R) CPU E5-2630 v4 @ 2.20 GHz and 500G of memory. The two node servers have two processors of Intel(R) Xeon(R) CPU E5-2630 v4 @ 2.20 GHz and 128G of memory and two processors of Intel(R) Xeon(R) CPU E5-2690 v3 @ 2.60 GHz and 128G of memory, respectively. The storage used for the pipeline was an HDD disk with Raid 6-10.

Table 2. Cluster Hardware Specifications

Cluster Type	No.	Processor	Memory
Head	1	2 x [Intel(R) Xeon(R) CPU E5-2630 v4 @ 2.20 GHz]	500G
Node	3	2 x [Intel(R) Xeon(R) CPU E5-2630 v4 @ 2.20 GHz]	128G
Node	1	2 x [Intel(R) Xeon(R) CPU E5-2690 v3 @ 2.60 GHz]	128G

4   Results

4.1   Different Methods Performance

We conducted a thorough analysis of a Whole Genome Dataset (WGS) dataset, using a variety of filters and tools to detect INDEL and SNP variants. The results of our analysis are presented in Table 3. This able also includes the FP.gt and FP.al values, FP.gt stands for number of genotype mismatches (alleles match, but different zygosity), and FP.al stands for number of allele mismatches (variants matched by position and not by haplotype). For SNPs the Ti/Tv ratio falls between [2.08-2.1], expecting a value of 2.08 for both samples.

Table 3. Comparison of Variant Calling Performance for Sample NA12878 and NA12877 at Different Coverages and Filters.

Sample ID	Coverage	Type	Filter	Recall	Precision	F1-Score	FP.gt	FP.al
NA12878	~50	Indel	GATK-ALL	0.960	0.967	0.964	2780	8497
			GATK-Hard Filter	0.960	0.969	0.965	2767	8011
			GATK-VQSL**	0.876	0.981	0.925	2008	4875
			DeepVariant	0.971	0.981	0.976	2153	5749
		SNP	GATK-ALL	0.993	0.991	0.992	2145	1701
			GATK-Hard Filter	0.970	0.995	0.982	1305	876
			GATK-VQSL***	0.888	0.999	0.940	284	137
			DeepVariant	0.993	0.999	0.996	1511	956
	~100	Indel	GATK-ALL	0.977	0.973	0.975	1542	7046
			GATK-Hard Filter	0.976	0.976	0.976	1524	6568
			GATK-VQSL**	0.905	0.986	0.944	918	4070
			DeepVariant	0.956	0.983	0.969	3058	4447
		SNP	GATK-ALL	0.994	0.991	0.993	1935	1727
			GATK-Hard Filter	0.969	0.996	0.982	1121	661
			GATK-VQSL***	0.889	0.999	0.941	253	99
			DeepVariant	0.967	0.998	0.982	4156	841
NA12877	~50	Indel	GATK-ALL	0.965	0.967	0.966	2555	8197
			GATK-Hard Filter	0.965	0.970	0.967	2544	7685
			GATK-VQSL**	0.884	0.982	0.930	1806	4637
			DeepVariant	0.973	0.982	0.977	1892	5642
		SNP	GATK-ALL	0.993	0.991	0.992	2229	1544
			GATK-Hard Filter	0.970	0.995	0.982	1465	838
			GATK-VQSL***	0.888	0.999	0.940	233	103
			DeepVariant	0.991	0.998	0.995	1611	964
	~100	Indel	GATK-ALL	0.978	0.975	0.976	1271	6529
			GATK-Hard Filter	0.977	0.978	0.978	1248	6080
			GATK-VQSL**	0.909	0.987	0.947	774	3756
			DeepVariant	0.960	0.984	0.972	2780	4359
		SNP	GATK-ALL	0.994	0.991	0.993	2076	1436
			GATK-Hard Filter	0.969	0.996	0.982	1259	633
			GATK-VQSL***	0.889	0.999	0.941	208	92
			DeepVariant	0.968	0.998	0.983	4104	878

The evaluation of four different variant calling methods for SNP and INDEL variants in the NA12877 and NA12878 genomes revealed that all methods achieved high recall and precision for SNP calling. However, there were differences in F1-score, with DeepVariant achieving the highest F1-score for both genomes and coverage depths.

All methods had lower recall than SNP calling for INDEL calling, with VQSL*** achieving the lowest recall for both genomes and coverage depths. Precision was generally high, with all methods achieving ¿0.96 precision for both genomes and coverage depths. F1-score was also generally high, with DeepVariant again achieving the highest F1-score for both genomes and coverage depths.

Although the methods tended to overestimate heterozygosity, the het/hom ratios were generally consistent across the different methods and coverage depths, values not present in the table. Overall, the evaluation results suggest that DeepVariant is a robust and reliable method for both SNP and INDEL calling, achieving high recall, precision, and F1-score across different coverage depths and genomes. Although, DeepVariant has higher numbers of FP.gt and FP.al in relation of GATK VQSL filter and decrease the quality when the coverage increase, that is the opposite of GATK.

4.2   Computational Time

The pipeline was tested with different combinations of samples, coverage, and tools, and the running times are reported in Table 4. For four samples and 50x coverage, the pipeline took 64 h to complete using GATK and hard filtering. For two samples with 200x coverage, the pipeline took around four days to complete using the same tool. When using GATK and VQSR, the pipeline did not report any issues, and the running time varied according to the sample size and coverage. Finally, DeepVariant was used to call variants, and the running times were less than those obtained with GATK.

Table 4. Running times for variant calling pipelines on different sample sizes and coverage

Pipeline	Samples	Coverage	Time
GATK-Hard Filter	4	50	64 h
GATK-Hard Filter	2	100	∼4 days
GATK-VQSR	4	50	64 h
GATK-VQSR	2	100	∼4 days
DeepVariant	4	50	∼60 h
DeepVariant	2	100	∼3.5 days

In conclusion, the variant calling pipeline was successfully executed on the high-performance computing cluster, and different combinations of samples, coverage, and tools were tested. The running times varied depending on the sample size, coverage, and the type of tool used, with DeepVariant being the most time-efficient. Further pipeline optimisation could be performed to reduce the running time and improve its accuracy.

5 Discussion

This study evaluated various variant calling methods for SNP and INDEL variants in the NA12877 and NA12878 genomes. These genomes are widely recognized benchmark samples for assessing variant calling method accuracy. Additionally, we analyzed the computational time required for each method. Since these are gold standard samples, we can generalize to a larger dataset by performing optimisation tests on these samples.

Our findings demonstrate that all four methods exhibited high precision and recall for SNP calling. DeepVariant yielded the highest F1-score for both genomes and coverage depths, consistent with previous studies showing its superior accuracy and speed compared to other variant calling methods, such as GATK and FreeBayes [35,36]. DeepVariant's high accuracy is attributed to its utilization of a deep neural network, which can identify complex patterns of genomic variation from large training datasets.

Although the methods tended to overestimate heterozygosity, the het/hom ratios were relatively consistent across the different methods and coverage depths. This finding suggests that the methods are suitable for detecting heterozygous variants common in diploid genomes.

Regarding computational time, our results indicate that the pipeline can be computationally demanding, particularly for samples with high coverage and when using GATK and VQSR. DeepVariant achieved faster running times than GATK, indicating that it could be a viable alternative for variant calling in large-scale projects, such as the GDI project.

Furthermore, our work contributes to the optimization of the variant calling pipeline for large-scale projects with numerous samples. Through this, we address the computational complexity and data management challenges associated with such projects, thereby facilitating a more efficient and effective analysis of genetic variations within populations. A key feature of our optimized pipeline is the generation of gVCFs that can aggregate variants across cohorts, providing a comprehensive approach to understanding the complex landscape of genetic variations in diploid genomes. Additionally, all gVCFs generated in our pipeline come with annotations from the Variant Effect Predictor (VEP), which further enhances the biological interpretation of genetic variants. By incorporating VEP annotations, researchers can gain valuable insights into the potential functional consequences of identified variants, ultimately supporting more informed decision-making in various genomic applications, such as disease association studies and personalized medicine.

6 Conclusion

In this study, we aimed to develop a pipeline that generates high-quality variant calls and annotations, is adaptable to specific queries, and efficiently and accurately analyzes genomic data. To achieve this, we comprehensively evaluated two variant calling methods using benchmark genomes for NA12877 and NA12878 individuals. Our findings reveal that DeepVariant outperforms other methods in accuracy and efficiency. Specifically, DeepVariant yielded the highest F1-score for both SNP and INDEL calling while maintaining consistent het/hom ratios. The high accuracy is attributed to its utilization of a deep neural network, which can identify complex patterns of genomic variation from large training datasets, although the best results are for runs with 50 coverage. Our results suggest the implementation of a variant calling pipeline for the GDI project that should utilize DeepVariant as the primary tool for SNP and INDEL calling. The pipeline begins with read cleaning and mapping using Trimmomatic and BWA, respectively, followed by duplicate removal with Picard Tools. Next, the variant calling step can be performed using DeepVariant merging the gVCF files with GLnexus. The final step of the pipeline involves variant annotation using VEP. By implementing this pipeline, we can ensure the accurate and efficient detection of genomic variation, providing high-quality cohort VCFs for downstream analyses. Furthermore, the pipeline is adaptable and easily modified to include additional tools, because is made with Snakemake based on Snakemake wrappers. Optimizing the pipeline can reduce the computational time required for variant calling, allowing for faster analyses of larger datasets.

In conclusion, our study provides valuable insights into the importance of accurate and efficient detection of genomic variation in large-scale genomic studies and can guide the decision-making process for implementing a variant calling pipeline in such projects.

Acknowledgements. This work has received funding from the EC under grant agreement 101081813, Genomic Data Infrastructure.

Reproducibility. The repository for the replication of the results described in this paper are available at https://github.com/bioinformatics-ua/GDI_Pipeline.git.

References

1. US DOE Joint Genome Institute: Hawkins Trevor 4 Branscomb Elbert 4 Predki Paul 4 Richardson Paul 4 Wenning Sarah 4 Slezak Tom 4 Doggett Norman 4 Cheng Jan-Fang 4 Olsen Anne 4 Lucas Susan 4 Elkin Christopher 4 Uberbacher Edward 4 Frazier Marvin 4, RIKEN Genomic Sciences Center: Sakaki Yoshiyuki 9 Fujiyama Asao 9 Hattori Masahira 9 Yada Tetsushi 9 Toyoda Atsushi 9 Itoh Takehiko 9 Kawagoe Chiharu 9 Watanabe Hidemi 9 Totoki Yasushi 9 Taylor Todd 9, Genoscope, CNRS UMR-8030: Weissenbach Jean 10 Heilig Roland 10 Saurin William 10 Artiguenave Francois 10 Brottier Philippe 10 Bruls Thomas 10 Pelletier Eric 10 Robert Catherine 10 Wincker Patrick 10, Institute of Molecular Biotechnology: Rosenthal André 12 Platzer Matthias 12 Nyakatura Gerald 12 Taudien Stefan 12 Rump Andreas 12 Department of Genome Analysis, GTC Sequencing Center: Smith Douglas R. 11 Doucette-Stamm Lynn 11 Rubenfield Marc 11 Weinstock Keith 11 Lee Hong Mei 11 Dubois JoAnn 11, Beijing Genomics Institute/Human Genome Center: Yang Huanming 13 Yu Jun 13 Wang Jian 13 Huang Guyang 14 Gu Jun 15, et al. Initial sequencing and analysis of the human genome. nature, 409(6822):860–921, 2001

2. Sims, D., Sudbery, I., Ilott, N.E., Heger, A., Ponting, C.P.: Sequencing depth and coverage: key considerations in genomic analyses. Nat. Rev. Genet. **15**(2), 121–132 (2014)
3. Kaye, J., Heeney, C., Hawkins, N., De Vries, J., Boddington, P.: Data sharing in genomics-re-shaping scientific practice. Nat. Rev. Genet. **10**(5), 331–335 (2009)
4. Papageorgiou, L., Eleni, P., Raftopoulou, S., Mantaiou, M., Megalooikonomou, V., Vlachakis, D.: Genomic big data hitting the storage bottleneck. EMBnet J. **24** (2018)
5. Maojo, V., Martin-Sanchez, F., Kulikowski, C., Rodriguez-Paton, A., Fritts, M.: Nanoinformatics and DNA-based computing: catalyzing nanomedicine. Pediatr. Res. **67**(5), 481–489 (2010)
6. European Genomic Data Infrastructure (GDI). https://gdi.onemilliongenomes.eu. Accessed 24 Feb 2023
7. Beyond 1 Million Genomes. https://b1mg-project.eu. Accessed 24 Feb 2023
8. Danecek, P., et al.: The variant call format and VCFtools. Bioinformatics **27**(15), 2156–2158 (2011)
9. Martin, M., et al.: WhatsHAP: fast and accurate read-based phasing. BioRxiv, 085050 (2016)
10. Alipanahi, B., Delong, A., Weirauch, M.T., Frey, B.J.: Predicting the sequence specificities of DNA- and RNA- binding proteins by deep learning. Nat. Biotechnol. **33**(8), 831–838 (2015)
11. Zou, J., Huss, M., Abid, A., Mohammadi, P., Torkamani, A., Telenti, A.: A primer on deep learning in genomics. Nat. Genet. **51**(1), 12–18 (2019)
12. Johannes, K.: DNA-SEQ-GATK-variant-calling (2021). https://github.com/snakemake-workflows/dna-seq-gatk-variant-calling
13. Van der Auwera, G.A., et al.: From fastQ data to high-confidence variant calls: the genome analysis toolkit best practices pipeline. Curr. Protoc. Bioinform. **43**(1), 11–10 (2013)
14. O'Rawe, J., et al.: Low concordance of multiple variant-calling pipelines: practical implications for exome and genome sequencing. Genome Med. **5**(3), 1–18 (2013)
15. McKenna, A., et al.: The genome analysis toolkit: a mapreduce framework for analyzing next-generation DNA sequencing data. Genome Res. **20**(9), 1297–1303 (2010)
16. Poplin, R., et al.: A universal SNP and small-indel variant caller using deep neural networks. Nat. Biotech. **36**(10), 983–987 (2018)
17. Olson, N.D., et al.: PrecisionFDA truth challenge v2: calling variants from short and long reads in difficult-to-map regions. Cell Genomics **2**(5), 100129 (2022)
18. Li, H., et al.: The sequence alignment/map format and samtools. Bioinformatics **25**(16), 2078–2079 (2009)
19. Garrison, E., Marth, G.: Haplotype-based variant detection from short-read sequencing. arXiv preprint (2012). arXiv:1207.3907
20. Koboldt, D.C., et al.: VarScan: variant detection in massively parallel sequencing of individual and pooled samples. Bioinformatics **25**(17), 2283–2285 (2009)
21. Sadedin, S.P., Pope, B., Oshlack, A.: Bpipe: a tool for running and managing bioinformatics pipelines. Bioinformatics **28**(11), 1525–1526 (2012)
22. Seven Bridges. Publications and case studies (2021)
23. Schneider, V.A., et al.: Evaluation of GRCh38 and de novo haploid genome assemblies demonstrates the enduring quality of the reference assembly. Genome Res. **27**(5), 849–864 (2017)
24. Church, D.M., et al.: Modernizing reference genome assemblies. PLoS Biol. **9**(7), e1001091 (2011)

25. Chaisson, M.J., et al.: Resolving the complexity of the human genome using single-molecule sequencing. Nature **517**(7536), 608–611 (2015)
26. Sedlazeck, F.J., et al.: Accurate detection of complex structural variations using single-molecule sequencing. Nat. Methods **15**(6), 461–468 (2018)
27. Eberle, M.A., et al.: A reference data set of 5.4 million phased human variants validated by genetic inheritance from sequencing a three-generation 17-member pedigree. Genome Res. **27**(1), 157–164 (2017)
28. DePristo, M.A., et al.: A framework for variation discovery and genotyping using next-generation DNA sequencing data. Nat. Genet. **43**(5), 491–498 (2011)
29. Bolger, A.M., Lohse, M., Usadel, B.: Trimmomatic: a flexible trimmer for illumina sequence data. Bioinformatics **30**(15), 2114–2120 (2014)
30. Andrews, S., et al.: FastQC: a quality control tool for high throughput sequence data (2010)
31. Broad Institute. Picard Toolkit. http://broadinstitute.github.io/picard/. Accessed 24 Feb 2023
32. Ewels, P., Magnusson, M., Lundin, S., Käller, M.: MultiQC: summarize analysis results for multiple tools and samples in a single report. Bioinformatics **32**(19), 3047–3048 (2016)
33. Miller, R.R., Zhang, L., Chen, L.: Next-generation sequencing bioinformatics pipelines. Clin. Lab. News: A Publication of AACC, **46**(3), 12–15 (2020)
34. Krusche, P., et al.: Best practices for benchmarking germline small-variant calls in human genomes. Nat. Biotechnol. **37**(5), 555–560 (2019)
35. Barbitoff, Y.A., Abasov, R., Tvorogova, V.E., Glotov, A.S., Predeus, A.V.: Systematic benchmark of state-of-the-art variant calling pipelines identifies major factors affecting accuracy of coding sequence variant discovery. BMC Genomics **23**(1), 155 (2022)
36. Lin, Y.-L., et al.: Comparison of GATK and DeepVariant by trio sequencing. Sci. Rep. **12**(1), 1809 (2022)

Healthcare and Diseases

Improving Fetal Health Monitoring: A Review of the Latest Developments and Future Directions

Restuning Widiasih[1]([✉])(ID), Hasballah Zakaria[2](ID), Siti Saidah Nasution[3](ID), Saffan Firdaus[4], and Risma Dwi Nur Pratiwi[1]

[1] Faculty of Nursing, Universitas Padjadjaran, Bandung, Indonesia
`restuning.widiasih@unpad.ac.id`
[2] School of Electrical Engineering and Informatics, Institut Teknologi Bandung, Bandung, Indonesia
[3] Faculty of Nursing, Universitas Sumatra Utara, Medan, Indonesia
[4] Faculty of Engineering, Universitas Indonesia, Jakarta, Indonesia

Abstract. Devices for monitoring heart rate and fetal movement are becoming increasingly sophisticated with the latest technological advancements. However, there is a pressing need for a more comprehensive review and analysis of these tools. The objective of this literature review is to identify fetal health monitoring devices, evaluate the sensitivity of monitoring fetal heart rate and growth/movement, and determine the target users for these devices. Method, the search was conducted using PubMed and Scopus databases, with PICO-based keywords that included pregnant women or pregnancy as the population, fetal or heart rate monitoring tool or fetal movement tool with sensitivity or reactivity as the research interest. A total of 2077 papers were initially identified, with 25 selected after article screening using the PRISMA approach and critical appraisal with JBI. Results, the analysis classified into four categories: the development of monitoring devices for fetal well-being, algorithm for more accurate maternal-fetal FHR filtering, fetal well-being indicators, and target users. The monitoring technology applied wave detection through cardiography and myography. The devices demonstrated a signal sensitivity of more than 75% for both the mother and the fetus. Conclusions, the analysis of the 25 articles revealed that monitoring technology is rapidly evolving, but almost all devices are designed for use by health workers. Only the Piezo Polymer Pressure Sensor is intended for independent monitoring by mothers and families. The development and research of independent fetal monitoring are necessary to improve monitoring during the new adaptation period after the Covid-19 pandemic.

Keywords: Covid-19 · Fetal monitoring tools · Fetal Heart rate · Fetal movement · Pregnancy · Sensitivity

I. Rojas et al. (Eds.): IWBBIO 2023, LNBI 13920, pp. 89–109, 2023.
https://doi.org/10.1007/978-3-031-34960-7_7

1 Introduction

Fetal well-being monitoring during pregnancy is essential for ensuring a healthy birth and development. The aims of fetal monitoring are to ensure the birth of healthy babies, monitor fetal growth and development, detect early abnormalities and fetal distress, and determine appropriate delivery methods (Evans et al. 2022). Various abnormalities and risks to fetal health can be detected by observing fetal well-being using methods such as electronic fetal monitoring (EFM), which has been widely employed to prevent a significant proportion of neonatal encephalopathy and cerebral palsy (Evans et al. 2022), reduce discrepancies in obstetric intervention (Mayes et al. 2018), and identify developmental disorders that pose a risk of perinatal mortality due to hypoxemia, acidemia, and stillbirth (Baschat et al. 2022; Valderrama et al. 2020). Fetal health monitoring is a concern for health services during pregnancy because it is closely related to the prevention of fetal morbidity and mortality, prompting the continued development and refinement of monitoring tools for fetal well-being.

Fetal well-being monitoring technologies include both invasive and non-invasive approaches. Invasive procedures directly affect the body's tissues of the fetus, while non-invasive procedures are prenatal methods that pose no risk and cause no pain to the mother and fetus (Alfirevic et al. 2017; Hamelmann et al. 2020). Fetal well-being is observed using tools such as Internal Electronic Fetal Monitoring (ECG) and Internal Electronic Contraction Monitoring, cardiotocography (CTG), auscultation with manual (Pinard/Laennec Stethoscope, Fotoscope) and digital (FetalPhonoCardio-Gram/FPCG) stethoscopes, Ultrasonography (USG), and non-invasive abdominal ECG (FECG) (Kauffmann and Silberman 2023; Valderrama et al. 2020). These technologies vary and have specific functions, such as monitoring fetal heart rate, uterine contractions, and the growth and development of fetal organs.

Tools for monitoring fetal heart rate, which are part of Fetal-Wellbeing Indicators, include intermittent auscultation (IA) and electronic fetal monitoring (EFM) methods. IA is employed for short-term FHR monitoring, whereas EFM is utilized for longer monitoring, ranging from 10 to 60 min, and offers a more detailed description of FHR performance (Valderrama et al. 2020). In addition to FHR, pregnant women are encouraged to monitor fetal movements as part of the independent monitoring of fetal well-being. In developing countries, including Indonesia, the most widely used fetal heart rate measurement technique among health workers is the Fetal Doppler, which is based on Doppler ultrasound (Hamelmann et al. 2020; Valderrama et al. 2020). Monitoring fetal well-being using EFM equipment can be costly. The fMCG, an expensive method requiring specialized operator training and dedicated shielded rooms, has become the gold standard for monitoring fetal well-being in developed countries (Valderrama et al. 2020).

Previous studies have analyzed the effectiveness of various fetal health monitoring technologies developed in recent years, such as FECG, Doppler ultrasonography, and fetal monitoring technology using a thermohygrometer. These analyses have focused on cost-effectiveness, gestational age, operator training, and evidence supporting the usefulness of the equipment. However, limited analyses exist regarding the sensitivity of devices for fetal health monitoring, the effectiveness of the methods, and the target user. This study aims to identify fetal health monitoring devices, evaluate the sensitivity of fetal heart rate monitoring devices and fetal growth/movement, and determine the target users of these devices.

2   Methods

2.1   Study Design

A systematic review employs systematic and explicit methods to identify, select, and critically assess relevant research, address one or more research questions, and provide a systematic synthesis of existing research characteristics (Gupta et al. 2018; Siddaway et al. 2019). The methodology and systematic presentation aim to minimize subjectivity and bias. This systematic review will focus on reviewing fetal well-being monitoring devices to improve fetal well-being.

2.2   Search Strategy

PubMed and Scopus databases used in the literature search for this systematic review. Keyword searches performed using the PICO approach (Table 1). The systematic review research was not be limited by year to enable the identification of research developments related to fetal monitoring over time.

2.3   Inclusion Criteria

Table 1. PICO Search Strategy

PICO's Framework	Keywords
Population/Problem	Pregnant women OR Pregnancy OR Gestating OR Gravid
Intervention	Fetal monitoring tool OR Heart rate monitoring tool OR Fetal movement tool OR Pulse rate instrument OR Fetal well-being monitoring tool
Comparation	-
Outcome	Sensitivity OR Reactivity

2.4   Selection of Studies

The initial search of the databases yielded 2,077 articles, with 481 articles found in Pubmed and 1,678 articles in Scopus. Of these, 82 articles were duplicates in both databases. In the initial title screening, 97 articles were identified, with 72 articles from Pubmed and 25 articles from Scopus. Next, the abstracts were screened, resulting in 72 articles from Pubmed and 6 articles from Scopus. Content screening was then performed on the 25 articles selected, using the JBI checklist form to determine their feasibility and quality based on a Quasi Experiment and RCT design. The 25 articles were divided into two research designs, Quasi Experiments with an average JBI score of 7 and RCT with a score of 7.5.

The analysis process involved reading and understanding the content of the articles at least three times and creating a data extraction table. The table was then analyzed by identifying similarities, differences, and comparisons. The main categories were determined from the extraction table, resulting in three main categories: devices for monitoring fetal movement, devices for monitoring fetal heart rate, and devices for monitoring both fetal heart rate and fetal movement (Fig. 1).

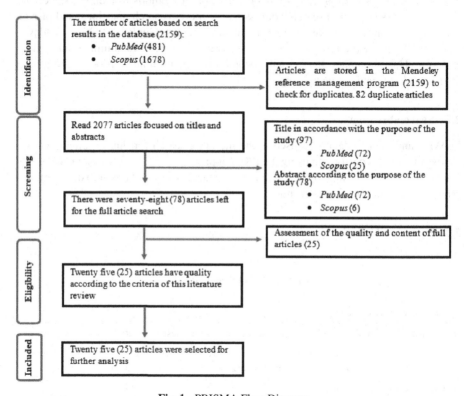

Fig. 1. PRISMA Flow Diagram

2.5 Data Analysis

The analysis of the articles was conducted in three stages. In the first stage, the full articles were read to understand their contents. The second stage involved extracting and summarizing the reading results in a table. Table 2 presents information on the article's title and author, research objectives, research locations, types of research, research samples, data collection methods, and a summary of research results. In the third stage, the researcher analyzed the similarities and differences in the content of each article and drew conclusions in the form of categories. The results of the analysis identified the development of device types, categories of fetal well-being observation, target users, and device sensitivity. The complete findings are presented in the findings and discussion section (Table 3).

3 Result

Table 4 illustrates that most of the research was conducted in developed countries using a quasi-experimental design with a sample size of pregnant women >30 weeks' gestation. The number of samples ranged from 6 to 1001 pregnant women, with research conducted from 1990 to 2019. The types of devices used to measure fetal well-being included fetal magneto-cardiographic (fMCG), electronic uterine monitoring (EUM), external Doppler, fetal scalp electrode monitor electromyogram (EMG), Cardiotocography (CTG) Non-stress test (NST), Pinard stethoscope, Intrauterine ECG Signal, fetal ECG, Photoscope (AAT), Piezo Polymer Pressure Sensor, fetal cardiovascular magnetic resonance (CMR), and fetal electrocardiographic (FECG). Almost all of these devices are operated by health workers, with only one study examining the use of the device by pregnant women independently. Fetal welfare monitoring devices are divided into two types, electronic-based devices and manual-based devices (Pinard stethoscope). The components of fetal well-being studied included fetal heart rate and fetal movement. Five articles analyzed algorithmic methods to separate fetal heart rate from maternal heart rate, such as LMS-based algorithm and ACWD model. The sensitivity value was >75%.

4 Discussion

4.1 The Development of Monitoring Devices for Fetal Well-Being

Based on the results of the literature analysis, fetal heart rate (FHR) measuring tools are divided into manual and electronic devices. Electronic-based tools are developing rapidly and are becoming increasingly sophisticated. Tools that have been developed continue to innovate and update their technology, such as CTG in (1984, 1992, 1998, 2003, 2004, 2009, 2011, 2012, 2014), FECG in (2012, 2012, 2017, 2018), AECG in (2002, 2008), EUM in (2002, 2019), FMCG in (1998, 2005, 2007, 2015, 2018), and Doppler ultrasound in (1993, 1998, 2018). On the other hand, the development of manual FHR monitoring tools is limited, even though research results had showed high sensitivity (Ashwal et al. 2019; Daniels and Boehm 1991). The use of fetal well-being monitoring devices requires trained health workers. Technological developments are the driving force for the development of digital health devices and are closely tied to mobile technology, which is widely used today.

Research in the UK strengthens the development of Health Technology towards digitization to meet community needs, enhance usability and interoperability, develop capacity for handling, processing, and analyzing data, address privacy and security concerns, and encourage digital inclusivity (Sheikh et al. 2021). Furthermore, the development of electronic health devices and digital-based health services creates future opportunities to improve integrated health services between health service facilities, health records,

Table 2. JBI Critical Appraisal for Quasi-Experimental

No.	JBI Components	Articles																							
		1	2	3	4	5	6	7	8	9	10	11	12	13	14	15	16	17	18	19	20	21	22	23	24
1	Is it clear in the study what is the 'cause' and what is the 'effect' (i.e. there is no confusion about which variable comes first)?	✓	✓	✓	✓	✓	✓	✓	✓	✓	✓	✓	✓	✓	✓	✓	✓	✓	✓	✓	✓	✓	✓	✓	✓
2	Were the participants included in any comparisons similar?	✓	✓	✓	✓	✓	✓	✓	✓	✓	✓	✓	✓	✓	✓	✓	✓	✓	✓	✓	✓	✓	✓	✓	✓
3	Were the participants included in any comparisons receiving similar treatment/care, other than the exposure or intervention of interest?	✓	✓	✓	✓	✓	✓	✓	✓	✓	✓	✓	✓	✓	✓	✓	✓	✓	✓	✓	✓	✓	✓	✓	✓
4	Was there a control group?	-	-	-	-	-	✓	-	-	-	-	-	-	-	-	-	-	-	✓	-	-	-	✓	-	✓
5	Were there multiple measurements of the outcome both pre and post the intervention/exposure?	-	-	-	-	-	-	-	-	-	-	-	-	-	-	-	-	-	✓	✓	-	-	-	-	-
6	Was follow up complete and if not, were differences between groups in terms of their follow up adequately described and analyzed?	✓	✓	✓	✓	✓	✓	✓	✓	✓	✓	✓	✓	✓	✓	✓	✓	✓	✓	✓	✓	✓	-	✓	✓
7	Were the outcomes of participants included in any comparisons measured in the same way?	✓	✓	✓	✓	✓	✓	✓	✓	✓	✓	✓	✓	✓	✓	✓	✓	✓	✓	✓	✓	✓	✓	✓	✓
8	Were outcomes measured in a reliable way?	✓	✓	✓	✓	✓	✓	✓	✓	✓	✓	✓	✓	✓	✓	✓	✓	✓	✓	✓	✓	✓	✓	✓	✓
9	Was appropriate statistical analysis used?	✓	✓	✓	✓	✓	✓	✓	✓	✓	✓	✓	✓	✓	✓	✓	✓	✓	✓	✓	-	✓	✓	✓	✓

Table 3. JBI Critical Appraisal for Randomized Controlled Trial

No.	JBI Components	Article
1	Was true randomization used for assignment of participants to treatment groups?	✓
2	Was allocation to treatment groups concealed?	✓
3	Were treatment groups similar at the baseline?	-
4	Were participants blind to treatment assignment?	✓
5	Were those delivering treatment blind to treatment assignment?	-
6	Were outcomes assessors blind to treatment assignment?	✓
7	Were treatment groups treated identically other than the intervention of interest?	✓
8	Was follow up complete and if not, were differences between groups in terms of their follow up adequately described and analyzed?	-
9	Were participants analyzed in the groups to which they were randomized?	✓
10	Were outcomes measured in the same way for treatment groups?	-
11	Were outcomes measured in a reliable way?	✓
12	Was appropriate statistical analysis used?	✓
13	Was the trial design appropriate, and any deviations from the standard RCT design (individual randomization, parallel groups) accounted for in the conduct and analysis of the trial?	-
Total		**8**

care providers, invest in health data science research, generate real-world data, develop artificial intelligence and robotics, facilitate public-private partnerships, and increase individual and family health awareness (Iyamu et al. 2021). The combination of development healthcare devices and digitalization refers to the technical process of converting analog records to digital data. Digitalization refers to the integration of digital technologies into public health operations, while digital transformation describes a cultural shift that pervasively integrates digital technologies and reorganizes services based on the health needs of the public.

The analysis of the article showed that there are limited technological inventions, especially related to monitoring fetal well-being aimed at self-observation by pregnant women or their families. A cultural shift that empowers pregnant women and their families to independently monitor fetal well-being is necessary to improve fetal well-being, prevent delays in problem identification, and prevent various disorders. Pregnant women and their families can be empowered to monitor their fetal well-being with valid indicators, which needs to be supported by future research on innovative tools for routine self-checking.

4.2 Algorithm for More Accurate Maternal-Fetal FHR Filtering

The results of the article analysis revealed that various algorithm approaches are used to increase the accuracy of FHR measurements by filtering maternal and fetal heart rate waves. Kuei-Chiang and Shynk (2002) used the (SC) algorithm to separate fetal and maternal heartbeats from an intrauterine electrocardiogram (IuECG) signal. Taralunga et al. (2008) employed an empirical mode decomposition (EMD) approach with vector machine (SVM) support, and the results showed that the fetal beats overlapped with the maternal QRS complex, but ESC can completely separate the mECG. Marchon et al. (2018) utilized the Linear Phase Sharp Transition BPF (LPST) to separate FECG from maternal ECG recording measured on the maternal abdomen using Independent Component Analysis (ICA), and identified the fetal P wave and T wave in 21 cases. Only one case extracted FECG using the wavelet theory based method, which will become a powerful tool for the differential diagnosis of fetal arrhythmias.

Filtering is necessary because the FECG is contaminated by numerous noise signals, including abdominal muscle electromyogram (EMG), electrohysterogram (EHG), maternal ECG (mECG), baseline wander due to maternal respiration, and power line interference. Further noise is caused when the signal passes through a multi-layer dielectric biological medium before being received by the electrodes (Signorini et al. 2003). In the future, FHR and MHR filtering has the potential to develop new fetal well-being monitoring technologies that are portable and can be operated independently by pregnant women and their families, particularly in low-middle income countries with various health service constraints, such as high morbidity and fetal death, pregnancy complications, and limited access to health services. Supporting health facilities for detailed and sophisticated monitoring of fetal well-being are not yet available in some regions of the country (Dahab and Sakellariou 2020; Mahomed et al. 1992). The need for a tool that can independently monitor fetal well-being is also highlighted in the research by Widiasih et al. (2021), where exploratory descriptive research was conducted during the COVID-19 pandemic, a situation that limited access to health services. Pregnant women and their families must be able to independently observe the well-being of their fetus for a long time, especially in situations that have limited knowledge or support facilities. The development of a portable device for monitoring fetal well-being targeting pregnant women and their families has the potential to improve the well-being of the unborn fetus, as part of efforts to prevent fetal morbidity and mortality.

4.3 Fetal Well-Being Indicators

The results of the analysis of fetal well-being monitoring devices showed that most of the articles focus on fetal heart rate, such as (Ashwal et al. 2019; Daniels and Boehm 1991; Kording et al. 2018; Mahomed et al. 1992; Mochimaru et al. 2004; Peters et al. 2012; Signorini et al. 2003; Velayo et al. 2017; Wakai 2004). Other devices for observation include monitoring fetal movement (Govindan et al. 2011), observation of intrauterine growth to prevent IUGR (Menéndez et al. 1998), and fetal response to sound stimulation (Sarno Jr.

et al. 1990). The technology for monitoring fetal heart rate is continuously developing with increasingly sophisticated and detailed examination, including the automatic detection of cardiac wave boundaries that works on fetal signals reconstructed from fMCG (Comani et al. 2005). Moreover, the sensitivity and effectiveness tests of FHR monitoring devices were also carried out with sensitivity values above 70% (Arias-Ortega et al. 2010; Daniels and Boehm 1991; Karlsson et al. 2000; Magenes et al. 2014).

Monitoring the fetal heart rate during labor can help identify changes in normal heartbeat patterns, and steps can be taken to solve underlying problems if changes are found. Monitoring the fetal heart rate can also help in preventing unnecessary treatments, and the normal fetal heartbeat may be sufficient to reassure patients and healthcare professionals that labor is safe. If there is an abnormal pattern of heart rate during pregnancy, healthcare professionals will first try to find the cause, and steps can be taken to help the fetus get more oxygen, such as changing positions. Next, if these procedures do not work, or if further test results suggest that the fetus has a problem, healthcare professionals may decide to deliver right away. In this case, the delivery is more likely to be by cesarean birth or with forceps or vacuum-assisted delivery (The American College of Obstetricians and Gynecologysts 2023).

On the other hand, there are still limited monitoring devices with indicators outside of FHR, such as monitoring fetal movements, especially in Low-Middle income countries, where reliance on the feelings of pregnant women is high, and this cannot be done for a long time. The influence of subjectivity on the part of the mother is also high. The development of fetal movement monitoring devices is necessary because decreased fetal movement might be a sign of potential fetal impairment or risk and may warrant further evaluation. Women are often taught by healthcare providers to be aware of fetal movements (Huecker et al. 2023).

4.4 Target Users

Most of the articles show that monitoring devices for fetal well-being are operated by health workers, and only one article aims to provide a portable fetal heart rate monitor, which will permit mothers to perform the fetal nonstress test in their homes (Zuckerwar et al. 1993). The development of this self-monitoring technology is increasing rapidly, even though its development and use are still limited in developed countries, and none have been used in developing countries, especially in Indonesia. A systematic review in Indonesia was carried out by Madiuw et al. (2019); the review analyzed fetal health monitoring tools published in various reputable international journals. In this technological era, various health technologies including media, instruments, and tools were created in the field of maternal and infant health. However, comprehensive information about these technology products is limited. Two databases, namely PubMed and CINAHL, found 16 articles on health technology. The technology was divided into two criteria, namely the development of questionnaires and medical devices. There is no technology originating from Indonesia. All foreign technologies have demonstrated effectiveness in detecting risks, and in interventions to improve prenatal care. In Indonesia, the latest invention related to fetal health monitoring is TeleCTG. TeleCTG is a portable device for monitoring uterine contractions and fetal heart rate to see whether there are any disturbances in the baby before or during delivery based on Telemedicine technology used by midwives

Table 4. Article Analysis

No.	Author (year), and Country	Methods	Devices	Objectives	Result
1	(Govindan et al. 2011) USA	Design: Quasi- Experimental Sample: 39 pregnant women (30–37 weeks)	Fetal magnetocar diographic (fMCG)	Identifying fetal movement based on fMCG signal	The fMCG recorded simultaneously with ultrasound from a single subject and show its improved performance over the QRS-amplitude based approach in the visually verified movements
2	(Ashwal et al. 2019) Sweden	Design: Quasi- Experimental Sample: 33 pregnant women	• Electronic Uterin Monitoring (EUM) • External Doppler • Fetal scalp electrode monitor electromyogram (EMG)	Comparing four devices in the accuracy of FHR	EUM accuracy was significantly higher than external Doppler (99.0% versus 96.6%, $p < .001$ for fetal heart rate < 110 bpm or > 160 bpm, PPA, significantly greater sensitivity than external Doppler ($p. < 0.001$)
3	(Mahomed et al. 1992) Zimbabwe	Design: Quasi- Experimental Sample: 200 pregnant women	• Cardiotocography (CTG) Sonicaid FM5L • the Pinard stethoscope	Comparing the accuracy of FHR using electronic trace and the Pinard stethoscope	The correlation coefficient two between tools observers was 0.8. The Pinard manual trace sensitivity was 75%

(continued)

Table 4. (*continued*)

No.	Author (year), and Country	Methods	Devices	Objectives	Result
4	(Kuei-Chiang and Shynk, 2002) California	Design: Quasi- Experimental Sample: 500 pregnant women	Intrauterine ECG Signal	Separating fetal and maternal heartbeats from an intrauterine electrocardiogram (IuECG) signal using the (SC) algorithm	The SC (successive cancellation) algorithm was effective in separating the fetal heart rate from the mother's heart rate
5	(Taralunga et al. 2008) Canada	Design: Quasi- Experimental Sample: 15 pregnant women	Fetal ECG	Separating fetal and maternal heartbeats from an maternal electrocardiogram (mECG) signal using Event Synchronous Canceller (ESC) algorithm	Even though the fetal beats overlap with the maternal QRS complex, ESC can completely separate the mECG
6	(Krupa et al. 2011) Malaysia	Design: Quasi- Experimental Sample: 15 pregnant women	Cardiotocography (CTG)	FHR extraction based on empirical mode decomposition (EMD) coupled with vector machine (SVM) support	The five-fold cross-validation test yielded an accuracy of 86% while the overall geometric mean of sensitivity and specificity was 94.8%

(*continued*)

Table 4. (*continued*)

No.	Author (year), and Country	Methods	Devices	Objectives	Result
7	(Wakai 2004) Madison	Design: Quasi- Experimental Sample: 61 pregnant women	Fetal magnetocar diography (fMCG) FMCG actography (akurasi tinggi)	Describing the application of fetal magnetocardiography (fMCG) to fetal heart rate variability (FHRV) and fetal trunk movement	FMCG is more specific for fetal trunk movement and FHR than isolated limb movements. Collaborative examination of fMCG and FHRV in FHR assists in neurodevelopmental research
8	(Daniels and Boehm, 1991) Albuquer que	Design: Quasi- Experimental Sample: 91 pregnant women	Non-stress test (NST) and Fotoscope (AAT)	Determining difference between the electronically monitored nonstress test (NST) and the auscultated acceleration test (AAT)	The sensitivity (100%) and specificity (85.37%) of the AAT indicate that the test is valid in predicting NST outcomes and thus appears to be a valid screening tool for fetal well-being and may be a reliable alternative to NST
9	(Peters et al. 2012) Netherlands	Design: Quasi- Experimental Sample: 149 pregnant women	Abdominal fetal ECG (NEMO prototipe)	Increasing the usability of abdominal fetal ECG recordings	The sensitivity and positive predictive value of this method is reduced to about 90% for SNR $\leq$ 2.4. The evaluation on the generated signals demonstrated excellent results (sensitivity of 0.98 for SNR ≥ 1.5)

(*continued*)

Table 4. (*continued*)

No.	Author (year), and Country	Methods	Devices	Objectives	Result
10	(Devoe et al. 1994) Georgia	Design: Quasi- Experimental Sample: 884 pregnant women	Doppler Fetal using a Hewlett-Packard M1350A fetal monitor	To correlate measures of Doppler-detected fetal movements with standard fetal heart rate parameters and perinatal outcomes	All fetal gesture parameters were significantly increased in consecutive 10 min blocks and during periods with increased or normal fetal heart rate variability compared to periods with fetal heart rate variability
11	(Zuckerwar et al. 1993) Hampton	Design: Quasi- Experimental Sample: 6 pregnant women	A piozopolymer pressure Sensor	To provide a portable fetal heart rate monitor, which will permit mother to perform the fetal nonstress test, in her home	The sensor design conforms to the distinctive features of the fetal heart tone, namely, the acoustic signature, frequency spectrum, signal amplitude, and localization. The components of a sensor serve to fulfill five functions: signal detection, acceleration cancellation, acoustical isolation, electrical shielding, and electrical isolation of the mother. An in vivo test on patients within the last six weeks of term reveals that nonstress test recordings from the acoustic monitor compare well with those obtained from conventional ultrasound

(*continued*)

Table 4. (*continued*)

No.	Author (year), and Country	Methods	Devices	Objectives	Result
12	(Kording et al. 2018) Germany	Design: Quasi- Experimental Sample: 15 pregnant women	Fetal cardiovascular magnetic resonance (CMR) 1.5 T CMR Scanners-compatible DUS	To investigate the effectiveness of a newly developed Doppler ultrasound (DUS) device in humans for fetal CMR gating	The fetal heart was detected with a variability of 26 ± 22 ms and a sensitivity of trigger detection of $96 \pm 4\%$. High-quality dynamic fetal CMR was successfully performed using the newly developed DUS device (4-chamber, mitral valve, foramen ovale, atrial septum, systolic-diastolic, stroke volume)
13	(Velayo et al. 2017) Philippines and United Kingdom	Design: Quasi- Experimental Sample: 46 pregnant women	fetal electrocardi ographic (FECG)	To evaluate the ECG system is capable of extracting fetal ECG signals from IUGR foetuses, and to assess the abdominal FECG parameters could be identified of cardiac status during pregnancy in IUGR	QTc and QTc intervals in IUGR fetuses were markedly prolonged by a value of 0.017 and $p = 0.002$ compared to their normal values. ECG parameters and Doppler, as well as the rest of the indices, couldn't be correlated with IUGR prediction. Increased sensitivity, but decreased specificity as ECG parameters are prolonged, has been observed when producing cutoff values for IUGR detection

(*continued*)

Table 4. (*continued*)

No.	Author (year), and Country	Methods	Devices	Objectives	Result
14	(Sarno Jr. et al. 1990) Los Angeles	Design: Quasi- Experimental Sample: 201 pregnant women	A tocodynamometer and a Doppler fetal heart rate transducer (Corometrics, CT, USA)	To evaluate fetal acoustic stimulation in the early intrapartum period as a predictor of subsequent fetal condition	Fourteen of the 201 fetuses (7%) showed a nonreactive response to fetal acoustic stimulation, and those fetuses were at significantly greater risk of initial and subsequent abnormal fetal heart rate patterns, meconium staining, and cesarean delivery because of fetal distress and Apgar scores < 7 at both 1 and 5 min. Transient fetal heart rate decelerations after a reactive response occurred in 25% of patients
15	(Wacker-Gussmann et al. 2018) Germany	Design: Quasi- Experimental Sample: 15 pregnant women	Fetal electrocardi ogram (fECG) Monica Healthcare System	To evaluate precise fetal cardiac time intervals (fCTIs) using the fetal ECG device: Monica Healthcare System	FECG (Monica Healthcare System) can detect the fCTI from 32 weeks with P-wave and QRS-complex could be easily identified in most ECG patterns (97% for P-wave, PQ and PR interval and 100% for QRS-complex), while the T-wave was detectable only 41%
16	(Arias-Ortega et al. 2010) Argentina	Design: Quasi- Experimental Sample: 25 pregnant women	LMS-based algorithm	To monitor fetal and maternal heart rate in real time	For the performance measures, sensitivity and accuracy have been applied. Overall, maternal and fetal detections have a high sensitivity of 95.3% and 87.1% respectively

(*continued*)

Table 4. (*continued*)

No.	Author (year), and Country	Methods	Devices	Objectives	Result
17	(Marchon et al. 2018) India	Design: Quasi- Experimental Sample: 1001 pregnant women	Linear Phase Sharp Transition BPF (LPST)	To analyze (LPST) FIR band-pass filter which of overlap of maternal ECG (MECG) spectrum	The performance parameters of the FQRS detector such as sensitivity (Se), positive predictive value (PPV), and accuracy (F1) were found to improve even for lower filter order
18	(Signorini et al. 2003) Italy	Design: Quasi- Experimental Sample: 35 pregnant women	Cardiotoco graphy (CTG)	To analyze new methodological approach for the CTG monitoring, based on a multiparametric FHR	The FHR pattern is more irregular with respect to normal fetuses. This observation underlines the difference between signal variability and regularity. As a matter of fact, the increase of ApEn in potential pathological conditions corresponds to a significant decrease in FHR variance with respect to normal healthy conditions
19	(Magenes et al. 2014) Italy	Design: Quasi- Experimental Sample: 120 pregnant women	A Hewlett Packard CTG fetal monitor	To identify the discrimination of Normal and Intra Uterine Growth Restricted (IUGR) fetuses based on a small set of parameters computed on the FHR signal	Comparing the LR built with the best univariate predictor, covariance, the multiparametric strategy allows significantly improving the 10-fold prediction performance in terms of accuracy (from 84% to 92%), with a 10% improvement of sensitivity

(*continued*)

Table 4. (*continued*)

No.	Author (year), and Country	Methods	Devices	Objectives	Result
20	(Zhao and Wakai 2002) USA	Design: Quasi- Experimental Sample: 39 pregnant women	Actocardiography, foetal magnetocardiogram (fMCG)	To describe a new method of fetal magnetocardiogram (fMCG) actocardiography, based on the high sensitivity of the fMCG to fetal trunk movements	The beat-to-beat FHR variability often decreased at or near the start of FHR accelerations, and this occurrence was an accurate marker of fetal movement onset, even when fetal movement onset lagged FHR acceleration
21	(Murta et al. 2015) Italy	Design: Quasi- Experimental Sample: 28 pregnant women	Fetal magnetocariograms (FMCG)	The detection of signal abnormalities according to which the fMCG data are broken down into stationary segments and then processed with ICA is part of this techniqueSegICA	Results showed that the SNR of fetal signals affected by fetal movements improved with SegICA. The best measure to detect signal nonstationarities of physiological origin was signal polarity reversal at threshold level 0.9. The first statistical moment also provided good results at threshold level 0.6. SegICA seems a promising method to separate fetal cardiac signals of improved quality from nonstationary fMCG recordings affected by fetal movements
22	(Mochimaru et al. 2004) Japan	Design: Quasi- Experimental Sample: 30 pregnant women	Fetal electrocardiogram (FECG)	To separate FECG from maternal ECG recording measured on the maternal abdomen using Independent Component Analysis (ICA)	The fetal P wave and T wave could be identified in 21 cases. FECG was extracted in only one case. FECG obtained by the wavelet theory-based method will become a powerful tool for the differential diagnosis of fetal arrhythmias

(*continued*)

Table 4. (*continued*)

No.	Author (year), and Country	Methods	Devices	Objectives	Result
23	(Karlsson et al. 2000) Poland	Design: Quasi- Experimental Sample: 22 pregnant women	The DopFet System: Ultrasonic doppler and electromyography (EMG)	The DopFet Parameters	Good sensitivity on real-time ultrasound and detectable 96% rotational movement, 100% flexion, and 97% hand movement were observed
24	(Comani et al. 2005) Italy	Design: Quasi- Experimental Sample: 49 pregnant women	Fetal mangnetoc ardiography (fMCG)	To describe an analytical model (ACWD) for the automatic detection of cardiac waves boundaries that works on fetal signals reconstructed from fMCG by means of independent component analysis	ACWD performances on short and long rhythm strips were investigated. ACWD demonstrated to be a robust tool providing dependable estimates of cardiac intervals and their variability during the third gestational trimester, also in case of fetal arrhythmias
25	(Schiermeier et al. 2007) Germany	Design: RCT Sample: 5 pregnant women	Cardiotocography (CTG) Fetal magnetocardiography (FMCG)	To analyzed applied both methods CTG and FMCG sequentially and simultaneously in healthy pregnancies	CTG can be used as a measurement of saturation and pulse, especially in the delivery room. Fetal short-term HRV was estimated on the basis of RMSSD values for both methods

and ob-gyn doctors. Portable devices in Indonesia target users are still for health professionals only (Sehati Group 2023). The application of the Internet of Medical Things makes access to TeleCTG devices easier, with concise and secure data. Even though it is used in several separate units, patient data can easily be sent to the doctor who is in charge of the patient, when he is not there. Various health technologies in the field of infant health were identified, but most of these internet-based technologies were made available to health professionals. Limited internet-based fetal monitoring technology in the form of simple and practical tools or devices for assessing fetal health, with the target users being pregnant women and their families independently.

5   Conclusion

The availability of technical inventions for fetal health monitoring that allow pregnant women and their families to identify problems in a timely manner is very limited. Future research should focus on innovative tools for regular self-examination to prevent various diseases or conditions. Currently, there is a limited number of monitors with non-FHR metrics such as fetal movement monitoring, especially in low and middle-income countries where monitoring relies on pregnant women's emotions. Therefore, the development of wearable fetal health monitoring devices for pregnant women and their families is crucial to improve fetal health and prevent fetal morbidity and mortality. However, in Indonesia, wearable devices are still limited to healthcare workers only.

References

Alfirevic, Z., Stampalija, T., Dowswell, T.: Fetal and umbilical Doppler ultrasound in high-risk pregnancies. Cochrane Database Syst. Rev. **6**(6), Cd007529 (2017). https://doi.org/10.1002/14651858.CD007529.pub4

Arias-Ortega, R., Gaitán-González, M., Yañez-Suárez, O.: Implementation of a real-time algorithm for maternal and fetal heart rate monitoring in a digital signal controller platform. Paper presented at the 2010 Annual International Conference of the IEEE Engineering in Medicine and Biology (2010)

Ashwal, E., Shinar, S., Aviram, A., Orbach, S., Yogev, Y., Hiersch, L.: A novel modality for intrapartum fetal heart rate monitoring. J. Matern Fetal Neonatal Med. **32**(6), 889–895 (2019). https://doi.org/10.1080/14767058.2017.1395010

Baschat, A.A., et al.: The role of the fetal biophysical profile in the management of fetal growth restriction. Am. J. Obstet. Gynecol. **226**(4), 475–486 (2022). https://doi.org/10.1016/j.ajog.2022.01.020

Comani, S., Mantini, D., Alleva, G., Di Luzio, S., Romani, G.L.: Automatic detection of cardiac waves on fetal magnetocardiographic signals. Physiol. Meas. **26**(4), 459 (2005)

Daniels, S.M., Boehm, N.: Auscultated fetal heart rate accelerations: an alternative to the nonstress test. J. Nurse Midwifery **36**(2), 88–94 (1991). https://doi.org/10.1016/0091-2182(91)90057-v

Dahab, R., Sakellariou, D.: Barriers to accessing maternal care in low income countries in Africa: a systematic review. Int. J. Environ. Res. Public Health **17**(12), 4292 (2020). https://doi.org/10.3390/ijerph17124292. PMID: 32560132; PMCID: PMC7344902

Devoe, L., et al.: Clinical experience with the Hewlett-Packard M-1350A fetal monitor: correlation of Doppler-detected fetal body movements with fetal heart rate parameters and perinatal outcome. Am. J. Obstet. Gynecol. **170**(2), 650–655 (1994). https://doi.org/10.1016/S0002-9378(94)70243-8

Evans, M.I., Britt, D.W., Evans, S.M., Devoe, L.D.: Changing perspectives of electronic fetal monitoring. Reprod. Sci. **29**(6), 1874–1894 (2021). https://doi.org/10.1007/s43032-021-007 49-2

Govindan, R.B., et al.: A novel approach to track fetal movement using multi-sensor magneto-cardiographic recordings. Ann. Biomed. Eng. **39**(3), 964–972 (2011). https://doi.org/10.1007/s10439-010-0231-z

Gupta, S., et al.: Systematic review of the literature: best practices. Acad. Radiol. **25**, 1481–1490 (2018). https://doi.org/10.1016/j.acra.2018.04.025

Hamelmann, P., et al.: Doppler ultrasound technology for fetal heart rate monitoring: a review. IEEE Trans. Ultrason. Ferroelectr. Freq. Control **67**(2), 226–238 (2020). https://doi.org/10.1109/tuffc.2019.2943626

Huecker, B.R., Jamil, R.T., Thistle, J.: Fetal Movement. StatPearls; StatPearls Publishing. https://www.ncbi.nlm.nih.gov/books/NBK470566/. Accessed 16 Jan 2023

Iyamu, I., et al.: defining digital public health and the role of digitization, digitalization, and digital transformation: scoping review. JMIR Public Health Surveill **7**(11), e30399 (2021). https://doi.org/10.2196/30399

Karlsson, B., et al.: The DopFet system: a new ultrasonic Doppler system for monitoring and characterization of fetal movement. Ultrasound Med. Biol. **26**(7), 1117–1124 (2000)

Kauffmann, T., Silberman, M.: Fetal Monitoring. In: StatPearls. StatPearls PublishingCopyright © 2023, StatPearls Publishing LLC. Accessed 21 Jan 2023

Kording, F., et al.: Dynamic fetal cardiovascular magnetic resonance imaging using Doppler ultrasound gating. J. Cardiovasc. Magn. Reson. **20**(1), 17 (2018). https://doi.org/10.1186/s12 968-018-0440-4

Krupa, N., MA, M.A., Zahedi, E., Ahmed, S., Hassan, F.M.: Antepartum fetal heart rate feature extraction and classification using empirical mode decomposition and support vector machine. Biomed. Eng. Online **10**, 1–15 (2011)

Kuei-Chiang, L., Shynk, J.J.: A successive cancellation algorithm for fetal heart-rate estimation using an intrauterine ECG signal. IEEE Trans. Biomed. Eng. **49**(9), 943–954 (2002). https://doi.org/10.1109/TBME.2002.802010

Madiuw, D., Widiasih, R., Napisah, P.: Health technologies for detecting high risk conditions in pregnancy: a systematic review. J. Nurs. Care **2**(3), 212 (2019). https://doi.org/10.24198/jnc.v2i3.22343

Magenes, G., Bellazzi, R., Fanelli, A., Signorini, M.G.: Multivariate analysis based on linear and non-linear FHR parameters for the identification of IUGR fetuses. Paper presented at the 2014 36th Annual International Conference of the IEEE Engineering in Medicine and Biology Society (2014)

Mahomed, K., Gupta, B.K., Matikiti, L., Murape, T.S.: A simplified form of cardiotocography for antenatal fetal assessment. Midwifery **8**(4), 191–194 (1992). https://doi.org/10.1016/S0266-6138(05)80006-3

Marchon, N., Naik, G., Pai, K.: Linear phase sharp transition BPF to detect noninvasive maternal and fetal heart rate. J. Healthc. Eng. **2018**, 1–14 (2018)

Mayes, M.E., Wilkinson, C., Kuah, S., Matthews, G., Turnbull, D.: Change in practice: a qualitative exploration of midwives' and doctors' views about the introduction of STan monitoring in an Australian hospital. BMC Health Serv. Res. **18**(1), 119 (2018). https://doi.org/10.1186/s12913-018-2920-5

Menéndez, T., et al.: Prenatal recording of fetal heart action with magnetocardiography. Z Kardiol. **87**(2), 111–118 (1998). https://doi.org/10.1007/s003920050162

Mochimaru, F., Fujimoto, Y., Ishikawa, Y.: The fetal electrocardiogram by independent component analysis and wavelets. Jpn. J. Physiol. **54**(5), 457–463 (2004)

Murta, L.O., Guzo, M.G., Moraes, E.R., Baffa, O., Wakai, R.T., Comani, S.: Segmented independent component analysis for improved separation of fetal cardiac signals from nonstationary fetal magnetocardiograms. Biomed. Eng./Biomedizinische Technik **60**(3), 235–244 (2015). https://doi.org/10.1515/bmt-2014-0114

Peters, C.H.L., van Laar, J.O.E.H., Vullings, R., Oei, S.G., Wijn, P.F.F.: Beat-to-beat heart rate detection in multi-lead abdominal fetal ECG recordings. Med. Eng. Phys. **34**(3), 333–338 (2012). https://doi.org/10.1016/j.medengphy.2011.07.025

Sarno, A.P., Jr., Ahn, M.O., Phelan, J.P., Paul, R.H.: Fetal acoustic stimulation in the early intrapartum period as a predictor of subsequent fetal condition. Am. J. Obstet. Gynecol. **162**(3), 762–767 (1990)

Schiermeier, S., et al.: Fetal heart rate variation in magnetocardiography and cardiotocography–a direct comparison of the two methods. Z. Geburtshilfe Neonatol. **211**(5), 179–184 (2007)

Sehati Group. Sehati TeleCTG. https://sehati.co/en/product/. Accessed 23 Jan 2023

Sheikh, A., et al.: Health information technology and digital innovation for national learning health and care systems. Lancet Digit Health **3**(6), e383–e396 (2021). https://doi.org/10.1016/s2589-7500(21)00005-4

Siddaway, A.P., Wood, A.M., Hedges, L.V.: How to do a systematic review: a best practice guide for conducting and reporting narrative reviews, meta-syntheses. Ann. Rev. Psychol. **70**, 747–770 (2019)

Signorini, M.G., Magenes, G., Cerutti, S., Arduini, D.: Linear and nonlinear parameters for the analysis of fetal heart rate signal from cardiotocographic recordings. IEEE Trans. Biomed. Eng. **50**(3), 365–374 (2003)

Taralunga, D.D., Wolf, W., Strungaru, R., Ungureanu, G.: Abdominal fetal ECG enhancement by event synchronous canceller. In: Annual International Conference of the IEEE Engineering in Medicine and Biology Society 2008, pp. 5402–5405 (2008). https://doi.org/10.1109/iembs.2008.4650436

The American College of Obstetricians and Gynecologysts. Fetal Heart Rate Monitoring During Labor. https://www.acog.org/womens-health/faqs/fetal-heart-rate-monitoring-during-labor. Accessed 21 Jan 2023

Valderrama, C.E., Ketabi, N., Marzbanrad, F., Rohloff, P., Clifford, G.D.: A review of fetal cardiac monitoring, with a focus on low- and middle-income countries. Physiol. Meas. **41**(11), 11tr01 (2020). https://doi.org/10.1088/1361-6579/abc4c7

Velayo, C.L., Funamoto, K., Silao, J.N.I., Kimura, Y., Nicolaides, K.: Evaluation of abdominal fetal electrocardiography in early intrauterine growth restriction. Front. Physiol. **8**, 437 (2017)

Wacker-Gussmann, A., Plankl, C., Sewald, M., Schneider, K.-T.M., Oberhoffer, R., Lobmaier, S.M.: Fetal cardiac time intervals in healthy pregnancies–an observational study by fetal ECG (Monica Healthcare System). J. Perinat. Med. **46**(6), 587–592 (2018)

Wakai, R.T.: Assessment of fetal neurodevelopment via fetal magnetocardiography. Exp. Neurol. **190**(Suppl 1), S65-71 (2004). https://doi.org/10.1016/j.expneurol.2004.04.019

Widiasih, R., et al.: Self-fetal wellbeing monitoring and ante-natal care during the COVID-19 pandemic: a qualitative descriptive study among pregnant women in Indonesia. Int. J. Environ. Res. Public Health **18**(21), 11672 (2021). https://www.mdpi.com/1660-4601/18/21/11672

Zhao, H., Wakai, R.T.: Simultaneity of foetal heart rate acceleration and foetal trunk movement determined by foetal magnetocardiogram actocardiography. Phys. Med. Biol. **47**(5), 839 (2002)

Zuckerwar, A.J., Pretlow, R.A., Stoughton, J.W., Baker, D.A.: Development of a piezopolymer pressure sensor for a portable fetal heart rate monitor. IEEE Trans. Biomed. Eng. **40**(9), 963–969 (1993). https://doi.org/10.1109/10.245618

Deep Learning for Parkinson's Disease Severity Stage Prediction Using a New Dataset

Zainab Maalej$^{(\boxtimes)}$ ⓘ, Fahmi Ben Rejabⓘ, and Kaouther Nouiraⓘ

Institut Supérieur de Gestion de Tunis, Université de Tunis, Tunis, Tunisia
maalej.zaineb@gmail.com

Abstract. Parkinson's Disease (PD) is a progressive neurological disorder affecting the Basal Ganglia (BG) region in the mid-brain producing degeneration of motor abilities. The severity assessment is generally analyzed through Unified Parkinson's Disease Rating Scale (UPDRS) as well as the amount changes noticed in the BG size in Positron Emission Tomography (PET) images. Predicting patients' severity state through the analysis of these symptoms over time remains a challenging task. This paper proposes a Long Short Term Memory (LSTM) model using a newly created dataset in order to predict the next severity stage. The dataset includes the UPDRS scores and the BG size for each patient. This is performed by implementing a new algorithm that focuses on PET images and computes BG size. These computed values were then merged with UPDRS scores in a CSV file. The dataset created is fed into the proposed LSTM model for predicting the next severity stage by analyzing the severity scores over time. The model's accuracy is assessed through several experiments and reached an accuracy of 84% which outperforms the other state-of-the-art method. These results confirm that our proposal holds great promise in providing a visualization of the next severity stage for all patients which aids physicians in monitoring disease progression and planning efficient treatment.

Keywords: Parkinson's Disease · Severity · LSTM · UPDRS · BG

1 Introduction

Parkinson's Disease (PD), the second most common neurodegenerative disease after Alzheimer's Disease, is a chronic and progressive condition affecting the central nervous system. It is caused essentially by the loss of dopaminergic neurons located in the Basal Ganglia (BG) region in the mid-brain [13] which plays a crucial role in providing dopamine that enables movement coordination.

Supported by Université de Tunis, Institut Supérieur de Gestion de Tunis (ISGT), BESTMOD Laboratory.

The decrease of dopamine engenders a reduction in the communication between dopaminergic neurons to coordinate body movements. As this disease grows progressively over time, it is critical to study the symptoms' severity in order to delay progression and its negative impact on patients as much as possible. To this end, Positron Emission Tomography (PET) images provide a fundamental tool to monitor PD progression [15]. By injecting Flourodopa F18 (F-Dopa) into the body, we can visualize the amount of F-dopa uptake by dopaminergic neurons through these images. We conclude that the PD has progressed over time if a decline in F-Dopa uptake will appear which causes a volume decrease of BG [15]. Consequently, a decreased size of BG will be identified in PET images.

Furthermore, in addition to PET images, several scales have been developed for assessing PD. Among these scales, the Unified Parkinson's Disease Rating Scale (UPDRS) is considered the most common scale used by neurologists to detect PD patients and rate the PD severity [16]. The UPDRS score comprises four different sub-scales: Part I and Part II examine non-motor and motor experience aspects of daily living respectively. Part III estimates the severity of motor complications. And part IV measures motor fluctuations as well as dyskinesias. Based on these scores, multiple severity stages can be depicted. Hoehn and Yahr's scale is the most common and significant descriptive five-level staging scale that provides a global clinical function estimate in PD severity. According to [4], stage 1 is characterized by only unilateral involvement, usually with minimal functional incapacity. Whereas stage 2 is distinguished by bilateral involvement without balance impairment. Stage 3 is defined as a bilateral disease with moderate disability. The patient is physically independent. In stage 4 which is a severely disabling disease, the patient is still able to walk and stand unassisted. Finally, in stage 5, the patient becomes confined to a wheelchair or bed.

Although there are several treatments including deep brain stimulation and dopamine-related medication, PD patients still suffer a gradual progression in symptoms severity [18]. Nevertheless, this progression from one stage to another differs between patients. Some can take a long time to progress to the next stage, others may have a short duration. In such cases, physicians face many difficulties in monitoring the PD progression of patients over time and are unable to estimate the next severity stage of each patient. The lack of knowledge of the patient's future severity stage can lead to false drug prescriptions that do not improve their condition and do not support delaying the severity disease in the future. In this respect, deep learning, part of Artificial Intelligence technologies, has reached notable success and improved performance surpassing state-of-art in numerous health domains [17]. It can aid physicians in decision-making and provide significant results. Several researchers have gained from deep learning models' significance in monitoring PD progression. Despite the efficacy of proposed approaches in the literature, most researchers investigate a single type of symptom, either images, signals, or UPDRS scores. Although the BG size characteristic is very prominent for PD severity prediction [15], it has not been widely applied in the literature. Most researchers focused on only UPDRS data to predict the next severity stage and do not take into account the size of BG visualized in PET images. In addition, there is limited interest in analyzing the history of all patient severity scores related to each clinical visit over time to predict the next severity stage. Besides, no one can deny that the selection of the most appropriate deep learning model based on the data continue to be a challenging task.

Based on the aforementioned limitations, we explore in this paper the following hypotheses. First, we assume that merging the BG size computed from PET images and UPDRS scores constitutes useful support for predicting the next severity stage of each patient. Second, we suppose that using Long Short Term Memory (LSTM) model to analyze the patients' historical data of all clinical visits is the most suitable alternative since it is among the most accurate models in deep learning for analyzing data over time. Specifically, we aim at; (i) implementing a new algorithm that computes the BG regions size of all PET images, (ii) creating a new dataset that merges the UPDRS scores and the BG size, (iii) predicting the next severity stage of different patients by applying LSTM model through several hyperparameters combination and using the dataset constructed, (iv) evaluating the performance of the proposal and comparing it to a previously developed method in the literature.

Our article is organized as follows. Section 2 summarizes the related work of deep learning models for PD severity prediction. Section 3 details the materials and methods of our proposed approach. Section 4 describes the experimental results. Concluding remarks are provided in Section 5

2   Related Work

The huge improvement in deep learning encouraged many researchers toward developing algorithms for predicting PD severity automatically. Three different orientations have been taken in the literature. One orientation concentrates on predicting PD severity through images, the second orientation uses signals, and the third one applies values. Most of the research articles applied Convolutional Neural Networks (CNN) to predict PD severity as a classification problem. In [10], authors proposed a tremor assessment system that focuses on the use of the CNN model to distinguish the severity of symptoms. This model takes as input a set of two-dimensional images collected and transformed from tremor signals. They confirmed that the proposed method could be used for monitoring PD tremor symptoms in daily life as it achieved an accuracy of 85%. In contrast, using signal data, [1] focused on processing and fusing raw data of gait ground reaction force with the purpose to categorize PD severity. A CNN model was applied to learn automatically the spatiotemporal signals selected resulting in an efficient classification with an F1-score of 95.5%. Similarly, [3] proposed a novel PD severity prediction system using a 1D-CNN model to process 18 1D-signals from foot sensors which measures the vertical ground reaction force. Their proposal reached an accuracy of 85.3%. From another point of view, there is also a focus on applying values as data type. For example, [6] and [9] demonstrated the efficiency of the CNN model for detecting and classifying the motor evolution state of each patient. Despite the efficacy of the CNN model that was confirmed by multiple researchers in monitoring the PD, [19] combined CNN and Long Short-Term Memory (LSTM) to classify the severity of the disease. They employed values extracted through gait signals from 73 healthy control and 93 PD subjects. Coupled with Vertical Ground Reaction Force data, other data were included as age, gender, weight, height, and PD severity levels. The model proposed achieved high accuracy of 98.61%. However, [7] applied a Deep Neural Network (DNN) which provided a high potential for autonomously predicting patients' severity states. Table 1 summarizes the related work for predicting PD severity. Although the high achievements reached in these works mentioned above, some limitations were noticed. First, despite its importance in predicting PD progression, there is no integration of the BG size feature. Most researchers deal with UPDRS scores and little interest was depicted in signals and images. Second, we observed that there is a lack of analysis all the history status of all patients over time to predict the PD severity stage. Third, despite its notable efficacy in prediction tasks, it is worth noting that there is a restricted interest in using the LSTM model compared to CNN.

For this purpose, first, our main goal is to create a new dataset that combines not only UPDRS scores but also data extracted from PET images which constitute a crucial role in monitoring PD progression. A new algorithm is developed to extract the most accurate data through PET images. Second, as we aim to predict the next severity stage based on scores over time, we propose a new LSTM model thanks to its effectiveness in handling sequential data with long intervals. The proposed model analyzes all the historical severity statuses of patients (UPDRS scores and BG size) and predicts the future severity stage. We believe that performing this type of model with the new dataset created coupling UPDRS and data from PET images can leverage an efficient severity prediction.

Table 1. Related works on PD progression prediction

Authors	Input Data	Model	Performance
Kim et al. [10]	Image	CNN	Accuracy = 85%
Alharthi et al. [1]	Signal	CNN	F1-score = 98% /96%
El Maachi et al. [3]	Signal	CNN	Accuracy = 85.3 %
Goschenhofer et al. [6]	Value	CNN	Accuracy = 86.95%
Iakovakis et al. [9]	Value	CNN	AUC = 0.89
Zhao et al. [19]	Value	CNN+LSTM	Accuracy = 98.61%
Grover et al. [7]	Value	DNN	Accuracy = 81%/62%

3   Materials and Methods

In this section, we introduce our proposed approach for predicting the next severity stage with its overall scheme illustrated in Fig. 1. We define four main steps for our proposal: (1) Data acquisition, (2) Dataset construction, (3) Data pre-processing, and (4) LSTM model predictions.

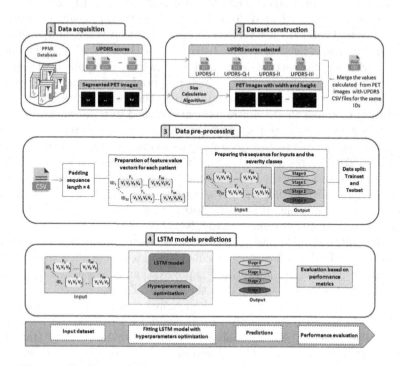

Fig. 1. The overall scheme of the proposed approach for predicting the severity stage

3.1   Data Acquisition

We collected the data from Parkinson's Progression Markers Initiative (PPMI) [12] database. PPMI is an international and large-scale clinical study. It is basically a collaborative effort of PD researchers that are experts in PD study implementation, biomarker development as well as data management to confirm biomarkers of PD progression. This database is comprised of study data presented in CSV files which are values characterizing various symptoms including patient characteristics, motor and non-motor assessments, medical history, and genetic data. Besides, PPMI covers image collections such as PET images, Magnetic Resonance Imaging (MRI), Computed Tomography (CT), and Single-Photon Emission Computed Tomography (SPECT).

Since we focus mainly on the PD evolution prediction, we identify only the relevant data which provide insight into the disease evolution. First, the UPDRS scores were selected specifically Part-I, Part-II, and Part-III. Then, PET images were identified in order to analyze the BG size. We selected only patients who have more than one PET image as we study the PD progression over time.

Thus, a total of 110 chronological PET images were collected from PPMI with 89 men and 21 women, and the age range is 33–76. A BG segmentation from PET images is a crucial step to delivering a meaningful visualization. In our previous work, we applied a new proposal based on the U-Net model for the BG region segmentation from PET images. The approach proposed reached significant performance. Consequently, we aim to use these segmented images in order to compute the size of BG for different patients.

3.2   Dataset Construction

This section provides the steps we followed to construct the new dataset for predicting the next severity stage. We initiate with the calculation of BG width and height from PET images followed by UPDRS scores.

PET Images. A size decrease of the BG region over time in PET images confirms the progression of PD. Hence, it is crucial to develop a new algorithm that aims to calculate the width and height of the left and right regions in each PET image. The implementation of our proposed algorithm has been performed in Python version 3.7.4. The multiple steps of this algorithm are illustrated in Algorithm 1. First, the dataset which includes all the segmented PET images from our previous work was fed into the algorithm proposed as input. The algorithm starts by reading each image as grayscale. Second, we applied a Canny edge detector for detecting all the edges of the two regions in each input image. Next, the Erosion filter is followed by Dilation. The erosion step is fundamental for removing white noises if exist. However, it may also shrink the regions. That is why we apply Dilation in the next step to dilate the edge region since the noise disappeared. This step is also crucial in order to join the majority of broken parts of the regions. As a next step, we save the contours of the resulting image and compute the number of existing pixels. A bounding rectangle with a minimum area is next drawn for each region in the image which promotes finally in calculating width and height through Euclidean distance. Figure 2 demonstrated an example of one PET image resulting from our algorithm. The width and height of BG regions of the left and right sides are labeled. After that, we saved the computed width and height of the left and right sides in a CSV file and we include other relevant features specific to each patient namely subject-ID, event-ID, age, gender, number of visits, and the size of left and right BG region which is equal to (width × height). This initial dataset contains 110 patients × 11 features.

UPDRS Data. Thanks to its ability in differentiating mild and severe PD, UPDRS is considered the most popular and effective tool to assess the progression of PD patients [5]. Therefore, we aim to add UPDRS scores in the previous CSV file created that comprises the information retrieved from PET images. From the PPMI database, we combine motor and non-motor aspects of experiences of daily living and motor complications severity. Using Python,

---

Algorithm 1. Size calculation algorithm

    Input: Dataset of PET images segmented
    Output: Dataset of PET images with width and height for each region
1: D_{in}="Define the path of input dataset"
2: D_{out}="Define the path of output dataset"
3: **for** each image $i \in D_{in}$ **do**
4:    Read i as grayscale
5:    I=Erode Filter(Dilate Filter(Canny Filter(i)))
6:    Save the resulted image I
7:    C = Find contours of I and store them in an array
8:    **for** $j \in C$ **do**
9:        $Area$ = ContourArea(j) ▷ Find the area of all $j \in C$ and store them in an array
10:   **end for**
11:   $Object = C[0]$
12:   B = Draw a bounding rectangle of $Object$ with minimum area
13:   B = Find 4 points that define the rectangle B
14:   Calculate how many pixels are there per centimeter
15:   Compute the width and height
16:   Label each region with width and height computed
17:   D_{out}= Save the resulted image comprising width and height
18: **end for**
19: **return** D_{out} ▷ The dataset D_{out} includes all images with width and height labeled for each region

---

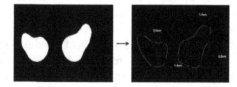

Fig. 2. Example of one image resulted from the proposed size calculation algorithm

we merged four distinct UPDRS datasets from PPMI with the dataset previously created which are MDS-UPDRS-Part-I, MDS-UPDRS-Part-I-Patient-Questionnaire, MDS-UPDRS-Part-II-Patient-Questionnaire, and MDS-UPDRS-Part-III. After that, we removed all the duplicates that emerged from rows and columns. The final dataset includes 116 patients × 89 features. It is observed that the number of patients compared to the first dataset increased from 110 to 116. The reason behind this increase is that for a single patient, we can detect more than one visit for UPDRS measurements. If so, one extra row will be added with the same feature values of BG size. Accordingly, we believe that our final dataset couples all the necessary data that promote efficiently the prediction of the severity stage.

3.3   Data Pre-processing

The multiple visits for each patient are represented in different rows in the dataset created. Each row contains all the feature values selected in the specific visit. However, this data representation is not suitable for the LSTM model to learn efficiently from the visit history of each patient. Hence, in order to have a more accurate representation, we changed the structure of this dataset. Fig 3 depicts the multiple steps followed. In the first step, taking into account the uniqueness of each patient's condition, there is a dissimilarity in the number of visits for all patients. Therefore, as illustrated in Fig. 3(A), we padded a sequence length of 4 for each patient in order to have a similarity of visit number between all the patients. Next, for all patients, we created a vector specific to each feature. Each value in the vector corresponds to one visit score. Thus, each feature vector presents a sequence of four different values related to the four visits as indicated in Fig. 3(B). Finally, we prepared the X-train and Y-train sets. The first three values are designed for X-train and the last value is for Y-train which is the target value to be predicted. This is depicted in Fig. 3(C).

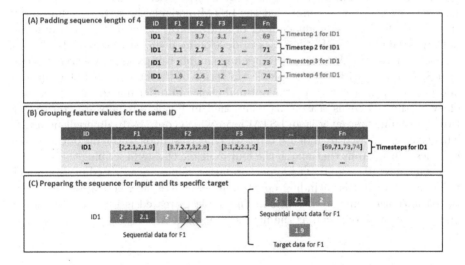

Fig. 3. The steps for pre-processing For clear illustration, one example of a patient is outlined. Considering that these steps are processed for all patients. "F" is denoted by Feature and "ID" by patient Identifier

3.4   Proposed LSTM Model

LSTM networks were introduced by [8] in 1997. It is a variation of the Recurrent Neural Network (RNN). As mentioned in [2], LSTM is a significant solution for modeling sequential data since they involve contextual information from numerous past inputs. It was proposed essentially to overcome the difficulty in learning long-term dependence in RNN and has made substantial advancements in applications [20] thanks to its memory cell that are modulated by both the input and forget gates [14]. Indeed, if these gates are closed, the contents of the memory cell will remain unchanged between the one-time step and the next one and the gating structure allows information to be conserved over various time steps [14]. These characteristics enable the LSTM model to overcome the issue of learning long-term dependence and also the vanishing gradient problem that appears with most RNN models. Hence, the LSTM model has gained much interest in numerous application domains namely speech recognition, handwriting recognition, generating sentences, and sequential data classifications and predictions.

It is worth mentioning that hyperparameter tuning is considered one of the main challenges when training deep learning models. For LSTM optimization performance, several decisions must be reached specifically the choice of epoch number, choice of activation function, choice of batch size, choice of the optimizer, the number of hidden layers, the number of neurons per layer, kernel initialization, and the choice of techniques for avoiding underfitting and overfitting of the model.

Since we aim to analyze a sequence of historical visits in order to predict the next severity stage, we assume that the LSTM model is the suitable solution for this issue. Choosing the right number of hidden layers as well as neuron number per layer remains a challenging task in developing deep learning models. As the dataset constructed does not contain a huge number of patients, we assume that an LSTM model with low complexity in terms of the number of hidden layers and neurons per layer will be the optimal choice in order not to fall into a case of overfitting. For this reason, as mentioned in Fig. 4, our network consists of two layers. The first layer is an LSTM layer which contains 10 different neurons. This layer is followed by a dense layer which corresponds to the output layer. It includes four neurons that match the four multiple classes describing the severity stages namely: (0, 1, 2, 3). The hyperparameter values are adjusted in this model through several trials including epoch number, batch size, activation function, kernel initializer, and optimization. The dataset created is fed into the model and the output is outlined as the next severity stage.

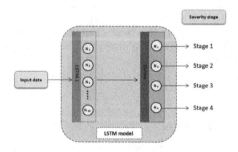

Fig. 4. The proposed LSTM model

4 Experiments

This section demonstrates the main process of the experiment displaying experimental results as well as results interpretation. Our proposed model is implemented in Python 3.7.4 using the Keras library and Tensorflow library at the backend. The performance of the model is tuned by conducting the experiments several times. Table 2 outlines the descriptions of the best hyperparameters value that are optimized for the proposed LSTM model. For the optimizer, we chose to apply Adam [11] thanks to its fast convergence as well as adaptive learning rate that requires less tuning [10]. Since they are adapted to multi-class predictions, we used Softmax as the activation function; and categorical cross-entropy as the loss function. We optimized a batch size to 4 and the number of epochs to 50.

Table 2. Description of the optimized hyperparameters

Hyperparameters	Values
Optimizer	Adam
Activation function	Softmax
Loss function	Categorical Crossentropy
Epoch	50
Batch size	4

In order to test the proposed model and find the best performance results, the created dataset was split randomly into three different splits as follows: First, we split the entire dataset into train and test sets in an 80/20 ratio. Then, the train set is split into train and validation sets in the ratio of 70/30. In this case, our LSTM model reached an accuracy equal to 71% and a loss of 0.40. Second, we changed the split into 70% for training and 30% for testing. We have maintained the same ratio of train and validation sets as 70/30. For this split, the LSTM accuracy slightly decreased to 70% and the loss increased to 0.41. Third, we applied a ratio of 60/40 for the train and test sets and we kept the same distribution for the train and validation sets at 70/30. In such a case, the accuracy model increases to 84% and the loss is equal to 0.40. Hence, it is noticed that the performance model increases when decreasing the train set to 60%. This may be due to overfitting when using a higher-size train set. Rising the number of data points in the train set may create a greater amount of noise in the data which leads to overfitting. Furthermore, we also tested the performance of our model when using a distribution of 50%–50% and 40%–60% for train and test sets respectively while retaining the same ratio of train and validation sets of 70/30. These reduced sizes of the training set produced a decline in the model accuracy of 81% and 70% respectively compared to the ratio of 60/40. This is justified by the fact that our model denotes an underfitting. With such train set sizes, the model may not have sufficient data points to make an efficient generalization to unseen data and can not learn the underlying distribution of the real data efficiently. Hence, the best distribution is depicted using a ratio of 60/40 for train and test sets. Its average accuracy and loss of training and validation sets are illustrated in Fig. 5. The training accuracy curve illuminates that the LSTM model can converge. In addition, the validation set accuracy increases as the training accuracy increases, which confirms that the LSTM model is learning meaningful features without nearly overfitting. Similarly, we can notice that training loss decreases progressively as well as the validation loss, which validates that the LSTM model performs predictions with low errors. Therefore, we can confirm that our proposed LSTM model achieved predictions with significant accuracy of 84% over a distribution of 60%-40% of train and test sets and a loss equal to 0.40. According to these results, we made a confusion matrix to visually display the accuracy of predicting severity stages. As we can see through Fig. 6, it is noteworthy that the LSTM succeeded in correctly predicting most of the severity classes.

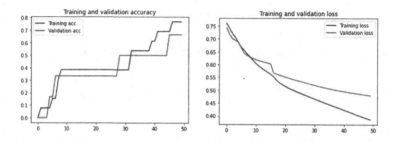

Fig. 5. Average accuracy and loss curve of training and validation sets

For comparison goals, given that we examined the efficacy of the LSTM model on a newly created dataset, no previous work has been published in the literature which applied the same dataset. Accordingly, we aim to compare our work with [7] that used similar features data for the prediction of PD severity stage as a classification problem. In this paper, authors applied DNN to predict the severity stages of multiple patients focusing on total UPDRS and motor UPDRS. The output layer is designed for "severe" and "not severe' classes. As illustrated in Table 3, it is observed that our LSTM model reached better performance. Their best accuracy rate is detected for motor UPDRS which is equal to 81% and 62% for total UPDRS whereas the best accuracy reached by our proposed model is 84%. Contrary to this work, our proposal focuses essentially on analyzing the past severity scores of several patients over time and then predicting the severity stage of each patient. The major finding of our proposal is that the LSTM model performed better when dealing with these data over time due to its capacity in maintaining the memory of all historical inputs. In addition, our created dataset contributed efficiently to the prediction task as it includes all the data symptoms needed for analyzing and predicting the PD

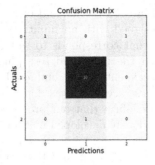

Fig. 6. Confusion matrix for multiclass classification using LSTM model

severity. Therefore, our proposal supported the initial hypotheses that the LSTM model could accurately estimate the severity stage from the dataset created. All these accomplishments produce the opportunity for widespread deployment in the clinical environment.

Nevertheless, due to the small number of patients in the PPMI database who hold more than one PET image, only 32 patients were identified and used to train and test the LSTM model. Since this model is based on deep learning, we assume that its performance will certainly be improved when adding more data to the dataset.

Table 3. Performance comparison with the previous study

Method	Accuracy
Grover et al. [7]	81%–62% (Motor UPDRS-Total UPDRS)
Our study	84%

5   Conclusion

In this study, we proposed a new LSTM model for the prediction of the next severity stage for several PD patients. Focusing on the PPMI database, we created a new dataset that includes all the data symptoms needed for analyzing the PD progression. We combined the UPDRS scores and the BG region sizes that were computed from PET images. First, we developed a new algorithm that takes as input all the BG-region images that were previously segmented from PET images and then calculates the width and height of each region in images. The output of this algorithm is a dataset that includes all patient images with width and height labeled for each region. Second, we created a dataset that comprises subject-ID, event-ID, age, gender, visits number, the calculated width and height values, the size of left and right BG regions, and the UPDRS scores.

All these data were merged into a CSV file and used to be analyzed in the new LSTM model proposed. Designed with an input layer with 10 neurons and a dense layer with 4 neurons that corresponds to the 4 different severity stages, the LSTM model performed the train and test phase using the constructed dataset where multiple experiments were conducted to identify the appropriate hyperparameters. The best accuracy was depicted using a data split ratio of 60/40 with 84%. We can confirm that our proposed model with the created dataset outperforms the previously developed method that aims also to analyze the UPDRS to predict the severity stage of PD patients. This may be clarified by the efficiency of the LSTM model which is the most common and significant deep learning model for analyzing and predicting data over time. Furthermore, coupling all the relevant information for PD severity prediction in one dataset contributes to enhancing the accuracy of the prediction. Therefore, we validate our initial hypothesis that our constructed dataset with the proposed LSTM model constitutes a significant tool to diagnose the progression of PD patients' stages in the future. This will allow undoubtedly physicians to take the necessary precautions for mitigating the advancement of this disease as much as possible. For future work, we will focus on augmenting the data points in the dataset constructed which can improve the model's accuracy.

References

1. Alharthi, A.S., Casson, A.J., Ozanyan, K.B.: Gait spatiotemporal signal analysis for Parkinson's disease detection and severity rating. IEEE Sens. J. **21**(2), 1838–1848 (2020)
2. Bouktif, S., Fiaz, A., Ouni, A., Serhani, M.A.: Optimal deep learning LSTM model for electric load forecasting using feature selection and genetic algorithm: Comparison with machine learning approaches. Energies **11**(7), 1636 (2018)
3. El Maachi, I., Bilodeau, G.A., Bouachir, W.: Deep 1D-convnet for accurate Parkinson disease detection and severity prediction from gait. Expert Syst. Appl. **143**, 113075 (2020)
4. Goetz, C.G., et al.: Movement disorder society task force report on the Hoehn and Yahr staging scale: status and recommendations the movement disorder society task force on rating scales for Parkinson's disease. Mov. Disord. **19**(9), 1020–1028 (2004)
5. Goetz, C.G., et al.: Movement disorder society-sponsored revision of the unified Parkinson's disease rating scale (MDS-UPDRS): scale presentation and clinimetric testing results. Mov. Disord. Official J. Mov. Disord. Soc. **23**(15), 2129–2170 (2008)
6. Goschenhofer, J., Pfister, F.M.J., Yuksel, K.A., Bischl, B., Fietzek, U., Thomas, J.: Wearable-based Parkinson's disease severity monitoring using deep learning. In: Brefeld, U., Fromont, E., Hotho, A., Knobbe, A., Maathuis, M., Robardet, C. (eds.) ECML PKDD 2019. LNCS (LNAI), vol. 11908, pp. 400–415. Springer, Cham (2020). https://doi.org/10.1007/978-3-030-46133-1_24
7. Grover, S., Bhartia, S., Yadav, A., Seeja, K., et al.: Predicting severity of Parkinson's disease using deep learning. Procedia Comput. Sci. **132**, 1788–1794 (2018)
8. Hochreiter, S., Schmidhuber, J.: Long short-term memory. Neural Comput. **9**(8), 1735–1780 (1997)

9. Iakovakis, D., et al.: Screening of parkinsonian subtle fine-motor impairment from touchscreen typing via deep learning. Sci. Rep. **10**(1), 1–13 (2020)

10. Kim, H.B., et al.: Wrist sensor-based tremor severity quantification in Parkinson's disease using convolutional neural network. Comput. Biol. Med. **95**, 140–146 (2018)

11. Kingma, D.P., Ba, J.: Adam: A method for stochastic optimization. In: Proceedings of International Conference Learning Representation (ICLR), pp. 1–15 (2015)

12. Marek, K., et al.: The Parkinson progression marker initiative (PPMI). Prog. Neurobiol. **95**(4), 629–635 (2011)

13. Mostafa, T.A., Cheng, I.: Parkinson's disease detection using ensemble architecture from MR images. In: 2020 IEEE 20th International Conference on Bioinformatics and Bioengineering (BIBE), pp. 987–992. IEEE (2020)

14. Patterson, J., Gibson, A.: Deep Learning: A Practitioner's Approach. O'Reilly Media Inc., Sebastopol (2017)

15. Pavese, N., Brooks, D.J.: Imaging neurodegeneration in Parkinson's disease. Biochim. Biophys. Acta (BBA)-Mol. Basis Dis. **1792**(7), 722–729 (2009)

16. Rascol, O., Goetz, C., Koller, W., Poewe, W., Sampaio, C.: Treatment interventions for Parkinson's disease: an evidence based assessment. Lancet **359**(9317), 1589–1598 (2002)

17. Ravì, D., et al.: Deep learning for health informatics. IEEE J. Biomed. Health Inform. **21**(1), 4–21 (2016)

18. Xiao, B., et al.: Quantitative susceptibility mapping based hybrid feature extraction for diagnosis of Parkinson's disease. NeuroImage Clin. **24**, 102070 (2019)

19. Zhao, A., Qi, L., Li, J., Dong, J., Yu, H.: A hybrid spatio-temporal model for detection and severity rating of Parkinson's disease from gait data. Neurocomputing **315**, 1–8 (2018)

20. Zhao, J., et al.: Do RNN and LSTM have long memory? In: International Conference on Machine Learning, pp. 11365–11375. PMLR (2020)

Improved Long-Term Forecasting of Emergency Department Arrivals with LSTM-Based Networks

Carolina Miranda-Garcia[1] (ID), Alberto Garces-Jimenez[1]([⊠]) (ID),
Jose Manuel Gomez-Pulido[1] (ID), and Helena Hernández-Martínez[2] (ID)

[1] Computing Sciences Department, Universidad de Alcala, Alcala de Henares, Spain
{lucia.miranda,jose.gomez}@uah.es, albertogarces0@gmail.com
[2] Nurse and Physiotherapy Department, Universidad de Alcala, Alcala de Henares, Spain
helena.hernandez@uah.es

Abstract. Patient admission to Emergency Departments suffers from a great variability. This makes the resource allocation difficult to adjust, resulting in an inefficient service. Several studies have addressed this issue with machine learning's regressors, time series analysis. This research proposes the use of improved recurrent neural networks that consider the dynamic nature of the data, introducing contextual variables that allow improving the predictability. Another important requirement from ED's administration is to have a wider predicting horizon for short- and long-term resource allocations. The results obtained using the data from one single Hospital in Madrid confirm that the use of deep learning with contextual variables improve the predictability to 6% MAPE for seven days and four months forecasts. As future research lines, the influence of special events, such as seasonal epidemics, pollution episodes, sports or leisure events, as well as the extension of this study to different types of hospitals' emergency departments.

Keywords: Emergency Department (ED) · Admissions Forecast · eHealth · Time Series · Long-Short Term Memory (LSTM) · Machine Learning · Deep Learning

1 Introduction

Emergency departments (ED) overcrowding is recognized as a growing global problem [1]. ED deterioration has numerous adverse consequences, such as increased interhospital referrals [2], increased waiting times [3], and less favorable patient outcomes [4]. The ability to predict the volume of ED visits could be of great help in achieving improved planning and management of hospital resources, both material and human.

The prediction of the number of ED visits has been widely studied in recent years. However, the problem is hard to generalize to the strict restrictions imposed to the medical data that make difficult to confirm the results with a variety of scenarios. In any case, some hospitals' ethics committees start releasing little by little the anonymized historical records necessary for this research.

I. Rojas et al. (Eds.): IWBBIO 2023, LNBI 13920, pp. 124–133, 2023.
https://doi.org/10.1007/978-3-031-34960-7_9

With respect to data models for prediction, most of the works found in the literature are based on time series techniques, typically Autoregressive Integrated Moving Average (ARIMA) and its different extensions; seasonal ARIMA or multivariate ARIMA [5], regression techniques [6]. Data mining-based models, like Artificial Neural Networks (ANNs) [7], Recurrent Neural Networks (RNNs) [8], and hybrid models [9] have been also tested with different results. An exhaustive review and analysis of forecasting methods in hospital ED has been proposed by Gul, M. et al. [10].

Some of the models have incorporated external variables that could be used in forecasting to improve their prediction performance. One type of these contextual variables come from the date, such as the importance of the day of the week variables [11]. Derived from date, it is possible to find the inclusion of holidays as an explanatory variable, although reports show contradictory results [12]. The date also entails the influence of influenza seasons. Another group of contextual variables come from the weather. The impact of temperature and other meteorological variables have certain predictive power on the number of EDs arrivals, but remain unclear, as these depend on geographic location and specific ED characteristics [13]. Weather is also limited for predicting beyond certain time horizon.

It is especially problematic predicting the number of admissions in abnormal situations such as the recently experienced Coronavirus Pandemic [14]. The historical evolution of the service cannot be used to train a model that will predict the coming admissions through these periods, characterized of unexpected behaviors. Other problem would be the sport events, mass concerts and similar happenings in the hospital's area of influence, as they are predictable and provided that the historical data used for training the models could be tagged considering them [15].

Deep Learning (DL) techniques seem to provide a good combination between the singularities brought by timing with the difficulties of extracting knowledge from the data. The aim of this paper is to show how the application of DL techniques improve the forecasting of patient's arrivals to the ED, in particular, the use of LSTM-based neural networks. Sudarshan et al. [16] showed the effectiveness of considering this type of networks in conjunction with weather and calendar variables, either from the previous three days or from future days and this work extends the horizon period.

It is interesting to observe which performances have been achieved so far to have a clear idea of what these systems are capable of. Articles forecasting daily ED presentations show achievements of MAPE between 5%–14% [17].

This study incorporates external characteristics obtained in previous days and programmed for the next days, framing the problem under the conceptual framework of encoder-decoder networks, which has been widely used in many other predictive tasks [18]. In order to compare the predictive ability of RNN and Machine Learning (ML) models, a Random Forest Regressor (RFR) is chosen as a reference model to test the performance of LSTM networks, since the promising results for predicting the number of emergency room's arrivals [19].

To evaluate the results obtained with these models, different prediction horizons have been set, according to the needs of the hospital planning team. First, a short-term (one and seven days) forecasting horizon is set, to facilitate daily and shift planning of staff and available resources. The Administrative Department have also manifested their interest in making long-term (four months) predictions, to anticipate planning policies, so that the changes potentially required can be addressed more efficiently.

This article is structured in sections. Next Section, Material & Methods, describes how the experiment is organized and justifies the techniques to obtain the Results, that are presented and discussed in the following section. Finally, in Conclusions, the results are commented, showing the novelty of this study, and describing the future work.

2   Materials and Methods

2.1   Data Acquisition

The database is provided by the ED of Gómez-Ulla Military Hospital, in Madrid, Spain. The database includes the arrival, admission, first medical attention, if the patient was under observation or was operated and the departure's dates and times of patients to its ED. The dataset also includes demographic information of the patients such as gender, date of birth, nationality, and place of residence.

The data is collected from January 1, 2015, to December 31, 2021, resulting 501,426 records. Due to the heavy impact caused on the EDs by COVID-19 pandemic [20], the information from January 2020 onwards is disregarded giving it up for ulterior studies. Because of this removal, the database contains 361,722 records.

The data is aggregated summing patients arriving on a daily basis. The date and time of admission allows the extraction of the number of the year's week, the day of the week and, according to the work calendar of the area, the national, regional, and local bank holidays with a Boolean variable. The year's week characterizes among others the seasonality of the data, while the day of the week, the special pattern observed between the labor days and the weekends. The influence of working shifts has not been considered at this point for a later more precise focus on this study.

As contextual variables, the dataset includes the information about the weather, which is retrieved from the historical data from the State Meteorological Agency of Spain (AEMET). Out of all the variables available from the Agency, only minimum and maximum daily temperatures are initially considered, because they show statistical significance correlation with daily arrivals to the ED, with p-values less than 0.00. However, since these variables show collinearity of Pearson $r = 0.94$, only maximum temperature is included as predictor.

2.2   Exploration

Figure 1 shows the number of daily arrivals to the ED since January 1, 2015 to December 31, 2019.

The ED hosts an average amount of 198.28 arrivals a day. The first observation is the rising trend of the number of arrivals. Fitting the signal with a linear a regression, the slope is an increment of 0.0246 patients per day (nine more every year). It is not that much, but it is important the model is able to update this information.

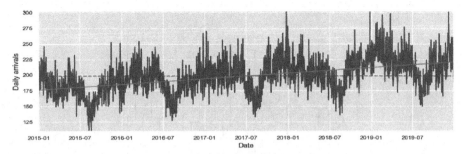

Fig. 1. Evolution of ED's daily arrivals (blue line), data trend (orange line) and average level (red dotted line) (Color figure online)

Another clear factor is the existence of a seasonal variation during some periods of the year. There is a valley of arrivals in the summer, i.e., time for seasonal holidays in the area and a peak around the end of the year, possibly caused for people's habits in Christmas. Figure 2 shows this variation in boxplots.

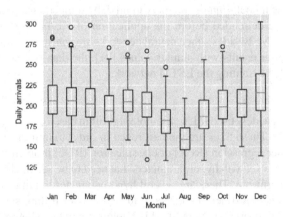

Fig. 2. Median and quartiles of the arrivals to the ED on each month

The ED demand nearly drops 20% with respect to the average in July and August, while rises 10% over the average in December.

Weekly seasonal pattern is also present as shown in Fig. 3.

Arrivals gradually grow at the weekend, reaching a maximum on Mondays. The arrivals decrease from Tuesday to Thursday both included, closing the weekly cycle.

2.3 Analysis Techniques

The experiment tests three predicting horizons, i.e., 1-day, 7-days and 120-days. Each one is performed with traditional ML's regressor, with a time series model and with a DL's network. The regressor is implemented with Random Forest (RFR), the time series analysis with SARIMAX, and the network with an encoder-decoder architecture of Long Short-Term Memory (LSTM) cells.

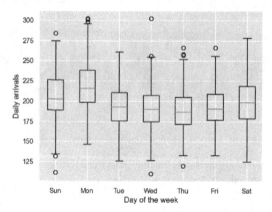

Fig. 3. ED's Arrivals on each day of the week

As said in the previous section, the periodic behavior observed is represented with the month and the day of the week. The bank holidays are also considered to improve the accuracy of the model. Finally, the maximum daily temperature is also considered.

Time series are detrended before the application of all the techniques, i.e. RFR, ARIMAX and LSTM. Besides, since neural networks are sensitive to unscaled data and time series get better results with similar values [21], the data is first normalized with 'min-max' before the training the LSTM network.

Random Forest

Random Forest is a widely used ML's ensemble algorithm based on bagging uncorrelated decision trees [22]. These trees are then averaged in order to minimize the variance. It is suitable for either regression (RFR) or classification (RFC) problems.

This technique is set with hyperparameters that need to be adjusted before balancing and avoid the underfitting. There are several hyperparameter tunning strategies for the optimization of this adjustment being the grid search over rolling subsets the chosen one. The following hyperparameters listed in Table 1 are obtained with temporal cross-validation for both the short-term and long-term horizon forecastings.

Table 1. Basic hyperparameters obtained after the tuning procedure

Hyperparameter	Short-term forecast horizon	Long-term forecast horizon
Number of trees	300	400
Maximum tree depth	23	28

On the other hand, the sequence of previous arrivals to the EDs are included as predictors in the RFR for the short-term forecast. A detailed description of the RFR's predictors is shown in Table 2.

Table 2. Chosen predictors for RFR with t-1 and t-7 previous days

Short-term forecast model	Long-term forecast model
Month (one-hot-encoded)	Month (one-hot-encoded)
Day of the week (one-hot-encoded)	Day of the week (one-hot-encoded)
Holiday (boolean)	Holiday (boolean)
Maximum temperature	Maximum temperature
t-1 & t-7 previous days arrivals	

SARIMAX

SARIMAX is an extension of the well-known Autoregressive Integrated Moving Average (ARIMA) model, introduced by George Box in 1970. Any SARIMA model is defined by seven parameters and its notation is (p, d, q)(P, D, Q)[s]. p is the number of autoregressive terms, d is the number of nonseasonal differences needed for stationarity, and q is the number of moving average terms. The parameters in capital letters P, D, Q are analogous to those in lower-case ones p, d, q and are referring to the seasonal part of the model with a lag, s. Residuals of the model are assumed to be white noise. In SARIMAX, external regressors are also included as predictors. The dataset for the SARIMAX includes maximum daily temperature, holidays and the month and day of the week are transformed to dummy variables.

The development of the model is based on Box-Jenkins methodology, which involves an iterative process of model identification, parameter estimation and residuals diagnostic. The final model is selected attending to the AIC criterion.

The selected model to predict the number of ED arrivals was SARIMAX (2, 0, 2)(1, 0, 0) [7]. Residuals were checked for normality, and both residuals and squared residuals were checked for lacking serial correlation.

Encoder-Decoder Architecture

Long Short-Term Memory (LSTM) networks were first introduced in the 90s by Hochreiter and Schmidhuber [21]. LSTMs are an improved type of RNN designed to avoid the vanishing gradient problem, and capable of learning long-term dependencies. LSTMs have gained popularity in recent years for their applications in time series problems, and data with sequential structure. Encoder/decoder networks have shown to be effective in many applications, including machine translation, speech transcription, and text summarisation [23, 24]. This architecture receives the input sequence in the encoder enabling the network to learn the underlying relationships between steps and features and thus generate an internal representation of the input sequence. The decoder transforms this internal representation of the output into an interpretable sequence. A time series problem can be handled with this architecture, where an input sequence of T time steps is transformed into a sequence of only one or several timesteps in the output.

The proposed architecture is depicted in Fig. 4.

The encoder network contains three stacked LSTM layers. If $\{x_t\}$ is the time series of the daily arrivals to the ED and $\{y_t\}$ is the multivariate time series of contextual data,

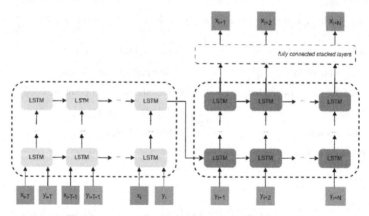

Fig. 4. LSTM-based encoder-decoder architecture

i.e., the maximum daily temperature, holidays, day of the week and month, then the inputs to the networks are $\{x_{i\text{-}T}..., x_i\}$ and $\{y_{i\text{-}T}..., y_i\}$, where i is the day and T is the taken number of timesteps backwards. The decoder network includes three stacked LSTM layers, initialized with the cell and hidden states of the encoder. Denoting N as the number of predicted timesteps, this network is fed with $\{y_{i+1}..., y_{i+N}\}$ for training.

Bayesian optimization method is applied to select the hyperparameter configuration of the network. In this model, the hyperparameters include batch size, number of layers and non-linear activation function in each one of the network components, and the number of neurons in each layer. Early stopping is used to control the number of epochs used in the training.

The proposed LSTM network works with a lookback-window of seven timesteps -days-.

Performance

The metrics considered to assess the accuracy of the predictions obtained by the proposed techniques are MAPE and RMSE.

a) MAPE (Mean Absolute Percentage Error): It is defined as:

$$MAPE = \frac{1}{n} \sum_{i=0}^{n} \left| \frac{\widehat{y_i} - y_i}{y_i} \right| .100[\%]$$

where y_i is the actual value, $\widehat{y_i}$ is the forecast value, and n is the number of fitted points. This statistic represents the mean percentage of the absolute errors. The smaller the MAPE is, the better the forecast is. MAPE is easy to understand and allows the comparison of results from different models applied to different data, since it is a scale-independent measure.

b) RMSE (root mean squared error) is defined as:

$$RMSE = \sqrt{\sum_{i=1}^{n} \frac{(\widehat{y_i} - y_i)^2}{n}}$$

where y_i is the actual value, $\hat{y}_i$ is the forecast value, and n is the number of fitted points. This statistic is the square root of the average of squared errors. Equally, the smaller the RMSE is, the better the forecast is. This metric is sensitive to outliers.

3   Results

This section presents the results obtained with each technique for both proposed forecast horizons. The models are trained with the first 90% records of the dataset and the metrics are evaluated with the remaining 10% records, i.e., since July, 2, 2019 to December, 31, 2019. Table 3 shows the MAPE values for the short-term horizon model, with one and seven days.

Table 3. Short-Term performance with one-day and seven-days forecasting horizons

Technique	Metric	One day	Seven days
LSTM	MAPE	5.84%	6.12%
	RMSE	15.68	16.43
SARIMAX	MAPE	6.31%	6.74%
	RMSE	16.55	17.91
RFR	MAPE	6.82%	7.21%
	RMSE	17.74	18.68

LSTM network outperforms SARIMAX and RFR both one-day ahead and seven-days ahead. The LSTM captures better than the other models the dynamics of the ED's stream. Previous works featuring neural networks including calendar and weather information report similar results with MAPE between 5.4%–8.04% when predicting one-day ahead [16, 25].

The accuracy is even better with other context variables, such as the results obtained by Zhao et al. [25], who achieved better MAPE, 4.5% for one-day prediction, and 4.71% for seven-days prediction. They included the air pollution, as well as weather and calendar-based information.

However, the performance obtained by Sudarshan, V.K. et al. [16] and Yousefi, M. et al. [26] predicting seven-days ahead yielded MAPEs of 6.3%–8.9%. This suggests that the predicting power of the proposed DL-based model for mid- and long-term forecasts is more promising. In fact, Table 4 shows MAPE and RMSE values for long-term (120 days) arrivals forecast giving similar performance to 7 days.

In this case it is possible to observe the similarities between the 7-days forecast with the 120-days forecast, obtained thanks to the long-term ability developed in the LSTM.

Table 4. Long-Term Performance with 120 days forecasting horizon

Technique	Metric	120 days
LSTM	MAPE	6.23%
	RMSE	16.32
SARIMAX	MAPE	7.14%
	RMSE	18.01
RFR	MAPE	7.31%
	RMSE	18.86

4 Conclusion

This study shows the suitability of the new approach, i.e., the use of LSTM encoder-decoder for predicting the ED' service demand in short- and long-term horizons.

In the future the model will be trialed with other EDs of different sizes, from the public and private systems for proving at what extent the model in generalizable.

Future lines of research also include assessing the prescription capacity of other contextual variables, such as irregular events, like sports, public demonstrations, mass concerts in the area, pollution and allergen levels or flu outbreaks. This will require the inclusion of soft decision-making strategies capable of learning from new situations.

Another research line that could lead to find representative features is the semantic analysis of social networks that could anticipate somewhat the requirements as per the identified keywords of the messages exchanged among the users.

It is important for the ED's administration to separate the demands from the shifts -morning, evening and overnight- as their patterns differ, and this will likely improve the accuracy of the model.

With regards to the supplied database, a more detailed coding of the ED services information will allow focused forecasts on specialized clinical areas.

References

1. Hoot, N.R., Aronsky, D.: Systematic review of emergency department crowding: causes, effects, and solutions. Ann. Emerg. Med. 52(2), 126–136 (2008)
2. Olshaker, J.S., Rathlev, N.K.: Emergency department overcrowding and ambulance diversion: the impact and potential solutions of extended boarding of admitted patients in the emergency department. J. Emerg. Med. 30(3), 351–356 (2006)
3. McCarthy, M.L., et al.: Crowding delays treatment and lengthens emergency department length of stay, even among high-acuity patients. Ann. Emerg. Med. 54(4), 492–503 (2009)
4. Bernstein, S.L., et al., Society for Academic Emergency Medicine, Emergency Department Crowding Task Force: The effect of emergency department crowding on clinically oriented outcomes. Acad. Emerg. Med. 16(1), 1–10 (2009)
5. Juang, W.C., Huang, S.J., Huang, F.D., Cheng, P.W., Wann, S.R.: Application of time series analysis in modelling and forecasting emergency department visits in a medical centre in Southern Taiwan. BMJ Open 7(11), e018628 (2017)

6. McCarthy, M.L., Zeger, S.L., Ding, R., Aronsky, D., Hoot, N.R., Kelen, G.D.: The challenge of predicting demand for emergency department services. Acad. Emerg. Med. **15**(4), 337–346 (2008)
7. Gul, M., Guneri, A.F.: Planning the future of emergency departments: forecasting ED patient arrivals by using regression and neural network models. Int. J. Ind. Eng. **23**(2), 137–154 (2016)
8. Harrou, F., Dairi, A., Kadri, F., Sun, Y.: Forecasting emergency department overcrowding: a deep learning framework. Chaos Solitons Fractals **139**, 110247 (2020)
9. Yucesan, M., Gul, M., Mete, S., Celik, E.: A forecasting model for patient arrivals of an emergency department in healthcare management systems. In: Intelligent Systems for Healthcare Management and Delivery, pp. 266–284. IGI Global (2019)
10. Gul, M., Celik, E.: An exhaustive review and analysis on applications of statistical forecasting in hospital emergency departments. Health Syst. **9**(4), 263–284 (2020)
11. Asplin, B.R., Flottemesch, T.J., Gordon, B.D.: Developing models for patient flow and daily surge capacity research. Acad. Emerg. Med. **13**(11), 1109–1113 (2006)
12. Rostami-Tabar, B., Ziel, F.: Anticipating special events in emergency department forecasting. Int. J. Forecast. **38**(3), 1197–1213 (2022)
13. Marcilio, I., Hajat, S., Gouveia, N.: Forecasting daily emergency department visits using calendar variables and ambient temperature readings. Acad. Emerg. Med. **20**(8), 769–777 (2013)
14. Etu, E.E., et al.: A comparison of univariate and multivariate forecasting models predicting emergency department patient arrivals during the COVID-19 pandemic. In: Healthcare, vol. 10, no. 6, p. 1120. MDPI (June 2022)
15. Kadri, F., Harrou, F., Chaabane, S., Tahon, C.: Time series modelling and forecasting of emergency department overcrowding. J. Med. Syst. **38**(9), 1–20 (2014)
16. Sudarshan, V.K., Brabrand, M., Range, T.M., Wiil, U.K.: Performance evaluation of emergency department patient arrivals forecasting models by including meteorological and calendar information: a comparative study. Comput. Biol. Med. **135**, 104541 (2021)
17. Whitt, W., Zhang, X.: Forecasting arrivals and occupancy levels in an emergency department. Oper. Res. Health Care **21**, 1–18 (2019)
18. Cui, Z., Zhou, Y., Guo, S., Wang, J., Xu, C.Y.: Effective improvement of multi-step-ahead flood forecasting accuracy through encoder-decoder with an exogenous input structure. J. Hydrol. **609**, 127764 (2022)
19. Gafni-Pappas, G., Khan, M.: Predicting daily emergency department visits using machine learning could increase accuracy. Am. J. Emerg. Med. **65**, 5–11 (2023)
20. Kuitunen, I., et al.: The effect of national lockdown due to COVID-19 on emergency department visits. Scand. J. Trauma Resusc. Emerg. Med. **28**(1), 1–8 (2020)
21. Hochreiter, S., Schmidhuber, J.: Long short-term memory. Neural Comput. **9**(8), 1735–1780 (1997)
22. Cho, K., Van Merriënboer, B., Bahdanau, D., Bengio, Y.: On the properties of neural machine translation: Encoder-decoder approaches. arXiv preprint arXiv:1409.1259 (2014)
23. Nallapati, R., Zhou, B., Gulcehre, C., Xiang, B.: Abstractive text summarization using sequence-to-sequence rnns and beyond. arXiv preprint arXiv:1602.06023 (2016)
24. Bérard, A., Pietquin, O., Servan, C., Besacier, L.: Listen and translate: A proof of concept for end-to-end speech-to-text translation. arXiv preprint arXiv:1612.01744 (2016)
25. Zhao, X., Lai, J.W., Ho, A.F.W., Liu, N., Ong, M.E.H., Cheong, K.H.: Predicting hospital emergency department visits with deep learning approaches. Biocybern. Biomed. Eng. **42**(3), 1051–1065 (2022)
26. Yousefi, M., Yousefi, M., Fathi, M., Fogliatto, F.S.: Patient visit forecasting in an emergency department using a deep neural network approach. Kybernetes **49**(9), 2335–2348 (2019)

High-Throughput Genomics: Bioinformatic Tools and Medical Applications

Targeted Next Generation Sequencing of a Custom Capture Panel to Target Sequence 112 Cancer Related Genes in Breast Cancer Tumors ERBB2 Positive from Lleida (Spain)

Iván Benítez[1] , Izaskun Urdanibia[1] , Xavier Matias-Guiu[2] , Ariadna Gasol[2] , Ana Serrate[1], Serafín Morales[2] , and Ana Velasco[2]([✉])

[1] Lleida Institute for Biomedical Research Dr. Pifarré Foundation (IRBLleida), 25198 Lleida, Spain
[2] Arnau de Vilanova University Hospital (HUAV), 25198 Lleida, Spain
avelasco@gss.cat

Abstract. Between 15–30% of invasive breast cancers have ERBB2 gene amplifications. Even though such homogeneous group, every patient has their own prognosis based on different features, some of which genetic involved. With that aim, we implemented a custom NGS panel comprising three probe subgroups for testing targeted mutations, copy number alteration and translocation in tumors with known HR and ERBB2 status previously assessed via immunohistochemistry and fluorescence in situ hybridization. DNA extracted from 47 primary breast cancers previously classified as ERBB2 positive were analyzed with a customized panel of 112 cancer related genes by targeted sequencing. Output data on fastq format was qualified, aligned and variant called trough different algorithms to find gene variations. A total of 20 different pathogenic mutations were found in 44% of tumors. Copy number analysis showed different levels of ERBB2 gene amplifications between tumors as so as different ERBB2 amplicon lengths. Additionally, the analysis of the raw data revealed the existence of two distinct mutation signatures. The identification of gene variations schemes that can yield distinct signatures holds the potential to accurately predict the subset of ERBB2-positive breast cancer patients who would respond best to treatment, specifically based on their pathological complete response (pCR).

Keywords: ERBB2 · Breast Cancer · NGS · Mutation Signatures · Trastuzumab

1 Introduction

Breast cancer is the most common cancer in women from Spain, with over 34,000 new cases diagnosed in 2020. Approximately 30% of cancers in women originate in the breast, making it the leading cause of cancer death among women in the country. The incidence rate for breast cancer is estimated to be 132 cases per 100,000 inhabitants, with a probability of 1 in 8 for women to develop the disease. The highest incidence

I. Rojas et al. (Eds.): IWBBIO 2023, LNBI 13920, pp. 137–150, 2023.
https://doi.org/10.1007/978-3-031-34960-7_10

occurs between the ages of 45–65, as hormonal changes during peri and post-menopause increase the risk. The incidence of cancer, including breast cancer, has been on the rise due to factors such as advances in early detection techniques and increased life expectancy. Aging is a major risk factor for developing cancer, along with modifiable factors such as sedentary lifestyles, tobacco and alcohol consumption, pollution exposure, and obesity. Improving prevention efforts and developing effective treatments remains critical in reducing the impact of breast cancer and other cancers on individuals and society [1].

The diagnosis, prognosis and treatment of breast cancer is based on clinical examination in combination with imaging and confirmed by pathology determination of histology, tumor grade, oestrogen receptor (ER), progesterone receptor (PgR), human epidermal growth factor receptor 2 (HER2/ERBB2) and proliferation marker Ki67 in core biopsies [2]. These biological markers are evaluated by immunohistochemical methods (IHC) using a standardized assessment methodology, e.g., Allred score or H-score, thereby HER2 is defined as positive by IHC (3+) when more than 10% of the cells harbor a complete membrane staining (protein is overexpressed), and by in situ hybridization (ISH) if the number of HER2 gene copies is >6. HER2 is part of the epidermal growth factor (EGF) family that stimulates cell growth and differentiation. Between 15–30% of invasive breast cancers have HER2 gene amplifications resulting in HER2 protein overexpression, which leads to an increase in tumor growth and aggressiveness [3]. Anti-HER drug Trastuzumab is a humanized monoclonal antibody targeting HER2 receptor, which was approved for use in 1998. Trastuzumab combined with chemotherapy (ChT) in patients with HER2 overexpression/amplification approximately halves the recurrence and mortality risk, compared with ChT alone [4]. The mechanisms of action of Trastuzumab have not been clearly defined, but likely include extracellular mechanisms involving antibody-dependent cellular cytotoxicity (ADCC), and intracellular mechanisms involving apoptosis and cell cycle arrest as well as inhibiting angiogenesis and preventing DNA repair following chemotherapy-induced damage. With the advancement of genomic studies, there are now more molecular targeted drugs available. Regrettably, certain patients do not react positively to treatments based on trastuzumab, and a few of them acquire secondary resistance following a remission of their illness. Although the exact mechanisms that lead to trastuzumab resistance are not completely comprehended, some researchers have suggested that the activation of the phosphoinositide 3-kinase (PI3K) pathway may play a role [5].

Precision medicine is a novel strategy that permits the personalization of medical care and illness prevention by utilizing the genomic makeup of an individual or disease, resulting in the customization of treatments that are specific to the patient's genetic profile. Recent research has demonstrated that cancers with identical pathological characteristics may actually be comprised of genetically distinct tumors [6, 7]. Thanks to advancements in technology, next-generation sequencing (NGS) now enables scientists to decode entire cancer genomes at a remarkable pace, transforming genomics research and providing an unprecedented level of insight into biological systems. By utilizing NGS to conduct clinical target sequencing of large-scale gene panels, researchers can identify the genetic drivers of disease and develop new molecular targeted drugs that are more effective. This approach not only allows for the selection of patients who are likely to respond well to treatment, but also reveals potential mechanisms of resistance to targeted therapies based on actionable driver mutations. Collectively, clinical target sequencing enables the differentiation of cancers based on anticipated therapeutic response and may ultimately serve as the catalyst for the realization and widespread adoption of precision medicine.

During the progression of cancer, various mutational processes caused by different factors are in play. These processes can be distinguished and pinpointed through the identification of mutational signatures, which are characterized by their distinct mutational patterns and specific effects on the genome [8].

The use of mutational signatures profiling has been successful in guiding oncological management and targeted therapies. For example, immunotherapy has been effective in treating mismatch repair deficient cancers of various types, while platinum and PARP inhibitors have been used to exploit synthetic lethality in homologous recombination deficient breast cancer. The first computational framework for deciphering mutational signatures from cancer genomics data was published in 2013 by Alexandrov and Stratton. They subsequently applied this framework to over 7,000 cancer genomes, creating the first comprehensive map of mutational signatures in human cancer. To date, over 100 mutational signatures have been identified across the spectrum of human cancer [8–10].

In order to find out some other druggable targets and features present in tumors diagnosed and treated with anti-HER2 therapy, the main objective of this study is to identify schemes on gene variations which could result in different tumor signatures that could help us to match each patient with the most effective treatment according to pathological complete response (pCR).

2 Methods

2.1 Breast Cancer Samples

The study population consisted of 47 patients (women) diagnosed and treated in Hospital Universitari Arnau de Vilanova (Lleida) whose clinical characteristics have been recorded in Oncology department. Some of these features are age at diagnosis, hormone status, HER2 status, treatment, pCR measured as tumor presence/absence after Trastuzumab treatment, relapse, or death.

All specimens contained >60% of tumor cells (as assessed before DNA extraction). IHC data included status for estrogen and progesterone receptors (ER and PR), P53 (positivity cut-off value of 1%), ERBB2 (0–3+ score, DAKO HercepTest kit scoring guidelines, defined as positive with 3+ and 2+ controlled by FISH according to ASCO guidelines), and Ki67 (positivity cut-off value of 20%).

The investigation was conducted following the guidelines of Good Clinical Practice and the Declaration of Helsinki by the World Medical Association. The patients gave their informed consent, and the research was approved by independent ethics committees.

2.2 Targeted Sequencing

DNA extracted from primary breast cancer tumors, previously classified as HER2 positive by IHC and FISH methods, was analyzed with a customized panel containing

49442 probes expanding exons from 112 cancer related genes [11] by targeted sequencing. Libraries for targets were generated with SureSelectQXT technology (Agilent) and sequenced on MiSeq v3 chemistry (Illumina).

2.3 Data Analysis

For descriptive purposes, an initial simple analysis was carried out using available tools and without statistical analysis as follows: Output data on fastq format was qualified, aligned and variant called with SureCall software (Agilent). Additional variant annotation was assessed with ANNOVAR [12] and the process for variant filtering was based on the selection of items annotated as pathologic or drug responsive, and variants which are expected to have similar impact through online data from population and disease public databases like SIFT, ExAC and ClinVar. Finally, copy number variations were rough estimated from read depth coverage [13] and read ratios between samples. Furthermore, HER2 gene expression analysis was performed by RT-PCR using TaqMan probes (Thermo Fisher Scientifics).

Secondary analysis involves several steps, including a) estimating the number of mutational signatures present in tumor samples; b) characterizing these signatures; c) comparing them with existing databases; d) evaluating the levels of exposure to these signatures in the samples; and e) assessing the potential associations between mutational signatures and treatment response. To be able to carry out these five steps, the sequencing data (fastq) was processed as follows: Output data on fastq format was qualified with FastQC [14], aligned and variant called with Tools from GATK4 [15] that includes Best Practices workflows for all major classes of variants for genomic analysis in gene panels.

In order to identify common patterns of mutations underlying biological processes, analysis was applied through the factorization of the non-negative matrix (NMF), which records the count of the N types of mutation (in this case, 96 combinations) present in K tumor samples [16]. First, the number of mutational signatures is estimated by selecting the model that maximizes the Bayesian Information Criterion (BIC). Then, the convergence of the parameters of the non-negative matrix factorization is verified by their stability in the process of resampling by Markov Chain Monte Carlo (MCMC). Once validated, the signatures are characterized and their similarity with signatures reported in the COSMIC catalog [17] is explored.

For this project, pre-surgical clinical staging after neoadjuvant chemotherapy treatment was used as a marker of clinical response. It was categorized as: a) No response: when no changes in tumor size are observed or tumor progression is observed in palpation and conventional imaging techniques; b) Partial response: when a decrease in size is evident in palpation and/or imaging; c) Complete response: when no tumor is identified in palpation and imaging.

To determine if there are differences in mutational signatures among tumors with varying therapeutic responses, the Kruskal-Wallis test was used. This test generates a set of p-values, $p(r)$, for each signature by evaluating the exposure matrix $E(r)$ for each sample. The median of the negative logarithm of these values defines the score for differential exposure (DES). Signatures with a DES ≥ 3 (equivalent to a p-value of 0.05) are considered to have differential exposure between groups. It is important to note that certain mutational signatures may be more prevalent in specific tumors, and this analysis aims to identify these differences.

3 Results

We have characterized the genomic landscape of ERBB2-positive breast cancer and investigated associations between genomic features and pCR. Sample collection was performed between 2013 and 2017 but more than 85% were gathered at years 2015 and 2016. Patients median age at sample collection time was 55.7 years (range from 29 to 86 years). 29 out of 47 patients were estrogen receptor positive (ER+) (61.7%) and 18 of them were ER- (38.29%). 17 of 47 (36.1%) presented pCR after anti-HER2 therapy. Survival intervals in patients who reached pCR oscillates between 16 and 75 months (average 40.2). 18 out of 47 patients (38.2%) presented partial response, 9 of them died with average survival of 27 months.

3.1 Targeted Sequencing

Raw data from parallel sequencing of 47 breast ERBB2 positive tumors (simple analysis) generated over 0.1 GB data per sample with approximately 115X depth on average. A total of 20 different pathogenic mutations were found in 44% of tumors, 64% of which were pathogenic variants in PIK3CA gene and 28.5% in TP53 gene. Other pathogenic variants have been found in HER2, RUNX1, ATM, SMAD4, CDH1, NTRK1 and CHEK2 genes (Fig. 1). Survival intervals in patients harboring pathogenic variants were of 33.6 months on average.

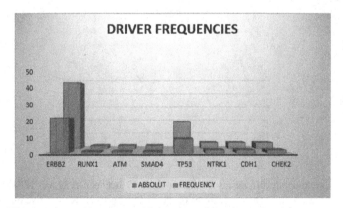

Fig. 1. Frequencies observed in a selected set of putative driver genes.

Copy number analysis showed different levels of HER2 gene amplifications between tumors as so as different ERBB2 amplicon lengths [18] (Fig. 2). In general, 24 out of 47 (51%) tumors have showed amplifications of some genes located on a region expanding 1Mbp at chr17q12-q21 cytogenetic band including CDK12, PPP1R1B, STARD3, PGAP3, HER2, GRB7, IKZF3, RARA and TOP2A. Although all tumors except one had in common the amplification of HER2 gene, 16 out of 24 (66.6%) presented amplification in a region expanding 262 Kbp from CDK12 to GRB7. RAD51, RPS6KB1, PPMID are downstream genes that presented amplification in some of the tumors in this series. Furthermore, this set of HER2 amplified tumors presented amplifications in other cytogenetics bands as chr5q22 (APC), chr6q22 (ROS1), chr8q23 (MYC), chr11q12 and chr11q22 (FGFR1, 8p11) amplification, also PPMID (17q23), and GNAS (20q13) were some other genes showing copy number variants.

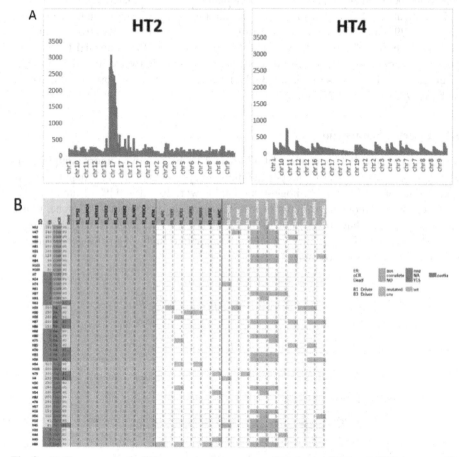

Fig. 2. A) Coverage depth differences in chromosome 17 between HT2 and HT4 tumours. B) Summary of biological and genomic features of the 47 sequences HER2 positive tumors.

3.2 Gene Signatures

Results obtained following GATK4 analysis showed 124,736 variants in 47 tumor samples, of which 111,524 were single nucleotide polymorphisms (SNPs), 1,740 were multiple nucleotide polymorphisms (MNPs), and 11,588 were insertions/deletions (indels). There were notable differences in mutational burden between cancer samples (Fig. 3).

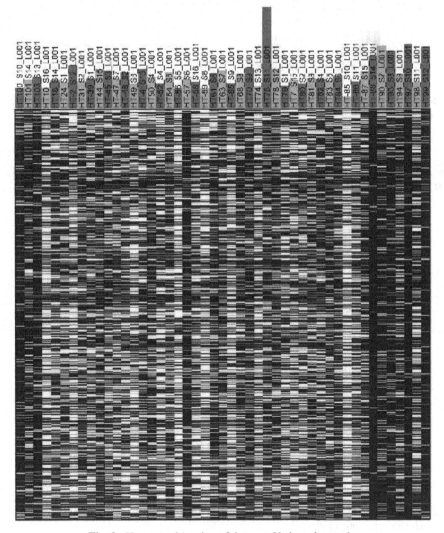

Fig. 3. Heatmap clustering of data set. Variants in purple.

Over 60% of the mutations identified were of the C > T and T > C type. Nearly half of the C > T variants were observed in CpG islands, which aligns with the most common mutation types in cancer GS14 (Fig. 4A). The study also identified 81 different sizes of indels, with smaller indels being the most frequent (Fig. 4B).

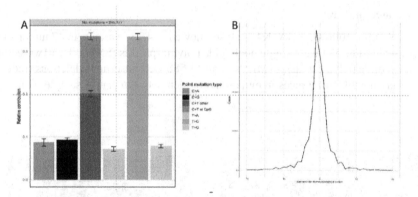

Fig. 4. A) Distribution of the type of mutation observed in the samples. B) Frequency of indels by size.

The identification of the number of signatures present in the samples was obtained by maximizing the median value of Bayesian information criterion (BIC) through the MCMC samples. The results identified two mutation signatures in the set of samples (Fig. 5).

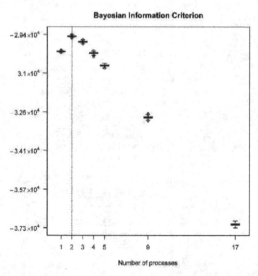

Fig. 5. Selection of number of mutational signatures. Returns the BIC value for the 1000 iterations for each number of processes.

Two mutation signatures (called S1 and S2) were identified in the set of 47 HER2+ breast tissue tumor samples. Both are characterized by the highest presence of C > T and T > C mutations, with the bases accompanying the central mutation being the differentiating parameter between them (Fig. 6).

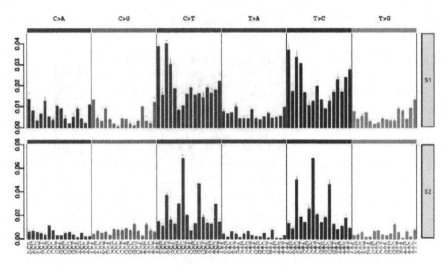

Fig. 6. Distribution of mutational signatures resulting from various mutational processes.

There is great availability in the loading of the signatures S1 and S2 among the experiments analyzed. In principle, it can be expected that a tumor that presents a greater burden of a mutational signature is more exposed to its consequences. Figure 7 shows tumors with exposure to S1 and S2 clearly higher than others.

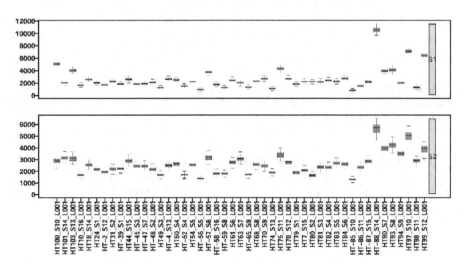

Fig. 7. Signature exposure per sample.

An unsupervised hierarchical clustering analysis of the samples based on their level of exposure to the obtained signatures (S1 and S2) shows a cluster that contains almost all the samples that did not obtain a clinical response (Fig. 8).

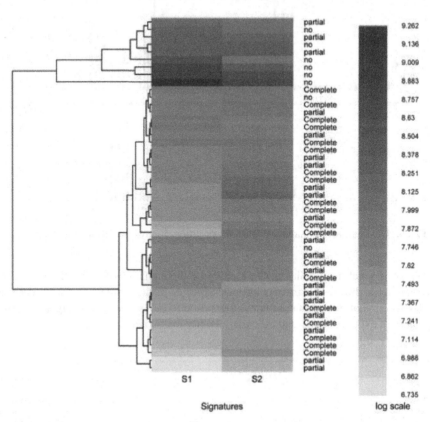

Fig. 8. Hierarchical classification based on the presence of mutational signatures, and their corresponding exposures for each genome sample. The dendrogram on the left illustrates the grouping of the samples according to their levels of exposure to the signatures.

The contribution of each sample to the constitution of the signatures is graphed (Fig. 9) as well as the level of exposure of the samples to each signature. All analyses support the hypothesis of a higher burden of S1 and S2 signatures (especially S1) in patients who did not respond to treatment.

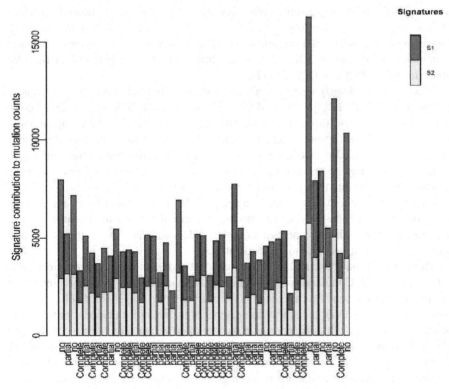

Fig. 9. Contribution of each sample to the constitution of the signatures.

4 Discussion

Our study delves into the genetic makeup of breast tumors in a specific group of patients with Her2-positive breast cancers who were treated with trastuzumab. Previously, it was believed that HER2 overexpression was indicative of a poor prognosis in breast cancer patients due to its correlation with a higher malignant potential. However, with the introduction of targeted anti-HER2 treatments, HER2 overexpression is now seen as a positive prognostic factor. Despite this, cancer cells can still develop resistance to these targeted therapies, making complete remission difficult to achieve. Therefore, investigating the mechanisms of resistance has become a significant area of research and development of genomic test that employ next-generation sequencers has enabled physicians to identify patients who are most likely to respond positively to specific drugs [19–21].

Owing to the fact that a significant number of patients with HER2-overexpressing breast cancer do not benefit equally from anti-HER2 therapy, personalized improved treatments could be considered in those cases harboring pathogenic variants that function as therapeutic targets (PIK3CA gene mutation or TOP2A gene amplification) [22] or as mechanisms for resistance to treatment (CCND1 and MYC gene amplifications) [23]. Thus, identification of these features suggest that it may be possible to predict, at the time of diagnosis, those ERBB2-positive breast cancer patients who could be treated with most effective therapy according to pathological complete response (pCR). Other possible druggable mutations included variants within HER-family members (including somatic mutations in HER2) or NTRK1 as so as amplifications in other potential therapeutic targets including ROS1 or CCND1 [11].

In addition to identifying nearly all genes previously implicated in breast cancer (PIK3CA, HER2, RUNX1, ATM, SMAD4, TP53, CDH1, NTRK and CHEK2), copy number analysis showed different levels of HER2 gene amplifications between tumors as so as different ERBB2 amplicon lengths. Studies have shown that patients with smaller ERBB2 amplicons tend to have better response rates to trastuzumab than those with larger amplicons. This may be because larger amplicons are associated with more complex genomic alterations that may confer resistance to Trastuzumab [24].

Overall, these genomic alterations may help to identify patients who are more likely to respond to trastuzumab and those who may require alternative treatment strategies. However, further studies are needed to better understand the role of these genomic alterations in patient response to trastuzumab [24].

The identification of two mutation signatures (S1 and S2) with the highest prevalence of C > T and T > C mutations did not yield a clear association with those annotated signatures from the somatic cancer mutation catalog from Sanger (COSMIC). A simulation study showed that the ability to decipher signatures of mutational processes is affected by the amount of noise present in the identified variants. In this project, the lack of paired normal tissue with the tumor sample may have added noise and limited the ability to identify pure mutational processes. Therefore, we believe that the identified signatures may include several mutational signatures, as some COSMIC signatures can be identified underlyingly. For instance, the similarity of signature S1 with signature SBS5 from the COSMIC catalog is notable. As described, the mutational burden of SBS5 increases in bladder cancer samples with ERCC2 mutations and in many types of cancer due to tobacco consumption. It should be noted that the lack of studies on mutational signatures in HER2+ breast cancer may be one of the reasons why signatures like these are not described in the literature.

There is a great variability in the burden of signatures S1 and S2 among the analyzed tumors. In principle, it is expected that a tumor with a higher burden of a mutational signature will have a greater exposure to its consequences. In Fig. 7, tumors with clearly higher exposure to S1 and S2 can be observed.

Due to this limited sample size, the study may not have had sufficient statistical power to identify differences between groups and although statistical associations between variables could not be analyzed due to the small sample size, however the study provides valuable insights into the topic under investigation and can serve as a basis for future research evaluating whether the S1 and S2 mutational signatures and the amplicon length are associated with a worse response to treatment, serving as biomarkers to classify patients and be able to offer other therapies.

References

1. ECIS - European Cancer Information System

2. Cardoso, F., et al.: Early breast cancer: ESMO clinical practice guidelines for diagnosis. Treat. Follow-Up. Ann. Oncol. **30**, 1194–1220 (2019). https://doi.org/10.1093/annonc/mdz173
3. Wang, L.: Early diagnosis of breast cancer. Sensors **17**, 1572 (2017)
4. Ross, J.S., Slodkowska, E.A., Symmans, W.F., Pusztai, L., Ravdin, P.M., Hortobagyi, G.N.: The HER-2 receptor and breast cancer: ten years of targeted Anti–HER-2 therapy and personalized medicine. Oncologist **14**, 320–368 (2009). https://doi.org/10.1634/theoncologist.2008-0230
5. Berns, K., et al.: A functional genetic approach identifies the PI3K pathway as a major determinant of trastuzumab resistance in breast cancer. Cancer Cell **12**, 395–402 (2007). https://doi.org/10.1016/j.ccr.2007.08.030
6. Katsnelson, A.: Momentum grows to make "personalized" medicine More "precise." Nat. Med. **19**, 249 (2013). https://doi.org/10.1038/nm0313-249
7. Yurkiewicz, B.Y.S.: The Prospects For, pp. 14–16 (2011)
8. Ng, C.K.Y., et al.: Intra-tumor genetic heterogeneity and alternative driver genetic alterations in breast cancers with heterogeneous HER2 gene amplification. Genome Biol. **16**, 107 (2015). https://doi.org/10.1186/s13059-015-0657-6
9. Alexandrov, L.B., Nik-Zainal, S., Wedge, D.C., Campbell, P.J., Stratton, M.R.: Deciphering signatures of mutational processes operative in human cancer. Cell Rep. **3**, 246–259 (2013). https://doi.org/10.1016/j.celrep.2012.12.008
10. O'Neil, N.J., Bailey, M.L., Hieter, P.: Synthetic lethality and cancer. Nat. Rev. Genet. **18**, 613–623 (2017). https://doi.org/10.1038/nrg.2017.47
11. Koboldt, D.C., et al.: Comprehensive molecular portraits of human breast tumours. Nature **490**, 61–70 (2012). https://doi.org/10.1038/nature11412
12. Wang, K., Li, M., Hakonarson, H.: ANNOVAR: functional annotation of genetic variants from high-throughput sequencing data. Nucleic Acids Res. **38**, e164–e164 (2010). https://doi.org/10.1093/nar/gkq603
13. Yoon, S., Xuan, Z., Makarov, V., Ye, K., Sebat, J.: Sensitive and accurate detection of copy number variants using read depth of coverage. Genome Res. **19**, 1586–1592 (2011). https://doi.org/10.1101/gr.092981.109
14. Andrews, S.: FastQC 0.11.4: a quality control tool for high throughput sequence data
15. Li, H., Durbin, R.: Fast and accurate long-read alignment with burrows-wheeler transform. Bioinformatics **26**, 589–595 (2010). https://doi.org/10.1093/bioinformatics/btp698
16. Blokzijl, F., Janssen, R., van Boxtel, R., Cuppen, E.: MutationalPatterns: comprehensive genome-wide analysis of mutational processes. Genome Med. **10**, 33 (2018). https://doi.org/10.1186/s13073-018-0539-0
17. COSMIC Mutational Signatures: Single Base Substitution (SBS) Signatures. https://cancer.sanger.ac.uk/cosmic/signatures/SBS/
18. Sahlberg, K.K., et al.: The HER2 amplicon includes several genes required for the growth and survival of HER2 positive breast cancer cells. Mol. Oncol. **7**, 392–401 (2013). https://doi.org/10.1016/j.molonc.2012.10.012
19. Bonadonna, G., et al.: Combination chemotherapy as an adjuvant treatment in operable breast cancer. N. Engl. J. Med. **294**, 405–410 (1976). https://doi.org/10.1056/NEJM197602192940801
20. Perou, C.M., et al.: Molecular portraits of human breast tumours. Nature **406**, 747–752 (2000). https://doi.org/10.1038/35021093
21. Brower, V.: NCI-MATCH pairs tumor mutations with matching drugs. Nat. Biotechnol. **33**, 790–791 (2015). https://doi.org/10.1038/nbt0815-790
22. Järvinen, T.A.H., Tanner, M., Bärlund, M., Borg, Å., Isola, J.: Characterization of topoisomerase iiα gene amplification and deletion in breast cancer. Genes Chromosom. Cancer **26**, 142–150 (1999). https://doi.org/10.1002/(SICI)1098-2264(199910)26:2%3c142::AID-GCC6%3e3.0.CO;2-B

23. Balázs, M., et al.: Genomic and gene expression characterization of a novel trastuzumab-resistant breast cancer cell line. Eur. J. Cancer Suppl. **6**, 119 (2008). https://doi.org/10.1016/s1359-6349(08)71628-8

24. Morrison, L.E., et al.: Effects of ERBB2 amplicon size and genomic alterations of chromosomes 1, 3, and 10 on patient response to trastuzumab in metastatic breast cancer. Genes, Chromosom. Cancer **46**, 397–405 (2007). https://doi.org/10.1002/gcc.20419

An Accurate Algorithm for Identifying Mutually Exclusive Patterns on Multiple Sets of Genomic Mutations

Siyu He, Jiayin Wang[✉], Zhongmeng Zhao, and Xuanping Zhang

School of Computer Science and Technology, Xi'an Jiaotong University, Xi'an 710049, China
wangjiayin@mail.xjtu.edu.cn

Abstract. In cancer genomics, the mutually exclusive patterns of somatic mutations are important biomarkers that are suggested to be valuable in cancer diagnosis and treatment. However, detecting these patterns of mutation data is an NP-hard problem, which pose a great challenge for computational approaches. Existing approaches either limit themselves to pair-wise mutually exclusive patterns or largely rely on prior knowledge and complicated computational processes. Furthermore, the existing algorithms are often designed for genotype datasets, which may lose the information about tumor clonality, which is emphasized in tumor progression. In this paper, an algorithm for multiple sets with mutually exclusive patterns based on a fuzzy strategy to deal with real-type datasets is proposed. Different from the existing approaches, the algorithm focuses on both similarity within subsets and mutual exclusion among subsets, taking the mutual exclusion degree as the optimization objective rather than a constraint condition. Fuzzy clustering of the is done mutations by method of membership degree, and a fuzzy strategy is used to iterate the clustering centers and membership degrees. Finally, the target subsets are obtained, which have the characteristics of high similarity within subsets and the largest number of mutations, and high mutual exclusion among subsets and the largest number of subsets. This paper conducted a series of experiments to verify the performance of the algorithm, including simulation datasets and truthful datasets from TCGA. According to the results, the algorithm shows good performance under different simulation configurations, and some of the mutually exclusive patterns detected from TCGA datasets were supported by published literatures. This paper compared the performance to MEGSA, which is the best and most widely used method at present. The purities and computational efficiencies on simulation datasets outperformed MEGSA.

Keywords: Gene mutation · Mutual exclusion · Clustering · Membership degree · Fuzzy c-means algorithm

1 Introduction

Cancers are driven by mutations, according to the research of cancer genomics, some mutations showed mutually exclusive patterns [1–5]. That is, if mutation A and mutation B are mutually exclusive, when A has occurred, it will significantly inhibit the occurrence

I. Rojas et al. (Eds.): IWBBIO 2023, LNBI 13920, pp. 151–164, 2023.
https://doi.org/10.1007/978-3-031-34960-7_11

of B, and vice versa. For example, it has been proved that TP53 is mutually exclusive of MDM2 or MDM4 [6], and the mutations of four genes (EGFR, KRAS, ERBB2, BRAF) in the EGFR-RAS-RAF signal transduction pathway in lung cancer are usually mutually exclusive [7]. Identifying mutually exclusive patterns is an important problem in cancer genomics, which is suggested to facilitate the research on identifying driver mutations, understanding cancer heterogeneities, guiding a more accurate diagnosis and treatment of cancer, etc. [8, 9]. For instance, in clinical manifestations, the expression of IFI44L is significantly reduced in Hepatocellular carcinoma (HCC) tumor tissues. That demonstrates that IFI44L is a novel tumor suppressor that affects cancer stemness, metastasis, and drug resistance via regulating the MET/SRC signaling pathway in HCC and is an important prognostic marker [10].

Although mutually exclusive patterns were discovered decades ago, recent research has emphasized their importance. This is because with advances in high-throughput sequencing technologies and decreases in the cost of sequencing in recent years have enabled TCGA and ICGC to collect large amounts of cancer genome mutation data, and the mutually exclusive patterns highlight their importance in diagnosis and treatment through data mining. However, the processing tools for large-scale data are relatively backward. There are two types of mutual exclusivity that the existing approaches mainly focus on: mutually exclusive patterns within a subset and mutually exclusive patterns between subsets. For patterns within a subset, the existing approaches are supposed to cluster one or multiple subsets of genes, where the genes in the same subset are mutually exclusive. For pattern among subsets, the existing approaches are supposed to cluster two subsets of genes, where the genes in the different subsets are mutually exclusive. That is to say, the mutual exclusion problem between sets is transformed into a pair-wise sets problem, which is completely different from the algorithm described in this paper.

Pathscan [11] and MEMCovery [12] take prior knowledge and gene penetrance as the search basis for mutually exclusive subsets, and effectively find the target subsets by reducing the search space. However, incomplete prior knowledge and inaccurate gene penetrance will directly affect the integrity and accuracy of the target subsets. RME algorithm is used to identify mutually exclusive patterns in tumor function modules (*de novo* sequencing data) [13]. RME uses a mutation as a classifier and the remaining mutations array as a trainer. After the mutex weight is determined by online fast flipping, the RME module is retrieved by greedy combination. Experiments show that due to the low mutation rates and coverages, 30% to 70% of the samples in a single mutation are not covered by the corresponding RME module, and the coverages can be improved only after deep filtering of the input data. In addition, the classification method of RME determines that the time for processing data will increase exponentially with the increase of data scale. MEMo algorithm is used to systematically identify the carcinogenic pathway module [14]. Based on prior-knowledge, MEMo uses filters to screen mutations with copy number variation and cooperative RNA expression, and searches for mutually exclusive subsets by looking for gene pairs on the same pathway. Since the pseudo discovery rate is used as an important parameter of the filter, the mutually exclusive weights of each subset need to meet the uniform distribution between 0 and 1. However, in practice, the weights do not obey the above distribution characteristics, so MEMo cannot effectively control the error rate. The Dendrix [3] algorithm was proposed for independent mutations, the maximum weight submatrix was determined by greedy algorithm and Markov chain Monte Carlo algorithm. In algorithm, the calculations of combination weight use the genes with high frequency mutations to determine the mutually exclusive signals. When most of the coverages of the gene set come from the same gene, there will be deviation in weight calculation. Muex [15] constructs a full probability model with coverage, non-exclusive and error rates (false positive and false negative rates) as parameters, and then assume that the gene mutation is independent, and finally build a model to search for the best mutually exclusive subset. However, the correctness of the assumption that mutation independence, and the correctness of likelihood probability obeying the standard normal distribution under the assumption that mutation independence remains to be discussed. MEGSA [9] algorithm, which uses likelihood ratio test as a statistic to test whether mutation subsets are mutually exclusive. And then, using multiple-path search algorithm and permutation test to test the global null hypothesis. Finally, the model is used to select the optimal mutually exclusive subsets. MEGSA is widely used in mutual exclusion detection.

The main difference among the above approaches lies in how to score the mutual exclusion, and how to design the process to identify the optimal mutually exclusive subsets, and most of them define metrics based on high coverage and approximate mutually exclusive principle. Some more rigorous and accurate methods are considered from the perspective of statistical testing. However, all these approaches use complex mathematical formulas and probabilistic methods for fitting and optimization. The processes of these approaches are difficult to understand and the calculations are inefficient, making them difficult to handle large-scale data.

Moreover, all the above approaches use genotype data as constraints to judge whether mutations are mutually exclusive or not. However, the constraints dealing become the optimization targets when deal with genotype data with variant allelic frequency, and the approaches cannot identify the mutually exclusive degree among mutations. In other words, all the above approaches cannot deal with real type data. The main difference between the genotype data and the real type data is that the genotype data clearly expresses whether the gene is mutated, while the real type data indicates the degree of mutation. It is obvious that the real type data can better express the true situation of the mutation.

Furthermore, it has been found that multiple mutations are mutually exclusive of each other in actual detection, so the results obtained by the pair-wise sets method in existing algorithms are incomplete. When the pair-wise sets problem becomes a multiple sets problem, it will encounter great challenges in terms of computational complexity.

To overcome the above two problems, in this paper, we propose an algorithm for multiple sets with mutually exclusive patterns based on fuzzy strategy, which considers both synergistic patterns and mutually exclusive patterns among multiple mutations. The goal is to get the results that the mutations within the subset are similar and the number of mutations is the largest, and any subset is mutually exclusive and has the largest number of subsets. There are few methods that consider both synergistic patterns and mutually exclusive patterns in existing approaches. The advantage of this method is that it transforms the problem of traversal search mutually exclusive patterns into a clustering problem of multiple sets with mutually exclusive patterns, which simplifies the complexity of the problem. If synergy and mutual exclusion occur frequently in actual research and analysis, its role in cancer occurrence and development can be verified from the perspective of biology and clinical medicine, which has strong practical significance. In addition, fuzzy strategy has two advantages. One is that the membership degree can ensure the continuity of the mutual exclusion degree, thus more fitting the characteristics of real-type data. Another is that the iterative process of the fuzzy strategy can directly reuse the latest calculation data without generating a target subset, which greatly simplifies the calculation process and improves the calculation efficiency.

2 Method

The mathematical meaning of the consideration of both synergistic patterns and mutually exclusive pattern is: $Si + ME = 1$, where Si represents synergy, ME represents mutual exclusion, and $Si \in [0,1]$, $ME \in [0,1]$. In the dataset with sample size P and mutation size N, K subsets with mutually exclusive pattern are searched, and the value of K is uncertain. This problem is considered as an NP-hard problem.

Based on the above conditions, this paper proposes an algorithm for multiple sets with mutually exclusive pattern based on fuzzy strategy to deal with real datasets. In the algorithm, mutual exclusivity problems among mutations are defined as the mutually exclusive problems among columns in matrix, where each row represents a sample and each column represents a mutation, and multiple subsets clustering are performed on the mutations. The problem is specifically described as: it is assumed that the mutation data matrix consists of samples $SP = \{p_0, p_1, p_2 \ldots p_{P-1}\}$ and mutations $SN = \{n_0, n_1, n_2 \ldots n_{N-1}\}$, detect set $G = \{G_0, G_1, G_2, G_3, \ldots, G_{K-1}\}$ with mutually exclusive patterns in the complete set SN, the following conditions must be met:

- Any two mutations in arbitrarily set, $n_x, n_y \in G_K$, need to be met $f(n_x, n_y) < Si$;
- The number of mutations in any subset is the largest, that is $\max|G_x| = M$;
- Any $n_x, n_y \in N$ and they belong to two mutually exclusive subsets, need to be met $h(n_x, n_y) > me$;
- The number of subsets in G is the largest, that is $\max|G| = K$.

In the above conditions, $f(n_x, n_y)$ is the similarity metric function of n_x and n_y, $h(n_x, n_y)$ is the mutually exclusive metric function of n_x and n_y, si is the similarity coefficient threshold, me is the mutual exclusion coefficient threshold.

The time complexity of calculating si and me among mutations is polynomial. However, multiple sets of mutually exclusive patterns are combinatorial optimization problems. Since the number of combinations with mutually exclusive patterns is uncertain, the number of combinations with mutually exclusive patterns may be about $C_G^K C_{G_x}^M$, if the above problem is solved by traversal search, the time complexity of calculation is $\geq O(2^k)$.

In this paper, the similarity and mutual exclusion degree among mutations are measured first to obtain the similarity and mutual exclusion degree matrix by using Euclidean distance. Based on the probability meaning of real type data, this paper introduces the clustering method of fuzzy c-means. Compared with hard clustering, which strictly divides the objects into a certain class and has the characteristics of either one or the other, fuzzy clustering uses the membership degree to establish the uncertainty description of the mutation to the class, so that the mutation has a certain attribute to each cluster, and can better reflect the mutual relationship pattern between mutations in the real situation. Not only that, fuzzy clustering also ensures the continuity of the solution space of the target problem, and only optimizes the membership degree and clustering center in the iterative process, without generating intermediate results, greatly simplifying the calculation processes and reducing the complexity. Therefore, this paper uses the weighted probability search method to obtain some ideal initial clustering centers, and then assigns the initial value of membership degree to all mutations according to the distance between mutations and the initial clustering center. Finally, the fuzzy c-means algorithm is used to continuously iterate the clustering center and membership degree, and then the target subsets are obtained. The subsets obtained in this way will maximize the number of internal mutations.

2.1 Similarity and Mutual Exclusion Measure

The measure of similarity and mutual exclusion between two-dimensional real data is essentially the measure of similarity and mutual exclusion between two vectors. Cosine similarity and Euclidean distance are the most direct and effective and widely used methods to measure the relationship between two vectors. In vector space, Euclidean distance measures the linear distance between two points, indicating the similarity and mutual exclusion of two vectors in specific values. The cosine similarity is the included angle between two vectors, which indicates the closeness and distance of two vectors in the direction. According to the actual situation of this paper, the similarity and mutual exclusion of vector values are more consistent with the similarity and mutual exclusion of mutations, while the specific meaning of the spatial direction between mutations is not clear. Therefore, it is more appropriate to choose Euclidean distance to measure the similarity and mutual exclusion of real data. Moreover, in the existing application of fuzzy clustering strategies, Euclidean distance is the mainstream. Therefore, the Euclidean distance is used to measure the similarity and mutual exclusion between the points.

2.2 Weighted Probability Search

The core of weighted probability search lies in the mutation selection probability, the mutation accumulative probability of mutation, and the mutation selection strategy of mutation.

- The mutation selection probability. The mutation selection probability is calculated by the formula (1). x_i represents the current mutation, $f(x_i)$ represents the fitness function between the mutation and the nearest seed point, K represents the number of seed points. The meaning of seed point is the initial cluster center. That is, the fitness function values of all mutations and the nearest seed point are normalized.

$$P(x_i) = \frac{f(x_i)}{\sum_{j=1}^{K} f(x_j)}, i \in [0, N-1] \tag{1}$$

The fitness function is used to calculate the overall fitness value of the mutations. There are two issues with the function design: one is the design of the fitness function, that is, the objective function may not be directly used as the fitness function. In this paper, the selected probability cannot be directly used as the fitness function value. Another issue is that the difference among the fitness function values $f(x_1), f(x_2), \ldots, f(x_K)$ of the selected mutations may not be too large, that is, the probability of the selected mutations is almost the same, which may weaken the selection function of the weighted probability search, and the fitness function needs to be calibrated.

Combined with the above two mentioned problems and the data status in this paper, power law calibration is the most direct and effective method for solving the fitness function design and increasing the difference among mutation fitness function values at one time. The function expression is: $f' = f^a$, where when $a > 1$, the selection pressure increases, and when $a < 1$, the selection pressure decreases. Therefore, the fitness function in this paper is calculated using formula (2), where $ED(x_i)$ represents the Euclidean distance between the current mutation and the nearest seed point.

$$fx_i = ED(x_i)^a, a \geq 2 \tag{2}$$

- The mutation accumulation probability. The calculation formula of the mutation accumulative probability is shown in formula (3). The accumulative probability means that the selected probability value of each mutation is represented by line segments of different lengths, which are combined into a straight line with a length of 1. The longer the line segment (which accounts for a larger proportion of the straight line), the greater the probability of it being selected. The principle is to arbitrarily select an arrangement sequence of all mutations (the sequence has no sorting rule, as each line segment represents the selected mutation, and the length of the line segment represents the probability of the selected mutation); the accumulative probability of any mutation is the accumulative sum of the selected probability values of the first several mutations corresponding to the mutation, as shown in Fig. 1.

$$Q(x_i) = \sum_{j=0}^{i} P(x_j) i \in [0, N-1] \tag{3}$$

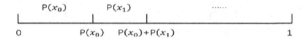

Fig. 1. Accumulative probability diagram

- The mutation selection strategy. The strategy involves generating a random number in the interval [0,1] and checking the number falls within that interval. It is clear that the length of the corresponding line segment is longer for the mutation with a higher probability of being selected, so the probability of randomly generated numbers falling in this interval is larger, and the probability of the mutation being selected is also larger. Conversely, for mutations with a smaller probability of selection, the length of the line segment will be relatively shorter, and the probability of random numbers in this interval is relatively smaller. However, there is still a chance of being selected, which avoids the problem of mutations with a small probability of selection being eliminated directly. This is also the reason why we must be random rather than simply taking the $P(x_i)$ with the largest measure.

By using the above method, we can identify as many seed points as possible, which maximizes the number of initial cluster centers and the K-value.

2.3 Algorithm Steps

The process of Algorithm for multiple sets with mutually exclusive patterns based on Fuzzy strategy is as follows:

(a) The similarity and mutual exclusion matrix is obtained using the Euclidean distance calculation formula.
(b) Randomly generate a site sequence number which will be the seed point (it is the first initial cluster center).
(c) Calculate the fitness function $f(x_i)$ of each point according to formula (2).
(d) Calculate the selected probability $P(x_i)$ of each point according to formula (1) and accumulate them to obtain the straight line with length of 1.
(e) Randomly generate a real number in the interval [0,1], determine that interval to which the random number belongs, and the site corresponding to the interval is the candidate.
(f) The steps c) to e) are repeated until the number of seed points K is stable.
(g) Initialize membership value u_{rc}, if the current site is a seed point, the membership degree to itself is 1, and the membership degree to other seed points is 0. If the site is not a seed point, there is certain degree of membership for each seed point, and the accumulative value is 1, that is:

$$\sum_{c=0}^{K-1} u_{rc} = 1, (r = 0, 1, 2, \ldots, N - 1) \tag{4}$$

Calculate the objective function value by using the clustering center iteration formula (5), the membership degree iteration formula (6), and the objective function iteration formula (7) until the objective function values is less than a certain threshold, or the difference between the current objective function value and the previous objective function value is less than a certain threshold, the iteration is terminated, and the objective function and its membership degree matrix are obtained.

$$c_r = \frac{\sum_{c=0}^{K-1}\left(x_c u_{rc}^m\right)}{\sum_{c=0}^{N-1} u_{rc}^m} \tag{5}$$

$$u_{rc} = \frac{1}{\sum_{l=0}^{K-1}\left(\frac{\|x_c - c_r\|}{\|x_c - c_l\|}\right)^{\frac{2}{m-1}}} \tag{6}$$

$$J = \sum_{r=0}^{N-1}\sum_{c=0}^{K-1} u_{rc}^m \|x_c - c_r\|^2 \tag{7}$$

The target subset is obtained by the target membership matrix.

The pseudo code of the algorithm for multiple sets with mutually exclusive pattern based on Fuzzy strategy is shown in Algorithm 1 (Table 1).

Algorithm 1

Input: Real data matrix
Output: K mutually exclusive subsets

1. Calculate the similarity and mutual exclusion matrix among each point.
2. Randomly generate an initial seed point.
3. do
4.   fitness function $f(x_i)\leftarrow$ calculate according to formula (2)
5.   site selection probability $P(x_i)\leftarrow$ calculate according to formula (1)
6.   randomly generate the next seed point
7. while max$(P(x_i) - \text{limit}) > \text{Min}$
8. initial membership $\mu \leftarrow$ calculate according to formula (4)
9. do
10.   initial cluster center $c\leftarrow$ calculate according to formula (5)
11.   iterative the membership $\mu_{rc} \leftarrow$ calculate according to formula (6)
12.   the objective function $J \leftarrow$ calculate according to formula (7)
13. while current objective function value - last iteration objective function value $\geq$ given threshold

Table 1. The definitions for the symbols

S_i	Synergy	ME	Mutual Exclusion
P	Sample size	N	Mutation size
K	The maximum number of subsets with mutually exclusive patterns in the mutation data matrix	SN	Mutations in the mutation data matrix
M	The maximum number of mutations in a mutually exclusive subset	$f(n_x, n_y)$	Similarity metric function
$h(n_x, n_y)$	Mutually exclusive metric function	si	Similarity coefficient threshold
me	Mutual exclusion coefficient threshold	$P(x_i)$	The mutation selection probability
u_{rc}	The value of membership	c_r	Clustering center
J	The objective function		

3   Discussion

The simulation experiment data are generated by the self-developed ME simulation program and the ms simulation software [16].

3.1   ME Simulation Principle and Evaluation of Result

Based on the data onto 2020 global cancer report, the ME simulation program in this paper determines that the mutation coverage accounts for 4.5–7.5% of the total sample, and the number of mutations contained in each subset is a random number between 2 and 5. In addition, the simulation data also takes full account of the interference among in subsets, the interference within subset, the random interference outside the exclusive set and other information. Finally, the generated data is scrambled for columns and rows to conform to the real situation as much as possible.

Because the detection algorithms involved in this paper are multi clustering algorithms, the evaluation indexes for the dichotomous clustering algorithm are no longer applicable. Therefore, the evaluation of experimental results consists of Purity and Silhouette Coefficient, Purity is a kind of external evaluation index, and Silhouette Coefficient is a kind of internal evaluation index.

Purity is to calculate the percentage of correct clustering by comparing with the data labels provided externally. The sample category with the highest frequency in each subset is used as the correct clustering, and then divided by the total number of samples in the subset. The purity value range is [0–1]. The larger the value, the better the clustering effect.

SC (Silhouette Coefficient) has no external label as a reference. It evaluates the clustering results by considering two factors: cohesion and separation. The process is to calculate the average distance $a(i)$ from the current site to other sites in the same subset, calculate the average distance $b(i)$ from the current site to all sites in other subsets, calculate the Silhouette Coefficient $S(i)$ according to formula (8). The value range of Silhouette Coefficient is $[-1 \sim 1]$. If it is close to 1, the site clustering is reasonable. If it is close to -1, the site should be clustered to other subsets. If it is close to 0, the site is on the boundary of the two subsets.

$$S(i) = \frac{b(i) - a(i)}{\max\{a(i), b(i)\}} \tag{8}$$

3.2 The Result and Analysis of ME Simulation Experiment

The number of samples is fixed at 500. When K is equal to 5, 7 and 10 respectively, 10, 50, 100, 150 and 200 data sets are randomly generated and analyzed with the algorithm. The evaluation results are shown in Fig. 2. Both the Purity and the Silhouette Coefficient are stable between 0.95 and 1, indicating that the clustering results are ideal no matter compared with the set results, or according to the compact density within the cluster and the dispersion between clusters.

Finally, when K values are 5, 7 and 10 respectively, data sets with sample sizes of 200, 400, 800, 1600 and 3200 are randomly generated to test the processing capacity of the algorithm when the sample size increases. The results are shown in Fig. 3. Experiments show that the computing time is basically in a linear relationship with the data size, indicating that when the data size increases and the dimension increases, its computing performance will not decline sharply.

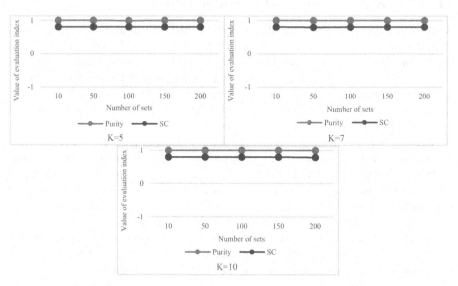

Fig. 2. Clustering evaluation of the Algorithm when p = 500

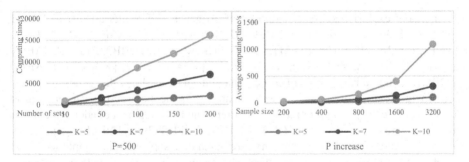

Fig. 3. Computing time of Algorithm

3.3 Ms Simulation Data

Ms simulation data is generated by ms simulation software. The ms simulation software is widely used to simulate population genomics, and its simulation principle follows Wright-Fisher neutral model. The parameters are set to:/ ms 15 1000 −t 10.004 −r 100.0 2501. Specifically, the default value of ms simulation software is used to set the seed points. Each sample randomly generates several mutations, generating a total of 1000 datasets in this paper. Randomly select 10, 50, 100, 150, 200 and 500 datasets from the above for mutual exclusion pattern detection. The evaluation results are shown in Fig. 4. Because ms simulation software cannot generate a comparison table, the results cannot be evaluated by Purity, but only by Silhouette Coefficient. The results show that the algorithm for real type data still has a better performance for genotype data generated by ms simulation software.

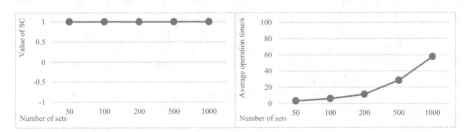

Fig. 4. Performance of the Algorithm in ms simulation data

3.4 Real Datasets

We applied our approach on a real dataset consisted of 589 samples, and the mutations of each sample ranges from several to tens of thousands. The mutations were obtained from clinical tests of NPC (Nasopharyngeal Cancer) and NSCLC (non-small-cell lung cancer). We located the mutations into the genes by comparing them with typical transcripts, 393 genes were finally obtained. After that, we filtered the noise according to coverage rates of mutations. Coverage means the proportion of mutation in all samples. Too high or too low coverage rate will be considered as noise. Then we obtained

11 mutually exclusive subsets composed of 36 genes, the results are list as below: {TEKT4, TERC}, {TMEM127, TMPRSS2, TNFAIP3}, {TNFRSF14, TNFRSF19}, {TOP1, TOP2A, TP53, TP63, TPMT, TSC1}, {TSHR, TTF1, TUBB3, TYMS, U2AF1}, {UGT1A1, VEGFA, VHL, WAS, WRN, WT1}, {XPA}, {XPC}, {XRCC1}, {YAP1}, {ZNF2}. Comparative analysis of subsets and clinical cases shows that genes in subsets can synergistically promote the occurrence and development of a certain cancer, such as TEKT4 and TERC jointly promote the occurrence and development of adenocarcinoma in sample with case number 234168. TMEM127, TMPRSS2 and TNFAIP3 jointly promoted undifferentiated non keratinizing carcinoma of nasopharynx in sample with case number 313577. The genes among subsets show a mutually exclusive pattern in sample, such as any two or more genes in XPA, XPC, XRCC1, YAP1, and ZNF2 do not appear in the same sample at the same time with large numbers. Usually, only one gene appears in one sample. This result verifies the mutual exclusion pattern of different subclones in the same cancer.

4   Comparative Experiments and Results

MEGSA [9] is the best and most widely used method among the existing approaches for detecting mutually exclusive patterns in genotype data. Therefore, MEGSA is used as the comparison target in this experiment and compared with the algorithm in the case of genotype data. There are m mutually exclusive subsets which generation by the MEGSA. The mutations in the subset are mutually exclusive, and the subsets are mutually exclusive as much as possible. Therefore, the comparison experiment is difficult to be carried out accurately, and can only be roughly compared from the two aspects of clustering result evaluation and computing time. The experiment uses ME simulation data, in which $K = 5$ and $P = 200$. The results are roughly compared in terms of evaluation purity and time efficiency by using multiple sets of data sets. After detected and analyzed, the average purities and average time efficiencies of the two algorithms are shown in Fig. 5. Both purity evaluation and the average time efficiencies of the algorithm in this paper are better than that of MEGSA.

Secondly, when the sample sizes are increased, the efficiencies of the two algorithms are analyzed, as shown in Fig. 6. In the case of the same data set, the algorithm in this paper and MEGSA show a linear relationship between the computing time and the sample size, but the time of the algorithm in this paper is much less than that of MEGSA.

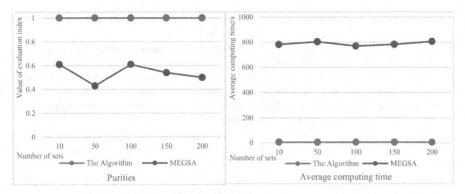

Fig. 5. Performances of the algorithm and MEGSA in ME data

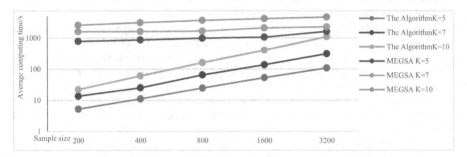

Fig. 6. Average computing time of the algorithm and MEGSA when P increases

5 Conclusion

The mutual exclusion pattern detection for real types data of gene mutation is a new type pattern proposed in this paper, and no similar work has been found at present. In this paper, we give the definition of mutually exclusive patterns in real type data from the perspective of mutually exclusive degree, and give the method to detecting. The experiments prove that the algorithm is correct and effective. Through the research, we can more clearly describe the synergy and mutually exclusive relationship among mutation mutations. Therefore, the algorithm will help to find hidden important information of tumors, thus assisting in accurate diagnosis and treatment, and then revealing the genetic code behind human tumors from an updated perspective.

References

1. Liu, S., Liu, J., et al.: MEScan: a powerful statistical framework for genome-scale mutual exclusivity analysis of caner mutations. Bioinformatics **37**(9), 1189–1197 (2021)
2. Yeang, C., Frank, M., Arnold, L.: Combinatorial patterns of somatic gene mutations in cancer. FASEB J. **22**(8), 2605–2622 (2008)
3. Fabio, V., Eli, U., Benjamin, J.: De novo discovery of mutated driver pathways in cancer. Genome Res. **22**(2), 375–385 (2012)

4. Cui, Y., Wang, T.: A statistic model of identifying mutual exclusivity mutations in cancer pathway. Taiyuan Shanxi Medical University (2016)
5. Wu, H.: Algorithm for detecting driver pathways in cancer based on mutated gene networks. Chin. J. Comput. **41**(6), 1180–1194 (2018)
6. Cancer Genome Atlas Research Network: Comprehensive genomic characterization defines human glioblastoma gene and core pathways. Nature **455**(7216), 1061–1068 (2008)
7. Hiromasa, H., Hisayuki, S., et al.: PIK3CA mutations and copy number gains in human lung cancers. Cancer Res. **68**(17), 6913–6921 (2008)
8. Yang, C., Zheng, T., et al.: A greedy algorithm for detecting mutually exclusive patterns in cancer mutation data. In: International Work-Conference on Bioinformatics and Biomedical Engineering (2019)
9. Xing, H., Paula, L., Jing, H., et al.: MEGSA: a powerful and flexible framework for analyzing mutual exclusivity of tumor mutations. Am. J. Hum. Genet. **98**(3), 442–455 (2016)
10. Huang, W., Tung, S., Chen, Y., et al.: IFI44L is a novel tumor suppressor in human hepatocellular carcinoma affecting cancer stemness, metastasis, and drug resistance via regulating met/Src signaling pathway. BMC Cancer **18**(1), 609 (2018)
11. Michael, C., John, W., Ling, L., et al.: PathScan: a tool for discerning mutational significance in groups of putative cancer genes. Bioinformatics **27**(12), 1595–1602 (2011)
12. Yoo-Ah, K., Cho, D., Phuong, D., et al.: MEMCover: integrated analysis of mutual exclusivity and functional network reveals dysregulated pathways across multiple cancer types. Bioinformatics **31**(12), i284–i292 (2015)
13. Christopher, A., Stephen, H., Erik, P., et al.: Discovering functional modules by identifying recurrent and mutually exclusive mutational patterns in tumors. BMC Med. Genomics **4**, 34 (2011)
14. Giovanni, C., Ethan, C., Chris, S., et al.: Mutual exclusivity analysis identifies oncogenic network modules. Genome Res. **22**(2), 398–406 (2012)
15. Ewa, S., Niko, B.: Modeling mutual exclusivity of cancer mutations. PLoS Comput. Biol. **10**(3), e1003503 (2014)
16. Richard, R.: Generating samples under a wright-fisher neutral model of genetic variation. Bioinformatics **18**(2), 337–338 (2002)

A 20-Year Journey of Tracing the Development of Web Catalogues for Rare Diseases

João Rafael Almeida[1,2(✉)] and José Luís Oliveira[1]

[1] IEETA / DETI, LASI, University of Aveiro, Aveiro, Portugal
[2] Department of Computation, University of A Coruña, Coruña, Spain
{joao.rafael.almeida,jlo}@ua.pt

Abstract. Rare diseases are affecting over 350 million individuals on a worldwide scale. However, studying such diseases is challenging due to the lack of individuals compliant with the study protocols. This unavailability of information raises some challenges when defining the best treatments or diagnosing patients in the early stages. Multiple organizations invested in sharing data and resources without violating patient privacy, which resulted in several platforms focused on aggregating information. Despite the benefits of these solutions, the evolution of data regulations leads to new challenges that may not be fully addressed in such platforms. Therefore, in this paper, we proposed an enhanced version of one of the identified open-source platforms for this purpose. With this work, we were able to propose different strategies for aggregating and sharing information about rare diseases, as well as to analyse the technological evolution when producing tools for biomedical data sharing, namely by analysing the evolution of the selected tool over the last two decades.

Keywords: Rare diseases · Biomedical semantics · Data integration · Interoperability · DiseaseCard

1 Introduction

A rare disease is defined as any disease that affects at most one person out of 2000 people [1]. Despite this low incidence at this scale, when considering worldwide populations, the number of subjects affected by such diseases is considered high, namely more than 350 million people [2]. Besides these numbers, such diseases have some impact since they result in significant sources of morbidity and mortality [2]. Of all diseases classified as rare, approximately 80% are exclusively genetic or have genetic subtypes. To demonstrate these values, a review was conducted by Bick *et al.* [3] based on 3 800 rare diseases listed by Orphanet.

One of the major challenges for biomedical researchers when studying such diseases is the lack of individuals that are compliant with the study protocol [4]. Conducting a trial with low number of subjects is challenging and the study may

I. Rojas et al. (Eds.): IWBBIO 2023, LNBI 13920, pp. 165–179, 2023.
https://doi.org/10.1007/978-3-031-34960-7_12

lead to results with low impact [4,5]. This unavailability of information raises another challenges, namely when defining the best treatments or diagnosing patients in yearly stages [6].

Over the last decades, the computational system had aided researchers to conduct medical studies, including the possibility of also analysing rare diseases. With the increase of computational power and methodologies for data sharing, the process of studying such diseases was simplified [7]. For instance, when different institutions are studying the same rare diseases, the researchers can share information, following well-defined regulations, to increase the number of subjects in the study. As result, these studies can produce more impactful findings [8].

These strategies resulted from initiatives focused on improving people's lives by supporting activities such as training, data access infrastructures, and the definition of data standards, among others [9]. In Europe, rare diseases have been recognised as a research field that can benefit from the coordination of European entities [10]. Currently, multiple organizations have the goal of sharing data and resources, namely the European Partnership on Rare Diseases, the European Joint Programme on Rare Diseases (EJP RD)[1], the Diseases Research Consortium (IRDiRC)[2] and the Eurordis[3], without violating the patient's privacy.

Over these years, efforts from initiatives resulted in several platforms for aggregating information collected in these studies. The main goal of these platforms was the simplification of sharing data with the scientific community, namely, the information originated from decades of medical studies that were kept at most in scientific articles. Examples of such platforms are the National Organisation for Rare Disorders (NORD)[4] or Online Mendelian Inheritance in Man database (OMIM)[5]. These are well-known for collecting and filtering information focused on rare diseases. Similar to these, other platforms were proposed, focused on collecting and aggregating all existing data related to certain diseases [11]. The objective of this latter was to simplify the data visualisation of different data types regarding rare diseases. In this paper, we revised the current state-of-the-art of these solutions and proposed improvements in one of the identified open-source solutions. As a result of this work, we were able to produce an enhanced version of this solution, as well as, analysed the evolution of these platforms over the last two decades.

2   Related Work

There are multiple platforms focused on exporting metadata about health datasets. Some of these are focused on rare diseases, including the study of genes, genetic phenotypes, proteins, genomes, enzymes, DNA and each one of

---

[1] https://www.ejprarediseases.org.

[2] https://irdirc.org/.

[3] https://www.eurordis.org/.

[4] https://rarediseases.org/.

[5] https://www.omim.org/.

possible known variations or mutations. These aspects are crucial to understand the impacts and discover possible treatments of any disease. As rare diseases are less frequent, being able to collect all available information is essential not only for the scientific community but also for patients and health professionals. Studying them is important since there are approximately 7 000 existing rare diseases of which are still poorly understood [12].

Orphanet has been the global reference source of information on rare diseases for the last two decades [13,14]. This platform was supported by the European Commission aiming to generate knowledge and assist in the identification of patients with rare diseases. The platform is based on three main aspects: i) equal access to knowledge by patients and the scientific community; ii) improvement of the visibility of rare diseases through a common language; and iii) data collection from various sources to generate knowledge [14]. The Orphanet global network is managed by the Orphanet Coordinating team at the French National Institute of Health and Medical Research (INSERM) and it is composed of 41 countries. National teams are responsible for gathering information on specialised centres, medical laboratories, ongoing research, and patient organisations. This makes Orphanet a platform that offers a wide range of free services, namely the inventory and encyclopedia of rare diseases. Orphadata is another component of this platform that provides databases about rare diseases and known drugs. Orphanet uses a nomenclature for rare diseases developed by the internal teams, for all diseases, subtypes, and groups of diseases on the platform. The nomenclature references other terminologies and databases such as ICD-11 [15], OMIM [16], Medical Language System (UMLS) [17], MeSH [18], and MedDRa [19]. Orphanet also analyzes the genes associated with rare diseases, collecting information from HUGO Gene Nomenclature Committee (HGNC) [20], OMIM [16], UniProt Knowledgebase (UniProtKB) [21], Ensembl [22], Reactome [23] and IUPHAR [24].

Malacards is another platform that aims to solve the challenge of integrating multiple sources of information about diseases [25]. Malacards standardizes knowledge by combining different perspectives on the same disease through the use of data-mining algorithms modelled from the GeneCards database. The platform organizes information into disease cards, each containing a main canonical name, a list of aliases, and several sections containing information on clinical features, genetics tests, anatomical context, drugs and therapeutics, and more. The platform offers several features, including a search engine that allows users to search for a particular disease by name, gene name, or keyword. It combines lexical heterogeneity to make possible the standardization of knowledge, combining different perspectives on the same disease explored in different sources. It enables researchers to understand and explore a vast amount of information on diseases [25].

Disease Ontology is a platform that aims to integrate a high number of sources of information about diseases by defining an ontology for the biomedical domain [26]. The ontology is open-source and focused on representing common and rare disease concepts from multiple biomedical databases, providing an interface between them. The concepts integrated into this ontology are extracted based on the UMLS [17] which includes diseases from ICD-11 [15], NCI Thesaurus [27], SNOMED-CT [28] and MeSH [18]. It also includes disease terms directly extracted from other sources, including OMIM [16], Orphanet [14] and NORD [29]. Disease Ontology is an open-source project, and most of its resources and results are fully available to the scientific community. The ontology and databases are available under a license that allows copy, redistribution and adaptation of these resources for any purpose [26].

The previous platforms exposes information mainly to researchers and health care professionals, with a high level of expertise. Genetic and Rare Disease Information Center (GARD) is another well-establish platform with the goal of providing reliable, timely and easy-to-understand information about rare diseases [30]. This platform also enables the interaction with information specialists, that can redirect patients to existing materials available on the platform or be oriented in terms of locating additional information in other resources [31]. The platform enables searching for information on more than 6 700 genetic rare diseases. These can be found through a simple search engine or through a browser engine that allows finding a disease alphabetically or by its category [32].

NORD platform is another valuable resource that provides hard-to-find information about rare diseases. The platform offers four distinct databases: i) the *Index of Rare Diseases*, which provides an alphabetical list of rare diseases; ii) the *Rare Disease Database*, which contains detailed information on 1 150 rare diseases along with associated reports and articles, some of which may require payment to access; iii) the *Index of Organizations*, which includes an alphabetical list of over 2 200 organizations and other sources of support for patients and families; and iv) the *Organizational Database*, which allows users to search for information about specific organizations associated with a particular disease [29].

National Center for Biotechnology Information (NCBI) provides organism names and classifications for every sequence in their nucleotide and protein sequence databases [33]. This platform allows access to all existing data through a simple search on its platform, namely associated with other branches of knowledge. One of the goals is to increase genomic, genetic and biomedical knowledge of the community. The results obtained from each search process contain a comprehensive set of links for each database [33].

DiseaseCard[6] is an open-source platform aiming to provide access to a network of the most relevant rare diseases scientific resources. The platform intents to deliver a lightweight holistic perspective on rare genetic diseases. However, instead of replicating the information already collected from others sources, the platforms redirect to these resources. With this strategy, it is possible to discover new platforms covering a specific disease and to view the information of each one, without leaving the DiseaseCard ecosystem [34]. This aggregation is made considering OMIM's morbid map as the root of the DiseaseCard knowledge base. Through OMIM entries and their conjugation with information from the HGNC, this network is expanded to another 20 resources [11]. This open-source plat-

---

[6] https://bioinformatics.ua.pt/diseasecard/.

form built in Java uses one of the first versions of COEUS [35], a semantic web application framework. By doing so, it allows for improving the semantic data integration and interoperability features of DiseaseCard. The knowledge base associated with this platform can be accessed through its API which includes two main data access alternatives: SPARQL Protocol and RDF Query Language (SPARQL) endpoints or LinkedData interfaces. These options facilitate the integration of data from this platform with external ones, in part due to the flexible output formats provided by the API [11].

As we described, there are some well-established platforms focused on aggregating information about rare diseases. However, the FAIR (Findable, Accessible, Interoperable, and Reusable) principles have emerged as a set of guiding principles to improve the management and sharing of data, making it more accessible to both humans and machines [36]. With this and the increase in concerns regarding data privacy, new challenges arose. Instead of developing a new platform compliant with these principles, we studied the most used platforms for rare genetic disease research and tried to improve one of them. Based on a detailed analysis conducted in previous work [37], we have selected DiseaseCard for this proposal.

3   Methods

The primary purpose of DiseaseCard[7] was the creation of a centralised platform to collect, connect and simplify the access of the most relevant scientific resources on rare diseases.

3.1   Implementation

The implementation of this version of the system was initially limited by the original technologies selected for its first developments. With the increase of new requirements, more challenges were raised, which forced the migration of some components to more recent technologies. Therefore, the main goal of the DiseaseCard restructuring process was to reuse as many existing components as possible, allowing to focus on providing a good foundation to fulfil the current functional requirements by introducing new components.

The system follows a client-server architecture. This new version includes Spring Boot framework in the backend, namely to support the Representational State Transfer (REST) Application Programming Interface (API). Spring Boot framework was adopted for being a Java-based framework, which is the programming language of this tool. As a result, the Java version was updated from version six to version eight. In addition, this framework through the Spring Spring Model-View-Controller (MVC) module facilitates the implementation of the API, following the standards for building web services. The client-side was

---

[7] This work relays on the latest version of DiseaseCard, in which the system core components were completely rebuilt.

completely rebuilt, keeping some of the original designs of the platform. For this layer of the system, new technologies were selected, since the original versions of DiseaseCard were implemented using Java Server Pages (JSP) pages. Figure 1 illustrates the five layers of the system and the technologies that support them. The client-side was implemented using ReactJS framework and Redux library.

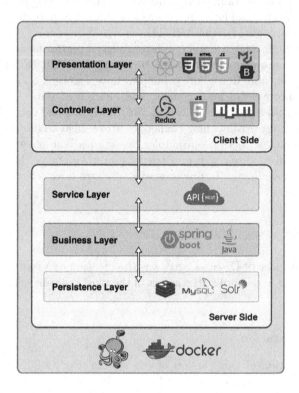

Fig. 1. Overview of the DiseaseCard layers and technologies.

Diseasecard's deployment strategy makes use of the Docker platform to containerise the various components that are part of the solution. The system deployment is composed of seven different docker containers. The persistence layer is structured in three independent docker containers, each running instances of the persistence technologies discussed above, namely an instance of MySQL, Redis and Solr. The core of the server side is running on a Tomcat server, capable of accessing the containers in the persistence layer. The client-side is composed of two distinct containers, one for the DiseaseCard and another for system administration. These containers are both running behind an Nginx server, which is robust and efficiently provides a server for web content.

3.2 Relationship Between Data Sources

DiseaseCard knowledge base is designed from an integration network that presents the OMIM's morbid map as the primary source of data. The process of importing and translating this data follows a warehousing data integration strategy, meaning that the extracted data is centralised in a single knowledge base, in opposition to real-time data gathering strategies. Internally, this process generates a set of triples that were created to represent the information of the resource itself and the relationships with resources from a different data source. The system relies on SCALEUS [38] to process the data from multiple and heterogenous data sources.

Figure 2 presents an overview of the DiseaseCard semantic structure. This diagram simplifies the identification of how each data source is introduced in the system, and how each resource correlates with the OMIM node, used as starting point. This interconnection is possible since most of the used databases contain a set of identifiers to other resources in external data sources. These identifiers are added into the system, which through complementary extraction methods, allows extracting data from another set of sources. The information from these last sources is associated with the identifier of the OMIM node.

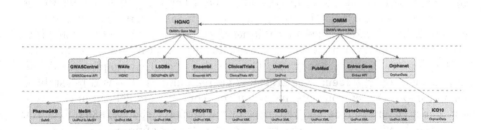

Fig. 2. Overview of resources are interconnected in DiseaseCard.

3.3 Semantic Structure

The relationships between resources are defined in an ontology. The ontology represents a shared, agreed and detailed model (or set of concepts) for this specific domain. An inherent advantage of using a domain ontology is the ability to define the semantic model of the data combined with the knowledge of the associated domain [39]. DiseaseCard's ontology is inherited from SCALEUS, which organises the relations into four different notions: i) seeds, ii) entities, iii) concepts and iv) resources.

All entities, concepts or resources in the ontology contain an associated node, denominated as seed. DiseaseCard defines a seed called `seed_DiseaseCard4`, and all the entities presented on DiseaseCard's ontology define the property `coeus:isIncludedIn` passing as value the DiseaseCard's seed name so that it can be associated with it. The ontology was created in a previous version of the tool [11], and the name of the original concepts was kept (for instance, the reference to Coeus in the nodes).

```
<owl:NamedIndividual rdf:about="&diseasecard;entity_Disease">
    <rdf:type rdf:resource="&coeus;Entity"/>
    <rdfs:label rdf:datatype="&xsd;string">entity_disease</rdfs:label>
    <dc:date rdf:datatype="&xsd;date">              </dc:date>
    <owl:versionInfo rdf:datatype="&xsd;string">4.2</owl:versionInfo>
    <rdfs:comment rdf:datatype="&xsd;string">Collects Disase entity knowledge.</rdfs:comment>
    <dc:description rdf:datatype="&xsd;string">Collects Disease entity knowledge.</dc:description>
    <dc:title rdf:datatype="&xsd;string">Disease</dc:title>
    <dc:creator rdf:datatype="&xsd;string">                        </dc:creator>
    <coeus:isEntityOf rdf:resource="&diseasecard;concept_OMIM"/>
    <coeus:isEntityOf rdf:resource="&diseasecard;concept_orphanet"/>
    <coeus:isIncludedIn rdf:resource="&diseasecard;seed_Diseasecard4"/>
</owl:NamedIndividual>
```

Fig. 3. Definition of the Entity Disease in the DiseaseCard Ontology.

Figure 3 shows the creation of an entity called `entity_Disease`, which defines the `coeus:isIncludedIn` property and some other properties related to the entity's creation context. However, the use of `coeus:isEntityOf` property is relevant for this context since it establishes the relation between entities and concepts. This property defines the creation of an entity that includes the definition of the property `coeus:isEntityOf`. This creation of concepts implies the definition of the property `coeus:hasEntity`. This reasoning is also used to establish relationships between concepts and resources (concepts are defined in the property `coeus:hasResource` and resources in the property `coeus:isResourceOf`).

In the definition of a resource, there are an extra set of properties that allows settling some properties, namely defining how the data extraction should be done. These properties are `coeus:method`, `coeus:endpoint`, `dc:publisher` and `dc:query`. The value of these properties is analysed internally and different operations are performed according to their value. The property that establishes the relation between entries of different resources is `coeus:isAssociatedTo`. This property is internally added during the data extraction process, allowing the creation of triples that are then used by DiseaseCard.

3.4 Data Readers

The implemented data readers aim to allow DiseaseCard to be able of supporting more data formats. This required adapting the existing ontology to include a new concept called "Parser". This concept details how to extract information from a data source, which is associated to a resource using the `dc:hasParser` and `dc:isParserOf` properties. Each parser has a set of extensible properties depending on the data format. Based on the available data sources, the need arose to focus development on the implementation of data readers that support data organised in CSV and XML formats.

CSV Parser. One of the most common formats among the available data sources is the CSV format. In this type of format, the data is organised in a very straightforward structure, as the data itself is delimited by commas. For this same reason, the process of extracting data in this type of format tends to be simpler than extracting from other existing formats.

The process of defining a *Parser* for this format starts by identifying the columns[8] that define the resource identifier and its relation with other entities. The column number of these identifiers is defined in the ontology using the `dc:resourceID` and `dc:externalResourceID` properties, respectively. In addition to defining the number of these columns, additional data extraction rules can be applied. These roles are established using regular expressions through the `dc:resourceRegex` and `dc:externalResourceRegex` properties. The first property refers to the resource identifier, and the second to the extended entity identifier.

XML Parser. The XML Parser was developed to be capable of handling more complex data. This node was defined in the ontology using the `dc:mainNode` property. The attribute where the information relates to the resource identifier is defined using the property `dc:resourceID`. In case the document to be analysed has information related to more than one identifier, it is necessary to define the `dc:uniqueResource` property.

Unlike resource identifiers which are unique, for each resource identifier, there can be more than one external identifier. In case there of multiple identifiers, it is necessary to define the `dc:uniqueExternalResource` property. Regardless of the number of identifiers of the entity that is being extended, the reader can extract the information from each node defined using the `dc:externalResourceNode` property. The name of the attribute where the identifier is defined can be defined using the `dc:externalResourceID` property. In case the identifier is not unique, the reader searches within the specified object using `dc:externalResourceNode` for multiple attributes with the name defined using the `dc:externalResourceID` property.

OMIM Parser. OMIM Parser allows the extraction and processing of all OMIM information. Each type of information has a specific property that allows specifying the number of the column where it is located. The name of the properties corresponds to the type of information that is associated with it using the underscore case.

4 Results

DiseaseCard's knowledge base comprises more than 2 million triples, that have been built from OMIM's maps combined with the rare genetic diseases network.

---

[8] In this context, a column corresponds to the position in which the value is found, with each comma incrementing the column's value. The column count starts at zero, so the first column corresponds to column zero.

Its engine indexes these triples, establishing more than 500 thousand connections with more than 100 thousand unique resources. On average, each unique resource is present in 5 single disease networks. Each rare disorder has on average 24 connections to external resources. OMIM's disease resources represent the largest proportion of individuals with more than 10 thousand HGNC entries and 20 thousand OMIM entries. Interestingly, MeSH and ICD terms are the least represented concepts in Diseasecard's knowledge base. All of this data is stored in Diseasecard's semantic knowledge base, which is accessible to end-users through an innovative in-context web workspace and to developers through an advanced semantic interoperability layer.

Besides these numbers, one of the results of this work was a new and fully operational platform for sharing information about rare diseases. One of the resulting features was the browser engine, in which consists of mapping all the diseases of the tool in an organized way. This allows users to deterministically explore all diseases, as this list includes information other than the disease name. The search engine module provides a visual interface to query all the data existing in DiseaseCard database. The main purpose of this mechanism is to allow the user to find a disease based on the name, OMIM ID or the name/ID of a related external endpoint. As the user enters any input into the search box, a list of auto-populated entries is suggested. The information that supports both of these engines is stored in a Redis database, to ensure that the system is responsible when processing queries.

The Disease Page aggregates all the information existing in the system about the selected disease, as shown in Fig. 4. This page can be divided into three different sections. The first consists of a fixed panel that contains a list of all external entries associated with the selected disease. The second consists of two possible graph views of these same endpoints, organised according to the data sources from which they were extracted. These graphs are interactive and allow the user to easily navigate through the high level of external endpoints that may be associated with the disease under analysis. Lastly, the third section is the space reserved for mirroring the external endpoint to the user, so that the user does not have to be redirected outside the DiseaseCard to be able to view the data presented in other data sources.

5   Discussion

One of the biggest challenges in the workflow behind rare disease data agglomeration platforms is the fact that the information is scattered across multiple different sources. In the process of collecting information from these sources, it is necessary to contemplate several aspects, such as which data source is most suitable for the type of data sought; what are the existing methods for obtaining them; and how to better organise them considering filtering methods.

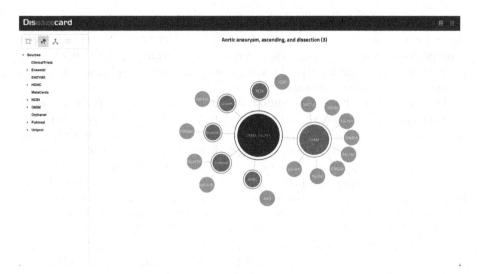

Fig. 4. DiseaseCard interface for a disease with all nodes associated.

The choice of which databases to use should consider whether or not there are methods available to access them, since some of the data sources are protected and require access permissions. Those that allow retrieving their information, usually provide the option to select specific target sets, allowing the automatic disposal of data that is not needed in a given use case. This option reduces the amount of data to fetch, which is important because as a general rule, these sources have high dimensions. However, when this option is not provided, it is important to clean the data before applying any type of internal processing of the system, because the smaller the amount of data to be processed, the less computational effort is necessary. Part of the cleaning process should focus on eliminating data redundancy between different sources, a recurring problem on some of the platforms previously analysed.

Another relevant aspect to consider in the development of this type of platform is the strategy for data source management, whether in terms of adding, removing or updating them. This process must be as easy and accessible as possible. This is one of DiseaseCard's current problems, as there is no easy way to manage its sources. The DiseaseCard platform depends on the URL of endpoints from external sources to be able to display them, so when they change it is not possible to present this data. To avoid this problem, a new component was developed, responsible for checking the endpoint associated with each resource, and alerting the team when they are not valid.

According to the current DiseaseCard architecture, it is necessary to interconnect the existing sources, associating them with OMIM entries. This allows the user to analyse the existing relations when consulting the data presented, which may lead to a better understanding of the concepts. This process is currently done by combining the OMIM IDs presented in its database and the

OMIM IDs existing in the HGNC database. This allows extracting the rest of the current associations through relations between internal IDs, genes or even through chromosomal locations. However, identifying valid relations between the different sources is not a simple task, as it sometimes requires knowledge about them and the data they contain.

DiseaseCard has undergone significant changes and evolved from its initial implementation to a modern and robust system over the last 20 years. Initially, the system was built using JavaServer Pages, which served as the primary technology for generating dynamic web pages. However, in the recent version, we implemented ReactJS, a modern front-end library that provided an enhanced user interface and improved performance. Alongside this, we introduced Spring Boot Framework, a powerful back-end framework that allowed for better management and scalability of the system. This evolution has resulted in a more robust and efficient system that is better suited to meet the needs of end-users and researchers in the medical domain. The choice of the set of tools to be used was strongly dependent on the type of technological advancements existing at the time of the start of the project. As new technologies emerge, in order to improve the development of new systems, it may make sense to technologically update an older system.

Besides these new updates, some lessons were learned during the process. In the beginning, DiseaseCard was developed as a collaborative application where a group of experts could collaborate online in order to share and disseminate their knowledge about genetic rare diseases [35]. At this stage, sharing data was the main goal and the technologies for web platforms were limited. Later, a new version was proposed, that already integrated web semantic [11]. This new concept supports the integration of distinct data sources following a semi-automatic strategy in a semantic web environment. With this evolution, the tool can expose the aggregated semantic rare diseases network through an advanced API. Although this version was a great evolution for the aggregation of rare diseases, at this time, concepts like FAIR were not discussed, and neither the data privacy was so concerning as today. This new version addressed some of the challenges that emerged and it could benefit from new frameworks for web development. Biomedical data sharing is still challenging for many reasons, and exposing the metadata of data sets can help researchers to know more about diseases, as well as, to better identify data sources of interest for further analysis.

6   Conclusions

Rare diseases are a significant global health concern, that affects millions of individuals worldwide. Studying then is still challenging due to the lack of individuals compliant with study protocols. To address these challenges, multiple organizations invested in sharing data and resources, resulting in several platforms focused on aggregating information. Despite the benefits of these solutions, the evolution of data regulations has led to new challenges that may not be fully addressed in such platforms. Therefore, we have studied these platforms to understand how DiseaseCard can be improved to better support researchers.

The proposed version of DiseaseCard can expose the aggregated semantic rare diseases network through a more automatic strategy, with a semantic web environment. This new version addressed some of the challenges that emerged over the years, such as the need to follow FAIR principles and ensure data privacy. As result, we recognise that biomedical data sharing is still challenging for many reasons, and exposing the data or metadata about data sets can help researchers to know more about diseases. In conclusion, DiseaseCard is a valuable tool for discovering information about genetic rare diseases.

Acknowledgements. This work has received funding from the EC under grant agreement 101081813, Genomic Data Infrastructure. J.R.A is funded by the FCT - Foundation for Science and Technology (national funds) under the grant SFRH/BD/147837/2019.

References

1. Schaaf, J., Sedlmayr, M., Schaefer, J., Storf, H.: Diagnosis of rare diseases: a scoping review of clinical decision support systems. Orphanet J. Rare Dis. **15**(1), 1–14 (2020)
2. Ferreira, C.R.: The burden of rare diseases. Am. J. Med. Genet. A **179**(6), 885–892 (2019)
3. Bick, D., Jones, M., Taylor, S.L., Taft, R.J., Belmont, J.: Case for genome sequencing in infants and children with rare, undiagnosed or genetic diseases. J. Med. Genet. **56**(12), 783–791 (2019)
4. Stoller, J.K.: The challenge of rare diseases. Chest **153**, 1309–1314 (2018)
5. Griggs, R.C., et al.: Clinical research for rare disease: opportunities, challenges, and solutions. Mol. Genet. Metab. **96**, 20–26 (2009)
6. Roessler, H.I., Knoers, N.V., van Haelst, M.M., van Haaften, G.: Drug repurposing for rare diseases. Trends Pharmacol. Sci. **42**, 255–267 (2021)
7. Almeida, J.R., Silva, L.B., Bos, I., Visser, P.J., Oliveira, J.L.: A methodology for cohort harmonisation in multicentre clinical research. Inform. Med. Unlocked **27**, 100760 (2021)
8. Bos, I., et al.: The EMIF-AD multimodal biomarker discovery study: design, methods and cohort characteristics. Alzheimer's Res. Ther. **10**(1), 64 (2018)
9. Lochmüller, H., et al.: The international rare diseases research consortium: policies and guidelines to maximize impact. Eur. J. Hum. Genet. **25**(12), 1293–1302 (2017)
10. Moliner, A.M., Waligora, J.: The European union policy in the field of rare diseases. In: Posada de la Paz, M., Taruscio, D., Groft, S.C. (eds.) Rare Diseases Epidemiology: Update and Overview. AEMB, vol. 1031, pp. 561–587. Springer, Cham (2017). https://doi.org/10.1007/978-3-319-67144-4_30
11. Lopes, P., Oliveira, J.L.: An innovative portal for rare genetic diseases research: the semantic diseasecard. J. Biomed. Inform. **46**, 1108–1115 (2013)
12. Litzkendorf, S., et al.: Use and importance of different information sources among patients with rare diseases and their relatives over time: a qualitative study. BMC Public Health **20**, 1–14 (2020)
13. Maiella, S., Rath, A., Angin, C., Mousson, F., Kremp, O.: Orphanet and its consortium: where to find expert-validated information on rare diseases. Rev. Neurologique **169**, S3-8 (2013)

14. Weinreich, S.S., Mangon, R., Sikkens, J., Teeuw, M.E., Cornel, M.: Orphanet: a European database for rare diseases. Ned. Tijd. Geneeskd. **152**(9), 518–519 (2008)
15. Harrison, J.E., Weber, S., Jakob, R., Chute, C.G.: ICD-11: an international classification of diseases for the twenty-first century. BMC Med. Inform. Decis. Mak. **21**(6), 1–10 (2021)
16. Amberger, J.S., Bocchini, C.A., Schiettecatte, F., Scott, A.F., Hamosh, A.: OMIM. org: online mendelian inheritance in man (OMIM®), an online catalog of human genes and genetic disorders. Nucleic Acids Res. **43**(D1), D789–D798 (2015)
17. Bodenreider, O.: The unified medical language system (UMLS): integrating biomedical terminology. Nucleic Acids Res. **32**(suppl_1), D267–D270 (2004)
18. Lipscomb, C.E.: Medical subject headings (MeSH). Bull. Med. Libr. Assoc. **88**(3), 265 (2000)
19. Mozzicato, P.: MedDRA: an overview of the medical dictionary for regulatory activities. Pharm. Med. **23**, 65–75 (2009)
20. Povey, S., Lovering, R., Bruford, E., Wright, M., Lush, M., Wain, H.: The HUGO gene nomenclature committee (HGNC). Hum. Genet. **109**, 678–680 (2001)
21. Boutet, E., et al.: UniProtKB/Swiss-Prot, the manually annotated section of the UniProt knowledgebase: how to use the entry view. In: Edwards, D. (ed.) Plant Bioinformatics. MMB, vol. 1374, pp. 23–54. Springer, New York (2016). https:// doi.org/10.1007/978-1-4939-3167-5_2
22. Howe, K.L., et al.: Ensembl 2021. Nucleic Acids Res. **49**(D1), D884–D891 (2021)
23. Jassal, B., et al.: The reactome pathway knowledgebase. Nucleic Acids Res. **48**(D1), D498–D503 (2020)
24. Harding, S.D., et al.: The IUPHAR guide to immunopharmacology: connecting immunology and pharmacology. Immunology **160**(1), 10–23 (2020)
25. Espe, S.: MalaCards: the human disease database. J. Med. Libr. Assoc. JMLA **106**(1), 140 (2018)
26. Kibbe, W.A., et al.: Disease ontology 2015 update: an expanded and updated database of Human diseases for linking biomedical knowledge through disease data. Nucleic Acids Res. **43**, D1071–D1078 (2015)
27. Sioutos, N., de Coronado, S., Haber, M.W., Hartel, F.W., Shaiu, W.L., Wright, L.W.: NCI thesaurus: a semantic model integrating cancer-related clinical and molecular information. J. Biomed. Inform. **40**(1), 30–43 (2007)
28. Donnelly, K., et al.: SNOMED-CT: the advanced terminology and coding system for eHealth. Stud. Health Technol. Inform. **121**, 279 (2006)
29. Phillips, K.: Alone we are rare, together we are strong: a review of the national organization for rare disorders (NORD®). J. Cons. Health Internet **26**(4), 444–451 (2022)
30. Kosakovsky Pond, S.L., Posada, D., Gravenor, M.B., Woelk, C.H., Frost, S.D.: GARD: a genetic algorithm for recombination detection. Bioinformatics **22**(24), 3096–3098 (2006)
31. Lewis, J., Snyder, M., Hyatt-Knorr, H.: Marking 15 years of the genetic and rare diseases information center. Transl. Sci. Rare Dis. **2**(1–2), 77–88 (2017)
32. Zhu, Q., Nguyen, D.T., Grishagin, I., Southall, N., Sid, E., Pariser, A.: An integrative knowledge graph for rare diseases, derived from the genetic and rare diseases information center (GARD). J. Biomed. Semant. **11**, 13 (2020)
33. Schoch, C.L., et al.: NCBI Taxonomy: a comprehensive update on curation, resources and tools. Database **2020** (2020)
34. Lopes, P., Oliveira, J.L.: COEUS: "semantic web in a box" for biomedical applications. J. Biomed. Semant. **3**, 1–19 (2012)

35. Oliveira, J.L., et al.: DiseaseCard: a web-based tool for the collaborative integration of genetic and medical information. In: Barreiro, J.M., Martín-Sánchez, F., Maojo, V., Sanz, F. (eds.) ISBMDA 2004. LNCS, vol. 3337, pp. 409–417. Springer, Heidelberg (2004). https://doi.org/10.1007/978-3-540-30547-7_41

36. Wilkinson, M.D., et al.: The FAIR guiding principles for scientific data management and stewardship. Sci. Data **3**(1), 1–9 (2016)

37. Sequeira, M., Almeida, J.R., Oliveira, J.L.: A comparative analysis of data platforms for rare diseases. In: 2021 IEEE 34th International Symposium on Computer-Based Medical Systems (CBMS), pp. 366–371 IEEE (2021)

38. Sernadela, P., González-Castro, L., Oliveira, J.L.: SCALEUS: semantic web services integration for biomedical applications. J. Med. Syst. **41**(4), 54 (2017). https://doi.org/10.1007/s10916-017-0705-8

39. Munir, K., Sheraz Anjum, M.: The use of ontologies for effective knowledge modelling and information retrieval. Appl. Comput. Inform. **14**(2), 116–126 (2018)

Unsupervised Investigation of Information Captured in Pathway Activity Score in scRNA-Seq Analysis

Kamila Szumala, Joanna Polanska[ID], and Joanna Zyla[(✉)][ID]

Department of Data Science and Engineering, Silesian University of Technology, Akademicka 16, 44-100 Gliwice, Poland
kamiszu970@student.polsl.pl, {joanna.polanska, joanna.zyla}@polsl.pl

Abstract. With the introduction of single cell RNA sequencing, research on cell, tissue and disease heterogeneity has a new boost. Transforming gene levels to explainable pathways via single-sample enrichment algorithms is a leading analysis step in understanding cell heterogeneity. In this study, eight different single-sample methods were investigated and accompanied by gene level outcomes as reference. For all, their ability to cell separation and clustering accuracy was tested. For this purpose, six scRNA-Seq datasets with labelled cells and their various numbers were collected. PLAGE method shows the best cell separation with statistically significant differences to gene level and to six other tested methods. The clustering accuracy analysis also indicates that PLAGE is the leading technique in single-sample enrichment methods in scRNA-Seq. Here the worst performance was observed to JASMINE algorithm which, in contrary to PLAGE, was designed to analyse the scRNA-Seq data. Moreover, Louvain clustering shows the best results regarding cell division regardless of the tested single-sample method. Finally, the results of clustering given by PLAGE reveal T cell subtypes initially not labelled, showing the great potential of this algorithm in heterogeneity investigation.

Keywords: pathway activation score · single-cell RNA sequencing · single-sample algorithms · algorithms effectiveness

1 Introduction

The rapid growth of technology is well observed in molecular biology. Since introducing microarrays, fast measuring thousands of genes in different cells/samples is possible. Single-cell RNA sequencing (scRNA-Seq) is a new powerful technology to study the transcriptomic profile of disease and normal tissues [1–3]. This technique allows for revealing heterogeneity of cellular populations at high resolution. With the development of scRNA-Seq, new bioinformatical tools were developed for identifying marker genes, visualising high dimensions and clustering cells [4, 5]. Despite it, biological interpretation of the clustering results and cell heterogeneity exploration remains a big challenge.

© The Author(s), under exclusive license to Springer Nature Switzerland AG 2023
I. Rojas et al. (Eds.): IWBBIO 2023, LNBI 13920, pp. 180–192, 2023.
https://doi.org/10.1007/978-3-031-34960-7_13

One widely used technique to encode transcriptional heterogeneity and classify cell subtypes is transforming gene level signal to explainable pathway/gene sets level. For this purpose, the Gene Set Analysis (GSA) techniques developed in the microarrays era are used. In a group of GSA methods, the single-sample enrichment approaches are used for cell heterogeneity investigation. The general concept of those methods is to transform transcripts expression from a particular pathway (e.g. apoptosis) and one cell into one representative value. This process is then repeated to each cell/sample and further across all investigated pathways. As a result, the pathway activity score (PAS) matrix is generated. Several algorithms in this group were designed for microarrays or bulk RNA-Seq analysis and are commonly used in scRNA-Seq (e.g. GSVA [6]). In [7–9], authors show that applying GSA algorithms designed for microarray analysis into bulk RNA-Seq does not impact their effectiveness. Nonetheless, in scRNA-Seq, data sparsity and normalisation impact are much larger than in other high-throughput molecular biology techniques. Thus new single-sample pathway enrichment approaches are proposed for scRNA-Seq investigation (e.g. Padoga2 [9]).

In this work, eight different single-sample enrichment algorithms were tested. Three of those methods were designed for scRNA-Seq, while the rest were previously proposed for microarrays analysis. For those methods, two characteristics were investigated: (i) cell type separation measured Silhouette index and (ii) clustering accuracy represented by normalised mutual information. Both represent the ability of tested algorithms to detect cell heterogeneity. Finally, obtained results were accompanied by biological interpretation.

2   Materials and Methods

2.1   Data Acquisition and Pre-processing

The study collected six publicly available scRNA-Seq datasets from different human organs and phenotypes. The datasets were selected to contain various cells (samples) with assigned cell types. All experiments were performed on Chromium 10x platform, and only originally labelled cells were extracted. Transcripts with zero and low variance expressions were filtered out in each dataset using Gaussian-mixture model clustering [10]. The same procedure was repeated for the cells, discarding those marked by fewer than 2,500 genes. Genes without annotation or whose Ensembl ID was shown to be duplicated (higher variance kept) were also removed. Finally, transcript expressions were log normalised using the Seurat package [11]. The first and smallest dataset comes from the breast cancer (BC) patient [12] and is available under accession number GSE75367. It comprises 244 cells and 16,639 genes labelled into five cell types representing various breast cancer types (Normal tissue, Luminal A, Luminal B, TNBC and HER2+). Next dataset comes from peripheral blood mononuclear cells (PBMC) containing 3,222 cells representing the nine labelled types and 15,817 genes. Data represent experiment 1 A from [13] and are stored in the Single Cell Portal of Broad Institute. The Bone marrow (BM) dataset consists of 4,390 cells and 18,875 genes, marked into 14 types. The dataset was presented in [14] and is stored under accession number E-MTAB-9067. Further, a dataset of COVID patients' blood consisting of 4,903 cells identified as six different types with 15,390 measured genes was collected. The dataset was published in [15] and is available under number E-MTAB-9221. Another experiment was taken from the HumanMousePancreas project and contained 12,400 cells (including 7,660 from the human pancreas, represented by ten distinct cell types) and 16,075 measured genes [16]. The largest dataset, consisting of 15,000 liver cells, in which the expression of 15,755 genes was measured, and 13 different groups of cells were marked. The dataset was published in [17] and is available under the number E-MTAB-10553.

Next, the gene sets/pathways collection was extracted from publicly available databases. For each dataset, separate gene sets containing each organ's characteristics, cell signatures and biological processes were filtered. First, the KEGG database [18] provided signalling pathways from the "Organismal Systems; Immune system" category for all immune datasets. Next, cell specific pathways were extracted from C8, C5 and C2 categories (cell type signature gene sets, ontology gene sets and curated gene sets, respectively) from MSigDB [19]. Next, specific for single-cell experiments, cell signatures were downloaded from the Cell Marker database [20], CYBERSORT [21] and PanglaoDB [22] - the latter two using the *tmod* R package [23]. Moreover, gene sets corresponding to each immunological cell type in each dataset were selected from the Chaussabel et al. [24] (DC) and Li et al. [25] (LI) collections, also available in the *tmod* R package [23]. The gene sets/pathways collection gathered to each dataset is available upon request. A quantitative summary of the sampled gene sets for each dataset is summarised in Table 1. Finally, a summary of an analysed dataset and assigned gene sets is presented in Table 2.

Table 1. Summary of extracted gene sets form each database to analysed datasets.

Database	Bone Marrow	Breast Cancer	COVID	Liver	Pancreas	PBMC
CellMarker	6	2	14	6	5	14
CIBERSORT	27	0	33	45	3	42
KEGG	29	27	29	29	7	29
MSigDB	22	18	12	34	80	12
PanglaoDB	15	0	6	13	9	7
DC	2	0	1	2	0	3
LI	36	0	40	32	0	45
Total	137	47	135	161	104	152

2.2   Tested Pathway Activity Transformation Algorithms

For the collected dataset and respective pathway collection, transformations of gene expression levels to predefined signalling pathways were performed using eight widely known single-sample enrichment algorithms. Below is a brief description of each tested algorithm and its approach to gene-pathway level transformation.

AUCell from the R package of the same name [26] is a dedicated method for scRNA-Seq enrichment analysis. All transcripts in the dataset are ranked in ascending order

Table 2. Overview of all datasets used in study with assigned pathways.

Dataset	# of genes	# of cells	# of cell types	# of pathway
Bone Marrow	18,875	3,918	14	137
Breast Cancer	16,501	232	5	47
COVID	15,390	4,903	6	135
Liver	15,573	14,852	12	161
Pancreas	16,077	7,660	10	104
PBMC	15,817	3,222	9	152

according to their expression level separately for each cell. Genes whit the same expression level are shuffled randomly at the relevant stage. For further analysis, the top 5% of transcripts are taken. The area under the curve (AUC) is calculated for the relation between a number of genes in the gene set under a particular rank. The AUC value is standardised by the maximum possible AUC for the used 5% of transcripts. The AUC score is given to each pathway and cell creating PAS.

Coincident Extreme Ranks in Numerical Observations (CERNO) algorithm [23] was designed to enrichment analysis between phenotypes in RNA microarray and bulk sequencing. However, it can also be readily applied for single-sample analysis of scRNA-Seq. The CERNO method ranks transcript by its expression, from giving the lowest value to the most expressed one. Further, each pathway's specific genes are extracted, and AUC is calculated using the Mann-Whitney approach. Similarly, to the AUCell, the AUC value makes the PAS matrix.

The Jointly Assessing Signature Mean and Inferring Enrichment (JASMINE) algorithm [27] is a rank-based method dedicated to the enrichment analysis of scRNA-Seq data. It starts by creating a ranking of the entire dataset from the smallest to the highest level of transcript expression. Moreover, tied ranks are applied. Then, PAS for a particular cell and pathway is calculated as a mean of the transcript ranks in a particular pathway and dividing them by the number of genes in the whole dataset. Finally, the received matrix is row normalised (for each pathway) to make results comparable between cells.

Vision, like previous methods, is dedicated to scRNA-Seq analysis [28]. At first, transcripts expression is normalised by the z-score approach. Next, the average expression of transcripts in particular pathways and cell is calculated. Further, a random set with the exact dimensions is generated as a reference in the calculation for each gene set separately. Finally, the randomly generated pathway's mean value and standard deviation are used to standardise the average transcript expression of the pathway.

The combined z-score [29] approach was initially implemented to analyse RNA microarray data, but because the algorithm analyses each sample separately, it can be used for scRNA-Seq without modification. First, z-score normalisation is performed on the entire transcript matrix, and then the PAS s for each cell j is calculated using Stouffer's integration method [30] as follows:

$$s_j = \frac{\sum_{i=1}^{n} z_i}{\sqrt{n}} \tag{1}$$

where z_i is a standardized count of gene i and n is a number of genes in particular pathway.

The next tested method was Pathway Level Analysis of Gene Expression (PLAGE) [31], created initially for RNA microarray analysis. The base of this method is the singular value decomposition (SVD) which is performed on the expression level matrix of genes belonging to a particular pathway. Here Principal Component Analysis (PCA) was performed for genes in pathways, and their first component corresponding to the highest total variation was taken, which is equivalent to SVD.

The penultimate algorithm was Gene Set Variation Analysis (GSVA) [6], designed for microarrays and bulk RNA sequencing but commonly used in scRNA-Seq. The transcript expression matrix at the first step is standardised using a discrete Poisson kernel. Next, genes are ranked by their expression in descending order and symmetrised around zero as follows:

$$r_{sym} = \left| \frac{N}{2} - r \right| \tag{2}$$

where N is the total number of analysed genes and r is the rank of transcript in a particular cell. Further, within each cell and particular pathway, a commonly known GSEA algorithm [8] is applied. Finally, the enrichment score is calculated as the difference in empirical cumulative distribution functions of transcript expression ranks inside and outside the gene set (modified Smirnov-Kolmogorov statistic).

Finally, the single-sample Gene Set Enrichment Analysis (ssGSEA) method was tested [32]. This algorithm is equivalent of GSVA without discrete Poisson kernel standardisation and with the application of simply tied ranks to transcript expression translation.

Both GSVA and ssGSEA were run from the *GSVA* R package (version 1.46.0). AUCell was run from the R package of the same name (version 1.20.2). All other methods are self-implemented due to their simplicity. PAS matrices were obtained for each algorithm and each dataset (8 algorithms x 6 datasets). Moreover, the top 20% of transcripts with the highest variance in each dataset were extracted and will be treated as reference PAS to single-sample enrichment algorithms in the testing process.

2.3 Algorithm's Evaluation

The single-sample enrichment methods and their PASs were tested regarding several characteristics important in scRNA-Seq analysis and results interpretation.

The first characteristic was the Silhouette Index (SI), which was calculated separately for each cell using the original labels and the post-transformation matrix (PAS). Mean SI values within the each cell type (cluster) were then taken to obtain a comparison between methods. This procedure was repeated for each dataset and cell type within it. SI will represent the separation between defined cell types based obtained by PASs. Briefly, it estimates the biological information consistency of PASs given by various single-sample algorithms. In [37], authors present similar approach, however, SI was measured for tSNE and UMAP transformation on PAS. Such approach does not refer directly to PAS algorithms itself and both dimensionality reductions does not always preserved correlations.

Second, tested characteristic was dedicated to measure clustering accuracy based on PASs matrices. Four different clustering algorithms were applied to the PASs obtained on tested single-sample algorithms and datasets. Clustering methods include state-of-the-art techniques, i.e. k-means [34] and hierarchical clustering (Euclidian distance and Ward agglomeration) [35]. For both, the optimal number of clusters was selected by Within-Cluster-Sum of Squared Error metric (WSS). K-means and hierarchical clustering algorithms were run using *factoextra* R package version 1.0.7. The following two methods are based on graphs and k-nearest neighbours. The Louvain method for community detection [36], commonly applied in scRNA-Seq analysis, and PARC method [37] were used. In both methods, the *a priori* number of clusters is not a required parameter and is fitted by algorithms based on graph separation. Based on the obtained results from different clustering techniques, the normalized mutual information (NMI) was calculated based on obtained clusters and original cell types in each dataset. This process was performed on all tested PASs from single-sample enrichment approaches.

The ANOVA was performed to check the differences in the performance of SI and NMI between tested PASs results. To assess significant differences between pairs of methods, the Tuckey *post-hoc* was applied. In both ANOVA and Tuckey tests, the significance level was set to $\alpha = 0.05$. Moreover, comparisons were made between PAS methods as well as between clustering approaches. Finally, the biological interpretation of obtained results from clustering was performed to justify the level of information captured in PASs matrices.

3   Results

3.1   Cell Type Separation

At first, for generated PASs by single-sample methods applied on all tested datasets, the Silhouette index was calculated. Next, the mean SI value within cell types was extracted. Moreover, the same procedure was performed on the top 20% of transcripts with the highest variance as a reference on the gene level. Using the ANOVA test, the differences in SI between tested methods are observed (p <0.0001). Results given by SI are presented in Fig. 1.

As shown in Fig. 1, the poorest cell separation is observed when the gene level is analysed (median SI = 0.03). Using single-sample algorithms and their PASs improve this characteristic. Out of the tested methods, the PLAGE algorithm shows a statistically significant difference compared to the majority of tested methods. Similar results were presented in [33], where cell separation was calculated on UMAP space generated on PASs. Here, we prove those findings but based only on PAS space. ssGSEA and z-score approach statistically does not differ from PLAGE, however, this is also true to other methods. Nevertheless, both show higher median SI values than the other. Surprisingly, methods dedicated to the scRNA-Seq analysis like AUCell, JASMINE and Vision have poor results in terms of cell separation compared to PLAGE, ssGSEA or simple z-score integration. GSVA algorithm is commonly used in scRNA-Seq research due to its discrete Poisson kernel transformation, here with median SI = 0.06 being the worst outcome in single-sample algorithms. Its poor cell separation results were also shown in [33]. Finally, it could be observed that SI vary across tested datasets. The best cell

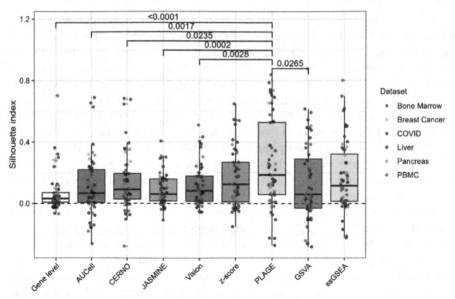

Fig. 1. The distribution of the Silhouette index for each tested PAS. Dots represent cell types in a particular dataset. P-values are the result of the Tuckey post-hoc test for ANOVA.

separation was obtained for the COVID dataset where for PLAGE algorithm median SI = 0.72 (as reference gene level median SI = 0.18). In contrary, the bone marrow dataset of similar cell number and immunological cell labels as COVID gives the worst results.

3.2 Clustering Accuracy

The second testing characteristic was clustering accuracy. Various clustering methods were applied for each single-sample method PASs of the analysed dataset. Next, cluster labels were compared to the original cell types, and NMI was calculated. Results across different single-sample enrichment methods are presented in Fig. 2 panel A, where the ANOVA testing shows significant differences between the tested methods ($p < 0.0001$).

As can be observed in Fig. 2. Panel A, clustering accuracy is the highest for gene level information and PLAGE (median NMI equals 0.68 and 0.69, respectively). Both gene level and PLAGE obtained statistically significantly better results than JASMINE and Vision. Moreover, PLAGE has better results than GSVA. Those results show that the PLAGE algorithm has great potential in scRNA-Seq analysis on both cluster accuracy and cell separation (Fig. 1). Poor results of Vision were caused due to PARC clustering. Figure 2 panel C shows that regardless of the dataset, NMI for PARC clustering drops for the Vision approach. When this clustering was removed, statistical significance between methods was only observed between pair PLAGE and JASMINE ($p = 0.0052$) and pair gene level and JASMINE ($p = 0.0096$). Moreover, based on Fig. 2 panel C, we can observe that Louvain community detection clustering has the most stable results regardless of the tested dataset. Thus, testing for reliable clustering for PASs was investigated. For this purpose, NMI was tested within clustering methods regardless

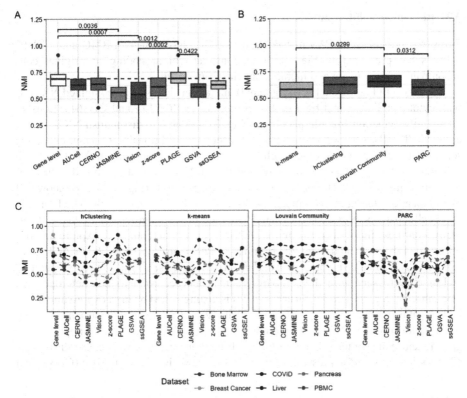

Fig. 2. Results of clustering accuracy expressed as NMI index. Panel A shows distribution of NMI for tested PASs regardless of clustering approaches. Dashed black line represent median value for gene level clustering result. Panel B represents NMI distribution regardless of PASs but on different clustering methods. Panel C shows detailed NMI results for applied clustering and PAS methods with distinction of datasets. P-values are result of Tuckey *post-hoc* test for ANOVA.

of single-sample enrichment approaches. Results of this exploration are presented in Fig. 2 panel B. ANOVA testing between clustering methods shows statistically significant differences with p = 0.0103. The Tuckey post-hoc analysis reveals that Louvain clustering shows better results compared to k-means and PARC methods. Out of other techniques used, hierarchical clustering can be distinguished. The results of hierarchical clustering and the Louvain community detection method on the PLAGE algorithm were further investigated by given biological implications.

3.3 Biological Validation

The tSNE projection of PAS matrix from PLAGE algorithm for PBMC and COVID datasets are presented in Fig. 3. Projections has colour coding of original cells accompanied by Louvain and hierarchical clustering results.

As shown in Fig. 3 panel A (right panel), hierarchical clustering combined NK cells with cytotoxic T cells and both groups of monocytes (CD14+ and CD16+). Louvain community detection (middle panel A of Fig. 2) was able to distinguish NK from the cytotoxic T cells, yet monocytes were still grouped. Additionally, Louvain clustering separates the group of cytotoxic T cells. Such an outcome may show that not all T cell subgroups [38] were labelled initially, and the PLAGE method combined with Louvain clustering may detect them. The COVID dataset results (Fig. 3 panel B) show that hierarchical clustering merged platelet cells with erythroid lineage cells. The Lou-vain approach separates those cell types and detects additional subgroups of T cells, and cor-rectly identifies other cell types, similar to PBMC dataset outcomes. To explore findings observed in the COVID dataset, the heatmap of PAS matrix given by the PLAGE algo-rithm was generated and is presented in Fig. 4. Louvain Clustering al-lowed to separate T cell subtypes by separating them into different clusters according to their heterogeneity. The subgroup represented by cluster 7 ("c7") is distinguished by high expression of sig-nalling pathways characteristic of Gamma Delta ($\gamma\delta$) T Cells. This indicates the presence of a rare form of the TCR receptor complex (responsible for binding to antigens) with gamma and delta protein chains in their cell membrane. Moreover, cells constituting c7 show low levels of expression of pathways characteristic of Cytotoxic (CD8+) and

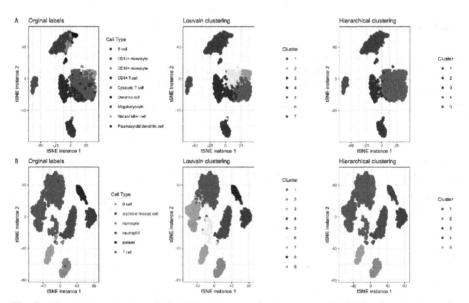

Fig. 3. The tSNE projection of PLAGE pathway activation score with labels given by Louvain and hierarchical clustering accompanied by original cell labels. Panel A show results for PBMC dataset. Panel B show results for COVID dataset.

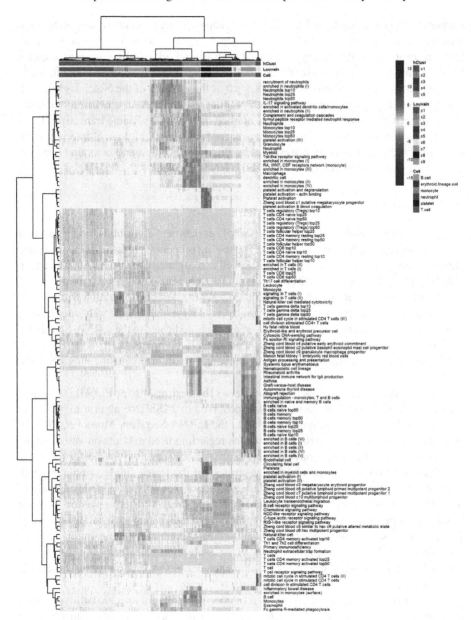

Fig. 4. The heatmap with hierarchical clustering dendrogram of PLAGE pathway activation score matrix on COVID dataset. Column colours represents original cells label, Louvain and hierarchical clustering.

Helper (CD4+) T Cells for which the TCR receptor usually occurs as a protein complex with alpha and beta (αβ) chains. Cells rep-resenting cluster 3 ("c3") present high levels of expression of pathways typical of both γδ and αβ T Cells, This may indicate that in some T cells, genes specific for both receptor types may be expressed, while their phenotype depends on other factors, such as the strength of the Notch signal. [39]. Cluster 6 (c6) represents a subset of T cells similar to B cells at the transcriptomic level. Moreover, when Louvain clustering was applied on top 20% variable transcripts T cell cluster was also divided but into mixed groups of not shared biological interpretation. In summary observed subgroups revealed by PLAGE algorithm and Louvain clustering enrich biological interpretation of analysed scRNA-Seq data.

4   Conclusions

The comparison of eight different single-sample enrichment algorithms, widely used or designed for scRNA-Seq was presented. The testing process include cell separation and clustering accuracy which are key points in cell heterogeneity investigation. Moreover, in testing process datasets of different cell number (dimensionality) were used.

Out of tested algorithms, PLAGE method shows the best cell separation with statistically significant differences to gene level and to six other tested methods. Those outcomes support the one presented in [37]. Moreover, methods dedicated to the scRNA-Seq analysis like AUCell, JASMINE and Vision have poor results compared to PLAGE. The worst cell separation is obtained on gene level showing importance of using single-sample methods.

The clustering accuracy analysis also shows the best results for PLAGE out tested methods. Here the worst performance was observed to JASMINE algorithm, which in contrary to PLAGE, was designed to analyse the scRNA-Seq data. Moreover, PLAGE algorithm sows the lowest variation of results regarding used datasets on various clustering techniques. Next, the Louvain community detection method for clustering shows the best results regarding cell division regardless of the tested single-sample method. Thus, this technique is suggested for cluster detection not only on gene level analysis of scRNA-Seq but also on pathway activation score level. Clusters given by Louvain technique on PLAGE algorithm identify several T cell subgroups not labelled originally. Those clusters show reasonable biology and indicate larger cell heterogeneity than initially observed. In summary, in this work PLAGE algorithm has the best results out of tested methods. The algorithm itself based on standard PCA, and outperform methods relied on gene ranking like JASMINE.

Unfortunately, there is still a lack of reference gene sets for specific cell types from different tissues. Therefore, their selection has to be done manually using different databases and own biological knowledge. For this reason, this study should also test cases where the manual selection of signalling pathways was omitted and a much larger number of pathways were included in the calculations without taking into account the presence of potential cell types. Furthermore, investigation on larges collection of dataset and more complex testing process is required to confirm those finding.

Acknowledgements. This work was financed by Silesian University of Technology grant for maintaining and developing research potential (KM, JP, JZ).

References

1. Tang, F., et al.: mRNA-Seq whole-transcriptome analysis of a single cell. Nat. Methods **6**(5), 377–382 (2009)
2. Method of the Year 2013. Nature Methods **11**(1) (2014)
3. Svensson, V., Vento-Tormo, R., Teichmann, S.A.: Exponential scaling of single-cell RNA-seq in the past decade. Nat. Protoc. **13**(4), 599–604 (2018)
4. Hie, B., Peters, J., Nyquist, S.K., Shalek, A.K., Berger, B., Bryson, B.D.: Computational methods for single-cell RNA sequencing. Annual Review of Biomedical Data Science **3**, 339–364 (2020)
5. Chen, G., Ning, B., Shi, T.: Single-cell RNA-seq technologies and related computational data analysis. Frontiers in Genetics, 317 (2019)
6. Hänzelmann, S., Castelo, R., Guinney, J.: GSVA: gene set variation analysis for microarray and RNA-seq data. BMC Bioinformatics **14**, 1–15 (2013)
7. Zyla, J., Leszczorz, K., Polanska, J.: Robustness of pathway enrichment analysis to transcriptome-wide gene expression platform. In: Panuccio, G., Rocha, M., Fdez-Riverola, F., Mohamad, M.S., Casado-Vara, R. (eds.) PACBB 2020. AISC, vol. 1240, pp. 176–185. Springer, Cham (2021). https://doi.org/10.1007/978-3-030-54568-0_18
8. Geistlinger, L., et al.: Toward a gold standard for benchmarking gene set enrichment analysis. Brief. Bioinform. **22**(1), 545–556 (2021)
9. Lake, B.B., et al.: Integrative single-cell analysis of transcriptional and epigenetic states in the human adult brain. Nat. Biotechnol. **36**(1), 70–80 (2018)
10. Mrukwa, A., Marczyk, M., Zyla, J.: Finding significantly enriched cells in single-cell RNA sequencing by single-sample approaches. In: Bioinformatics and Biomedical Engineering: 9th International Work-Conference, IWBBIO 2022, Maspalomas, Gran Canaria, Spain, June 27–30, 2022, Proceedings, Part II, pp. 33–44. Springer International Publishing, Cham (2022)
11. Hao, Y., et al.: Integrated analysis of multimodal single-cell data. Cell **184**(13), 3573–3587 (2021)
12. Jordan, N.V., et al.: HER2 expression identifies dynamic functional states within circulating breast cancer cells. Nature **537**(7618), 102–106 (2016)
13. Ding, J., et al.: Systematic comparison of single-cell and single-nucleus RNA-sequencing methods. Nat. Biotechnol. **38**(6), 737–746 (2020)
14. Ranzoni, A.M., et al.: Integrative single-cell RNA-seq and ATAC-seq analysis of human developmental hematopoiesis. Cell Stem Cell **28**(3), 472–487 (2021)
15. Silvin, A., et al.: Elevated calprotectin and abnormal myeloid cell subsets discriminate severe from mild COVID-19. Cell **182**(6), 1401–1418 (2020)
16. Baron, M., et al.: A single-cell transcriptomic map of the human and mouse pancreas reveals inter-and intra-cell population structure. Cell Syst. **3**(4), 346–360 (2016)
17. Wang, Z.Y., et al.: Single-cell and bulk transcriptomics of the liver reveals potential targets of NASH with fibrosis. Sci. Rep. **11**(1), 19396 (2021)
18. Kanehisa, M., Furumichi, M., Tanabe, M., Sato, Y., Morishima, K.: KEGG: new perspectives on genomes, pathways, diseases and drugs. Nucleic Acids Res. **45**(D1), D353–D361 (2017)
19. Liberzon, A., Subramanian, A., Pinchback, R., Thorvaldsdóttir, H., Tamayo, P., Mesirov, J.P.: Molecular signatures database (MSigDB) 3.0. Bioinformatics **27**(12), 1739–1740 (2011)
20. Zhang, X., et al.: Cell Marker: a manually curated resource of cell markers in human and mouse. Nucleic Acids Res. **47**(D1), D721–D728 (2019)
21. Chen, B., Khodadoust, M.S., Liu, C.L., Newman, A.M., Alizadeh, A.A.: Profiling tumor infiltrating immune cells with CIBERSORT. In: von Stechow, L. (ed.) Cancer Systems Biology. MMB, vol. 1711, pp. 243–259. Springer, New York (2018). https://doi.org/10.1007/978-1-4939-7493-1_12

22. Franzén, O., Gan, L.M., Björkegren, J.L.: PanglaoDB: a web server for exploration of mouse and human single-cell RNA sequencing data. Database **2019** (2019)

23. Zyla, J., Marczyk, M., Domaszewska, T., Kaufmann, S.H., Polanska, J., Weiner, J., 3rd.: Gene set enrichment for reproducible science: comparison of CERNO and eight other algorithms. Bioinformatics **35**(24), 5146–5154 (2019)

24. Chaussabel, D., et al.: A modular analysis framework for blood genomics studies: application to systemic lupus erythematosus. Immunity **29**(1), 150–164 (2008)

25. Li, S., et al.: Molecular signatures of antibody responses derived from a systems biology study of five human vaccines. Nat. Immunol. **15**(2), 195–204 (2014)

26. Aibar, S., et al.: SCENIC: single-cell regulatory network inference and clustering. Nat. Methods **14**(11), 1083–1086 (2017)

27. Noureen, N., Ye, Z., Chen, Y., Wang, X., Zheng, S.: Signature-scoring methods developed for bulk samples are not adequate for cancer single-cell RNA sequencing data. Elife **11**, e71994 (2022)

28. DeTomaso, D., Jones, M.G., Subramaniam, M., Ashuach, T., Ye, C.J., Yosef, N.: Functional interpretation of single cell similarity maps. Nat. Commun. **10**(1), 4376 (2019)

29. Lee, E., Chuang, H.Y., Kim, J.W., Ideker, T., Lee, D.: Inferring pathway activity toward precise disease classification. PLoS Comput. Biol. **4**(11), e1000217 (2008)

30. Stouffer, S.A., Suchman, E.A., DeVinney, L.C., Star, S.A., Williams, R.M. Jr.: The American Soldier, Vol. 1: Adjustment during Army Life. Princeton University Press, Princeton (1949)

31. Tomfohr, J., Lu, J., Kepler, T.B.: Pathway level analysis of gene expression using singular value decomposition. BMC Bioinformatics **6**, 1–11 (2005)

32. Barbie, D.A., et al.: Systematic RNA interference reveals that oncogenic KRAS-driven cancers require TBK1. Nature **462**(7269), 108–112 (2009)

33. Zhang, Y., et al.: Benchmarking algorithms for pathway activity transformation of single-cell RNA-seq data. Comput. Struct. Biotechnol. J. **18**, 2953–2961 (2020)

34. MacQueen, J.B.: Some methods for classification and analysis of multi-variate observations. In: Le Cam, L.M., Neyman, J. (eds.) Proceedings of the Fifth Berkeley Symposium on Mathematical Statistics and Probability, vol. 1, pp. 281–297. University of California Press, California (1967)

35. Bridges, C.C., Jr.: Hierarchical cluster analysis. Psychol. Rep. **18**(3), 851–854 (1966)

36. Blondel, V.D., Guillaume, J.L., Lambiotte, R., Lefebvre, E.: Fast unfolding of communities in large networks. J. Stat. Mech: Theory Exp. **2008**(10), P10008 (2008)

37. Stassen, S.V., Siu, D.M., Lee, K.C., Ho, J.W., So, H.K., Tsia, K.K.: PARC: ultrafast and accurate clustering of phenotypic data of millions of single cells. Bioinformatics **36**(9), 2778–2786 (2020)

38. Golubovskaya, V., Wu, L.: Different subsets of T cells, memory, effector functions, and CAR-T immunotherapy. Cancers **8**(3), 36 (2016)

39. Sherwood, A.M., et al.: Deep sequencing of the human TCRγ and TCRβ repertoires suggests that TCRβ rearranges after αβ and γδ T cell commitment. Sci. Transl. Med. **3**(90), 90ra61 (2011)

Meta-analysis of Gene Activity (MAGA) Contributions and Correlation with Gene Expression, Through GAGAM

Lorenzo Martini[iD], Roberta Bardini[iD], Alessandro Savino[iD],
and Stefano Di Carlo[(✉)][iD]

Control and Computer Engineering Department, Politecnico di Torino,
10129 Turin, Italy
stefano.dicarlo@polito.it
https://www.smilies.polito.it

Abstract. It is well-known how sequencing technologies propelled cellular biology research in the latest years, giving an incredible insight into the basic mechanisms of cells. Single-cell RNA sequencing is at the front in this field, with Single-cell ATAC sequencing supporting it and becoming more popular. In this regard, multi-modal technologies play a crucial role, allowing the possibility to perform the mentioned sequencing modalities simultaneously on the same cells. Yet, there still needs to be a clear and dedicated way to analyze this multi-modal data. One of the current methods is to calculate the Gene Activity Matrix, which summarizes the accessibility of the genes at the genomic level, to have a more direct link with the transcriptomic data. However, this concept is not well-defined, and it is unclear how various accessible regions impact the expression of the genes. Therefore, this work presents a meta-analysis of the Gene Activity matrix based on the Genomic-Annotated Gene Activity Matrix model, aiming to investigate the different influences of its contributions on the activity and their correlation with the expression. This allows having a better grasp on how the different functional regions of the genome affect not only the activity but also the expression of the genes.

Keywords: Multimodal single-cell data · Gene Activity Matrix · Bioinformatics

1 Introduction

Next Generation Sequencing (NGS) technologies are the backbone of the latest cellular biology research. With their incredible power to investigate fundamental cell mechanisms, NGS technologies enable the study of cellular states with high resolution, which is crucial to investigate cellular heterogeneity.

The single-cell RNA sequencing (scRNA-seq) technology is the most widely employed technology to study thousands of single-cell transcriptional profiles and investigate cellular heterogeneity based on gene expression [4]. In addition,

I. Rojas et al. (Eds.): IWBBIO 2023, LNBI 13920, pp. 193–207, 2023.
https://doi.org/10.1007/978-3-031-34960-7_14

single-cell assays for transposase-accessible chromatin sequencing (scATAC-seq) is becoming popular. Thanks to its ability to probe the whole genome and assess the accessible chromatin regions, scATAC-seq provides a complementary insight into the fundamental process of gene regulation [3] and expression [2].

These two faces of the same medal give an unprecedented way to investigate these complex mechanisms through their joint analysis. So, it is not surprising that multi-modal technologies, which allow simultaneously assessing both scRNA-seq and scATAC-seq from the same cells, are becoming crucial when investigating cell-related phenomena, including heterogeneity [6]. However, the intrinsic difference in data type between the two technologies poses some caveats to a proper joint analysis [9,19].

In general, it is not trivial to correlate the accessibility of a particular region of the genome to gene expression, given the incredible and complex machinery involved in gene regulation. This means that scATAC-seq datasets are built considering genes as prominent features. In contrast, scATAC-seq datasets consider genomic regions as features, making integrating these two data difficult.

The concept of Gene Activity (GA) is a viable approach to correlate accessibility with gene expression [17]. The GA summarizes the genomic accessibility information in a form where the features are genes instead of genomic regions, representing how much the gene is accessible and potentially transcribed. It enables the translation of scATAC-seq data into a matrix formally similar and directly comparable to the scRNA-seq matrix, allowing for a direct investigation of the correlation between the two biological levels. However, no clear definition exists of how to model the relationship between accessible regions and genes.

A promising approach to solve this problem is the Genomic-Annotated Gene Activity Matrix (GAGAM) approach [15,16]. In GAGAM, the association between genomic regions and accessible genes relies on a genomic model based on genomic annotations. This model constructs a Gene Activity Matrix (GAM) consisting of three different contributions associated with different functional genomic regions (i.e., promoters, exons, and enhancers). This should better model the gene regulatory landscape crucial to understanding gene expression, not just the simple gene body accessibility. GAGAM has, therefore, the peculiarity of creating a simple model that integrates different scATAC-seq signals, thus understanding and treating differently the functional information related to the accessible regions. However, in GAGAM, the complexity of the gene regulation mechanism remains hidden, and the relationship and interaction between specific regulatory elements and the gene bodies are still not represented, and, more specifically, how their accessibility impacts the actual expression remains implicit. This modeling limitation consequentially restrains the ability of GAGAM to represent the entire gene regulation mechanism accurately.

This work presents a complete meta-analysis of the GA contributions of GAGAM, starting from a multi-modal dataset to pave the way for more effective analysis. The analysis uses these data to understand better the correlation between GA and expression. Results presented in this paper help improve the definition of GA models, accurately represent the role of gene regulatory mechanisms, and support the investigation of the complex relations between DNA accessibility and gene expression. Eventually, it will also help the single-cell study of rising multi-modal datasets.

2   Background

2.1   Single-Cell Sequencing Technologies

To fully understand the proposed analysis, it is crucial to introduce the basic technologies involved in this work. First, a quick explanation of the scATAC-seq data helps understand the derived concept of GA. scATAC-seq is a technology to provide information on the epigenomic state of the cells by probing the whole genome, which leverages the Tn5-transposase to detect all the regions where the chromatin is open, and the DNA sequence is accessible [20]. Through that, it is possible to investigate not only the genes (as for scRNA-seq) but also various functional elements, like enhancers and promoters, [11], that are scattered all over the genome but are crucial for gene regulation [7]. While scRNA-seq data have genes as features, scATAC-seq data use peaks, i.e., short genomic regions described by their coordinates on the chromosomes. This intrinsic difference poses a considerable hurdle when correlating the two biological levels. One way to overcome this is to transform the peaks into gene-like data and compare the two technologies. As mentioned in Sect. 1, GA is one way to do so [6].

However, the current models to define GA tend to oversimplify the relationship between a gene and the accessibility of its genomic region. Specifically, some approaches like GeneScoring [13] and Signac [18] look indiscriminately to the peak signals overlapping the gene body regions without distinction between coding and non-coding regulatory elements. On the other hand, Cicero [6] defines its GA in a more structured way, considering other regulatory regions but collapsing the gene region to a single base. These methods generally retain little biological information from the raw scATAC-seq data, usually only related to gene coding regions, even if they represent only a tiny percentage of the whole signal [5]. Beyond these simplistic models, other approaches aim to comprise more accessible genomic regions and their impact on overall GA. Specifically, this work employs GAGAM since it uses curated genomic annotations to functionally label the peaks and, consequentially, associates them with the genes through a simple model [15].

2.2   GAGAM

GAGAM comprises information on several DNA regions, particularly from exons and non-coding regions with a regulatory role (i.e., the promoter and the enhancers of genes). This represents the main strength of GAGAM, which tries to analyze this biological information from scATAC-seq data and put it together in a model-driven method, more representative of the biological level, and could better support the cellular heterogeneity study. Therefore, GAGAM poses the basis for a more detailed investigation of the relationship between accessibility and expression in single-cell data. Its modular structure lets us control which contributions to consider in the analysis and study their relation to expression levels individually. The latter is important, especially when considering the role of regulatory regions, whose relation with gene expression is challenging to investigate. Systematically analyzing their accessibility to gene expression could build new ways to study the complex matter of gene regulation. For this reason, this work proposes a meta-analysis of the general relationship between promoter, exon, and enhancer contributions and their prevalence in the GAGAM model and a study of the correlation of each of them with gene expression.

Let us briefly recall the workflow required to construct GAGAM, starting from the raw data, to provide the reader with the necessary background. One everyday use of single-cell data is to study cellular heterogeneity, that is, to highlight the key changing features that characterize the different types of cells. GAGAM also goes in this direction. All methods to analyze single-cell data require properly preprocessing the initial data. Indeed, before constructing GAGAM itself, one preprocesses both parts of the multi-modal data (i.e., the scRNA-seq and scATAC-seq matrices). This preprocessing includes (i) Quality Control check, (ii) normalization and standardization, (iii) Principal Component Analysis (PCA), (iv) Uniform Manifold Approximation and Projection (UMAP) dimensionality reduction, (v) clustering, and (vi) Differential Expression (DE) analysis [14]. In particular, clustering and DE are the most relevant results, providing a reliable figure of the dataset's cell states/cell types, giving ground for the following investigations.

Algorithm 1. GAGAM construction

1: **for** p in P **do**
2:     Overlap p to the genomic annotation
3:     Assign label to p (prom, enhD, exon)
4:     Map p to genes
5: **end for**
6: Construct **prom** activity matrix $\mathbf{A}^{prom}_{|G| \times |C|}$
7: Construct **enhD** activity matrix $\mathbf{A}^{enhd}_{|G| \times |C|}$
8: Construct **exon** activity matrix $\mathbf{A}^{exon}_{|G| \times |C|}$

9: $\mathbf{GAGAM} = w_p \cdot \mathbf{A}^{prom}_{|G| \times |C|} + w_{en} \cdot \mathbf{A}^{enhd}_{|G| \times |C|} + w_{ex} \cdot \mathbf{A}^{exon}_{|G| \times |C|}$

GAGAM, works on the preprocessed scATAC-seq data alone, organized in the form of a matrix $\mathbf{D}_{|P| \times |C|}$, where P is the set peaks in the dataset, and C is the set of available cells. As shown in Algorithm 1 (lines 1 to 5), first, GAGAM employs the UCSC Genome Browser [12] to obtain genomic annotations and label all the peaks $p \in P$ overlapping with regions of interest (i.e., promoters, genes' exons, and enhancers), assigning to them the respective labels prom, exon, enhD. In this way, it is possible to understand the function of the accessible peaks and, consequentially, relate the function to the genes. Then it constructs three label-specific matrices (Algorithm 1, lines 6 to 8), which constitute the three contributions investigated in this work, denoted as $\mathbf{A}^l_{|G| \times |C|}$ where G is the set of genes with at least one peak mapping to them and $l \in \{prom, enhd, exon\}$. Finally, GAGAM sums all the contributions weighted by model-specific weights to obtain the final GA matrix. In the GAGAM implementation introduced in [15], simple binary weights are used, but a better understanding of the contribution of the three matrices could help fine-tune them. The final model of GAGAM activity translates the simple accessibility data from scATAC-seq into a score representing the overall accessibility of the gene and its potential to be expressed. This simple structure allows for investigating each contribution individually in a direct way, which is the core of this work and a vital element of the potentiality of GAGAM itself.

3   Meta-analysis

A multi-modal dataset is required to jointly analyze Gene Activity *per se* and gene expression. The dataset of choice is an open-access dataset from the 10X Genomics platform, consisting of 10,691 cells from adult murine peripheral blood mononuclear cell (PBMC) [1]. The scATAC-seq part of the dataset has a total of 115,179 peaks as features, while the scRNA-seq part has 36,601 genes. The tools employed to process and elaborate the data are GAGAM (the focus of this paper, accessible from [15]) and Seurat [4]. The latter is one of the most well-known and highly-utilized single-cell pipelines. All the code employed for this work is available at https://github.com/smilies-polito/MAGA, including all the supplementary material and figures.

Before starting with the actual meta-analysis, it is worth noting that scRNA-seq and scATAC-seq detect only a tiny fraction of the actual signal from each cell (around 10–45% for scRNA-seq and only 1–10% for scATAC-seq [5]). This translates into considerable sparsity for the data. For each cell, the dataset contains several zero entries that could be false negatives [10]. This characteristic introduces noise when trying to correlate accessibility and expression. For this reason, this work explores the idea of performing the analysis based on the concept of *aggregated cell* behavior. Specifically, it aggregates cells from the same clusters obtained from preprocessing the raw scRNA-seq data, representing the average over groups of cells instead of single cells. This way, these clusters should, with high probability, represent the cell types [4]; thus, exploring them could be relevant for cellular heterogeneity studies.

The procedure to calculate gene activity and expression of the aggregated cells is described in Algorithm 2. For the subsequent analyses, let us denote with $\mathbf{A}_{|G| \times |AC|}$ the aggregated cells activity matrix and with $\mathbf{E}_{|G| \times |AC|}$ the aggregated cells expression matrix, where G is the set of genes and AC is the set of aggregated cells.

The meta-analysis focuses on three separate investigations reported in the following subsections.

Algorithm 2. Aggregated cells definition

1: Initialize activity matrix $\mathbf{A}_{|G|\times|AC|}$
2: Initialize expression matrix $\mathbf{E}_{|G|\times|AC|}$
3: **for** i in AC **do**   ▷ $i \rightarrow$ is a single cluster
4:     **for** g in G **do**   ▷ $g \rightarrow$ is a single gene
5:         Compute $\mathbf{A}_{g,i}$, the average activity of g $\forall c \in i$
6:         Compute $\mathbf{E}_{g,i}$, the average expression of g $\forall c \in i$
7:         Compute variance of g activity $\forall c \in i$
8:         Compute variance of g expression $\forall c \in i$
9:     **end for**
10: **end for**

3.1   Peaks Information

After processing the data and obtaining the GAGAM contributions, the first investigation focuses on the three labels (i.e., prom, exon, enhD), assigned to peaks during the GAGAM construction (Algorithm 1 line 3), and the information they carry on. Indeed, understanding what type of information and how much is retained from the raw scATAC-seq data is crucial for creating models from them. As discussed in Sect. 2, while other GA methods look into gene regions only, GAGAM focuses on more regions of the genome. This type of analysis supports the correctness of this choice. Moreover, showing the non-equal distribution of the labels justifies the separate investigation of the three contributions performed in the following sections.

Let us focus on the proportions between non-labeled, prom, exon, and enhancers labels. This proportion gives direct knowledge of how much information GAGAM retains from the raw scATAC-seq, which is not trivial [20]. Furthermore, it is relevant to investigate the peak-to-gene assignments. From GAGAM computation, it is also possible to retrieve the link between peaks and genes; thus, it is straightforward to study how many and which labeled peaks the model assigns to the genes. This simple analysis gives insight into the general GA model constitution, investigating how much accessibility information relates to each gene, which is crucial for the improvement of the model itself.

Figure 1 shows some information about peak labeling. Of the 115,179 peaks, 92,100 (80%) received one of the labels. GAGAM does not necessarily employ all labeled peaks since it filters out the ones that do not associate with a gene (namely, for some promoter and enhancer peaks.). Among the labeled peaks, there is a clear predominance of the enhancer peaks (about 34% of all peaks) against a limited portion of promoter peaks (about 13% of all peaks), remarking that the epigenetic information from the scATAC-seq data comes from distal regions and not just near the gene coding regions.

Figure 2 presents the number of labeled peaks assigned to genes for each label type. The vast majority of genes (86%) have only one promoter peak set to them, which appears to be the average case. However, a few genes have multiple promoter peaks mapping to them, which is entirely unexpected. Directly examining the UCSC genome browser, it becomes clear that the multiple promoters map

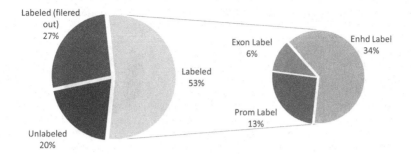

Fig. 1. The figure shows the distribution of labels among the peaks. Additionally, the labeled peaks are divided into three different labels. Most peaks receive a label, even though some data is not considered. Most labels are enhancers, showing how much of the information and potentiality of the scATAC-seq data comes from non-gene-related regions.

to different isoforms of the same genes. This shows the capability of GAGAM to have detailed resolution on the GA, meaning it can independently explore several isoforms of genes.

Regarding the exon peaks, most genes do not have any mapping to them. One of the reasons stems from the small number of exon peaks (namely, 7046, only 6% of the total peaks), so they only cover some genes.

On the other hand, the genes tend to have many enhancer peaks linked to them, also because, unlike the other labeled peaks, a single enhancer can map to multiple genes. The latter highlights the relevance of creating a reliable model to connect them to genes, which is central to future developments.

3.2 Activity-Expression Correlation

Our previous work [15] qualitatively addressed this type of investigation to broadly study Activity-Expression (A-E) patterns for specific genes. This paper applies a quantitative approach. Pearson coefficient [8] computed between the activity and expression of each gene on aggregated cells is used to quantify the A-E correlation, and this information is visualized through a set A-E plots. The investigation is performed independently on each peak label from the previous section. This enables us to investigate how each accessibility label affects the gene expression and better tune the GA models.

The promoter peaks are expected to be most correlated with the expression [18,20]. The correlation is relatively high for most genes, with 71% correlating higher than 0.5 with statistically significant p-values ($p < 0.05$). Moreover, focusing on the top 1,000 genes with the lowest variance, 94% of them correlate higher than 0.5, strengthening the previous result. It is also interesting to notice that considering the list of DE genes, the correlation is still relevant, with 91% being over the mentioned threshold. Therefore, it is fair to state that the accessibility of the gene's promoter results in its detectable and subsequent expression.

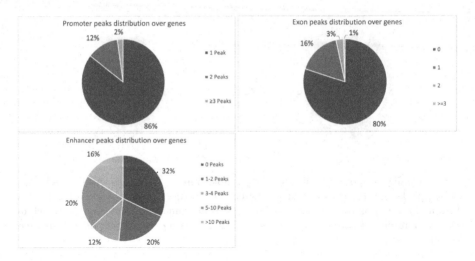

Fig. 2. Distribution of the number of peaks mapping to genes, divided by label types. The percentages represent the percentage of all genes with n peaks mapping to them.

Besides a purely numerical analysis, it is relevant to visualize the correlation. For each aggregated cell, it is possible to plot each gene as a point in the A-E space and explore the general trend of the genes inside the clusters. This way, one can investigate the difference between clusters that could convey pertinent information for the cellular heterogeneity study. For the sake of space and clarity, only the first four A-E plots (denoted as CL0, CL1, CL2, and CL3) are reported in Fig. 3. Information written in these plots is representative of the overall results. All remaining plots and figures are available at https://github.com/smilies-polito/MAGA. The points are on a log space to allow better visualization, while the colors represent meaningful information on each gene. The black points are all the genes, while the red points represent the differentially expressed genes for the cluster. The latter comes from the DE analysis performed on the clusters, precisely the top specific marker genes per cluster. Moreover, the plots contain green dashed lines defining the genes' mean activity or expression in that specific aggregated cell.

Figure 3 shows that most points are in the top-right area, meaning high expression and activity, demonstrating the correlation between the two characteristics. Most of the DE genes from that specific cluster (points in red) lie in this area, highlighting their relevant high activity other than expression. The least populated area is the top-left area, representing the low activity high expression area, while the opposite is pretty populated. This observation is not trivial, and it highlights that the promoter accessibility of a gene can implicate various levels of expression. However, when the accessibility is low or null, the genes are rarely expressed.

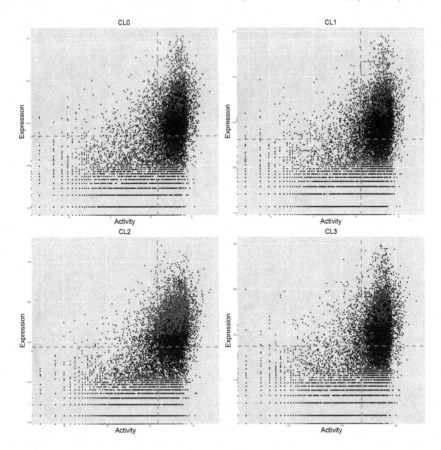

Fig. 3. Activity-Expression plots. Each point represents a gene, with its expression and promoter activity from the aggregated cells. Here are present only the first four aggregated cells. The red dots represent the DE genes from the cluster (Color figure online)

The exon contribution is the smallest. Only 7,046 genes have at least one exon peak linked to them. Nonetheless, their specific information can be informative to the model. The Pearson correlation on the aggregated cells reveals that only about 55% of the remaining genes correlate greater than 0.5, with the percentage lowering to about 41% when considering the lowest variance genes. This is more clear from the A-E plots (Fig. 4), where one can see that the points tend to occupy the right-most part consistently, indicating the high activity area. Differently from the promoter activity, which was spread out and ranged on different levels, the exon activity appears more set in a binary-like fashion. The exon contribution to the activity seems to have a relatively high contribution or not to be present at all. This also explains the lack of a high correlation with the expression since the exon contribution does not appear to influence it strongly.

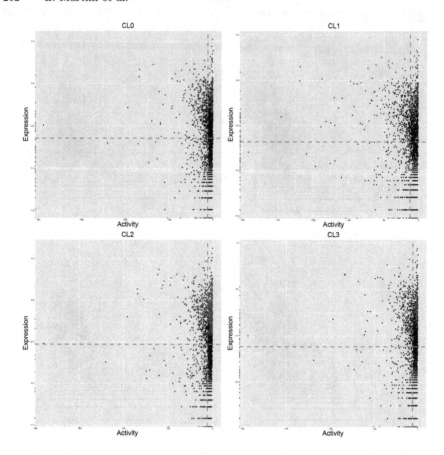

Fig. 4. A-E plots. Each point represents a gene, with its expression and exon activity from the aggregated cells. Here are present only the first four aggregated cells. The red dots represent the DE genes from the cluster. (Color figure online)

The enhancer contribution is the most complex and potentially important to explain gene regulation. Many different enhancer peaks can map to a gene, and each can map to various other genes, making the correlation less trivial to study. The Pearson correlation confirms that about 63% of the genes correlate with enhancer activity and expression higher than the threshold of 0.5. It is lower than the promoter contribution but higher than the exon one. When considering the lowest variable genes, differently from the exon case, this percentage rises to 70%. Eventually, the correlation calculated for the DE genes improves, with 72% of them being over the threshold.

The results in Fig. 5 differ slightly from promoters and exons. The points still predominate in the top-right area but spread out more, showing that the correlation is less predominant. This is prominent in specific clusters like the "CL2", where many genes have discordant activity and expression, meaning that the enhancer activity contribution is less necessary for the expression than

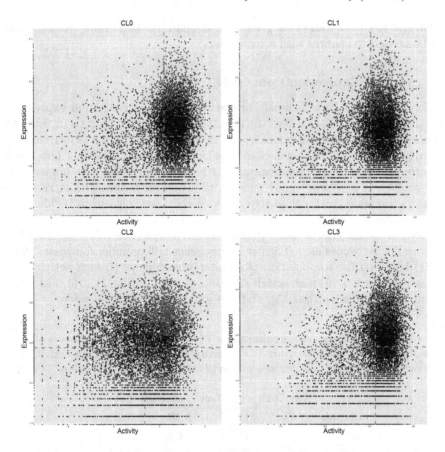

Fig. 5. A-E plots. Each point represents a gene, with its expression and enhancer activity from the aggregated cells. Here are present only the first four cluster aggregated cells. The red dots represent the DE genes from the cluster (Color figure online)

the other contributions. The red points, representing the DE genes, still reside in the right-top area, even if more to the top-left.

Intriguingly, the Pearson correlation on the DE genes remains relatively high for all peak labels. This can be ascribed to the fact that these genes, which have, by definition, high expression variability among the clusters, must also show a similar activity variability at all levels. Therefore, the concept of marker genes, well-known in transcriptomic analysis studies, could even be applied to epigenomic studies. However, it is curious that sometimes these genes are expressed despite a low or even null activity, as in the enhancer case. At first glance, it might appear counterintuitive to have expression and no GA, which could stem from the noise caused by the different depths of the technologies. However, the epigenetic and transcript levels work on very different regulation time scales, meaning a change in accessibility does not immediately propagates to the expression. Therefore, the case could represent an informative dynamical change undetected from the static view given by the separated data. More investigations will be needed to confirm such a hypothesis.

3.3   Activity-Expression Coherence

Section 3.2 discussed the general correlation between activity and expression. Yet, there are some cases where the causal relationship between them could be more complex (e.g., for the enhancer contribution). Therefore, it is intriguing to investigate the coherence between expression and activity, meaning how much the presence of activity of genes implicates their expression in a binary way. This analysis studies how many genes with an activity greater than zero also have an expression greater than zero and vice-versa (see Algorithm 3). The algorithm reduces the activity (a) and expression (e) in the $\mathbf{A}_{|G| \times |AC|}$ and $\mathbf{E}_{|G| \times |AC|}$ matrices to a binary form (lines 1–14), considering positive values as one while negative or null values as 0. Then it counts genes falling under four cases (lines 15–31) based on the combination of a and e of each gene: (i) high activity and expression (line 19), (ii) high activity but low expression (line 21), (iii) low activity but high expression (line 25), and (iv) low activity and expression.

Table 1. Activity-Expression coherence: The table shows the percentages of genes in the four cases. Namely, they are High activity High expression, High activity Low expression, Low activity High expression, and Low activity Low expression.

Label	High-High	High-Low	Low-High	Low-Low
Promoter	83.8%	14.9%	0.4%	0.9%
Exon	79.4%	13.2%	5.6%	1.8%
Enhancer	60.0%	7.8%	24.1%	8.1%

This approach highlights the interesting non-trivial case where activity and expression exhibit low or null coherence. Table 1 reports all the results as the average percentage of genes in the four cases, distinguished by peak label.

The A-E coherence for the promoter contribution is in line with the correlation results as about 83% belong to the high-high class and about 15% to the high-low case. It remarks the fact that promoter accessibility is necessary but not sufficient for the expression.

Unlike the correlation results, in the case of exons, the high-high case still includes most of the genes, despite being in a lower percentage than previously. It is relevant to notice the increase in the low-high case, which could support the hypothesis of the possible dynamical transcription changes. Indeed, the exon accessibility in gene regulation could follow different timings than promoters and be a better probe for these dynamical changes.

Finally, the A-E coherence is the most surprising for the enhancer contribution. The high-high count in the enhancer case lowers to 60% in favor of a significant increase in the low-high case. This differs from the 80% of all genes

Algorithm 3. Coherence calculation

```
 1: for a ∈ A_{|G|×|AC|} do
 2:     if a > 0 then
 3:         a = 1
 4:     else
 5:         a = 0
 6:     end if
 7: end for
 8: for e ∈ E_{|G|×|AC|} do
 9:     if a > 0 then
10:         e = 1
11:     else
12:         e = 0
13:     end if
14: end for
15: for a ∈ A_{|G|×|AC|} do
16:     for e ∈ E_{|G|×|AC|} do
17:         if a = 1 then
18:             if e = 1 then
19:                 High_High = High_High + 1
20:             else if e = 0 then
21:                 High_Low = High_Low + 1
22:             end if
23:         else if a = 0 then
24:             if e = 1 then
25:                 Low_High = Low_High + 1
26:             else if e = 0 then
27:                 Low_Low = Low_Low + 1
28:             end if
29:         end if
30:     end for
31: end for
```

identified during correlation analysis. The low-high case was almost nonexistent when considering correlation, but now it includes about 24% of the genes. Therefore, the enhancer activity seems not a strictly necessary condition for the expression, although, when present, it positively correlates with it. Many reasons may justify this behavior. The modeling of enhancers on gene regulation is far from trivial, and for clarity, the GAGAM model simplifies it. As previously mentioned, one enhancer peak can map to many genes and influence their activity simultaneously. At the same time, in a single cell, it is likely to impact only a subset of them at a time. Moreover, the current model cannot distinguish between enhancers and silencers, which affects the sign of the contribution to the expression. Finally, the timescales involving the enhancers in gene regulation are probably even more dilated than the previous contributions. The low-high correlation could represent a dynamic change in the gene expression, as mentioned before. Despite these limitations, the results show the model can retrieve a significant correlation between enhancer contribution and expression.

In general, the A-E coherence results emphasize that the accessibility of a gene is not a sufficient condition for the expression, and in some cases, neither is necessary. The reason behind this not-trivial behavior could be different and exciting to investigate. Specifically, the enhancer contribution seems the most complex but potentially informative for gene regulation, and its understanding and fine-tuned interpretation could unravel crucial details. However, this is not part of this work and is left for future development.

3.4   Conclusions

This work presented a meta-analysis of GAGAM informative content and, precisely, how its building blocks correlate with the expression on a multimodal single-cell dataset. The results are pretty revealing. First, from the peak information analysis (Sect. 3.1), one can immediately understand that the information retained by GA from the raw scATAC-seq data is limited and diversified. Indeed, only about half of the original peaks are considered, with them being unevenly distributed between the three labels. In particular, there is an evident predominance of enhancer peaks, which highlights the limitations of other GA methods (Sect. 2) since they focus only on promoter and gene body regions, which cover only a tiny portion of the epigenetic data information. Understanding that an optimal GAM should retain as much information from the raw data as possible is fundamental, especially for accurately modeling the epigenetic level.

Regarding the A-E correlation, this meta-analysis focused on studying how the three contributions of GAGAM correlate with the actual expression. First, the promoter contribution shows the most linear behavior, meaning that genes with active promoters also consistently display expression. On the other hand, the exon contribution already has less clear and trivial results; namely, the exon accessibility has a remarkably lower correlation with the expression than the other two. Lastly, the enhancers show the most complex results. In particular, the A-E plots and the A-E coherence highlights the significant number of discordant genes (i.e., the genes with High activity Low expression and Low activity High expression), which emphasizes the intricate relationship between gene expression and activity.

In general, the incoherencies between gene activity and expression are interesting. Indeed, the different time scales at which transcriptomic and epigenomics work could cause these discrepancies, giving insight into the dynamic changes that are part of gene regulation. This fact highlights the power and relevance of studying multimodal data through the GAM, which could help go beyond the intrinsic static nature of single-cell data.

In any case, this analysis is crucial to improving the underlying GA model. Comprehending the relationship between specific genomic regions' activity and gene expression can help fine-tune each contribution's weights on the final matrix. Moreover, better models also go toward a more accurate representation of the gene regulatory mechanism, opening the possibility of investigating

in new ways. Lastly, this type of meta-analysis can become an extra tool for studying the increasingly popular multimodal datasets and help the joint analysis of scRNA-seq and scATAC-seq.

References

1. 10XGenomics: 10k peripheral blood mononuclear cells (PBMCs) from a healthy donor single cell multiome ATAC + gene expression dataset by cell ranger arc 2.0.0, 10x genomics (2021)
2. Baek, S., Lee, I.: Single-cell ATAC sequencing analysis: from data preprocessing to hypothesis generation. Comput. Struct. Biotechnol. J. **18**, 1429–1439 (2020)
3. Buenrostro, J.D., et al.: Integrated single-cell analysis maps the continuous regulatory landscape of human hematopoietic differentiation. Cell **173**(6), 1535–1548 e16 (2018)
4. Chen, G., Ning, B., Shi, T.: Single-cell RNA-seq technologies and related computational data analysis. Front. Genet. **10** (2019)
5. Chen, H., et al.: Assessment of computational methods for the analysis of single-cell ATAC-seq data. Genome Biol. **20**(1), 241 (2019)
6. Chen, S., Lake, B.B., Zhang, K.: High-throughput sequencing of the transcriptome and chromatin accessibility in the same cell. Nat. Biotechnol. **37**, 1452–1457 (2019)
7. Danese, A., Richter, M.L., Chaichoompu, K., et al.: EpiScanpy: integrated single-cell epigenomic analysis. Nat. Commun. **12**(D1), 5228 (2021)
8. Freedman, D., Pisani, R., Purves, R.: Statistics (international student edition) 4th edn. W. W. Norton & Company, New York (2007)
9. Hao, Y., et al.: Integrated analysis of multimodal single-cell data. Cell **184**(13), 3573–3587 (2021)
10. Hwang, B., Lee, J.H., Bang, D.: Single-cell RNA sequencing technologies and bioinformatics pipelines. Exp. Mol. Med. **50**, 1–14 (2018)
11. Kelsey, G., Stegle, O., Reik, W.: Single-cell epigenomics: recording the past and predicting the future. Science **358**(6359), 69–75 (2017)
12. Kent, J., et al.: The human genome browser at UCSC. Genome Res. **12** (2002)
13. Lareau, C.A., Duarte, F.M., Chew, J.G., et al.: Droplet-based combinatorial indexing for massive-scale single-cell chromatin accessibility. Nat. Biotechnol. **37**, 916–924 (2019)
14. Luecken, M.D., Theis, F.J.: Current best practices in single-cell RNA-seq analysis: a tutorial. Mol. Syst. Biol. **15**(6), e8746 (2019)
15. Martini, L., Bardini, R., Savino, A., Di Carlo, S.: GAGAM v1.2: an improvement on peak labeling and genomic annotated gene activity matrix construction. Genes **14**(1) (2023)
16. Martini, L., et al.: GAGAM: a genomic annotation-based enrichment of scATAC-seq data for gene activity matrix. In: Rojas, I., et al. (eds.) Bioinformatics and Biomedical Engineering, vol. 13347, pp. 18–32. Springer, Cham (2022). https://doi.org/10.1007/978-3-031-07802-6_2
17. Pliner, H.A., et al.: Cicero predicts cis-regulatory DNA interactions from single-cell chromatin accessibility data. Mol. Cell **71**, 1–14 (2018)
18. Stuart, T., et al.: Single-cell chromatin state analysis with Signac. Nat. Methods **18**, 1333–1341 (2021)
19. Subramanian, I., et al.: Multi-omics data integration, interpretation, and its application. Bioinform. Biol. Insights **14**, 1177932219899051 (2020)
20. Yan, F., et al.: From reads to insight: a hitchhiker's guide to ATAC-seq data analysis. Genome Biol. **21**(22) (2020)

Predicting Papillary Renal Cell Carcinoma Prognosis Using Integrative Analysis of Histopathological Images and Genomic Data

Shaira L. Kee[1] , Michael Aaron G. Sy[1,2] , Samuel P. Border[3,4] ,
Nicholas J. Lucarelli[3,4] , Akshita Gupta[4,5] , Pinaki Sarder[3,4,6] ,
Marvin C. Masalunga[7] , and Myles Joshua T. Tan[1,2,4,6(✉)]

[1] Department of Natural Sciences, University of St. La Salle, Bacolod, Philippines
mylesjoshua.tan@medicine.ufl.edu
[2] Department of Electronics Engineering, University of St. La Salle, Bacolod, Philippines
[3] J. Crayton Pruitt Family Department of Biomedical Engineering, Herbert Wertheim College of Engineering, University of Florida, Gainesville, FL, USA
[4] Division of Nephrology, Hypertension and Renal Transplantation – Quantitative Health Section, Department of Medicine, College of Medicine, University of Florida, Gainesville, FL, USA
[5] Department of Health Outcomes and Biomedical Informatics, College of Medicine, University of Florida, Gainesville, FL, USA
[6] Department of Electrical and Computer Engineering, Herbert Wertheim College of Engineering, University of Florida, Gainesville, FL, USA
[7] Medical Education Unit, School of Medicine, Southwestern University - PHINMA, Cebu, Philippines
mcmasalunga.swu@phinmaed.com

Abstract. Renal cell carcinoma (RCC) is a common malignant tumor of the adult kidney, with the papillary subtype (pRCC) as the second most frequent. There is a need to improve evaluative criteria for pRCC due to overlapping diagnostic characteristics in RCC subtypes. To create a better prognostic model for pRCC, we proposed an integration of morphologic and genomic features. Matched images and genomic data from The Cancer Genome Atlas were used. Image features were extracted using CellProfiler, and prognostic image features were selected using least absolute shrinkage and selection operator and support vector machine algorithms. Eigengene modules were identified using weighted gene co-expression network analysis. Risk groups based on prognostic features were significantly distinct ($p < 0.05$) according to Kaplan-Meier analysis and log-rank test results. We used two image features and nine eigengene modules to construct a model with the Random Survival Forest method, measuring 11-, 16-, and 20-month areas under the curve (AUC) of a time-dependent receiver operating curve. The integrative model (AUCs: 0.877, 0.769, and 0.811) outperformed models trained with eigengenes alone (AUCs: 0.75, 0.733, and 0.785) and morphological features alone (AUCs: 0.593, 0.523, 0.603). This suggests that an integrative prognostic model based on histopathological images and genomic features could significantly improve survival prediction for pRCC patients and assist in clinical decision-making.

I. Rojas et al. (Eds.): IWBBIO 2023, LNBI 13920, pp. 208–221, 2023.
https://doi.org/10.1007/978-3-031-34960-7_15

Keywords: Renal cell carcinoma (RCC) · prognostic model · genomic features · histopathological images · The Cancer Genome Atlas

1 Introduction

In the past decade, localized renal cell carcinoma (RCC) has increased in prevalence and incidence substantially [1]. It is the most common malignant neoplasm arising from the adult kidney [2] and is responsible for ~95% of all cases [3]. RCCs are malignant tumors of the renal cortex displaying distinct clinical, morphologic, and genetic characterizations [4, 5]. Currently, there are 20 different RCC variants [6]. Classically, the Heidelberg classification system categorizes RCC into the following histologic subtypes: clear cell, papillary, chromophobe, collecting duct, and unclassified RCC [7]. Papillary renal cell carcinoma (pRCC) is the second most commonly identified subtype of RCC (10%-15% of cases), following the clear cell subtype (ccRCC) [8]. pRCC is distinguished from ccRCC morphologically by the presence of basophilic or eosinophilic cells in a papillary or tubular form [9]. pRCC compared to ccRCC, has been reported with a greater male predominance [9]. Clear cell papillary renal cell carcinoma (ccpRCC) is a distinct histotype that progresses in a more indolent manner [4]. pRCC tumors possess immunohistochemical and genetic profiles that are distinct from ccRCC and ccpRCC [9]. While frontline surgical extirpation of suspected malignant localized RCC remains the standard of care, recent increases in renal mass biopsies for risk stratification indicate a growing preference of both patients and surgeons for the characterization of tumor at the outset to inform treatment decisions. Precise pathologic information coupled with emerging molecular tools remains the best course of pretreatment risk stratification [3].

Pathologists use immunohistochemical staining to increase contrast between specific cell types in biopsy specimens for tumor evaluation [10]. With the growing digitization of Whole-Slide Images (WSIs) [11], computational image analysis has shown great potential in diagnosis and discovery of new biomarkers for multiple types of cancers, such as breast [12], colon [13], and lung [14]. Morphological interpretation of histologic images is the basis of pathological evaluation. WSIs are a rich source of biological information as this level of resolution facilitates detailed assessment of the relationship of cancer cells with other cells and tumor microenvironment (TME), all of which are referred to as "hallmarks of cancer" [15]. Accurate and reproducible models can be made to assess prognosis through automatically quantifying morphological features. A pipeline is developed which automatically segments cancer images and generates quantitative features [16]. The majority of previous work in quantitative pathology has required laborious image annotation by skilled pathologists, which is prone to human error [17]. Automation using image analysis and machine learning (ML) methods has significantly corrected inconsistencies that result from histologic preparation [12, 13], as even expert human eyes have difficulty in distinguishing some granular image features [18].

In addition to histopathologic images, information on molecular alteration has also been widely adopted for predicting cancer clinical outcomes [19, 20]. Cancers are diverse, with varying genetics, phenotypes, subtypes, and outcomes. In the past decades, molecular stratification of tumors using gene expression microarrays has been considered an important field in cancer research [21], with an increase in application of integrative genomics or panomics approaches [22, 23]. This approach is used to identify biomarkers for stratification of patients into groups with different clinical outcomes. Solid tumors are heterogeneous tissues composed of a mixture of cancer and normal cells that further complicates the interpretation of their molecular profiles [24]. Cancer, immune, and stromal cells form an integral part of the TME; however, their admixture poses challenges for molecular assays, especially in large scale analyses [24]. Alternatively, Molecular profiles yield quantifiable results via computational and statistical inferences on big data. For instance, studies have shown that lymphocytic infiltration can be inferred from gene expression profiles [25], as well as cellularity from single-nucleotide polymorphism (SNP) information [26]. However, these approaches are indirect and strongly rely on statistical assumptions [27]. To leverage the richness of histopathological information and quantitative results of computational analyses, a systematic approach that integrates histopathology and genomics was developed. An image-based approach was used to increase the power of molecular assays and to complement them with each other to unveil prognostic features otherwise invisible in molecular data [2]. Recent studies have also highlighted the contribution of gene expression and morphologic phenotypes to cancer growth and progression [12, 28]. The integration extends recent approaches that only identified morphological features prediction of survival by image analysis [12]. One study ventured into an integrative approach for predicting ccRCC prognosis or time-to-event outcomes [2]. The study utilized both histopathologic images and eigengenes to predict patient outcome. Eigengenes are the summarized genetic module expression profile used to reduce a gene co-expression network involving thousands of genes [29]. The model developed from the study was able to generate risk indices that correlated strongly with survival of ccRCC patients, outperforming predictions based on morphologic features or eigengenes separately. Cheng et al. [2] showed significant correlation according to Bonferroni correction between image features and eigengenes, with results suggesting that low expression of some eigengene were related to poor prognostic outcome, implying impaired renal function; while high expression of some eigengene observed to coexpress in multiple types of cancers [30] indicated aggressiveness and was negatively related to patient prognosis [2]. The integration of multi-modal data for prognosis estimation has led to new insights into the influence of tissue genotype on phenotype [31, 32].

Random Survival Forest (RSF) is an ensemble tree-based ML model for the analysis of survival data. The method has been used in several studies showing superior predictive performance to traditional strategies in low and high-dimensional settings [33, 34]. RSF can capture complex relationships between predictors and survival without prior specification requirements. Risk prediction models play a vital role in personalized decision-making, especially for time-to-event outcomes of cancer patients [34]. Traditional prediction models often ignore the longitudinal nature of medical records, using only baseline information. The use of risk prediction models incorporates longitudinal information to produce updated survival predictions during follow-up check-ups. This allows for high accuracy in the automated prediction of cancer prognosis, showing significant promise in improving the quality of care where pathologists are scarce. This study aims to overcome the difficulty of sub-classifying pRCC using morphological features alone and contribute to better prognosis methods for pRCC patients.

The study is limited to the risk index calculation based on the cellular morphological features and eigengenes of the pRCC cohort. Due to the lack of other large cohorts of pRCC with matched histologic image and genomic data, out-of-bag method and cross-validation were used in the algorithms. Since the large number of genes posed a challenge to obtaining sufficient statistical power, instead of focusing on individual genes, a gene coexpression network analysis (GCNA) was used to cluster genes into coexpressed modules or highly correlated genes, where each module was summarized as an eigengene.

This paper is organized into four sections. Section 2 provides the methodology of the development of the prognostic models, detailing the image processing, image feature extraction and elimination, analysis of eigengenes, patient stratification, and prognostic model development. In Sect. 3, generated prognostic features, gene enrichment analysis, and comparison of prognostic models are discussed. The clinical and biochemical implications of the results are discussed in Sect. 4.

2 Materials and Methods

2.1 Dataset Overview

The pRCC patient samples used in the study include matched hematoxylin and eosin (H&E)-stained WSIs, transcriptome, somatic mutation, and clinical information. The patient data were acquired from The Cancer Genome Atlas (TCGA, https://portal.gdc.cancer.gov/) [35] data portal of the National Cancer Institute Genomic Data Commons. Microscopic images (20 × and 40 × magnification) were obtained from the TCGA, which can be visualized at the Cancer Digital Slide Archive (CDSA, http://cancer.digitalslidearchive.net/) [36]. Gene expression profiles with 20,531 Entrez-identified sequences were taken from Broad Genome Data Analysis Center (GDAC) Firehose. Prior to preprocessing, the dataset included 287 pRCC cases. Patients with missing data were excluded from the study, leaving us with 277 patients.

2.2 Histopathological Image Processing

Image Preprocessing and Image Feature Extraction. WSIs were first chopped into equal-sized patches (1000 × 1000 × 3) using Openslide. The resulting image patches were then assessed for tissue content using pixel intensity statistics. The generated sub-images were then filtered according to quality. Sub-images with >50% white background were excluded. To do so, a range value for the whole dataset was determined, where images from one randomly chosen patient were grouped according to the cell coverage of their image patches. The mean and standard deviation of the pixel value of each image were calculated, and were averaged to determine the pixel value range for each group. Then, 20 sub-images were randomly selected for the next step to eliminate sample selection bias and reduce computing load. Since there were varying sources of the WSIs, the images were color normalized to prepare for downstream analysis. Macenko color normalization [37] was used. Then, H&E stains were separated from the original images to facilitate nuclear extraction for downstream feature calculation. Quantitative performance evaluation of nuclei segmentation was performed using manually annotated subset of image patches.

CellProfiler version 4.2.1 [38, 39] software was used to extract image features from each sub-image. Before extraction of image features, ground truth masks for the TCGA tumor biospecimen samples were generated using CellProfiler 4.2.1 and GNU Image Manipulation Program (GIMP) version 2.10.42. Using GIMP, tumoral nuclei were annotated by encircling the object using red outlines, while blue outlines were drawn outside of the encircled objects to differentiate the background from the foreground. The objects segmented were verified to be tumoral nuclei by a pathologist. The outlines generated using GIMP were then exported to CellProfiler to create the label images or the object masks. The label images were used to assess the performance of the pipeline for the automated segmentation of the nuclei. Individual cell segmentation was conducted using builtin modules in CellProfiler and included Otsu Thresholding followed by morphological postprocessing. Here, we measured object intensity, object size, object shape, image granularity, and image texture. The measurement of object size and shape includes features such as area, Zernike shape, perimeter, formfactor, solidity, Euler's number, and orientation. The Zernike shape features comprise a set of 30 shape characteristics that are derived from Zernike polynomials ranging from order 0 to order 9. Finally, we extracted 448 image features from each sub-image and calculated the mean value of the 20 representative sub-images for each patient.

Feature Elimination. To obtain prognosis-related features, an R implementation of the support vector machine's recursive feature elimination (SVM-RFE) and Lasso Regression algorithms were employed to filter the prognostic image features most correlated with pRCC prognosis. A 5-fold cross validation in both the SVM-RFE and Lasso Regression algorithms were applied. A total of 441 image features were filtered using the SVM-RFE and Lasso Regression. Feature reduction pipeline was built using the glmnet and caret libraries. Data for this model was split using a 70:30 ratio. The image features were also filtered using the e1071 multiple SVM-RFE (mSVM-RFE) [40] as adapted from https://github.com/johncolby/SVM-RFE. SVM-RFE is a ML method with a backward feature SVM. Feature filtration using SVM-RFE was performed using a method performed by Li et al. [38]. The SVM-RFE was used to rank the pathological image features in a descending order of significance, iteratively eliminating the minimum features and training the model with the remaining image features until all features are removed. The maximal cross-validated accuracy was adopted as the evaluation index to select the optimal feature subset related to the prognosis. Finally, intersection of the optimal subset of features from the SVM-RFE model and LASSO regression was used to obtain the most relevant pathological features to the prognosis.

2.3 Gene Coexpression Analysis

The profiles of mRNA expression for the pRCC tumors in TCGA were transformed from Illumina HiSeq 2000 RNA-seq V2 read counts to normalized transcripts per million (TPM). TPM fulfills the invariant average criterion that reads per kilobase of transcript per million reads mapped (RPKM) does not account for. By definition, TPM and RPKM are proportional, thus are closely related as shown by this equation:

$$\text{TPM} = 10^6 \times \frac{RPKM}{\text{Sum}(RPKM)} \tag{1}$$

Expression data were then scaled with the natural logarithm operation [41]:

$$X_{\text{input}} = \log(X_{\text{original}} + 1) \tag{2}$$

where X_{original} was the genetic data or the non-negative RNA sequencing expression values (Illumina Hi-Seq RNA-seq v2 RSEM normalized), and X_{input} was the input covariate vector for the coexpression network analysis.

Weighted gene co-expression network analysis (WGCNA) package [42] was used to cluster genes into coexpressed modules and each module was summarized as an eigengene using the protocol described in the study [42]. This approach allows for substantial improvement in statistical power and for more focus to be placed on important biological processes and genetic variations related to coexpressed gene modules. We adopted code from https://github.com/Lindseynicer/WGCNA_tutorial. The generated module eigengenes were then correlated with prognostic image features. Biological relevance of the module eigengenes was obtained through Metascape (https://metasc ape.org/).

2.4 Risk Categorization

Each prognostic feature was correlated with survival status to determine the direction of relationship. For each prognostic feature, we divided the patients into two groups (low and high-risk groups) where the median of each prognostic feature was used as a cut-off point. Depending on the relationship and feature values relative to the median, prognostic features of each patient were categorized as high or low risks. Finally, we assigned a patient's risk level based on the dominant risk level among their features.

2.5 Prognosis Prediction Model

The overall workflow is shown in Fig. 1. RSF algorithm was used to construct an integrative prognostic model. The method will be implemented using the R package random-ForestSRC, tailoring based on the different ensemble parameters that affect building of survival trees. The number of splits in each candidate variable can reduce computation time compared to testing all possible split points for each covariate. A 7:3 ratio and 10-fold cross validation was used during model development. The RF model estimated survival risk for each patient and determined their risk scores, which was plotted on a survival curve. As predicted by the learned model, we compared 11-month survival differences between the two groups using Kaplan-Meier (KM) analysis and log-rank test. KM analysis was used for patient stratification, while the log-rank test calculated the p-value, where $p < 0.05$ is considered significant.

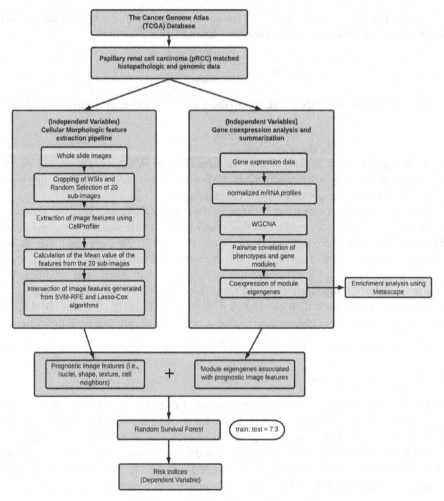

Fig. 1. Data analysis and integration workflow

3 Results

3.1 Patient Characteristics

277 pRCC patients (72 female and 205 male) were included (Table 1). Histopathological images, mRNA expression data, and clinicopathological information were downloaded from TCGA, Broad GDAC Firehose, and cBioportal.

3.2 Prognosis-related Image Features and Co-expression Gene Module Selection

441 image features were used for the data dimension algorithms used, specifically LASSO and mSVM-RFE. The optimal subset of features determined by the feature

Table 1. Demographic and clinical characteristics of pRCC patients.

Characteristics		Total ($n = 277$)	Train ($n = 194$)	Test ($n = 83$)
Gender	Male	205	141	64
	Female	72	53	19
Events	Alive	235	164	71
	Dead	42	30	12

elimination of the mSVM-RFE algorithm obtained six features while LASSO regression identified 20. We then found the intersection of the results of the two algorithms to obtain two pRCC prognostic image features (Image Granularity feature and Zernike shape feature). To identify the prognostic co-expression gene modules, WGCNA was applied to evaluate the relationship between the two prognostic image features and eigengene modules. The most significant positive association with the Zernike feature was the dark red module, while it was the turquoise module for the Image Granularity feature. Examples of selected histopathological sub-images in both high-risk and low-risk groups are presented in Fig. 2.

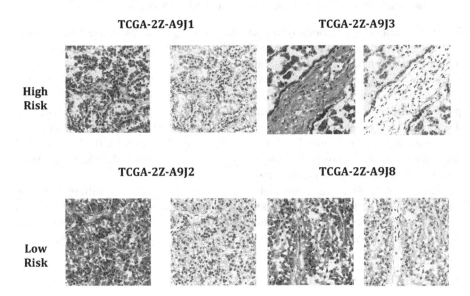

Fig. 2. H&E and eosin-stained histopathological sub-images in high- and low-risk groups

3.3 Enrichment Analysis of the Key Gene Modules

There were 471 genes in the purple module, 147 genes in the dark red module, and 3500 genes in the turquoise module, with other significant genes shown in Fig. 3. These indicate that there are significant intrinsic associations among the biological functioning of

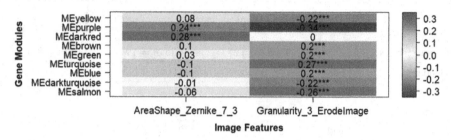

Fig. 3. Significant associations with the image features from nine modules (salmon, dark turquoise, blue, turquoise, green, brown, dark red, purple, and yellow). (Color figure online)

the selected genes in each module. Most genes were enriched in the biological processes such as metabolic process, response to stimulus, biological regulation, developmental process, localization, growth, immune system response, signaling, and cellular process. More specific associations include nuclear factor erythroid 2–related factor 2 (NRF2) pathway, glucuronidation, cellular response to DNA damage, and vascular endothelial growth factor (VEGF), particularly the VEGFA-VEGFR2 signaling pathway.

3.4   Construction and Evaluation of the Integrative Prognostic Model

The pRCC patients were randomly divided into a training ($n = 194$) and test set ($n = 83$). 30 trees were used in the integrative model. The time-dependent ROC curve can better demonstrate the model's predictive ability over time, as it incorporates both survival state and survival time in the results. In the test set shown in Fig. 4, the 11-, 16-, and 20-month areas under the curve (AUCs) were 0.877, 0.769, and 0.811 respectively. The predictive accuracy of the test set remained at a good level, especially at the 11-month discrimination. The integrative model performed significantly better than the standalone models as seen in Fig. 5, with 11-, 16-, and 20-month AUCs of 0.75, 0.733, and 0.785 respectively for the Genomic prognostic model test set; and 0.593, 0.523, and 0.603 respectively for the Phenotypic prognostic model test set.

The KM curve of the overall unweighted data showed a significantly different survival rate between the risk groups (p <0.0001). Also, the test results of the KM analysis with weights from Random Forest demonstrated that the survival rate of low-risk score patients was significantly better than that of high-risk score patients ($p = 0.00071$).

4   Discussion and Conclusion

Our study was able to identify two significant image features with the prognosis of pRCC through feature elimination, including a Zernike shape and a Granularity image. We conclude that the texture and morphology of pathological images may be correlated to pRCC prognosis. Apart from predicting prognosis, the variation in the pathological image features may incur different cell arrangements which may cause variations in the invasion of various potential tumors. We were able to conclude that histopathological image features have a certain ability to predict patient survival, and the combination of genomics and clinical data could further improve the prognosis prediction of pRCC. In the enrichment analysis, associations with the NRF2 pathway and metabolic processes were observed. Recent studies found that NRF2 indeed exhibits an aberrant activation in cancer [43]. Evidence shows that nuclear factor (erythroid-derived 2)-like 2/Kelch-like ECH-associated protein 1 (NRF2/KEAP1) signaling pathway is associated with cancer cell proliferation and tumorigenesis through metabolic reprogramming. This correlates with the established notion that RCC is known as a metabolic disease [44] as seen from the diverse array of metabolic defects and perturbations occurring as a result of the genetics driving the tumors. Similarly, study of the genetic information in pRCC denotes significant relation to the metabolic process. In the enrichment analysis, associations with the NRF2 pathway and metabolic processes were observed. Recent studies found that NRF2 indeed exhibits an aberrant activation in cancer [43]. Evidence shows that NRF2/KEAP1 signaling pathway is associated with cancer cell proliferation and tumorigenesis through metabolic reprogramming. This correlates with the established notion that RCC is known as a metabolic disease [44] as seen from the diverse array of metabolic defects and perturbations occurring as a result of the genetics driving the tumors. Similarly, study of the genetic information in pRCC denotes significant relation to the metabolic process.

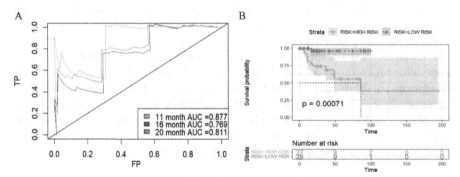

Fig. 4. (A) Test Group Survival Analysis AUC for Genomic and Phenotypic Data. (B) KM Curve of Genomic and Phenotypic Data with weight from Random Forest

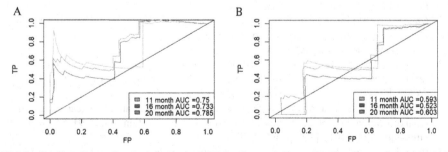

Fig. 5. (A) Test Group Survival Analysis AUC for Genomic Data. (B) Test Group Survival AUC for Phenotypic Data.

Moreover, glucuronidation also suggests an important relation to the genetic prognostic features of pRCC patients. UDP-glucuronosyltransferases (UGT) encode enzymes that regulate the glucuronidation pathway in humans. UGTs are important metabolic enzymes responsible for approximately 40–70% of endo and xenobiotic reactions [45], which include anti-cancer drugs. UGT enzymes are highly expressed in metabolic tissues such as the liver, intestine, and kidney, consistent with their role in facilitating the elimination of certain metabolites; however, the expression of most UGT members also extends to many other organs and blood cells. Generally, UGTs are anchored to the luminal side of the endoplasmic reticulum (ER), explaining why glucuronidation reactions generally occur in the lumen of the ER [46]. Moreover, UGTs, along with other drug-metabolizing enzymes and transporters, can participate in the inactivation of xenobiotics. Drug inactivation by UGTs is emerging as an important mechanism of drug resistance in cancer [46].

Typically, kidney damage is prone to trigger ER stress. In the kidney, ER stress and unfolded protein response (UPR) participate in acute and chronic histological damages, as it is linked to the molecular basis of progression of Chronic kidney diseases; however, contradictorily, it also promotes cellular adaptation and nephroprotection [47]. UPR pathway is launched by the ER, as it is involved in maintaining ER homeostasis. Dysregulation of the UPR pathway is linked to kidney disease symptoms. Experimental models have also revealed that disruption of the UPR causes podocyte injury and albuminuria as a mouse grows older [48]. This may suggest why pRCC predominantly occurs in the older generation.

Overall, the model deepened the cognition about the genomic and histopathological image information of pRCC, which could potentially aid in clinical decision-making and treatment of the disease. Moreover, further exploration of biological mechanisms of the histopathological image features and the integration of other data sources, such as clinical data, can lead to a more comprehensive understanding of the disease and improve patient outcomes. Future studies can also explore the generalizability and reproducibility of the model on larger and more diverse patient cohorts to validate its clinical utility.

References

1. Siegel, R.L., Miller, K.D., Jemal, A.: Cancer statistics, 2020. CA. Cancer J. Clin. **70**(1), 7–30 (2020). https://doi.org/10.3322/caac.21590
2. Cheng, J., et al.: Integrative analysis of histopathological images and genomic data predicts clear cell renal cell carcinoma prognosis. Cancer Res. **77**(21), e91–e100 (2017). https://doi.org/10.1158/0008-5472.CAN-17-0313
3. Filippou, P., Shuch, B., Psutka, S.P.: Advances in the characterization of clear cell papillary renal cell carcinoma: identifying the sheep in wolf's clothing. Eur. Urol. **79**(4), 478–479 (2021). https://doi.org/10.1016/j.eururo.2021.01.023
4. Morlote, D.M., Harada, S., Batista, D., Gordetsky, J., Rais-Bahrami, S.: Clear cell papillary renal cell carcinoma: molecular profile and virtual karyotype. Hum. Pathol. **91**, 52–60 (2019). https://doi.org/10.1016/j.humpath.2019.05.011
5. Rysz, J., Franczyk, B., Ławiński, J., Gluba-Brzózka, A.: Characteristics of clear cell papillary Renal Cell Carcinoma (ccpRCC). Int. J. Mol. Sci. **23**(1), 151 (2021). https://doi.org/10.3390/ijms23010151

6. Shuch, B., et al.: Understanding pathologic variants of renal cell carcinoma: distilling therapeutic opportunities from biologic complexity. Eur. Urol. **67**(1), 85–97 (2015). https://doi.org/10.1016/j.eururo.2014.04.029

7. Kovacs, G., et al.: The heidelberg classification of renal cell tumours. J. Pathol. **183**(2), 131–133 (1997). https://doi.org/10.1002/(SICI)1096-9896(199710)183:2%3c131::AID-PATH931%3e3.0.CO;2-G

8. Akhtar, M., Al-Bozom, I.A., Al Hussain, T.: Papillary Renal Cell Carcinoma (PRCC): an update. Adv. Anat. Pathol. **26**(2), 124–132 (2019). https://doi.org/10.1097/PAP.0000000000000220

9. Mendhiratta, N., Muraki, P., Sisk, A.E., Shuch, B.: Papillary renal cell carcinoma: review. Urol. Oncol. Semin. Orig. Investig. **39**(6), 327–337 (2021). https://doi.org/10.1016/j.urolonc.2021.04.013

10. Clark, I., Torbenson, M.S.: Immunohistochemistry and special stains in medical liver pathology. Adv. Anat. Pathol. **24**(2), 99–109 (2017). https://doi.org/10.1097/PAP.00000000000000139

11. Cooper, L.A., et al.: Digital pathology: data-intensive frontier in medical imaging. Proc. IEEE Inst. Electr. Electron. Eng. **100**(4), 991–1003 (2012). https://doi.org/10.1109/JPROC.2011.2182074

12. Beck, A.H., et al.: Systematic analysis of breast cancer morphology uncovers stromal features associated with survival. Sci. Transl. Med. **3**(108), 108ra113 (2011). https://doi.org/10.1126/scitranslmed.3002564

13. Gultekin, T., Koyuncu, C.F., Sokmensuer, C., Gunduz-Demir, C.: Two-tier tissue decomposition for histopathological image representation and classification. IEEE Trans. Med. Imaging **34**(1), 275–283 (2015). https://doi.org/10.1109/TMI.2014.2354373

14. Yu, K.-H., et al.: Predicting non-small cell lung cancer prognosis by fully automated microscopic pathology image features. Nat. Commun. **7**, 12474 (2016). https://doi.org/10.1038/ncomms12474

15. Hanahan, D., Weinberg, R.A.: Hallmarks of cancer: the next generation. Cell **144**(5), 646–674 (2011). https://doi.org/10.1016/j.cell.2011.02.013

16. Al-Lahham, H.Z., Alomari, R.S., Hiary, H., Chaudhary, V.: Automating proliferation rate estimation from Ki-67 histology images. In: Medical Imaging, 2012 Computer-Aided Diagnosis SPIE, pp. 669–675 (2012).https://doi.org/10.1117/12.911009

17. Mulrane, L., Rexhepaj, E., Penney, S., Callanan, J.J., Gallagher, W.M.: Automated image analysis in histopathology: a valuable tool in medical diagnostics. Expert Rev. Mol. Diagn. **8**(6), 707–725 (2008). https://doi.org/10.1586/14737159.8.6.707

18. Bartlett, J.M., et al.: Evaluating HER2 amplification and overexpression in breast cancer. J. Pathol. **195**(4), 422–428 (2001). https://doi.org/10.1002/path.971

19. Gulati, S., et al.: Systematic evaluation of the prognostic impact and intratumour heterogeneity of clear cell renal cell carcinoma biomarkers. Eur. Urol. **66**(5), 936–948 (2014). https://doi.org/10.1016/j.eururo.2014.06.053

20. Maroto, P., Rini, B.: Molecular biomarkers in advanced renal cell carcinoma. Clin. Cancer Res. **20**(8), 2060–2071 (2014). https://doi.org/10.1158/1078-0432.CCR-13-1351

21. Haury, A.-C., Gestraud, P., Vert, J.-P.: The influence of feature selection methods on accuracy, stability and interpretability of molecular signatures. PLoS ONE **6**(12), e28210 (2011). https://doi.org/10.1371/journal.pone.0028210

22. Bastien, R.R., et al.: PAM50 breast cancer subtyping by RT-qPCR and concordance with standard clinical molecular markers. BMC Med. Genomics **5**, 44 (2012). https://doi.org/10.1186/1755-8794-5-44

23. He, S., et al.: Aurora kinase A induces miR-17-92 cluster through regulation of E2F1 transcription factor. Cell. Mol. Life Sci. CMLS **67**(12), 2069–2076 (2010). https://doi.org/10.1007/s00018-010-0340-8

24. Yuan, Y., et al.: Quantitative image analysis of cellular heterogeneity in breast tumors complements genomic profiling. Sci. Transl. Med. **4**(157), 157ra143 (2012). https://doi.org/10.1126/scitranslmed.3004330

25. Calabrò, A., et al.: Effects of infiltrating lymphocytes and estrogen receptor on gene expression and prognosis in breast cancer. Breast Cancer Res. Treat. **116**(1), 69–77 (2009). https://doi.org/10.1007/s10549-008-0105-3

26. Assié, G., LaFramboise, T., Platzer, P., Bertherat, J., Stratakis, C.A., Eng, C.: SNP arrays in heterogeneous tissue: highly accurate collection of both germline and somatic genetic information from unpaired single tumor samples. Am. J. Hum. Genet. **82**(4), 903–915 (2008). https://doi.org/10.1016/j.ajhg.2008.01.012

27. Neuvial, P., Bengtsson, H., Speed, T.P.: Statistical analysis of single nucleotide polymorphism microarrays in cancer studies. In: Lu, H.H.-S., Schölkopf, B., Zhao, H. (eds.) Handbook of Statistical Bioinformatics. Springer Handbooks of Computational Statistics, pp. 225–255. Springer, Berlin, Heidelberg (2011). https://doi.org/10.1007/978-3-642-16345-6_11

28. Oh, E.-Y., et al.: Extensive rewiring of epithelial-stromal co-expression networks in breast cancer. Genome Biol. **16**, 128 (2015). https://doi.org/10.1186/s13059-015-0675-4

29. Langfelder, P., Horvath, S.: Eigengene networks for studying the relationships between co-expression modules. BMC Syst. Biol. **1**, 54 (2007). https://doi.org/10.1186/1752-0509-1-54

30. Zhang, G., Xu, S., Yuan, Z., Shen, L.: <p>Weighted gene coexpression network analysis identifies specific modules and hub genes related to major depression</p>. Neuropsychiatr. Dis. Treat. **16**, 703–713 (2020). https://doi.org/10.2147/NDT.S244452

31. Colen, R., et al.: NCI workshop report: clinical and computational requirements for correlating imaging phenotypes with genomics signatures. Transl. Oncol. **7**(5), 556–569 (2014). https://doi.org/10.1016/j.tranon.2014.07.007

32. Martins, F.C., et al.: Combined image and genomic analysis of high-grade serous ovarian cancer reveals PTEN loss as a common driver event and prognostic classifier. Genome Biol. **15**(12), 526 (2014). https://doi.org/10.1186/s13059-014-0526-8

33. Mogensen, U.B., Ishwaran, H., Gerds, T.A.: Evaluating random forests for survival analysis using prediction error curves. J. Stat. Softw. **50**(11), 1–23 (2012). https://doi.org/10.18637/jss.v050.i11

34. Pickett, K.L., Suresh, K., Campbell, K.R., Davis, S., Juarez-Colunga, E.: Random survival forests for dynamic predictions of a time-to-event outcome using a longitudinal biomarker. BMC Med. Res. Methodol. **21**, 216 (2021). https://doi.org/10.1186/s12874-021-01375-x

35. Tomczak, K., Czerwińska, P., Wiznerowicz, M.: The Cancer Genome Atlas (TCGA): an immeasurable source of knowledge. Contemp. Oncol. Poznan Pol. **19**(1A), A68-77 (2015). https://doi.org/10.5114/wo.2014.47136

36. Gutman, D.A., et al.: Cancer digital slide archive: an informatics resource to support integrated in silico analysis of TCGA pathology data. J. Am. Med. Inform. Assoc. JAMIA **20**(6), 1091–1098 (2013). https://doi.org/10.1136/amiajnl-2012-001469

37. Macenko, M., et al.: A method for normalizing histology slides for quantitative analysis. In: 2009 IEEE International Symposium on Biomedical Imaging: From Nano to Macro, Boston, MA, USA, pp. 1107–1110. IEEE (2009). https://doi.org/10.1109/ISBI.2009.5193250

38. Li, H., Chen, L., Zeng, H., Liao, Q., Ji, J., Ma, X.: Integrative analysis of histopathological images and genomic data in colon adenocarcinoma. Front Oncol 11 (2021). Accessed: Aug. 09, 2022. https://www.frontiersin.org/articles/https://doi.org/10.3389/fonc.2021.636451

39. Soliman, K.: CellProfiler: novel automated image segmentation procedure for super-resolution microscopy. Biol. Proced. Online **17**(1), 11 (2015). https://doi.org/10.1186/s12575-015-0023-9

40. Duan, K.-B., Rajapakse, J.C., Wang, H., Azuaje, F.: Multiple SVM-RFE for gene selection in cancer classification with expression data. IEEE Trans. Nanobioscience **4**(3), 228–234 (2005). https://doi.org/10.1109/tnb.2005.853657

41. Huang, Z., et al.: Deep learning-based cancer survival prognosis from RNA-seq data: approaches and evaluations. BMC Med. Genomics **13**(Suppl 5), 41 (2020). https://doi.org/10.1186/s12920-020-0686-1

42. Langfelder, P., Horvath, S.: WGCNA: an R package for weighted correlation network analysis. BMC Bioinformatics **9**, 559 (2008). https://doi.org/10.1186/1471-2105-9-559

43. Song, M.-Y., Lee, D.-Y., Chun, K.-S., Kim, E.-H.: The role of NRF2/KEAP1 signaling pathway in cancer metabolism. Int. J. Mol. Sci. **22**(9), 4376 (2021). https://doi.org/10.3390/ijms22094376

44. Rathmell, W.K., Rathmell, J.C., Linehan, W.M.: Metabolic pathways in kidney cancer: current therapies and future directions. J. Clin. Oncol. **36**(36), 3540–3546 (2018). https://doi.org/10.1200/JCO.2018.79.2309

45. Mano, E.C.C., Scott, A.L., Honorio, K.M.: UDP-glucuronosyltransferases: structure, function and drug design studies. Curr. Med. Chem. **25**(27), 3247–3255 (2018). https://doi.org/10.2174/0929867325666180226111311

46. Allain, E.P., Rouleau, M., Lévesque, E., Guillemette, C.: Emerging roles for UDP-glucuronosyltransferases in drug resistance and cancer progression. Br. J. Cancer **122**(9), 1277–1287 (2020). https://doi.org/10.1038/s41416-019-0722-0

47. Gallazzini, M., Pallet, N.: Endoplasmic reticulum stress and kidney dysfunction. Biol. Cell **110**(9), 205–216 (2018). https://doi.org/10.1111/boc.201800019

48. Cybulsky, A.V.: Endoplasmic reticulum stress, the unfolded protein response and autophagy in kidney diseases. Nat. Rev. Nephrol. **13**(11), 681–696 (2017). https://doi.org/10.1038/nrneph.2017.129

Image Visualization and Signal Analysis

Medical X-ray Image Classification Method Based on Convolutional Neural Networks

Veska Gancheva[1]([⊠]) [ID], Tsviatko Jongov[1], and Ivaylo Georgiev[2]

[1] Technical University of Sofia, Kliment Ohridski 8, 1000 Sofia, Bulgaria
vgan@tu-sofia.bg
[2] Stephan Angeloff Institute of Microbiology, Academy of Georgi Bonchev, 1113 Sofia, Bulgaria
ivailo@microbio.bas.bg

Abstract. Artificial intelligence and machine learning, including convolutional neural networks are increasingly entering the field of healthcare and medicine. The aim of the study is to optimize the learning process of convolutional neural networks through X-ray images pre-processing. A model for optimizing the overall architecture of a classifying convolutional neural network of chest X-rays by reducing the total number of convolutional operations is presented. The experimental results prove the successful application of the optimization process on the training of classification convolutional networks. There is a significant reduction in the training time of each epoch in the optimized convolutional networks. The optimization is of the order of 25% for the network with an input layer size of 124 × 124 and about 27% for the network with an input layer size of 122 × 122. The method can be applied in any field of image classification in which the informative image regions are grouped and subject to segmentation.

Keywords: Artificial Intelligence · Classification · Convolutional Neural Networks · COVID-19 · Deep Learning

1 Introduction

Artificial intelligence and machine learning are increasingly entering the field of healthcare. The analyzes applied with the innovative technology are becoming more accurate and applicable in various fields of medicine, assisting for faster and sometimes more accurate clinical diagnosis.

It is generally accepted that artificial intelligence tools will enhance and facilitate human work rather than taking the position of doctors and other healthcare professionals. Artificial intelligence is ready to assist healthcare workers with activities like administrative workflow, clinical documentation, patient outreach, and more specialized support like image analysis, medical device automation, and patient monitoring [1]. The prediction and identification of health cases, disease populations, disease states, and immune responses have advanced significantly because of recent advancements in artificial intelligence and machine learning technology. The use of these methods in

I. Rojas et al. (Eds.): IWBBIO 2023, LNBI 13920, pp. 225–244, 2023.
https://doi.org/10.1007/978-3-031-34960-7_16

healthcare settings is expanding quickly. The usage of supervised, unsupervised, and reinforcement learning is demonstrated in a quick introduction of machine learning-based learning methodologies and algorithms. Radiology, genetics, electronic health records, and neuroimaging are just a few of the healthcare fields that use artificial intelligence and machine learning-based methods. The risks and challenges of their application in healthcare, such as system privacy and ethical issues are discussed, and suggestions for future applications are examined [2].

The best trial sample can be found using healthcare data, which can also be used to increase the number of data points collected, evaluate trial participants' ongoing data, and correct databased errors. The use of ML-based techniques can help identify early signs of an epidemic or pandemic, offer a variety of treatment options and customized care, boost hospital and healthcare system efficiency overall, and reduce healthcare costs [3]. To achieve the greatest results, developing clinical decision support, sickness detection, and individualized treatment techniques will be essential.

The application of diagnostic imaging in healthcare include: 1) Diagnosis of cardiovascular abnormalities and diseases; 2) Diagnosis of fractures; 3) Diagnosis of non-neurological diseases; 4) Diagnosis of cancer; 5) Diagnosis of respiratory diseases.

COVID-19 is a viral disease that mainly affects the respiratory system. A wide variety of medical professionals in various fields continue to seek different options for diagnosing and treating the disease. Much of the researchers' efforts have been focused on the radiological manifestations of COVID-19 in the lungs [4].

The analysis of X-ray images for the diagnosis of respiratory diseases, and in particular pneumonia caused by infection with COVID-19, is extremely relevant [5–8]. This type of diagnostic system is based on artificial intelligence and in particular on convolutional neural networks - a branch of artificial intelligence dealing with image analysis. Some of the most popular applications of convolutional neural networks are [9–12]: 1) Detection of an image or part of an image; 2) Image segmentation; 3) Image classification.

Image classification finds its direct application in the diagnosis of respiratory diseases and the detection of pneumonia caused by COVID-19 disease. The training relies on a sufficient amount of pre-classified X-ray images and a serious system and time resource required for the network training process. Deep learning algorithms may quickly identify pneumonia and Covid-19 from chest X-rays or CT images [13]. Using various deep learning model architectures that have been developed or altered, research has been done on the application of deep learning approaches aimed at detecting community-acquired pneumonia, viral pneumonia, and Covid-19 from chest X-ray and CT images. The diagnosis of respiratory illnesses and the detection of pneumonia brought on by the disease COVID-19 are two areas where image classification is directly applied. The network training process requires a sizable system and time resources in addition to a sufficient number of pre-classified X-ray pictures.

In order to quantitatively compare studies, five deep learning architectures were optimized and development techniques were put into place [14]. COVID-19 local CT image dataset and the global chest X-ray dataset are utilized in a comparative investigation of the performance of four widely used deep transfer learning (DTL) models, including

VGG16, VGG19, ResNet50, and DenseNet [15]. This review study covers the main conclusions, including the AI technique utilized, the type of classification carried out, the datasets used, the results in terms of accuracy, specificity, sensitivity, score, etc., along with the limitations and future work for COVID-19 detection. A deep-learning model, specifically a convolutional neural network with pre-trained weights, which allows to use transfer learning to obtain new retrained models to classify COVID-19, pneumonia, and healthy patients is presented in [16].

The aim of the study is to explore the possibilities for optimizing the learning process of this type of systems through pre-processing of incoming X-ray images. The paper describes a novel method based on combining two types of convolutional neural networks and optimizing the training of the CNNs by reducing the active area used for the training. The method can be applied on any image dataset for image classification, which has redundant areas at the edges of the images.

2 Image Classification Optimization

The image classification process aims to divide a group of images into predefined classes. To this end, a convolution network with many hidden layers of convolution, maxpooling and dropout type is used. It is a mandatory condition that the last layers be of a fully connected type, with a last layer having a number of neurons equal to the number of classes. In the classifying neural network, the first layer is convolutional. The size is constant and in case the input images have different sizes, they should be uniform and equal to the size of the input convolution layer. The learning process of this type of classifying convolutional neural networks is resource intensive and depends on many factors, such as input volume, selected input convolution layer size, number and type of layers, number of filters for each layer, configuration of fully connected layers, and etc. It is customary for the first convolutional layer to have the highest resolution, with all subsequent convolutional layers decreasing in size until the fully bonded layers are reached. Following this logic, it is clear that the total number of convolution operations depends on the size of each convolution layer, as well as on the number of the filters required to extract the various properties. In addition, the total number of convolutional operations in a neural network is directly related to its productivity and resource intensity in the process of training and work. Reducing the total number of convolutional operations leads to a reduction in the time for training and operation of the network. On the other hand, choosing too small convolutional input layer may reduce the accuracy of the final network classification.

The present study examines a model for optimizing the overall architecture of a classifying convolutional neural network of chest X-rays, by reducing the total number of convolutional operations in the network. The total number of convolution operations can be reduced by reducing the size of the input convolution layer.

Chest X-ray is one of the most commonly used X-ray examinations. It is performed in order to diagnose pathological changes in the organs located in the chest area, as well as the surrounding anatomical structures. X-rays are used to create images of the heart, lungs, airways, blood vessels and lymph nodes.

When diagnosing pneumonia caused by COVID-19, the active region to be examined in an X-ray image is the area of the lung that does not usually occupy the full size of the image. Only this region is important in the classification of images with and without pneumonia. It is possible to use a pre-segmentation module to delineate the boundaries of the lung in the X-ray image. Thus, segmented images can be reduced together with the first input convolution layer, which will reduce the total number of convolution operations and the ability to optimize the learning process. The segmentation module can be implemented with a segmentation convolutional network. To achieve the main objectives of the study, it is necessary to perform the following tasks:

A) Design of a module for lung segmentation in an X-ray image. The module aims to generate images with the outline of the lungs in an X-ray image.
B) Design of a module for analysis of segmented images and reduction of X-ray images, based on the analysis of segmentation.
C) Design of a module for classification of X-ray images into two categories - I) X-ray images of healthy patients without diseases and II) X-ray images of patients with pneumonia caused by infection with COVID-19.
D) Comparison of system resources used in the design of a module for classification of normal X-ray images and reduced X-ray images.
E) Comparison of the quality of the modules for classification of normal X-ray images and of reduced X-ray images.

The design of the modules for classification of normal and reduced X-ray images are to be performed on the same system, in the same conditions for correct and informative comparison of the occupied resources and achieved quality of the two classification modules.

The following optimized model for X-ray image classification is proposed (Fig. 1).

A) Segmentation convolution network training
- Input data: X-ray images; Segmented images;
- Output data: Trained model for X-ray image segmentation.

The first module in the proposed classification model is the module performing the segmentation of X-ray images. This module is implemented with a segmenting convolutional neural network that needs to be trained. Chest X-rays and the corresponding manually segmented images outlining the position of the lung are used for input.

B) X-ray image segmentation
- Input data: X-ray images; Trained model for X-ray image segmentation;
- Output data: Segmented images.

The segmentation model, thus trained, is used to generate segments of non-segmented X-ray images, which will subsequently be used to optimize the learning process of the classifying neural network.

C) Analysis of the results of the segmented images
- Input data: Segmented images;
- Output data: Active regions of the segments.

The analysis of the segmented images shows the positions of the active regions in the images. Active regions indicate the parts of the images containing all the information needed for the next classification. Information outside the active regions is redundant and is to be removed.

D) Reducing the size of X-ray images
- Input data: X-ray images; Active regions of the images;
- Output data: Reduced X-ray images.

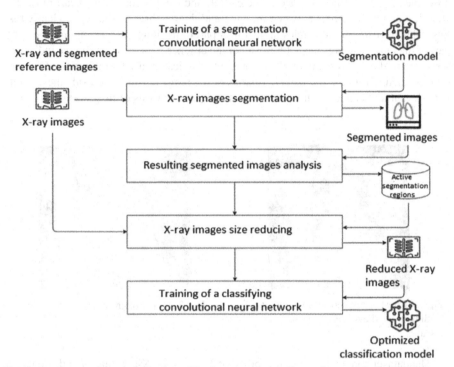

Fig. 1. Processes of an optimized model of a convolutional neural network for X-ray image classification.

The X-ray image reduction process removes redundant information and prepares the training data for the classifying neural network.

E) Training of a classifying convolutional network
- Input data: Reduced X-ray images;
- Output data: Optimized model of a convolutional neural network for image classification.

The X-ray images, thus reduced, are to be used as input for the optimized training of a classifying convolutional neural network. In addition to the learning process, the classification process is optimized for all images whose active regions after segmentation fall into the selected size at the input convolution level.

3 Segmentaion of X-ray Image

3.1 Conventional Diagnostics and Segmentation Data

The conventional manner of diagnosing pneumonia, through imaging, involves the analysis of a chest X-ray and the search for the presence or absence of a number of imaging indicators. Some of the most commonly used indicators when reading an X-ray to diagnose pneumonia are the presence of a clear left and right border of the heart muscle, a clear left and right border between the lung and the diaphragm, sharp and clearly distinguishable costophrenic angles in the lower part of the lung, no blurring in the central part of the lung. The presence of blurring or the absence of some of the above indicators serves in the diagnosis of pneumonia. Each of the indicators used for this type of diagnosis is located within the internal boundaries of the lung. The image outside these limits is irrelevant and cannot be used in the diagnosis and analysis of lung diseases (Figs. 2 and 3).

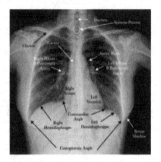

Fig. 2. Chest X-ray and the main points of reading in the diagnosis of various lung diseases

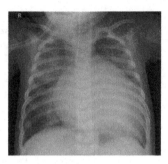

Fig. 3. X-ray of chest with pneumonia

The dataset contains x-rays and corresponding masks. X-ray images in this data set have been acquired from the tuberculosis control program of the Department of Health and Human Services of Montgomery County, MD, USA [17–19]. This set contains 138 posterior-anterior x-rays, of which 80 x-rays are normal and 58 x-rays are abnormal with manifestations of tuberculosis. All images are de-identified and available in DICOM format. The set covers a wide range of abnormalities.

The process of segmentation of the lung from an X-ray image is to delineate the boundaries of the lung and generate an output image - a mask. To create a segmentation module, it is necessary to use pre-manually segmented X-ray images (Fig. 4) [20]. The image pairs serve as input data for the training of a convolutional neural network. The purpose of such a trained network is to be able to generate segmented images at given input chest X-rays.

3.2 Creating and Training a Segmenting Convolutional Network

The input data for training is a set of image pairs - X-ray images and corresponding segmented images.

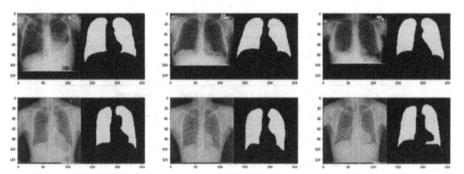

Fig. 4. X-ray input data and the corresponding segmented images of lung [17].

Input data parameters:

- Number of X-ray images: 704;
- Image format: PNG;
- Color channels: 1;
- Horizontal image size: varying (~3000 pxs);
- Vertical image size: varying (~3000 pxs);

A software implementation in the Python language has been developed to create and train a convolutional neural network for segmenting chest X-rays with anterior imaging. A function for loading test data used for validation of the neural network training process has been implemented. The test data are not used for network training, but serve for intermediate reporting of segmentation accuracy. The training and validation data lists are created according to the input resolution of the neural network. A function for estimating the loss in network training and a function for creating a convolutional neural network for segmentation have been implemented. The last layers are to be of convolutional type to achieve an image as an output of the network. The activation function of the convolution layers is of the ReLu (Rectified Linear Unit) type [21], which equates all negative values to zero while maintaining the positive values as they are input. The ReLU activation function has been introduced in order to optimize the learning process and the work process.

$$Re(x) = \max(0, x) \tag{1}$$

The designed convolutional neural network architecture is presented in Fig. 5.

The number of filters and each filter size are described in column "Output Shape". The first two numbers define the size of the filter in pixels, while the third number denotes the number of filters for each CNN layer. Each layer is described by layer type, layer size, number of filters and number of training parameters. The total number of parameters to be trained in the neural network is 7759521, with an input and output convolution layer with a size of 128 × 128 pixels and one color channel for brightness. Feedback functions, used for automatic recording of the model of the trained convolutional neural network when reaching optimal results of accuracy and termination of the learning process have been designed. Lists of data for training, validation and test have been generated. 10%

Model: "model"

Layer (type)	Output Shape	Param #	Connected to
input_1 (InputLayer)	[(None, 128, 128, 1)	0	
conv2d (Conv2D)	(None, 128, 128, 32)	320	input_1[0][0]
conv2d_1 (Conv2D)	(None, 128, 128, 32)	9248	conv2d[0][0]
max_pooling2d (MaxPooling2D)	(None, 64, 64, 32)	0	conv2d_1[0][0]
conv2d_2 (Conv2D)	(None, 64, 64, 64)	18496	max_pooling2d[0][0]
conv2d_3 (Conv2D)	(None, 64, 64, 64)	36928	conv2d_2[0][0]
max_pooling2d_1 (MaxPooling2D)	(None, 32, 32, 64)	0	conv2d_3[0][0]
conv2d_4 (Conv2D)	(None, 32, 32, 128)	73856	max_pooling2d_1[0][0]
conv2d_5 (Conv2D)	(None, 32, 32, 128)	147584	conv2d_4[0][0]
max_pooling2d_2 (MaxPooling2D)	(None, 16, 16, 128)	0	conv2d_5[0][0]
conv2d_6 (Conv2D)	(None, 16, 16, 256)	295168	max_pooling2d_2[0][0]
conv2d_7 (Conv2D)	(None, 16, 16, 256)	590080	conv2d_6[0][0]
max_pooling2d_3 (MaxPooling2D)	(None, 8, 8, 256)	0	conv2d_7[0][0]
conv2d_8 (Conv2D)	(None, 8, 8, 512)	1180160	max_pooling2d_3[0][0]
conv2d_9 (Conv2D)	(None, 8, 8, 512)	2359808	conv2d_8[0][0]
conv2d_transpose (Conv2DTranspo	(None, 16, 16, 256)	524544	conv2d_9[0][0]
concatenate (Concatenate)	(None, 16, 16, 512)	0	conv2d_transpose[0][0] conv2d_7[0][0]
conv2d_10 (Conv2D)	(None, 16, 16, 256)	1179904	concatenate[0][0]
conv2d_11 (Conv2D)	(None, 16, 16, 256)	590080	conv2d_10[0][0]
conv2d_transpose_1 (Conv2DTrans	(None, 32, 32, 128)	131200	conv2d_11[0][0]
concatenate_1 (Concatenate)	(None, 32, 32, 256)	0	conv2d_transpose_1[0][0] conv2d_5[0][0]
conv2d_12 (Conv2D)	(None, 32, 32, 128)	295040	concatenate_1[0][0]
conv2d_13 (Conv2D)	(None, 32, 32, 128)	147584	conv2d_12[0][0]
conv2d_transpose_2 (Conv2DTrans	(None, 64, 64, 64)	32832	conv2d_13[0][0]
concatenate_2 (Concatenate)	(None, 64, 64, 128)	0	conv2d_transpose_2[0][0] conv2d_3[0][0]
conv2d_14 (Conv2D)	(None, 64, 64, 64)	73792	concatenate_2[0][0]
conv2d_15 (Conv2D)	(None, 64, 64, 64)	36928	conv2d_14[0][0]
conv2d_transpose_3 (Conv2DTrans	(None, 128, 128, 32)	8224	conv2d_15[0][0]
concatenate_3 (Concatenate)	(None, 128, 128, 64)	0	conv2d_transpose_3[0][0] conv2d_1[0][0]
conv2d_16 (Conv2D)	(None, 128, 128, 32)	18464	concatenate_3[0][0]
conv2d_17 (Conv2D)	(None, 128, 128, 32)	9248	conv2d_16[0][0]
conv2d_18 (Conv2D)	(None, 128, 128, 1)	33	conv2d_17[0][0]

Total params: 7,759,521
Trainable params: 7,759,521
Non-trainable params: 0

Fig. 5. Architecture of a model for segmenting convolutional neural network

of the data is allocated for validation. The accuracy of the model is extremely important, as it is calculated by means of validation data, which are separated from the training data. The training process is started with a set of 50 epochs, and each epoch represents one cycle of training passing through all input data.

The achieved accuracy of the trained model on the validation data is 0.9802 and the loss in the analysis of the validation data is −0.9600 (Figs. 6 and 7).

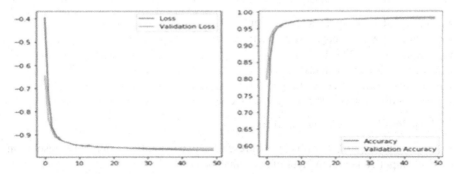

Fig. 6. Training results – loss, validation loss, accuracy, validation accuracy.

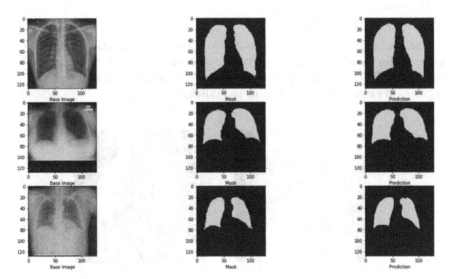

Fig. 7. Results of the trained neural network. "Base Image" – input x-ray images, "Mask" – input masks (segmented images), "Prediction" – predicted/computed mask.

4   Image Reduction and Neural Network Training

4.1   Generating of Segmented Images

Once trained, the convolutional neural network model can be applied to X-ray image segmentation. The input images are divided into 4 groups - training data, validation data, X-ray images with the presence of pneumonia caused by COVID-19 and X-ray images of a healthy lung. Input data parameters:

- Number of X-ray images for training: 150
- Number of X-ray images for validation: 40
- Image format: PNG, JPG
- Number of color channels: 3
- Horizontal image size: varying (~2000 pixels)
- Vertical image size: varying (~2000 pixels)

Functions have been developed for:

- Loading the input X-ray images and rescaling to a resolution equal to the resolution of the input convolution layer of the trained network.
- Generating of segmented images - masks with a trained model.
- Loading the model of the convolutional neural network, setting the resolution of the input data.
- Generating segment images - masks.

The used dataset is organized into 3 groups (train, test, validation) and contains subgroups for each image category (Pneumonia/Normal) [22, 23]. There are 5863 X-Ray images (JPEG) and 2 categories (Pneumonia/Normal). Generation is performed for all groups of input images - for training and validation (Fig. 8).

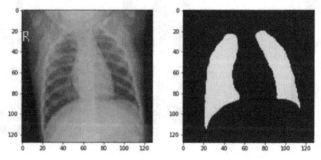

Fig. 8. X-ray image of the lung and the prepared segment image from the pre-trained model of a segmenting convolutional neural network [22].

4.2   Active Regions Computation

The active data region should be calculated for each segmented image. A segmented image I having one luminance channel, width W and height H can be considered as a two-dimensional vector. If the accepted index values h and take the values from 1 to the corresponding width and height of the two-dimensional vector, then:

$$h = \{1, 2, \dots H - 1, H\} \tag{2}$$

$$w = \{1, 2, \dots W - 1, W\} \tag{3}$$

The two-dimensional vector can be divided into a group of horizontal one-dimensional vectors corresponding to the rows in the image:

$$X_h = \left[x_{1,h}, x_{2,h}, ..x_{W-1,h}, x_{W,h} \right] \tag{4}$$

or a group of vertical one-dimensional vectors corresponding to the columns in the image:

$$Y_w = \left[y_{w,1}, y_{w,2}, ..y_{w,H-1}, y_{w,H} \right] \tag{5}$$

On the two groups of one-dimensional vectors, thus formed, a search can be performed for the four boundaries - upper, lower, left and right, separating the active region of the segmented image. For this purpose, two functions for searching for an active value in a one-dimensional horizontal vector $Ar(Xh)$ and a vertical vector $Ac(Yw)$ are defined:

$$Ar(X_h) = Ar(x_{i,h}) = \begin{cases} 1, x_{i,h} \geq 0,5 \\ Ar(x_{i+1,h}), x_{i,h} < 0,5 \end{cases} \tag{6}$$

$$Ac(Y_w) = Ac(y_{w,i}) = \begin{cases} 1, y_{w,i} \geq 0,5 \\ Ac(y_{w,i+1}), y_{w,i} < 0,5 \end{cases} \tag{7}$$

The functions applied to a horizontal or vertical vector assume a value of 1 in case a value higher than or equal to the average value that each element of the vector can accept is found. All values are normalized and accept values from 0 to 1. Thus, the search functions are activated when an open element with a value equal to or greater than 0.5 is activated.

Thus defined, search functions of active elements in one-dimensional horizontal and vertical vectors are applied in the definition of the functions for calculation of the four boundaries of the active region in the image - BT, BB, BL, and BR (upper, lower, left and right boundary).

$$B_T(I) = B_T(X_k) = \begin{cases} k, Ar(X_k) = 1 \\ B_T(X_{k+1}), Ar(X_k) \neq 1 \end{cases} \tag{8}$$

where $k = \{1, 2, ..H - 1, H\}$.

$$B_B(I) = B_B(X_k) = \begin{cases} k, Ar(X_k) = 1 \\ B_B(X_{k-1}), Ar(X_k) \neq 1 \end{cases} \tag{9}$$

where $k = \{H, H - 1, ..2, 1\}$.

The search is performed in ascending order of the horizontal vectors when searching for the upper active limit and in descending order when searching for the lower active limit. The two-dimensional horizontal vectors belong to the input two-dimensional vector of the image $Xk \in I$:

$$B_L(I) = B_L(Y_k) = \begin{cases} k, Ac(Y_k) = 1 \\ B_L(Y_{k+1}), Ac(Y_k) \neq 1 \end{cases} \tag{10}$$

where $k = \{1, 2, ..W - 1, W\}$.

$$B_R(I) = B_R(Y_k) = \begin{cases} k, Ac(Y_k) = 1 \\ B_R(Y_{k-1}), Ac(Y_k) \neq 1 \end{cases} \tag{11}$$

where $k = \{W, W - 1, ..2, 1\}$.

The search is performed in ascending order of the vertical vectors when searching for the left active boundary and in descending order when searching for the right active boundary. The two-dimensional vertical vectors belong to the input two-dimensional vector of the image $Yk \in I$.

Functions have been developed for:

- Calculation of active regions from the created segment images.
- Loading segmented images and rescaling to a preset size.
- Calculation of upper, lower, left and right limits of an active region.
- Calculation the boundaries of a list of segmented images and generate a list of active regions, widths and heights of regions.
- Setting image width and height of 128 pixels, load and transform input data.
- Normalization of brightness values in the range from -1 to $+1$.
- Calculation of active regions of the segmented images from the neural network training group for classification.
- Saving a list of active regions in a.csv file.

In the same way, the lists of active regions of the data for validation and testing after training are computed and recorded.

4.3 Analysis of the Segmented Image

The analysis of the resulting segmented images shows the presence of irrelevant data at the edges that can be removed. The following images visualize the active regions of the lung in the corresponding X-ray images. The active regions are defined by an upper, lower, left and right border, and no active pixels should fall outside the active regions (Fig. 9). The distribution of the widths and heights of the active regions in the X-ray images may be seen using functions, and the active regions can also be loaded from previously recorded images.CSV files; training data dispersion visualization.

From the distribution of the widths and heights of the active regions in the already rescaled images, it is noticed that the majority of the images have a width and height of the active region of 100 pixels (Figs. 10 and 11). The presented distributions of active regions prove that the reduction of X-ray images by active regions is possible and will not lead to loss of information necessary for the construction of the classification module.

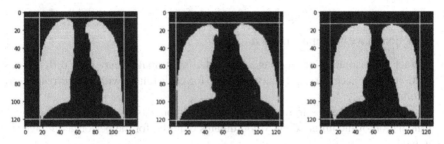

Fig. 9. Segmented images with visualization of the active regions.

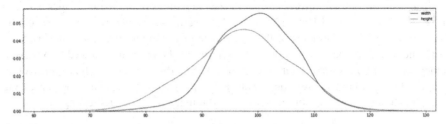

Fig. 10. Distribution of widths and heights of the active regions in images with a healthy lung.

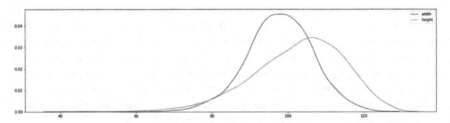

Fig. 11. Distribution of widths and heights of active regions of images with pneumonia.

4.4 Image Reduction

From the analysis of the distributions of the active regions, a reduction in the size of the X-ray images by 8% was chosen. The size reduction needs to be performed according to the respective sizes and positions of the active regions in each image. Information about the positions of the active regions is extremely important for the preservation of the informative data in the reduced images.

Input data parameters:

- Number of X-ray images for training: 150
- Number of X-ray images for validation: 40
- Data for active regions
- Image format: PNG, JPG
- Number of color channels: 3
- Horizontal image size: varying (~2000 pixels)

- Vertical image size: varying (~2000 pixels)

  Functions have been implemented for:

- Extract the resolution of an image. Since the input images may have different resolutions, it is necessary to reduce the size for each image with the corresponding resolution.
- Extract an active region from the saved active region lists.
- Calculation of the initial x and y coordinates required for the performed image size reduction.
- X-ray image size reduction.

The position and size information of the active regions of the X-ray images is stored in a relative image size of 128 x 128 pixels. This requires the conversion of the reduction coordinates to the coordinates of the actual size of the input image. The preservation of all the details in the image is the main priority of the procedure, in order to prevent underestimation of the quality of the classification module. PNG, which is a format that uses lossless graphic compression, is selected as the output file format. The processing is applied to all groups of input images - for training, validation and testing.

4.5   Training of a Classifying Convolutional Neural Network

The training process of a classifying convolutional neural network with reduced chest X-ray images has been developed in three main variants, the first presenting standard image training without reducing the number of convolutional operations, and the other two using reduced input convolutional layers. The implementation of the study in three variants is necessary for the correct measurement of the level of optimization, as well as for monitoring the quality of classification of the three models of neural networks. The training of a classifying convolutional neural network is without reducing the number of convolutional operations in the input convolutional layer [13].

Input data parameters:

- Number of X-ray images for training: 150
- Number of X-ray images for validation: 40
- Image format: PNG
- Number of color channels: 3
- Horizontal image size: varying (~1850 pixels)
- Vertical image size: varying (~1850 pixels)

The number of images for sequential processing is set, the size of the input convolutional layer and the number of epochs, corresponding to the number of cycles of submission of the full set of input data in training. The data from the input files are divided into training and validation groups. The data contain 1081 images with a healthy lung and 3104 images with the presence of pneumonia. The type of the convolutional network is sequential, and the following types of layers are used in the model:

- Input layer;
- Standard two-dimensional convolution layer (Conv2D);
- SeparableConv2D;

- MaxPool2D layer;
- Normalization layer (BatchNormalization);
- Dropout;
- Leveling layer (Flatten);
- Fully connected layer (Dense).

The last layer of the classification convolutional network is of the fully connected layer type and consists of only one neuron. The activation function is also different, in which case it is typical to use a Sigmoid type function. This provides initial values in the last neuron in the range of -1 to 1, which helps for correct classification. For each weight in the learning process, the gradient or the direction in which the overall function increases its value is calculated. Since the total value of the function represents the loss, it is necessary to change each weight gradually in the opposite direction to the gradient. And here the speed of change of each weight is extremely important. This speed is determined by the optimization algorithm and serves to overcome the small local extremes in search of the global extremum for each weight. The loss estimation function is binary cross-entropy, which calculates the loss at each stage of training.

$$L\left(Y, \hat{Y}\right) = -\frac{1}{N} \sum_{n=1}^{N} y_i . log \hat{y}_i + (1 - y_i) . \log(1 - \hat{y}_i) \tag{12}$$

where is vector of the output values, and Y is vector with the corresponding expected values.

Accuracy is calculated on the basis of the quotient of the neural network outputs, coinciding with the expected output values and the total number of values.

$$ACC(TP, N) = \frac{TP}{N} \tag{13}$$

where TP is the number of output values that match the expected values, and N is the total number of values.

The learning process is performed with set 25 epochs - a training cycle, passing through all input data. The training process is performed with predefined 25 epochs - a cycle of training passing through all input data. Thus created, the training model is applied to three types of classifying convolutional neural networks in order to assess the level of optimization and change in the levels of accuracy and loss.

5   Experimental Results and Discussion

The experimental training of neural networks is divided into three parts, in each part are trained: 1) Neural network with standard input convolution layer - without optimization. The size of the input convolution layer is 128 × 128. 2) Neural network with reduced input convolution layer. The size of the input convolution layer is 124 × 124. 3) Neural network with reduced input convolution layer. The size of the input convolution layer is 122 × 122. The training is carried out in 25 epochs, as in each epoch all input data for training are processed and temporary levels of precision and error are calculated based on the validation data. The training of the three proposed neural networks follows successively. The results are presented in Figs. 12, 13, 14, 15 and 16.

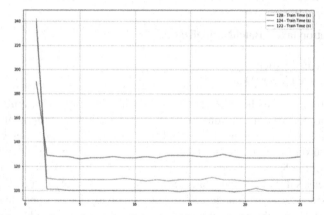

Fig. 12. Training time for each epoch - from 1 to 25 of the three types of convolutional neural networks - 128, 124 and 122.

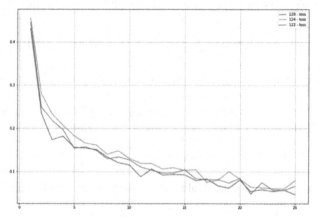

Fig. 13. Losses of the three types of convolutional neural networks calculated in training for each epoch on training data.

It is observed from the obtained results that the training time for each epoch of the optimized neural networks is less than the training time for each epoch of the non-optimized neural network. The accuracy of the training data progressively increases and the loss decreases, which shows good training of the model. Accuracy and loss in validation data follow the same direction but with a higher variance. In the different parts of the training process there is an advantage of each of the neural networks over the other two, taking into account the temporary levels of loss and accuracy.

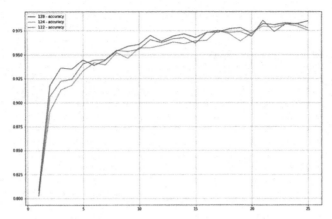

Fig. 14. Accuracy of the three types of convolutional neural networks, calculated during training for each age on the training data.

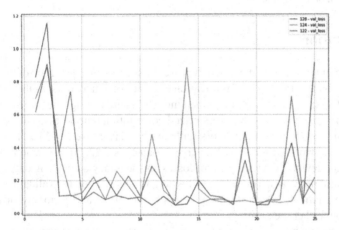

Fig. 15. Losses of the three types of convolutional neural networks, calculated during training for each epoch on the validation data.

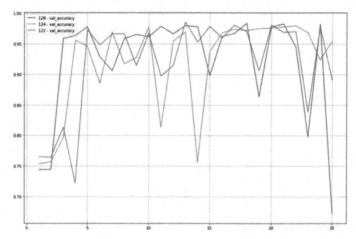

Fig. 16. Accuracy of the three types of convolutional neural networks, calculated during training for each epoch on the validation data.

6   Conclusion

The experimental results of the research prove the successful application of the optimization process on the training of classification convolutional networks, as the optimization does not affect the accuracy of the trained models. There is a significant reduction in the training time of each epoch in the optimized convolutional networks. The optimization is of the order of 25% for the network with an input layer size of 124 × 124 and about 27% for the network with an input layer size of 122 × 122. At the same time, there is no significant deviation in the values of losses and accuracy on training data of the three types of neural networks. The values of losses and accuracy on the validation data show significant variations, which do not give a significant advantage to any of the neural networks, but rather are arbitrary.

Three datasets have been used for conducting the research – one dataset for X-ray masks and two datasets for image classification. The images are in lossy jpeg or png format not of perfect quality, but are suitable for proving the point of the research.

The current study is applied to segmentation and classification of X-ray images of the lung, but the method can be applied in any field of image classification in which the informative image regions are grouped and subject to segmentation.

Of great importance is the stage of preliminary segmentation and analysis of the active regions, in which the distribution of the widths and heights of the obtained active regions is studied. This distribution is necessary to determine the effectiveness of the described optimization model.

Acknowledgements. This research was funded by National Science Fund, Bulgarian Ministry of Education and Science, grant number KP-06-N37/24, project "Innovative Platform for Intelligent Management and Analysis of Big Data Streams Supporting Biomedical Scientific Research".

References

1. Bohr, A., Memarzadeh, K.: The rise of artificial intelligence in healthcare applications. Artificial Intelligence in Healthcare, 25–60 (2020). https://doi.org/10.1016/B978-0-12-818438-7.00002-2. Epub 2020 Jun 26. PMCID: PMC7325854
2. Habehh, H., Gohel, S.: Machine learning in healthcare. Curr. Genomics **22**(4), 291–300 (2021). https://doi.org/10.2174/1389202922666210705124359. PMID: 35273459; PMCID: PMC8822225
3. Javaid, M., et al.: Significance of machine learning in healthcare: features, pillars and applications. Int. J. Intell. Netw. **3**, 58–73 (2022). ISSN 2666–6030. https://doi.org/10.1016/j.ijin.2022.05.002
4. Soffer, S., Ben-Cohen, A., Shimon, O., Amitai, M.M., Greenspan, H., Klang, E.: Convolutional neural networks for radiologic images: a radiologist's guide. Radiology **290**(3), 590–606 (2019). https://doi.org/10.1148/radiol.2018180547
5. Borkowski, A.A., Viswanadhan, N.A., Thomas, L.B., Guzman, R.D., Deland, L.A., Mastorides, S.M.: Using artificial intelligence for COVID-19 chest x-ray diagnosis. Federal Practitioner: for the health care professionals of the VA, DoD, and PHS **37**(9), 398–404 (2020). https://doi.org/10.12788/fp.0045
6. Reshi, A.A., et al.: An efficient CNN model for COVID-19 disease detection based on x-ray image classification. Complexity **2021**, Article ID 6621607, 12 (2021). https://doi.org/10.1155/2021/6621607
7. Wang, S., et al.: A deep learning algorithm using CT images to screen for Corona virus disease (COVID-19). Eur. Radiol. **31**(8), 6096–6104 (2021). https://doi.org/10.1007/s00330-021-07715-1
8. Arias-Garzón, D., et al.: COVID-19 detection in x-ray images using convolutional neural networks. Mach. Learn. Appl. **6**, 100138 (2021). ISSN 2666–8270.https://doi.org/10.1016/j.mlwa.2021.100138
9. Sorić, M., Pongrac, D., Inza, I.: Using convolutional neural network for chest x-ray image classification. In: 43rd International Convention on Information, Communication and Electronic Technology (MIPRO) (2020). https://doi.org/10.23919/MIPRO48935.2020.9245376
10. Nkwentsha, X., Nkwentsha, X., Hounkanrin, A., Hounkanrin, A., Nicolls, F.: Automatic classification of medical x-ray images with convolutional neural networks. In: 2020 International SAUPEC/RobMech/PRASA Conference. https://doi.org/10.1109/SAUPEC/RobMech/PRASA48453.2020.9041052
11. Yang, N., Niu, H., Chen, L., et al.: X-ray weld image classification using improved convolutional neural network. AIP Conference Proceedings **1995**, 020035 (2018). https://doi.org/10.1063/1.5048766
12. Yadav, S.S., Jadhav, S.M.: Deep convolutional neural network based medical image classification for disease diagnosis. J. Big Data **6**(1), 1–18 (2019). https://doi.org/10.1186/s40537-019-0276-2
13. Shah, A., Shah, M.: Advancement of deep learning in pneumonia/covid-19 classification and localization: a systematic review with qualitative and quantitative analysis. J. Chronic Dis. Transl. Med. **8**, 154–171 (2022). https://doi.org/10.1002/cdt3.17
14. Baltazar, L.R., et al.: Artificial intelligence on COVID-19 pneumonia detection using chest x-ray images (2021). https://doi.org/10.1371/journal.pone.0257884
15. Bhatele, K.R., Jha, A., Tiwari, D., et al.: COVID-19 detection: a systematic review of machine and deep learning-based approaches utilizing chest x-rays and CT scans. Cogn. Comput. (2022). https://doi.org/10.1007/s12559-022-10076-6
16. Luján-García, J.E., Moreno-Ibarra, M.A., Villuendas-Rey, Y., Yáñez-Márquez, C.: Fast COVID-19 and pneumonia classification using chest x-ray images. Mathematics **8**, 1423 (2020). https://doi.org/10.3390/math8091423

17. Chest Xray Masks and Labels, Pulmonary Chest X-Ray Defect Detection. https://www.kag gle.com/nikhilpandey360/chest-xray-masks-and-labels

18. Jaeger, S., et al.: Automatic tuberculosis screening using chest radiographs. IEEE Trans. Med. Imaging **33**(2), 233–245 (2014). https://doi.org/10.1109/TMI.2013.2284099. PMID: 24108713

19. Candemir, S., et al.: Lung segmentation in chest radiographs using anatomical atlases with nonrigid registration. IEEE Trans. Med. Imaging **33**(2), 577–590 (2014). https://doi.org/10. 1109/TMI.2013.2290491. PMID: 24239990

20. Jaegerm, S., Candemirm, S., Antanim, S., Wángm, Y.X., Lum, P.X., Thomam, G.: Two public chest x-ray datasets for computer-aided screening of pulmonary diseases. Quant. Imaging Med. Surg. **4**(6), 475–477 (2014). https://doi.org/10.3978/j.issn.2223-4292.2014.11.20. PMID: 25525580; PMCID: PMC4256233

21. Brownlee, J.A.: Gentle Introduction to the Rectified Linear Unit (ReLU). Machine Learning Mastery (2019)

22. Chest X-Ray Images (Pneumonia). https://www.kaggle.com/paultimothymooney/chest-xray-pneumonia

23. Kermany, D., et al.: Identifying medical diagnoses and treatable diseases by image-based deep learning. Cell **172**(5), P1122-1131 (2018). https://doi.org/10.1016/j.cell.2018.02.010

Digital Breast Tomosynthesis Reconstruction Techniques in Healthcare Systems: A Review

Imane Samiry[✉], Ilhame Ait Lbachir, Imane Daoudi, Saida Tallal, and Sayouti Adil

Engineering Research Laboratory, ENSEM, Hassan II University, Casablanca, Morocco
{imane.samiry.doc21,ilhame.aitlbachir,i.daoudi,
s.tallal}@ensem.ac.ma

Abstract. Digital Breast Tomosynthesis (DBT) images are widely used to increase breast cancer detection and reduce recall rates in healthcare systems for breast cancer detection. In the field of medical imaging, computer-aided diagnosis (CAD) systems are used to analyze this type of images. Generally, in order to achieve an early detection of breast cancer, these CAD systems start with the reconstruction part of the image, the pre-processing step and then the segmentation and classification. However, the post-acquisition techniques of DBT can impact the detection and diagnosis of breast cancer, and bias the final decision in computer-aided detection and diagnosis systems. Mainly, the reconstruction phase in computer aided detection systems, that helps prepare the DBT for further analysis, such as segmentation and classification of abnormalities. In this paper, we present a survey of different techniques for DBT reconstruction, that we compared theoretically in terms of advantages and drawbacks, particularly for healthcare systems dedicated to breast cancer detection.

Keywords: breast cancer · computer aided detection · digital breast tomosynthesis · healthcare systems · reconstruction

1 Introduction

Breast cancer is the most common malignancy in women [1]. In healthcare systems for breast cancer detection and diagnosis, full-field digital mammography (FFDM) is frequently employed [2]. However, mammography has an intrinsic limitation that leads it to miss some worrisome malignant tumors when tissue overlaps, especially in dense breasts [3].

A developing technique for detecting and diagnosing breast cancer called Digital Breast Tomosynthesis uses quasi-three-dimensional imaging to provide a detailed assessment of the dense tissue inside the breast. By reducing the impact of overlapping tissue on screening, DBT has surpassed digital mammography (DM) in terms of lesion identification, characterization, and diagnosis [4, 5].

Early detection of breast abnormalities with DBT can facilitate improved treatment and management of breast cancer [6, 7]. In medical imaging, computer-aided detection

© The Author(s), under exclusive license to Springer Nature Switzerland AG 2023
I. Rojas et al. (Eds.): IWBBIO 2023, LNBI 13920, pp. 245–255, 2023.
https://doi.org/10.1007/978-3-031-34960-7_17

(CADe) and computer-aided diagnostic (CADx) systems are used to analyze this type of images.

In fact, the goal of CAD systems is to assist radiologists in their interpretations and decision [8]. We present in Fig. 1 the block diagram of DBT processing in CAD systems.

Fig. 1. Block diagram of DBT processing in CAD systems

In the first step, the three-dimensional DBT image is reconstructed by combining the set of the projected views [9]. Several reconstruction methods have been proposed in the literature, and can be classified into four categories, namely, back-projection techniques, transform techniques, algebraic reconstruction techniques, and statistical reconstruction techniques [9].

In the second step, the quality of DBT images is improved by enhancing the contrast and eliminating the existing noise. The aim of this step is to prepare the reconstructed image to the following processes.

In the third step, the regions of interest (ROIs) likely to have masses, micro-calcifications or architectural distortions are extracted [10].

In the fourth step, the extracted ROIs in the previous step are classified into benign or malignant abnormalities.

In this paper, we focus on analyzing DBT reconstruction step. The main contribution of this paper is the proposition of a review of the different methods for DBT preprocessing techniques in the literature. We provide herein a qualitative comparison of those methods.

This paper is organized as follows: In Sect. 2, a comparison between DM and DBT is presented, highlighting the advantages of the latter. Section 3 reviews recent DBT reconstruction methods. Section 4 proposes an analytical comparison study of the existing reconstruction methods for DBT, and a discussion is performed. Finally, the conclusion is presented in Sect. 5.

2   Tomosynthesis Technology for Breast Imaging

Tomosynthesis is a new technology for breast imaging [11]. The tomosynthesis device has an x-ray tube that is not fixed. It can also make a circular arc of 15° to 45°, allowing several images to be taken from different angles. The computer then generates a three-dimensional (3D) image of the breast gland from the different planes obtained. The tomosynthesis device in Fig. 2 can also perform direct digital conventional mammograph.

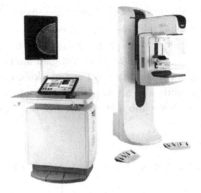

Fig. 2. Tomosynthesis device

For diagnostic purposes, tomosynthesis may be requested as an additional examination instead of compression. Radiologists suggest its use when mammography interpretation is difficult, particularly in women with dense breasts, young women and women at risk.

2.1 The Advantages of Tomosynthesis Compared to 2D Mammography

Currently, projection mammography (see Fig. 3a) [12] is the most popular imaging method for detecting occult breast cancer. The projection method, which creates 2D pictures of 3D breast anatomy, has a number of fundamental limitations that affect projection mammography [13]. In the projection image, normal tissues that are spatially isolated from one another might superimpose, creating artificially dense tissue patches that resemble lesions in appearance. True lesions may be covered up by normal tissue on projection mammograms, which is another problem. [14, 15].

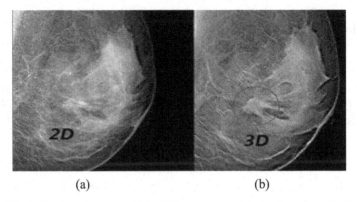

(a)   (b)

Fig. 3. Region of interest: (a) in (2D) mammography, (b) in tomosynthesis

Digital breast tomosynthesis (see Fig. 3b) is an emerging tomographic imaging modality. In DBT, tomographic pictures are reconstructed using a large number of x-ray projection images (usually 9 to 48 projections) that were collected over a limited angular range (15° to 60°). Tomosynthesis can mitigate the problems of superposition of nonadjacent tissue (false positive densities) and masking of real lesions observed in projection mammography, while also allowing 3D localization of lesions [13].

The following Table 1 shows the similarities and differences between these two breast imaging techniques:

Table 1. Comparison between tomosynthesis and mammography

Similarities	Differences
-Both are imaging techniques used to detect signs of breast cancer	-Tomosynthesis is a more advanced and detailed imaging technique than traditional mammography
-They are used both for annual exams and to check for breast cancer progression	-A traditional mammogram captures only a 2D image
	-Tomosynthesis can examine multiple layers of the breast in a 3D image
	-Tomosynthesis is more recommended for dense breast
	-Tomosynthesis allows doctors to see small lesions and other signs of breast cancer earlier than they would with a traditional mammogram

According to several studies, DBT can offer better diagnostic precision than conventional digital mammography [14–16].

3   Methods of Reconstruction Phase

3.1   The Importance of the Reconstruction Phase in CAD Systems

The DBT technique uses reconstruction algorithms to converts a number of projection images [17] that are acquired at limited view angles into a three-dimensional (3D) image set (see Fig. 4) [18].

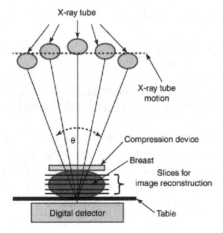

Fig. 4. A schematic presentation of the tomosynthesis principle

Four categories of reconstruction approaches have been proposed in the literature: back-projection techniques, transform techniques, algebraic reconstruction techniques, and statistical reconstruction techniques [9].

3.2 Back-Projection Algorithms

The Shift-And-Add Reconstruction (SAA).
The shift-and-add algorithm is the standard tomosynthesis reconstruction algorithm (SAA) [9]. Figure 5 provides an illustration of SAA's core concept. A voxel's value

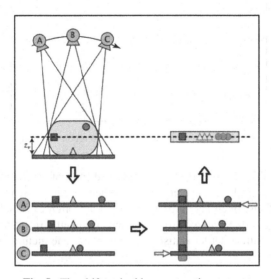

Fig. 5. The shift-and-add reconstruction concept

at position (x_v, y_v, z_v) is represented by:

$$V(x_v, y_v, z_v) = \frac{1}{N_p} \sum_{k=1}^{N_p} I_{(org,k)}(x_v + \xi_k(z_v), y_v + \eta k(z_v)) \qquad (1)$$

where $V(x_v, y_v, z_v)$ is a voxel of the reconstructed volume, $I_{(org,k)}$ is the kth projection image, N_p is the number of projections and $\xi(z_v)$ and $\eta(z_v)$ are the shift factors in x and y direction respectively.

As a result, SAA is a strictly linear operator and does not include any filtering steps. But the SAA technique is only effective when the X-ray source's direction is parallel to the detector (corresponding to a linear motion at a fixed height above the detector). Different features in a chosen plane cannot be simultaneously recognized by simply moving and adding the projections when a rotational tube motion, such as in DBT, is present.

The Simple Back-Projection (SBP).
The shift necessary to register an object point depends on where it is located in the plane. No matter how the X-ray tube is moved, an algorithm for back-projection (BP), also known as simple back-projection (SBP), accurately includes the imaging geometry. This method's equation is quite similar to Eq. (1) The extra scale factor $s_k(z_v)$ makes a difference. Therefore, the equation is:

$$V(x_v, y_v, z_v) = \frac{1}{N_p} \sum_{k=1}^{N_p} I_{(org,k)}(s_k(z_v) \cdot x_v + \xi_k(z_v), \; s_k(z_v) \cdot y_v + \eta k(z_v))$$

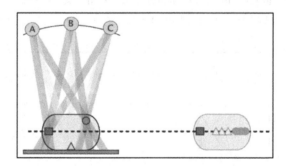

Fig. 6. The simple back-projection reconstruction

Consequently, the SAA approach is a condensed form of the SBP method. With the SBP approach, certain additional techniques have been used to lessen the streak artifacts (out-of-plane artifacts, interplane artifacts), which are often extremely strong in SBP reconstructed images (see Fig. 6). Strong artifacts frequently result from features with particularly high contrast. Order statistics operators are one method used to minimize these artifacts [9].

3.3 Transform Algorithms

Transform algorithms include the filtered back-projection (FBP) which is the analytical algorithm widely recognized for reconstructing tomosynthesis (Fig. 7).

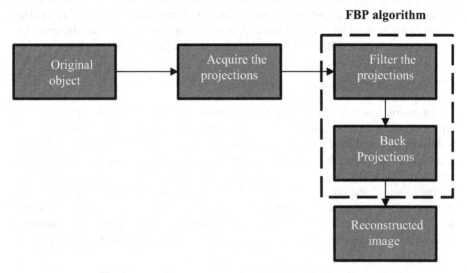

Fig. 7. Transform reconstruction with the FBP algorithm

In the FBP approach [19, 20], projection images are converted into the spatial frequency domain using the Fourier slice theorem. The 2D Fourier transform of a projection image captured at a specific angle translates, with a parallel beam approximation, to a slice sample in the spatial frequency domain at that angle. In order to acquire a discrete sampling of the whole spectrum of the observed volume, numerous projection images can be combined.

3.4 Algebraic Reconstruction Techniques

Many algebraic techniques have been researched for the general reconstruction problem. These methods, which generally set up a system of simultaneous linear equations and use iterative techniques to solve this linear algebra issue, include the algebraic reconstruction technique (ART) [21], the simultaneous iterative reconstruction technique (SIRT) [22], and the simultaneous algebraic reconstruction technique (SART) [23].

In ART, the linear attenuation coefficients are updated in a ray-by-ray manner. The difference between the detected and computed values updates each voxel along the ray being considered. The linear attenuation coefficient is updated using only one projection value at a time, so although ART has a quick convergence rate, it will eventually arrive at a (least squares) solution, which can be very noisy for severely ill-posed inverse problems like limited-angle tomosynthesis reconstruction [24].

To improve the ART method, variations on its implementation have been proposed. In SART [23], each voxel's linear attenuation coefficient is adjusted based on all rays passing through it. This has shown out to be a successful tradeoff between a noisy solution and a slow convergence rate.

Furthermore, Chen et al. [25] developed for DBT a new algebraic reconstruction technique called matrix inversion tomosynthesis (MITS) [25]. MITS was proven to work well with high frequency information and the removal of out-of-plane artifacts, but badly with low frequency information. [26] To address this, Chen et al. devised the Gaussian frequency blending reconstruction algorithm, which combines the good low frequency behavior of FBP with the high frequency response of MITS (GFB).

3.5 Statistical Reconstruction Techniques

The reconstruction algorithms that have been explored thus far are all deterministic techniques. Statistical reconstruction techniques, on the other hand, approach the inverse problem from a statistical perspective, considering the unknown attenuation coefficients as a random variable adhering to some certain probability distribution function. The maximum likelihood method (ML) is an illustration of this kind of approach. It is challenging to search the whole space of unknown sets for the ML answer since the logarithm probability function does not have an analytical solution. Iteratively maximizing the log-likelihood function has been studied using a variety of techniques, including the expectation maximization (EM) algorithm [23] convex algorithm [27], and gradient algorithm [28].

4 Analysis and Discussions

The following table compares the different categories of Digital Breast Tomosynthesis reconstruction methods (Table 2):

Image reconstruction is the basic operation performed in three-dimensional imaging. It enables the combination of a number of projected perspectives in order to recreate the original volume. Any pre-processing applied to the projected views should only be used to obtain a reversal of the projection process.

As already presented, several methods of reconstruction DBT images have been proposed in the literature, and can be classified into four categories, namely, back-projection techniques, transform techniques, algebraic reconstruction techniques, and statistical reconstruction techniques.

Algebraic reconstruction techniques have the ability to produce superior quality images with less radiation exposure compared to FBP. However, their practical application is limited due to the longer processing time required [17]. Furthermore, the researchers discovered that the BP algorithm provided the highest quality in-plane images for larger features with low contrast, but suffered from out-of-plane artifacts [29]. The FBP algorithm, as expected, performed better than BP for high frequency features.

To sum up each method has its strong points and other points that are weak. However, the latter can be covered in most cases by combining different methods in order to benefit from the advantages of each according to the case studied in the tomosynthesis images.

Table 2. Comparison of reconstruction methods

Method		Advantage	Limitation
Back-projection algorithms	SAA	Shift-and-add is a strictly linear operator and does not include any filtering steps	-The SAA technique is only effective when the X-ray source's direction is parallel to the detector -Different features in a chosen plane cannot be simultaneously recognized by simply moving and adding the projections when a rotational tube motion, such as in DBT, is present
	SBP	-No matter how the X-ray tube is moved, an algorithm for BP accurately includes the imaging geometry	-The streak artifacts (out-of-plane artifacts, interplane artifacts), are often strong in SBP reconstructed images
Transform algorithms	FBP	-This non-linear operation quite effectively reduces the streak artifacts in the resulting volume	-More cost for using the filter -Some authors found the reconstruction results to be very noisy and the structure details to be poorly visible
Algebraic reconstruction techniques	MITS	-Work well with high frequency information and the removal of out-of-plane artifacts	-Work badly with low frequency information

(continued)

Table 2. *(continued)*

Method		Advantage	Limitation
Statistical reconstruction techniques	ML	-Adaptable to a variety of algorithms (EM, convex and gradient algorithm)	-It is difficult to search the entire space of unknown sets to find the ML solution

5   Conclusion and Future Works

In conclusion, this paper reviewed several techniques for DBT reconstruction with the aim of merging multiple projected perspectives to reconstruct the original three-dimensional volume. Based on a theoretical comparison, it was concluded that a combination of these different methods, each offering its important role in improving the image reconstruction, can give a good result. The use of some reconstruction algorithms also helps to fill the gaps in the methods used. Robust reconstruction methods are crucial for improving image quality and have a significant impact on subsequent analyses. As a potential direction for future work, a segmentation method for DBT images based on new artificial intelligence technologies could be proposed.

References

1. Buda, M., et al.: Detection of masses and architectural distortions in digital breast tomosynthesis: a publicly available dataset of 5,060 patients and a deep learning model. arXiv preprint arXiv:2011.07995 (2020)
2. Nystrom, L., Andersson, I., Bjurstam, N., Frisell, J., Nordenskjold, B., Rutqvist, L.E.: Long-term effects of mammography screening: updated overview of the Swedish randomised trials. Lancet **359**, 909–919 (2002)
3. Carney, P.A., et al.: Individual and combined effects of age, breast density, and hormone replacement therapy use on the accuracy of screening mammography. Ann. Intern. Med. **138**, 168–175 (2003)
4. Michell, M.J., et al.: A comparison of the accuracy of film-screen mammography, full-field digital mammography, and digital breast tomosynthesis. Clin. Radiol. **67**, 976–981 (2012)
5. Haas, B.M., Kalra, V., Geisel, J., Raghu, M., Durand, M., Philpotts, L.E.: Comparison of tomosynthesis plus digital mammography and digital mammography alone for breast cancer screening. Radiology **269**, 694–700 (2013)
6. Lakshminarayanan, A.S., Radhakrishnan, S., Pandiasankar, G.M., Ramu, S.: Diagnosis of cancer using hybrid clustering and convolution neural network from breast thermal image. J. Test. Eval. **47**(6), 3975–3987 (2019)
7. Wu, W., Pirbhulal, S., Li, G.: Adaptive computing-based biometric security for intelligent medical applications. Neural Comput. Appl. **32**(15), 11055–11064 (2020). https://doi.org/10.1007/s00521-018-3855-9
8. Ait Ibachir, I., Es-salhi, R., Daoudi, I., Tallal, S., Medromi, H.: A survey on segmentation techniques of mammogram images. In: El-Azouzi, R., Menasché, D.S., Sabir, E., Pellegrini, F.D., Benjillali, M. (eds.) Advances in Ubiquitous Networking 2. LNEE, vol. 397, pp. 545–556. Springer, Singapore (2017). https://doi.org/10.1007/978-981-10-1627-1_43
9. Peters, G.: Computer-aided detection for digital breast tomosynthesis (Doctoral dissertation, Télécom ParisTech) (2007)
10. Fan, M., Zheng, H., Zheng, S., et al.: Mass detection and segmentation in digital breast tomosynthesis using 3D-mask region-based convolutional neural network: a comparative analysis. Front. Mol. Biosci. **7**, 599333 (2020)
11. http://www.depistagesein.ca/tomosynthese/#.Ywrl33bP3IX
12. https://moffitt.org/for-healthcare-professionals/clinical-programs-and-services/breast-oncology-program/treatments-services/digital-breast-tomosynthesis/
13. Kuo, J., Ringer, P.A., Fallows, S.G., Bakic, P.R., Maidment, A.D., Ng, S.: Dynamic reconstruction and rendering of 3D tomosynthesis images. In: Medical Imaging: Physics of Medical Imaging, vol. 7961, pp. 355–365. SPIE (2011)
14. Rafferty, E.A., Niklason, L., Halpern, E.: Assessing radiologist performance using combined full-field digital mammography and breast tomosynthesis versus full-field digital mammography alone: results of a multi-center, multi-reader trial. In: 93rd Scientific Assembly and Annual Meeting of the RSNA (2007)
15. Poplack, S.P., Tosteson, T.D., Kogel, C.A., Nagy, H.M.: Digital breast tomosynthesis: initial experience in 98 women with abnormal digital screening mammography. AJR Am. J. Roentgenol. **189**, 616–623 (2007)

16. Gennaro, G., et al.: Digital breast tomosynthesis versus digital mammography: a clinical performance study. Eur. Radiol. **20**, 1545–1553 (2010)
17. Zhu, F., et al.: Comparison and optimization of iterative reconstruction algorithms in digital breast tomosynthesis. Optik **203**, 164033 (2020)
18. Zackrisson, S., Houssami, N.: Evolution of mammography screening: from film screen to digital breast tomosynthesis. In: Breast Cancer Screening, pp. 323–346. Academic Press (2016)
19. Lauritsch, G., Haerer, W.: Theoretical framework for filtered back projection in tomosynthesis. In: Proceedings of the SPIE Medical Imaging, vol. 3338, pp. 1127–1137 (1998)
20. Mertelmeier, T., Orman, J., Haerer, W., Dudam, M.K.: Optimizing filtered backprojection reconstruction for a breast tomosynthesis prototype device. In: Proceedings of the SPIE Medical Imaging, vol. 6142, p. 61420F. SPIE (2006)
21. Gordon, R., Bender, R., Herman, G.T.: Algebraic reconstruction techniques (ART) for three-dimensional electron microscopy and x-ray photography. J. Theor. Biol. **29**, 471–481 (1970)
22. Gilbert, P.: Iterative methods for the three-dimensional reconstruction of an object from projections. J. Theor. Biol. **36**, 105–117 (1972)
23. Andersen, A.H., Kak, A.C.: Simultaneous Algebraic Reconstruction Technique (SART): a superior implementation of the ART algorithm. Ultrason. Imaging **6**(1), 81–94 (1984)
24. Zhang, Y., et al.: Tomosynthesis reconstruction using the simultaneous algebraic reconstruction technique (SART) on breast phantom data. In: Proceedings of the SPIE Medical Imaging, vol. 6142, p. 614249. SPIE (2006)
25. Chen, Y., Lo, J.Y., Dobbins, J.T., III.: Impulse response analysis for several digital tomosynthesis mammography reconstruction algorithms. Proc. SPIE **5745**, 541–549 (2005)
26. Chen, Y., Lo, J.Y., Baker, J.A., Dobbins, J.T., III.: Gaussian frequency blending algorithm with matrix inversion tomosynthesis (MITS) and filtered back projection (FBP) for better digital breast tomosynthesis reconstruction. Proc. SPIE **6142**, 61420E-61429E (2006)
27. Wu, T., et al.: Tomographic mammography using a limited number of low-dose cone-beam projection images. Med. Phys. **30**(3), 365–380 (2003)
28. Lange, K., Fessler, J.A.: Globally convergent algorithms for maximum a posteriori transmission tomography. IEEE Trans. Image Process. **4**(10), 1430–1438 (1995)
29. Sechopoulos, I.: A review of breast tomosynthesis. Part II. Image reconstruction, processing and analysis, and advanced applications. Med. Phys. **40**(1), 014302 (2013). https://doi.org/10.1118/1.4770281

BCAnalyzer: A Semi-automated Tool for the Rapid Quantification of Cell Monolayer from Microscopic Images in Scratch Assay

Aleksandr Sinitca[1]([✉])[ID], Airat Kayumov[2][ID], Pavel Zelenikhin[2][ID], Andrey Porfiriev[2][ID], Dmitrii Kaplun[1][ID], and Mikhail Bogachev[1][ID]

[1] Centre for Digital Telecommunication Technologies, St. Petersburg Electrotechnical University "LETI", St. Petersburg, Russia
{amsinitca,dikaplun}@etu.ru, rogex@yandex.com
[2] Institute for Fundamental Medicine and Biology, Kazan Federal University, 18 Kremlevskaya street, Kazan 420008, Tatarstan, Russia
kairatr@yandex.ru

Abstract. The scratch assay is a simple and low-cost approach to evaluate the speed and character of cell migration *in vitro*. The principle is based on the online imaging of the "scratch" in the cells monolayer being filled with new cells from both edges in real time. Thus, the scratch assay represents a model of cell migration during wound healing and is compatible with imaging of live cells during migration under various conditions. For the quantitative assessment of the scratch area in microscopic images, we suggest a simple semi-automated two-step algorithm based on the local edge density estimation, which does not require any preliminary learning or tuning, although with a couple of parameters directly controllable by the end user to adjust the analysis resolution and sensitivity, respectively. Using several representative examples of cell lines, we show explicitly the effectiveness of the image segmentation and quantification of the cells monolayer and discuss benefits and limitations of the proposed approach. A simple open-source software tool based on the proposed algorithm with on-the-fly visualization allowing for a straightforward feedback by an investigator without any specific expertise in image analysis techniques is freely available online at https://gitlab.com/digiratory/biomedimaging/bcanalyzer

Keywords: Cell culture · Cell migration · Scratch assay · Proliferation

1 Introduction

The *in vitro* scratch assay is a simple and low-cost approach to investigate cell migration. The principle of this approach is based on creating an artificial "scratch" in a preformed cell monolayer with consequent imaging of this area in

This work has been supported by the Russian Science Foundation (project No. 21-79-20219), https://rscf.ru/project/21-79-20219/.

regular time intervals to evaluate the cells migration rates from both edges of the "scratch" to its center [6]. Thus, selection of the continuously decaying cell-free area following its coverage by cells monolayer by microscopic image analysis methods as a function of time quantifies the relative speed of cells migration, this way modeling such processes as wound healing, cell-cell and cell-extracellular matrix interactions, respectively (see, e.g., [8] and references therein).

Conventional approaches to the scratch measurements typically require considerable time and workload of lab personnel. Generally, manual measurements of the scratch width at 10 different positions per image are performed using general purpose image analysis software, such as ImageJ / Fiji [10,11] or similar, followed by manual formation of a sample-exposition-effect table (an approach that is prone to technical errors) and averaging over them to estimate the cell migration rate, although some dedicated plug-ins aiming to take over part of this burden have been proposed [13]. Although technical errors could be eliminated at this stage by statistical analysis, this further increases workload on qualified lab personnel. As a prominent example of a detailed instruction including multi-step manual data collection and analysis procedures, we refer to [9] and references therein. Moreover, for an accurate assessment some authors suggest taking images as frequently as every 10 min [5], that would drastically increase the work volume, and thus require automated segmentation and quantification of the images.

Simple threshold based computer vision methods that are widely applied to fluorescent microscopy image analysis [1] appear inefficient in scratch assay quantification. Thus, there is neither specific intensity contrast between the cell monolayer in bright-field view nor staining is consistent with long-term experiments on viable cells. To overcome the above issues, methods and algorithms capable of extracting structural information solely by computer vision methods without relying upon any physical staining or contrasting techniques are of interest [15].

Several software tools are available to the date to quantify the areas of the "scratches" (or monolayer confluency, that is simply the inverse quantity). Among recent and prominent examples, [4] have presented an automatic analysis pipeline detecting scratch boundaries and measuring areas based on level sets for topology-preservation and use an entropy-based energy functional, while [2] offered an automatic PyScratch software to perform robust analysis of confluent cells. Another example is the TScratch algorithm that quantifies the scratch images by using edge-detection and the fast discrete curvelet transform to automate the measurement of the area occupied by cells in the images, allowing for a visual inspection of the analysis results and manual modification of analysis parameters [3]. Software implementation of an improved Robust Quantitative Scratch Assay algorithm claimed to overcome certain drawbacks of the above methodology has been reported in [16].

Here we suggest a simple semi-automated two-step algorithm based on the local edge density estimation, which does not require any preliminary learning or tuning, although with a couple of parameters directly controllable by the end user to adjust the analysis resolution and sensitivity, respectively. Using several representative examples of cell lines, we show explicitly the effectiveness of the image segmentation and quantification of the cells monolayer and discuss benefits and limitations of the proposed approach. A simple open-source software tool based on the proposed algorithm with on-the-fly visualization allowing for a straightforward feedback by a investigator without any specific expertise in image analysis techniques is freely available online at https://gitlab.com/digiratory/biomedimaging/bcanalyzer.

2   Materials and Methods

2.1   Sample Images Used for the Algorithm Demonstration

The image set contains color microscopic images of an *in vitro* scratch assay (see [6] for a the protocol description, as well as [8] for further details and an intuitive video guidance) made on A549 (human alveolar adenocarcinom), B16 (murine melanoma), and MCF7 (human breast adenocarcinom) cell lines. Briefly, a cell monolayer with 85% or more confluency was formed in 12-well culture plates and scraped in a straight line using a pipette tip ($200 \mu L$). Then, the medium in the wells was replaced and cultivation was followed for the next 24 h. Images of the scratches were captured immediately following the scratch formation and after 24 h of the cultivation. Images were obtained with the Zeiss Axio Observer 1.0 microscope (Carl Zeiss AG, Oberkochen, Germany) with 400x magnification. The original series of experiments with the respective cell lines have been reported earlier in [21].

2.2   Image Set for a Systematic Algorithm Performance Validation

For a systematic comparison of the effectiveness of the proposed approach we used the collective cell migration image set BBBC019v2 containing images of scratch assays on eight various cell lines which differ in both cells density and shapes, as well as imaging conditions. The dataset was obtained from the Broad Bioimage Benchmark Collection that is freely available online at [7].

In addition to the manual segmentation, the above image set contains also the results from its systematic analysis by three alternative methods reported in an earlier work [19]. This allowed us to further benchmark our approach against alternative methods, including tscratch [3], multiCellSeg [20], and topman [14] under the same imaging conditions.

3   Software Implementation

The software interface is shown in Fig. 1. For a better generalization two additional control sliders for the canny edge detection thresholds tuning have been added, although for the majority of considered scenarios using a single threshold led to acceptable results. Additionally, support of an alternative slope difference distribution (SDD) multi-thresholding algorithms [17] potentially useful for the analysis of images with multi-modal intensity distribution have been added.

The typical image processing workflow using the implemented software follows the pipeline below. In the first step, image(s) selected for the analysis are dragged & dropped in the main image visualization area (A) and appear listed in the panel (B) on the left of it. Next, the input channels (C) and the keynote algorithm settings (D) as well as algorithm parameters (E) are selected and tuned by the investigator, with instant visualization of the selection results by the inversion of the selected part of the sample image selected in the file list (B). For convenience, the sensitivity threshold and the resolution parameter (corresponding to the gliding window size) are displayed as fractions of the total image area, such that the investigator can observe both the selected area and at least to a first approximation its potential variation arising from discreteness effects displayed in the text boxes next to the upper two sliders, respectively. Once the parameters have been adjusted to a satisfactory level, that could be also cross-checked by fixing the adjusted parameters by *Apply to all* in panel (E) followed by selection of different images from the file list (B). Further processing of a series of images could be performed in a background mode using *File → Export*, with options either to save tabulated monolayer confluency

---

Data: $0 < \text{ws\_perc} < 50$;   /* Window relative size for local edges */
Data: $0 < cut < 255$;   /* Canny Upper Threshold */
Data: $0 < clt < 255$;   /* Canny Lower Threshold */
Data: $img \leftarrow int[c, w, h]$;   /* Original grayscale or RGB image */
Data: $remove\_bg$;   /* Flag for background removing */
Result: BI;   /* Quantificated image */
$W_s \leftarrow w * \text{ws\_perc}/100$;
if $c = 3$ then
$\quad | \quad GS_{ij} \leftarrow 0.299R_{ij} + 0.587G_{ij} + 0.114B_{ij}$;
else
$\quad | \quad GS_{ij} \leftarrow img_{ij}$
end
if $remove\_bg$ then
$\quad | \quad I_{bg} \leftarrow \text{blur}(GS)$;   /* Estimate background as blurred image */
$\quad | \quad I \leftarrow (GS - I_{bg}) - min(GS - I_{bg})$;
else
$\quad | \quad I \leftarrow GS$
end
$IC \leftarrow \text{canny\_det}(I, clt, cut)$;   /* Canny algorithm without filtering */
for ($i = W_s/2;\ i < w - W_s/2;\ i+ = 1$) {
$\quad$ for ($j = W_s/2;\ j < h - W_s/2;\ i+ = 1$) {
$\quad\quad | \quad D_{ij} \leftarrow \frac{1}{W_s^2} \sum_{x=i-\frac{W_s}{2}}^{i+\frac{W_s}{2}} \sum_{y=j-\frac{W_s}{2}}^{j+\frac{W_s}{2}} IC_{xy}.$
$\quad$ }
}
$BI \leftarrow \text{threshold\_alg}(D)$;   /* Binarized edge density map */

Algorithm 1: Proposed method algorithm

and/or scratch areas (both absolute values and relative fractions of the total image area, respectively) only, or also binary masks and/or visualization results as well.

The proposed Algorithm 1 takes as argument RGB or grayscale image and parameters from GUI, namely relative size of windows for estimation of local edge density (ws_perc), upper and lower thresholds for canny edge detector (*cut* and *clt*) and flag for background elimination (*remove_bg*). The result of algorithm is mask with same size as original image with a quantification. It is worth noting, that in the algorithm used canny edge detector without image filtering in the beginning.

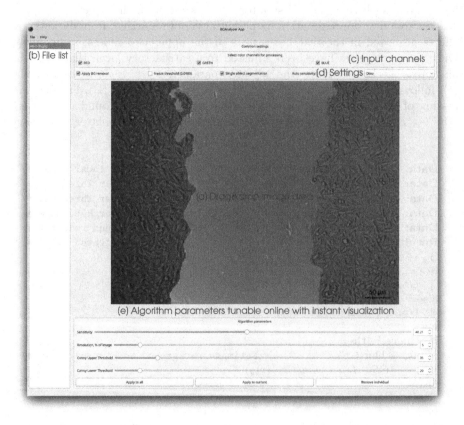

Fig. 1. The implemented software interface: (A) image drag & drop and visualization area; (B) list of files provided for analysis; common settings panel, including selection of (C) input channels and (D) keynote pipeline variants; (E) algorithm parameters tunable online to control sensitivity, resolution, as well as edge detector thresholds.

4 Results and Discussion

4.1 Prominent Examples of Cell Scratch Assay Analysis for Different Cell Lines

Figures 2, 3, 4 represent typical application examples of the reported algorithm to various scratch assay results performed on cells with various shapes, including the A549 human alveolar adenocarcinom cells, B16 murine melanoma cell line, as well as MCF7 human breast adenocarcinom cells, respectively. The algorithm was tuned on automatic choose of sensitivity and threshold by otsu approach (see Fig. 1D), followed with fine tuning of selection by correction of resolution and Canny lower and upper thresholds (see Fig. 1E).

The cell-free areas on a plates with of A549 line could be segmented and quantified with high accuracy in a fully automated mode because of high edges density on image areas with cell monolayer Fig. 2. By contrast, both B16 and MCF7 cells did not form a monolayer with tightly connected cells, leading to low edges density on image areas with cells (see Figs. 3, 4) and consequently leading to non-successful segmentation of the image in automated mode. To overcome the problem and select the cell-free area as present of Figs. 3, 4, the resolution (i.e. the gliding window size) was manually increased and both lower and upper Canny thresholds have been set to values allowing suitable selection.

Thus, these examples show explicitly that in all cases the scratch areas could be selected on the image. Consequently, the relative monolayer confluency and/or scratch areas could be calculated in two-step approach with automatic choose of initial parameters followed by the expert-driven settings of the algorithm parameters for the best result. Being optimized on the representative image of the set, a series of other images obtained under the same conditions can be analyzed with the same set of parameters as determined from the first one followed by minor corrections for each image, if desired. Finally, the results can be exported to the standard *.csv-file for the following statistical analysis.

4.2 Systematic Algorithm Validation Using Previously Reported Reference Image Sets

For a more systematic validation of the proposed algorithm, we next analyze the scratch assay image series obtained from [7] in a fully automated mode and compare the segmentation accuracy using the conventional Otsu thresholding and the slope difference distribution (SDD) multi-thresholding algorithms [17]. While we remain focused on the binary classification scenario, we test three different threshold values derived from SDD denoted as the lower, median and upper thresholds, and evaluate the segmentation accuracy for each of them, this way also comparing two different automated threshold adjustment strategies. Of note, while SDD thresholding method has been reported to outperform conventional methods reminiscent to the Otsu approach, as we show below, its particular performance largely depends on the imaging conditions and image quality. However, even despite certain potential drawbacks, another possible utilization of

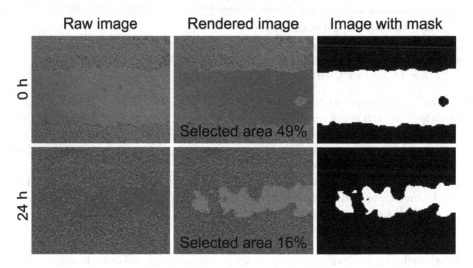

Fig. 2. Representative cell scratch assay images for the A549 human alveolar adenocarcinom cell line.

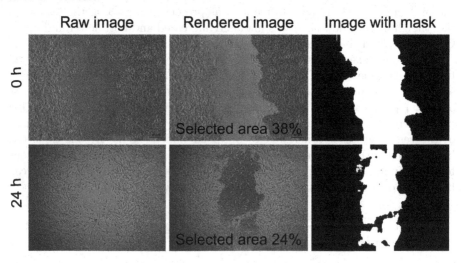

Fig. 3. Representative cell scratch assay images for the B16 murine melanoma cell line.

multi-thresholding method is simply providing the investigator with additional semi-automated threshold selection options, thus leading to certain compromise between performance and quality by leaving the particular choice of the particular threshold (out of several options proposed by the automatic algorithm) to the investigator who is typically an application domain expert.

Figures 5, 6 and 7 indicate that the proposed approach in a fully automatic thresholding analysis mode (assuming that the best suitable algorithm options including single object segmentation and background elimination, and a rele-

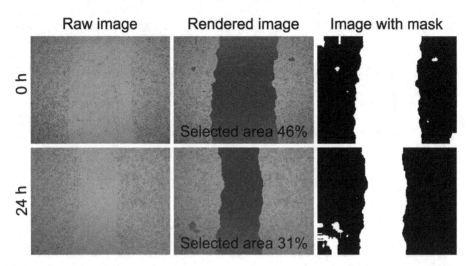

Fig. 4. Representative cell scratch assay images for the MCF7 human breast adeno-carcinom cell line.

vant threshold selection approach have been selected) for the majority of the analyzed image subsets either demonstrated comparable effectiveness with the best alternatives among three previously proposed approaches (multiCellSeg, topman, and tscratch) or in some cases even moderately outperformed the best of three alternatives, indicating its universality. In particular,

- for the HEK293 subset, the effectiveness was comparable with multiCellSeg, while clearly outperforming topman and tscratch algorithms;
- for the Init subset, the performance was comparable with multiCellSeg and topman, while slightly outperforming tscratch;
- for the MDCK subset, the proposed approach was comparable with multi-CellSeg, while outperforming topman and tscratch;
- for the Scatter subset, the proposed approach outperformed all three methods, with the best results achieved when using either median or upper SDD based thresholds, while the improvement was slightly less pronounced when using the Otsu threshold, although remaining rather *on par* with multiCellSeg and topman, while still outperforming tscratch with the lower SDD threshold;
- for the SN15 subset, the performance is comparable for all four considered methods.

There were only few exceptions from the above, in particular,

- for the Melanoma and Tscratch subsets, the performance was significantly reduced than observed for all three alternative methods, indicating that the proposed approach is not suitable for respective analysis conditions, that we attribute to the absence of congruent monolayer in the majority of images;
- for the Microfluidic subset, the situation was probably most non-trivial among all considered examples. The proposed approach significantly outperformed

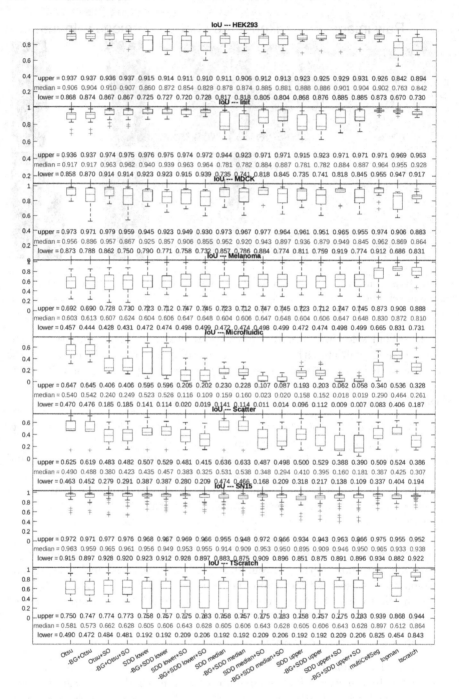

Fig. 5. Comparison of the segmentation effectiveness in a fully automated mode using Otsu thresholding, evaluated using the IoU metric, for the BBBC019v2 image set, tested both with and without background elimination and single object segmentation in all possible combinations. The results are compared against manual segmentation, as well as three alternative methods, including tScratch, multiCellSeg and topman. The results are provided for the resolution setting corresponding to the window size of 40 × 40 pixels.

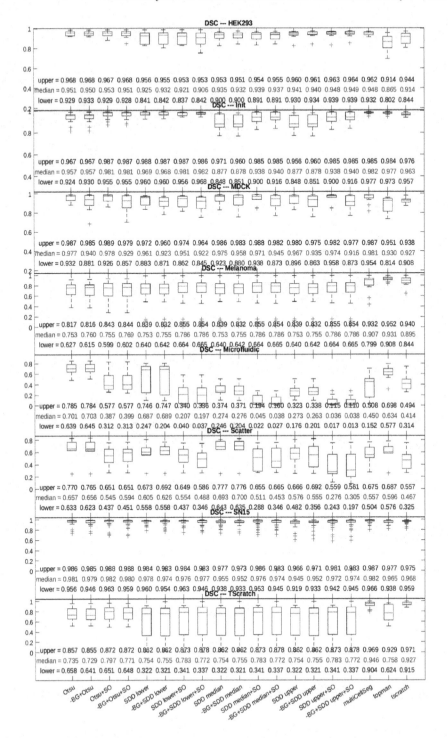

Fig. 6. CSimilar to Fig. 5 but evaluated using the Accuracy metric.

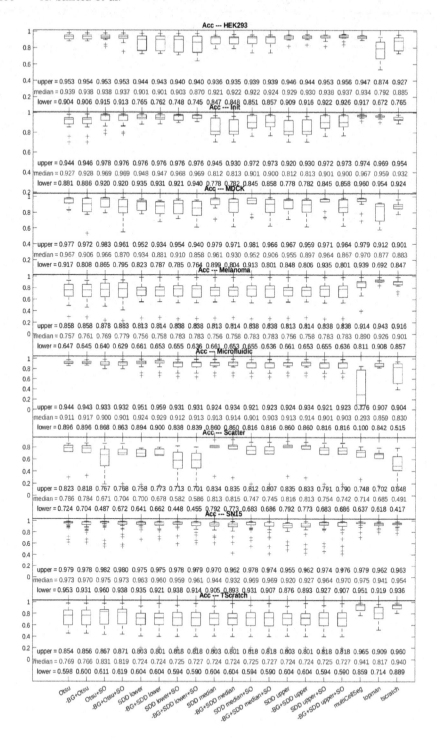

Fig. 7. Similar to Fig. 5 but evaluated using the Accuracy metric.

all three alternative methods in terms of Accuracy, while low values of the Dice-Sorensen coefficient and the IoU metric for all four considered methods indicate that the actual overlap is rather low, and high Accuracy values above 0.9 are rather due the overall statistical effect based on high correlations between the cell content and edge density in the entire analyzed image. Despite obvious drawbacks, this feature could be nevertheless useful when the total cell content is of interest, while particular cell localization is not essential in a given experimental setting.

Among recent approaches to the automated thresholding problem, although the recently reported slope difference distribution (SDD) based approach [17] indicates promising evidence from medical imaging applications [18], particular performance of various automated algorithms largely depends on a variety of imaging conditions, leading to certain limitations in the automatic method selection. However, one important feature of the above algorithm is the embedded opportunity of multi-thresholding based on data-driven estimates of threshold values. In particular, combining the local edge density estimate with the SDD thresholding approach leads to an improved segmentation accuracy under certain scenarios, especially in imbalanced cases where the conventional Otsu approach demonstrates limited effectiveness, although in more favorable imaging conditions, a more simple Otsu algorithm may still be preferential. While it seems that at the moment there is no universal solution to the fully automated selection of the algorithm and its optional parameters, online feedback by the domain expert remains the best way to their initial adjustment, while further analysis of large-batch image sets obtained under more or less similar imaging conditions could be performed in a fully automated mode.

5   Conclusion

To summarize, we have developed a semi-automatic algorithm and implemented it as a simple open-source software tool with only a couple of tuning parameters that are easily controllable online by the investigator with immediate visualization of the results this way allowing for a straightforward feedback. The software is available at https://gitlab.com/digiratory/biomedimaging/bcanalyzer, for the description of an earlier prototype, we also refer to [12]. To our opinion, the keynote advantage of the proposed algorithm and software tool is that, in marked contrast with manual approaches, our image analysis procedure is semi-automatic. Thus, while the initial settings are selected by the investigator who is typically a domain expert, with optional automated adjustment of thresholds based on objective statistical criteria, they could be easily further modified by manual adjustment of the algorithm parameters. This allows simultaneously providing complete visualization of the results at each step of the image processing, in order to enable expert control and online assessment of the analysis accuracy. Once the algorithm parameters have been optimized and approved by the investigator assessment based on either single or several representative sample image(s), the software is capable of further analysis of the entire image series following by the export of tabulated results in a standard *.csv format in a fully automatic mode, that could be easily done over a cup of coffee.

References

1. Bogachev, M.I., et al.: Fast and simple tool for the quantification of biofilm-embedded cells sub-populations from fluorescent microscopic images. PLOS ONE **13**(5), 1–24 (2018). https://doi.org/10.1371/journal.pone.0193267
2. Garcia-Fossa, F., Gaal, V., de Jesus, M.B.: PyScratch: an ease of use tool for analysis of scratch assays. Comput. Meth. Programs Biomed. **193**, 105476 (2020)
3. Gebäck, T., Schulz, M.M.P., Koumoutsakos, P., Detmar, M.: TScratch: a novel and simple software tool for automated analysis of monolayer wound healing assays: short technical reports. Biotechniques **46**(4), 265–274 (2009)
4. Glaß, M., Möller, B., Zirkel, A., Wächter, K., Hüttelmaier, S., Posch, S.: Cell migration analysis: segmenting scratch assay images with level sets and support vector machines. Pattern Recogn. **45**(9), 3154–3165 (2012)
5. Kauanova, S., Urazbayev, A., Vorobjev, I.: The frequent sampling of wound scratch assay reveals the "opportunity" window for quantitative evaluation of cell motility-impeding drugs. Front. Cell Dev. Biol. **9**, 640972 (2021)
6. Liang, C.C., Park, A.Y., Guan, J.L.: In vitro scratch assay: a convenient and inexpensive method for analysis of cell migration in vitro. Nat. Protoc. **2**(2), 329–333 (2007)
7. Ljosa, V., Sokolnicki, K.L., Carpenter, A.E.: Annotated high-throughput microscopy image sets for validation. Nat. Meth. **9**(7), 637 (2012)
8. Mouritzen, M.V., Jenssen, H.: Optimized scratch assay for in vitro testing of cell migration with an automated optical camera. JoVE (J. Visualized Exp.) (138), e57691 (2018)
9. Pinto, B.I., Cruz, N.D., Lujan, O.R., Propper, C.R., Kellar, R.S.: In vitro scratch assay to demonstrate effects of arsenic on skin cell migration. JoVE (J. Visualized Exp.) (144), e58838 (2019)
10. Schindelin, J., et al.: Fiji: an open-source platform for biological-image analysis. Nat. Meth. **9**(7), 676–682 (2012)
11. Schneider, C.A., Rasband, W.S., Eliceiri, K.W.: NIH image to ImageJ: 25 years of image analysis. Nat. Meth. **9**(7), 671–675 (2012)
12. Sinitca, A.M., Kayumov, A.R., Zelenikhin, P.V., Porfiriev, A.G., Kaplun, D.I., Bogachev, M.I.: Segmentation of patchy areas in biomedical images based on local edge density estimation. Biomed. Signal Process. Control **79**, 104189 (2023). https://doi.org/10.1016/j.bspc.2022.104189
13. Suarez-Arnedo, A., Figueroa, F.T., Clavijo, C., Arbeláez, P., Cruz, J.C., Muñoz-Camargo, C.: An image j plugin for the high throughput image analysis of in vitro scratch wound healing assays. PLoS ONE **15**(7), e0232565 (2020)
14. Topman, G., Sharabani-Yosef, O., Gefen, A.: A standardized objective method for continuously measuring the kinematics of cultures covering a mechanically damaged site. Med. Eng. Phy. **34**(2), 225–232 (2012)
15. Trizna, E., et al.: Brightfield vs fluorescent staining dataset - a test bed image set for machine learning based virtual staining. Sci. Data **10**, 160 (2023). https://doi.org/10.1038/s41597-023-02065-7
16. Vargas, A., et al.: Robust quantitative scratch assay. Bioinformatics **32**(9), 1439–1440 (2016)

17. Wang, Z.: A new approach for segmentation and quantification of cells or nanoparticles. IEEE Trans. Industr. Inf. **12**(3), 962–971 (2016)

18. Wang, Z.: Automatic localization and segmentation of the ventricles in magnetic resonance images. IEEE Trans. Circuits Syst. Video Technol. **31**(2), 621–631 (2020)

19. Zaritsky, A., Manor, N., Wolf, L., Ben-Jacob, E., Tsarfaty, I.: Benchmark for multi-cellular segmentation of bright field microscopy images. BMC Bioinform. **14**(1), 1–6 (2013)

20. Zaritsky, A., et al.: Cell motility dynamics: a novel segmentation algorithm to quantify multi-cellular bright field microscopy images. PLoS ONE **6**(11), e27593 (2011)

21. Zelenikhin, P., et al.: Bacillus pumilus ribonuclease inhibits migration of human duodenum adenocarcinoma HuTu 80 cells. Mol. Biol. **54**(1), 128–133 (2020)

Color Hippocampus Image Segmentation Using Quantum Inspired Firefly Algorithm and Merging of Channel-Wise Optimums

Alokeparna Choudhury, Sourav Samanta, Sanjoy Pratihar[(✉)],
and Oishila Bandyopadhyay

Computer Science and Engineering, Indian Institute of Information Technology,
Kalyani 741235, India
sanjoy@iiitkalyani.ac.in

Abstract. Color image segmentation is essential for medical image processing to figure out the cells, tissues, lesion areas, etc. The hippocampus
is an extension of the temporal lobe of the brain. This area of the brain
has been intensively studied for its clinical significance. It is the first and
most severely affected structure in neuropsychiatric conditions. Metaheuristic algorithm-based optimal segmentation is a widely accepted
method in the medical domain. In this work, a hybrid method called
the *quantum-inspired firefly algorithm (QIFA)* has been implemented
in a multi-core environment to perform color segmentation of the hippocampus images in a parallel manner. The parallel QIFA runs on three
different channels, *Red*, *Green*, and *Blue* of the input color image, and
a subsequent merging is applied. The correlation has been considered
as the objective function. Finally, a study has been carried out concerning various image segmentation evaluation parameters, and the proposed
method has been compared to other metaheuristic algorithms. The analysis of the results shows that the method is effective for medical image
segmentation. The speed-up of the technique has also been examined in
detail for various image sizes and color levels.

Keywords: Hippocampus images · Color image segmentation ·
Quantum inspired firefly algorithm (QIFA) · Medical image
segmentation

1 Introduction

In the *computer-aided diagnosis* (CAD) system, segmentation is a crucial phase
of medical image processing for the localization of a particular cell, tissue, lesion,
etc. The segmentation of medical images using meta-heuristics is a well-used
strategy. In recent years, hybrid metaheuristics have been utilized to improve
the performance of traditional meta-heuristic approaches. The hippocampus is

I. Rojas et al. (Eds.): IWBBIO 2023, LNBI 13920, pp. 270–282, 2023.
https://doi.org/10.1007/978-3-031-34960-7_19

a deep element in the brain's left and right medial temporal lobes. Most of the hippocampus's cells are pyramidal except for the basket cells, which are of different kinds. The pyramidal cells are connected to the basket cells in the different hippocampal layers. There are different layers - the deepest layer is named the *alveus*, which contains the pyramidal axons. The above layer is *stratum oriens*, which contains the basket cells. The next layer is the *pyramidal layer* which contains the pyramidal neurons. Above that, the layers are *stratum radiatum*, *stratum lacunose*, and *stratum moleculare*, and they contain the dendrites and some fibres [1]. The different fields of the hippocampus are $CA1$, $CA2$, $CA3$, and $CA4$, which are very important in forming hippocampal circuitry with several pyramidal neurons. Here CA is the *Cornu Ammonis*. The pyramidal neurons in the different fields of CA form synapse with each other in an efficient way [1]. In the hippocampus, the *dentate gyrus* also contains the pyramidal cells [1]. Generally, the hippocampus performs three primary functions: forming new memories, memory consolidation, and spatial navigation or spatial memory. Pyramidal neurons in the hippocampus are essential in maintaining the processes mentioned above. Injury to the hippocampus areas produces numerous cognitive memory-related problems, such as amnesia. While the damages occur in the hippocampal regions, the hippocampus volume significantly reduces, called hippocampal atrophy [2]. In the case of early Alzheimer's disease (AD), the hippocampal volume reduces by $15\% - 30\%$ [3], whereas in the patients with moderate AD, it has been seen that the volume reduces up to 50% [4]. As most of the pyramidal neurons are located in the $CA1$ region of the hippocampus and *dentate gyrus*, it has been shown in the brain postmortem analysis of AD patients that there is significantly less number of synapses between the neurons [5].

Thus, detecting the losses of neurons connects the hippocampus volume reduction due to several brain diseases. These observations lead to the early detection and diagnosis of the disorders where color image segmentation is the crucial preprocessing task. In this work, a hybrid method called the *quantum-inspired firefly algorithm (QIFA)* has been implemented in a multi-core environment to perform the color segmentation of the hippocampus images. The QIFA runs concurrently on different channels (*Red*, *Green*, and *Blue*), and a channel-wise merging is applied to obtain the final segmentation.

2 Literature Survey

A few contemporary works in medical image segmentation are highlighted briefly in this section. Chakraborty et al. [6] presented this segmentation technique for segmenting microscopic hippocampal images using a modified Cuckoo Search (CS) algorithm. Their optimization algorithm has considered Otsu's between-class variance, Kapur's entropy, and Tsallis entropy as the objective functions. The proposed strategy for optimizing threshold levels with the Tsallis entropy functioned well. Dey et al. [7] reviewed some medical image segmentation problems based on several metaheuristic algorithms that have been addressed in other works. In addition, they reported the outcomes of the overall segmentation techniques and the use of several metaheuristics in various medical image segmentation areas. Ghosh et al. [8] proposed the hybrid clustering algorithm, C-FAFCM, using a hybrid chaotic firefly algorithm clubbed with a modified fuzzy C-means (m-FCM) clustering. This clustering algorithm was utilized to automatically segment MRI volumetric datasets. The proposed strategy reduces the influence of image noise. In this study, the suggested C-FAFCM is compared to a number of fuzzy approaches, including FCM, BCFCM, FAFCM, and En-FAOFCM. The suggested method outperforms all other approaches according to several evaluation criteria, such as Accuracy, Similarity Coefficients (Dice). Giuliani [9] proposed a hybrid method combining the gaussian mixture model and the firefly algorithm. Fireflies are used in their proposed method to find the optimum cluster centroids. The authors experimented with the proposed model with various images, including blood cells and cervical vertebrae. The results established the algorithm's effectiveness with the test images mentioned above. In the paper proposed by Oliva et al. [10], multilevel thresholding-based studies on nine different metaheuristics algorithms (MA) are presented. In this survey, the authors conclude that Otsu and Kapur were the most often utilized objective functions in the analyzed MAs and the most prevalent performance indicators are the values of objective functions, CPU time, PSNR, SSIM, etc. The most commonly utilized average threshold levels are 2, 4, and 5. They have also demonstrated the potential future application of these algorithms for color image segmentation. Upadhyay et al. [11] presented a Crow Search Algorithm (CSA) based on Kapur's Entropy for optimizing multilevel threshold points. The performance of the method is compared to those of prominent existing metaheuristics such as Particle Swarm Optimization (PSO), Differential Evolution (DE), Grey Wolf Optimizer (GWO), Moth-Flame Optimization (MFO) and Cuckoo Search as the number of iterations grows (CS). Segmentation quality is quantified using PSNR, SSIM, and FSIM. The proposed technique produces superior results for higher threshold levels. Hernandez del Rio et al. [12] proposed a two-dimensional (2D) histogram non-local means-based segmentation approach along with the SCA and PSO algorithms to find the optimal threshold points for efficient segmentation. The images from the Berkeley Segmentation Dataset (BSDS300) and Benchmark were used in the experiment. The Rényi entropy has been used as the objective function in the work. The hybrid MFE-LFA was suggested by Pare et al. [13] for thresholding-based multilevel color image segmentation. The proposed algorithm uses threshold levels as 2, 5, 8, and 12. The method has been compared with metaheuristic algorithms like PSO, ABC, JADE, and CS. The authors used evaluation parameters like PSNR and MSE to judge the segmentation quality. He et al. [14] provide a different effective segmentation algorithm for color images. An efficient krill herd (EKH) technique is intended to identify the appropriate threshold locations for thresholding multilayer color images. Their work uses Otsu's approach, Kapur's entropy, and Tsallis' entropy as objective functions. Threshold levels 3, 4, 5, and 6 have been used to test ten color benchmark photos. Dhal et al. [15] developed a fuzzy image segmentation based on clustering. Three strategies are seen in the work: rough set-based population, random attraction, and local search. This approach has experimented on the dataset ALL-IDB2.

So, many metaheuristic algorithms have been proposed recently for color image segmentation. The solutions to the problem of segmentation of color medical images are still having scopes for improvement in terms of accuracy in segmentation, robustness, and CPU time. Our contributions to this work are listed below.

1. A modified quantum-inspired firefly algorithm (QIFA) has been proposed for color hippocampus image segmentation.
2. The proposed method uses concurrent execution of QIFA on different color channels, *Red*, *Green* and *Blue*, of the input color image, and merges the channel-wise optimums to get the final segmentation.
3. Concurrent execution of the QIFA on different color channels ensures speedy execution, and it is equivalent to applying on the gray-scale image.
4. The results has been analyzed using evaluation parameters like *correlation*, *SSIM*, *entropy*, *PSNR*, and *MSE* which establish the applicability of the proposed method for color hippocampus image segmentation.

The following sections present the details of the work. The proposed quantum-inspired firefly algorithm deployed in a multi-core architecture combined with the merging of channel-wise optimums has been presented in Sect. 3. Experimental setups have been discussed in Sect. 4. Finally, Sect. 5 analyses the results, and Sect. 6 presents concluding remarks.

3 Proposed Method

The quantum-inspired firefly algorithm (QIFA) proposed by Choudhury et al. [16] has been efficiently applied for threshold-based segmentation of gray-level hippocampus images. The structure of the previously proposed QIFA algorithm has been shown in Algorithm 1. This work proposes parallel QIFA, implemented in multi-core architecture. Initially, three color channels, R, G, and B, are extracted from the color hippocampus image and the minimum and maximum gray level values are identified from each channel. Then the three initial firefly populations are generated for three channels. Now, the QIFA algorithm is concurrently applied to three quantum populations. These three instances of QIFA are assigned to three cores for execution in parallel. Each QIFA algorithm targets to find the best firefly until the maximum iteration. The maximum iteration of value 300 is fixed to the same value for all channels. So, when the maximum iteration is reached, the three fireflies FF_r, FF_g, and FF_b are produced and sorted. Finally, as shown in Algorithm 2, a merging procedure results in the color-segmented image. The goal of the merging algorithm is to limit the number of color levels used to a value equal to κ. All algorithm parameters are defined at Line 1. The parameters include Randomization parameter (α), Attractiveness coefficient (β), Light absorption coefficient (γ), and Qauntum Rotation Angle for Update (δ). The length of the quantum firefly is calculated at Line 2 (considering each threshold is of 8-bit length), and it randomly initializes the position of

---

Algorithm 1: QUANTUM-INSPIRED-FIREFLY-ALGORITHM

Input: $\mathfrak{J}^{rgb}$ ⟵ Original Color Image
$\quad\quad\kappa$ ⟵ Number of Color Thresholds
Output: $\mathfrak{J}^{rgb}_{seg}$

1 Define QIFA parameters α, β, γ, and δ;
2 Define the dimensions of Quantum firefly $\mathfrak{D} = 8 \times \kappa$;
3 Initialize population of Quantum fireflies $\psi^{quantum}_{pop}$ of length $\mathfrak{D}$ using Hadamard gate for each color component;
    // p loop will be executed parallelly
4 **for** $p \leftarrow 1$ **to** 3 **do**
5   **while** $t \leq$ *Max Iteration* **do**
6   **for** $i \leftarrow 1$ **to** n **do**
7   **for** $j \leftarrow 1$ **to** n **do**
8   Measure the $\psi^{quantum}_{pop}$ to get ψ^{bin}_{pop};
9   Convert ψ^{binary}_{pop} to get $\psi^{decimal}_{pop}$;
10   Segment $\mathfrak{J}$ with each quantum firefly and calculate the light intensity $\mathfrak{F}$;
11   Sort the $\mathfrak{F}$ to get $\mathfrak{F}'$;
12   According to $\mathfrak{F}'$ quantum fireflies are ranked and the current best $\psi^{quantum}_{global}$ is found;
13   Find the Best $\psi^{decimal}_{global}$ from $\psi^{decimal}_{pop}$;
14   **if** $(\mathfrak{F}_j \geq \mathfrak{F}_i)$ **then**
15   Firefly i moves in all $\mathfrak{D}$ dimensions towards Firefly j ;
16   **else**
17   Update the quantum firefly according to $\psi^{quantum}_{global}$ using quantum update operaion;
18   β varies using $e^{-\gamma r}$;
19   Obtain new solutions and update brightness values;
20             $j \leftarrow j + 2$;
21         $i \leftarrow i + 2$;

22 $\mathfrak{J}^{rgb}_{seg} \leftarrow$ MERGE-COMPONENTS $(\mathfrak{J}^{rgb}, \kappa^r_{optimum}, \kappa^g_{optimum}, \kappa^b_{optimum})$

---

the quantum firefly using Hadamard gate, which preserves the quantum superposition of solution as shown in Line 3. The three quantum firefly populations of size n for three color channels are generated. QIFA will be executed parallel to segment the three color channels as mentioned at Line 4. In each QIFA, the quantum firefly population is measured and converted to binary and decimal populations. The decimal firefly is used to segment the image and then rank each firefly according to segmentation quality from as per Line 8 to Line 13. Suppose the condition at Line 14 is satisfied. In that case, Quantum firefly is updated according to firefly movement operation at Line 15 otherwise updated by quantum update operation at Line. Then β is modified according to the classical firefly. Also, the new solution is updated. The loop is executed up to the

---

Algorithm 2: MERGE-COMPONENTS

Input: $\mathfrak{I}^{rgb}$, $\kappa^r_{optimum}$, $\kappa^g_{optimum}$, $\kappa^b_{optimum}$
Output: $\mathfrak{I}^{rgb}_{seg}$

1   **for** $c \leftarrow 1$ **to** $\kappa - 1$ **do**

2     $Level^r_c \leftarrow \left(\kappa^r_{optimum}[c] + \kappa^r_{optimum}[c+1]\right)/2$

3     $Level^g_c \leftarrow \left(\kappa^g_{optimum}[c] + \kappa^g_{optimum}[c+1]\right)/2$

4     $Level^b_c \leftarrow \left(\kappa^b_{optimum}[c] + \kappa^b_{optimum}[c+1]\right)/2$

5   $c \leftarrow \kappa$

6   $Level^r_c \leftarrow max\left(\mathfrak{I}^{rgb}(:,:,1)\right)$

7   $Level^g_c \leftarrow max\left(\mathfrak{I}^{rgb}(:,:,2)\right)$

8   $Level^b_c \leftarrow max\left(\mathfrak{I}^{rgb}(:,:,3)\right)$

9   **for** $i \leftarrow 1$ **to** h **do**

10   **for** $j \leftarrow 1$ **to** w **do**

11   **for** $c \leftarrow 1$ **to** κ **do**

12       $ED^{rgb}_c \leftarrow$
$$\sqrt{(\mathfrak{I}^{rgb}(i,j,1) - Level^r_c)^2 + (\mathfrak{I}^{rgb}(i,j,2) - Level^g_c)^2 + (\mathfrak{I}^{rgb}(i,j,3) - Level^b_c)^2}$$

13      $[idrgb_{min}] \leftarrow min\left(ED^{rgb}_c\right)$

14      $\mathfrak{I}^r_{seg}(i,j) \leftarrow \kappa^{r\,idrgb_{min}}_{optimum}$

15      $\mathfrak{I}^g_{seg}(i,j) \leftarrow \kappa^{g\,idrgb_{min}}_{optimum}$

16      $\mathfrak{I}^b_{seg}(i,j) \leftarrow \kappa^{b\,idrgb_{min}}_{optimum}$

17   $\mathfrak{I}^{rgb}_{seg} \leftarrow combine\left(\mathfrak{I}^r_{seg}, \mathfrak{I}^g_{seg}, \mathfrak{I}^b_{seg}\right)$

18   **return** $\mathfrak{I}^{rgb}_{seg}$

---

maximum iteration. After execution of parallel for loop of Line 4, three optimum solutions (one from each color component) are assigned into $\kappa^r_{optimum}$, $\kappa^g_{optimum}$, and $\kappa^b_{optimum}$ respectively. Then MERGE-COMPONENTS (Algorithm 2) is invoked to get the final segmented color image $\mathfrak{I}^{rgb}_{seg}$ of κ color levels. Algorithm 2 takes the original color image and the three lists of the optimum threshold level of size κ for each color component. Then boundary levels are calculated at Line 2–4. The last (κ-th) boundary levels are assigned according to maximum pixel intensity of each color component of $\mathfrak{I}^{rgb}$ at Line 6–8. The Euclidean distance between boundary level values is calculated for each color pixel, as shown in Line 12. The minimum distance among all the boundary values is calculated at Line 13, and finally, color threshold values are assigned according to the minimum index value at Line 14–16. Finally, the segmented color image $\mathfrak{I}^{rgb}_{seg}$ is obtained.

4   Experimental Setup

In this experiment, color images of rat hippocampus have been considered. The results have been shown on the two sample images. The image size is 2352×3136 for both Sample-1 and Sample-2. The experiment has been carried out in $i7$ laptop (hexacore) and Matlab 2016a. Matlab utility "parfor" has been utilized to

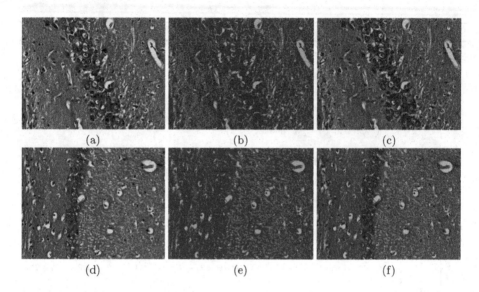

Fig. 1. Segmentation using QIFA and subsequent merging of channel-wise bests: (a) `Sample-1`; (b)-(c) Segmented `Sample-1` with $\kappa = 3$ and $\kappa = 5$; (d) `Sample-2`; (e)-(f) Segmented `Sample-2` with $\kappa = 3$ and $\kappa = 5$;

perform the segmentation of each channel in parallel. The correlation between the original image and the segmented image determines the segmentation quality as shown in Eq. 1. Here, A and $\hat{A}$ are the original and segmented images, respectively. We have used the parameters *SSIM*, *entropy*, and *PSNR* to understand the segmentation quality obtained from the proposed method. A brief introduction of the parameters is given below in Eqs. 2, 3, 4, and 5 respectively.

$$Correlation\left(A, \hat{A}\right) = \frac{\sum_m \sum_n (A_{mn} - A_\iota)(\hat{A}_{mn} - \hat{A}_\iota)}{\sqrt{\left(\sum_m \sum_n (A_{mn} - A_\iota)^2\right)\left(\sum_m \sum_n (\hat{A}_{mn} - \hat{A}_\iota)^2\right)}} \tag{1}$$

$$SSIM\left(A, \hat{A}\right) = \frac{\left(2\mu_A\mu_{\hat{A}} + c_1\right)\left(2\sigma_{A\hat{A}} + c_2\right)}{\left(\mu_A^2 + \mu_{\hat{A}}^2 + c_1\right)\left(\sigma_A^2 + \sigma_{\hat{A}}^2 + c_2\right)} \tag{2}$$

$$Entropy = -\sum_{i=1}^{n} p_i log_2 p_i \tag{3}$$

$$PSNR = 20log_{10}\left(\frac{max_f}{\sqrt{MSE}}\right) \tag{4}$$

$$MSE = \frac{1}{m \times n}\sum_m \sum_n \left(A_{mn} - \hat{A}_{mn}\right)^2 \tag{5}$$

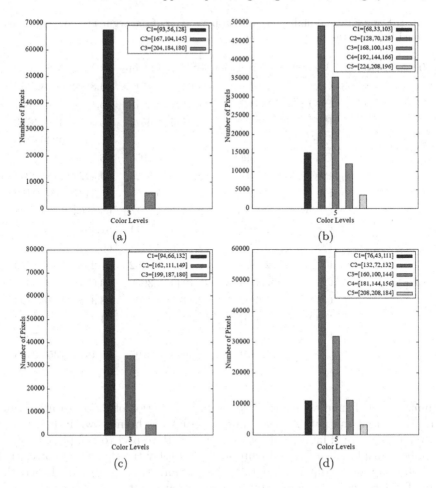

Fig. 2. Histograms of segmented images (shown in Fig. 1) using the proposed method: (a)-(b) Sample-1; $\kappa = 3$ and $\kappa = 5$; (c)-(d) Sample-2; $\kappa = 3$ and $\kappa = 5$.

5   Result and Analysis

The result of the experiment will be discussed in this section. The two sample images are shown in Fig. 1-(a) and (d). The segmentation results shown in Fig. 1 (b), (c) and (e), (f) are for $\kappa = 3$ and $\kappa = 5$, corresponding to the input images. The histograms of the segmented images concerning Sample-1 and Sample-2 are shown in Fig. 2 for $\kappa = 3$ and $\kappa = 5$. The value of different segmentation quality evaluation parameters is demonstrated in Table 1. Each value represents the

Table 1. Segmentation quality evaluation parameters from segmented output images obtained by using the proposed method.

Image	κ	Algorithm	Correlation	SSIM	Entropy	PSNR	MSE
Sample-1	$\kappa = 3$	GA	2.3478	2.0408	3.5517	54.8862	4338.6713
		PSO	2.3465	1.9851	3.5818	53.8331	5109.8641
		FA	2.3447	1.9955	3.6577	56.8558	4547.5783
		QIFA	2.3507	2.0349	3.6273	56.4494	4240.6193
	$\kappa = 5$	GA	2.7396	2.4878	5.7055	65.8194	2132.8281
		PSO	2.7156	2.41	5.6522	64.7729	2185.3244
		FA	2.72	2.4635	5.807	65.4137	2007.3167
		QIFA	2.7425	2.4895	5.8014	65.6964	1754.8488
Sample-2	$\kappa = 3$	GA	2.3905	2.1254	3.4371	61.4917	3596.9291
		PSO	2.3712	2.1229	3.2451	60.9346	3698.4936
		FA	2.3863	2.1701	3.3282	61.285	3166.2579
		QIFA	2.3775	2.1243	3.3038	60.6244	3963.1392
	$\kappa = 5$	GA	2.6844	2.4473	5.3491	68.321	1476.5103
		PSO	2.6497	2.395	5.1786	64.3631	2200.8482
		FA	2.7312	2.5161	5.3539	71.1501	1462.7987
		QIFA	2.7191	2.5268	5.4297	69.4649	1354.2406

summation of channel-wise parameters values. Variation of CPU execution time for a sequential and parallel version of the QIFA has been shown in Fig. 5. The convergence curve of individual color channels with respect to Sample-1(a) & (e) have been displayed in Fig. 3. Figure 3(a)–(f) display the changes of correlation with iteration under different κ for Sample-1. Similarly, Fig. 3(g)–(l) display the same for Sample-2. A comparative segmentation result set has been shown in Fig. 4. Figure 5(a) displays the changes in execution time concerning image size. The original image was 2352 × 3136, then the original image has been scaled down to 10%, 20%, 30%, 40%, 50% 60%, 70%, 80%, and 90%. Finally, the scaled-down and original images are executed for both versions of the algorithms under the fixed color level of $\kappa = 8$. Figure 5(b) shows the execution time of both algorithms for various values of $\kappa = 3, 5$, and the image size is fixed to 12.5%. A comparison on execution time with other mentioned algorithms is also shown in Fig. 6 for $\kappa = 3$ and $\kappa = 5$.

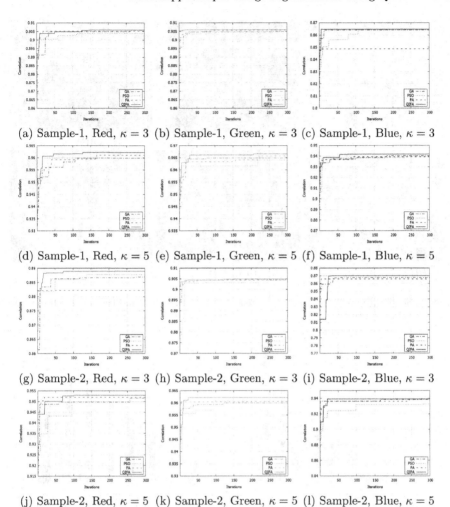

(a) Sample-1, Red, $\kappa = 3$ (b) Sample-1, Green, $\kappa = 3$ (c) Sample-1, Blue, $\kappa = 3$

(d) Sample-1, Red, $\kappa = 5$ (e) Sample-1, Green, $\kappa = 5$ (f) Sample-1, Blue, $\kappa = 5$

(g) Sample-2, Red, $\kappa = 3$ (h) Sample-2, Green, $\kappa = 3$ (i) Sample-2, Blue, $\kappa = 3$

(j) Sample-2, Red, $\kappa = 5$ (k) Sample-2, Green, $\kappa = 5$ (l) Sample-2, Blue, $\kappa = 5$

Fig. 3. Convergence curves (Iterations vs Correlation) of `Sample 1` and `Sample-2` corresponding to all the channels and different κ values.

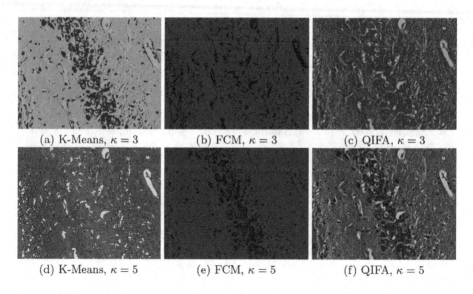

(a) K-Means, $\kappa = 3$ (b) FCM, $\kappa = 3$ (c) QIFA, $\kappa = 3$

(d) K-Means, $\kappa = 5$ (e) FCM, $\kappa = 5$ (f) QIFA, $\kappa = 5$

Fig. 4. Segmented images (`Sample-1`) using various methods and the proposed method.

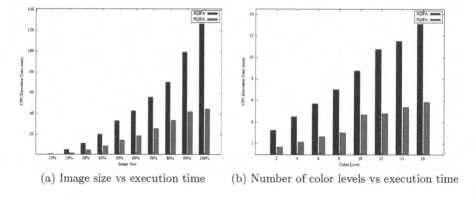

(a) Image size vs execution time   (b) Number of color levels vs execution time

Fig. 5. Comparison of CPU times to execute sequential and parallel QIFA (concurrent for three channels) – Violet: sequential QIFA, Green: parallel QIFA. (Color figure online)

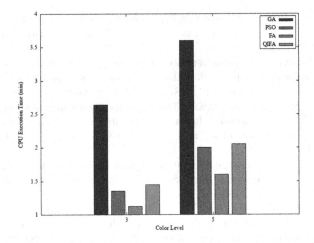

Fig. 6. Execution time: Proposed QIFA with channel merging vs other algorithms.

6 Conclusion

In this work, the QIFA algorithm has been deployed in the multi-core architecture to make it parallel. The method successfully produced good-quality color-segmented images. Detailed experimentation has been carried out to investigate the performance of the proposed parallel algorithm on various image sizes and color threshold levels. The segmentation results have been compared with several other algorithms. It corroborates that the QIFA can generate satisfactory color image segmentation when applied to microscopic images by simply merging channel-wise optimums. The proposed method can be further extended to other modalities of medical images, like MRI images, histopathological images, CT images, etc.

References

1. Lange, N., et al.: Two macroscopic and microscopic brain imaging studies of human hippocampus in early Alzheimer's disease and schizophrenia research. Stat. Med. **23**(2), 327–350 (2004). https://doi.org/10.1002/sim.1720
2. Mesejo, P., Ibáñez, Ó., Cordón, Ó., Cagnoni, S.: A survey on image segmentation using metaheuristic-based deformable models: state of the art and critical analysis. Appl. Soft Comput. **44**, 1–29 (2016). https://doi.org/10.1016/j.asoc.2016.03.004
3. Heckers, S.: Neuroimaging studies of the hippocampus in schizophrenia. Hippocampus **11**(5), 520–528 (2001). https://doi.org/10.1002/hipo.1068
4. Heckers, S., et al.: Impaired recruitment of the hippocampus during conscious recollection in schizophrenia. Nat. Neurosci. **1**(4), 318–323 (1998). https://doi.org/10.1038/1137
5. Ragland, J.D.: Effect of schizophrenia on frontotemporal activity during word encoding and recognition: a PET cerebral blood flow study. Am. J. Psychiatry **158**(7), 1114–1125 (2001). https://doi.org/10.1176/appi.ajp.158.7.1114

6. Chakraborty, S., et al.: Modified cuckoo search algorithm in microscopic image segmentation of hippocampus. Microsc. Res. Tech. **80**(10), 1051–1072 (2017). https://doi.org/10.1002/jemt.22900

7. Dey, N., Ashour, A.S.: Meta-heuristic algorithms in medical image segmentation. In: Advancements in Applied Metaheuristic Computing, pp. 185-203. IGI Global (2018). https://doi.org/10.4018/978-1-5225-4151-6.ch008

8. Ghosh, P., Mali, K., Das, S.K.: Chaotic firefly algorithm-based fuzzy C-means algorithm for segmentation of brain tissues in magnetic resonance images. J. Vis. Commun. Image Representation **54**, 63–79 (2018). https://doi.org/10.1016/j.jvcir.2018.04.007

9. Giuliani, D.: A grayscale segmentation approach using the firefly algorithm and the gaussian mixture model. Int. J. Swarm Intell. Res. **9**(1), 39–57 (2018). https://doi.org/10.4018/ijsir.2018010103

10. Oliva, D., Abd Elaziz, M., Hinojosa, S.: Multilevel thresholding for image segmentation based on metaheuristic algorithms. In: Metaheuristic Algorithms for Image Segmentation: Theory and Applications. SCI, vol. 825, pp. 59–69. Springer, Cham (2019). https://doi.org/10.1007/978-3-030-12931-6_6

11. Upadhyay, P., Chhabra, J.K.: Kapur's entropy based optimal multilevel image segmentation using crow search algorithm. Appl. Soft Comput. **97**, 105522 (2020). https://doi.org/10.1016/j.asoc.2019.105522

12. Hernandez del Rio, A.A., Cuevas, E., Zaldivar, D.: Multi-level image thresholding segmentation using 2D histogram non-local means and metaheuristics algorithms. In: Oliva, D., Hinojosa, S. (eds.) Applications of Hybrid Metaheuristic Algorithms for Image Processing. SCI, vol. 890, pp. 121–149. Springer, Cham (2020). https://doi.org/10.1007/978-3-030-40977-7_6

13. Pare, S., Bhandari, A.K., Kumar, A., Singh, G.K.: A new technique for multilevel color image thresholding based on modified fuzzy entropy and Lévy flight firefly algorithm. Comput. Electr. Eng. **70**, 476–495 (2018). https://doi.org/10.1016/j.compeleceng.2017.08.008

14. He, L., Huang, S.: An efficient krill herd algorithm for color image multilevel thresholding segmentation problem. Appl. Soft Comput. **89**, 106063 (2020). https://doi.org/10.1016/j.asoc.2020.106063

15. Dhal, K.G., Das, A., Ray, S., Gálvez, J.: Randomly attracted rough firefly algorithm for histogram based fuzzy image clustering. Knowl. Based Syst. **216**, 106814 (2021). https://doi.org/10.1016/j.knosys.2021.106814

16. Choudhury, A., Samanta, S., Pratihar, S., Bandyopadhyay, O.: Multilevel segmentation of Hippocampus images using global steered quantum inspired firefly algorithm. Appl. Intell. **52**(7), 7339–7372 (2021). https://doi.org/10.1007/s10489-021-02688-6

Breast Cancer Histologic Grade Identification by Graph Neural Network Embeddings

Salvatore Calderaro[1]([✉]) [iD], Giosué Lo Bosco[1] [iD], Filippo Vella[2] [iD],
and Riccardo Rizzo[2] [iD]

[1] Department of Mathematics and Computer Science, University of Palermo,
90123 Palermo, Italy
{salvatore.calderaro01,giosue.lobosco}@unipa.it
[2] Institute for High-Performance Computing and Networking,
National Research Council of Italy, 90146 Palermo, Italy
{filippo.vella,riccardo.rizzo}@icar.cnr.it

Abstract. Deep neural networks are nowadays state-of-the-art methodologies for general-purpose image classification. As a consequence, such approaches are also employed in the context of histopathology biopsy image classification. This specific task is usually performed by separating the image into patches, giving them as input to the Deep Model and evaluating the single sub-part outputs. This approach has the main drawback of not considering the global structure of the input image and can lead to avoiding the discovery of relevant patterns among non-overlapping patches. Differently from this commonly adopted assumption, in this paper, we propose to face the problem by representing the input into a proper embedding resulting from a graph representation built from the tissue regions of the image. This graph representation is capable of maintaining the image structure and considering the relations among its relevant parts. The effectiveness of this representation is shown in the case of automatic tumor grading identification of breast cancer, using public available datasets.

Keywords: Histology images · Graph Neural Networks · Breast Cancer

1 Introduction

According to Arnold et al. [3], breast cancer is the most commonly diagnosed cancer type, with 1 in 8 cancer diagnoses worldwide and counting, in 2020, over 2.3 million new cases. The tumor diagnosis is obtained, in the first step, using mammography or ultrasound scans to identify suspicious regions of the breast, and then a tissue biopsy to determine the presence and assess the cancer grade. During the biopsy examination, the pathologist searches for specific features that can help formulate a disease prognosis and assess how the cancer is growing or

I. Rojas et al. (Eds.): IWBBIO 2023, LNBI 13920, pp. 283–296, 2023.
https://doi.org/10.1007/978-3-031-34960-7_20

spreading. These features, taken as a whole, determine the spread of cancer at the diagnosis time and have been formalized by the "Nottingham Grading System" [9]. Looking at the sample with a microscope, the pathologist analyses features of the cancer cells, such as their similarity with the normal ones, and decides a score for each feature that will be summarised in a global score called Grade. The Grade varies from 1 to 3, where Grade 1 labels well-differentiated cells and usually a less aggressive cancer, Grade 2 labels the cancer of intermediate nature, and Grade 3 regards the less differentiated cells, usually more aggressive cancers. The Grade value is used as a parameter in the decision for post-surgery treatment. The identification of the most relevant characteristics can be time-consuming and Machine Learning (ML) methods for image classification can help the pathologist speed up the diagnosis process. Deep convolutional neural networks (DCNN) are state-of-the-art methodologies for image classification, and they could profitably be adopted as standard post-surgery image classification systems [4–6]. Unfortunately, their complexity is strictly related to the input size of the image, so in order to provide efficiency, the input needs to be reduced from the whole image to a set of its parts (patches). Consequently, every DCNN could be prone to fail in identifying large structural features or relationships between distant image patterns. We have tried to avoid such drawbacks in this paper, proposing a graph representation of the histopathological image built using the tissue regions of the image. Furthermore, we propose a classification system based on a graph neural network, fast to train, and whose performances are comparable with other state-of-the-art methods. We have tested our methodology on two publicly available datasets of breast invasive ductal carcinoma [8, 27]. The paper is structured as follows: in Sect. 2, we report a brief description of the state-of-the-art approaches for the Breast Cancer Histologic Grade Identification; in Sect. 3, we describe the dataset used for the experiments, the graph neural networks theory, the technique used for building the graph, the experimental setup and the obtained results; in Sect. 6 the conclusions are given.

2 Related Works

At the moment, few studies address the problem of the Grade evaluation in cancer images and this lack is even more evident for breast cancer cases. Here we review the methodologies adopting the datasets we employ in this paper, also mentioning two other recent approaches working on similar datasets.

One of the contributions in this field is the work of Dimitropoulos et al. [8], which introduced a linear dynamical system approach and a dataset used to test proposed methodologies. This dataset, that is also adopted in this paper, is characterized by a limited number of images (in total, 300) with resolution 1280×960, corresponding to 21 patients with invasive ductal carcinoma. The images are obtained from regions affected by the tumor and are captured using a Nikon digital camera with a magnification factor of 40×. The scarcity of images in this dataset implies a difficulty in the design of performing models. The histopathological images were divided into patches using different strategies: overlapping, non-overlapping, and random selections. These patches are considered as multidimensional signals that are modeled using a higher-order linear dynamical system and encoded using the VLAD (Vector of Locally Aggregated Descriptors) to obtain the final input descriptors. The Grade identification is performed by a support vector machine, which achieves a 95.8% of accuracy using as inputs the descriptors of overlapping image patches sized at 8×8.

The same dataset was used in a study by Nanni et al. [18]. They proposed a solution based on the ensemble of ResNet50 networks differentiated in terms of the learning algorithm used. An accuracy of 94.33% is reached using 14 models trained with the stochastic gradient descent algorithm and 21 others with variants of the ADAM algorithm.

Li et al. [16] addressed the Grade identification in the dataset mentioned above using a model based on the Xception network. It is used for feature extraction, and other transfer-learned models are adopted for Grade identification. The result achieved in terms of accuracy is 93.01 ± 1.52.

Senousy et al. [22] propose an entropy-based elastic ensemble DCNN model named 3E-Net. The model consists of several DCNNs designed and implemented concerning the image input size and the patch number. As a feature extractor for patches, a DenseNet-16 [11] is used. The 3E-Net is an ensemble method where the final predictions of several CNNs are weighted by the uncertainty, measured by Shannon entropy, of the predictions of the related DCNN. The performance of 3E-Net is evaluated using a score that does not consider the images on which the ensemble is uncertain. Following this paradigm, which discards some of the images in the dataset, the model achieves an accuracy of 99.50.

A second dataset of graded cancer images, called the PathoIDCG dataset, also enrolled in this paper, has been introduced by Yan et al. [27] together with a deep-learning method for Grade identification. The dataset is composed of 3644 images of invasive ductal carcinoma of the breast, acquired at two different magnification factors: $20\times$ and $40\times$. NGNet is the deep method they propose, which processes two kinds of input: the original image and the image segmented by DeepLabV3+ net The purpose of this segmentation is to introduce the nucleus region in the input image. An attention mechanism is also adopted. To evaluate the performance of the NGNet, the authors use 80% of the images to train and validate the model and the remaining 20% for the testing. The result obtained in terms of accuracy is 93.4.

Another paper by the same authors [28], with the same dataset, proposes a "divide-and-attention" network named DANet that learns representative pathological image features w.r.t. different tissue structures and adaptively focuses on the most important ones. The basic idea is similar to the previous methodology, with the difference that DANet receives three inputs: the original image, the segmented image with DeepLabV3+ net which contains only the nuclei, and the image without the nuclei. The authors test the method on three different histopathological image datasets. We only report the results obtained for the dataset introduced in [27]. The dataset was augmented performing zoom, flip and contrast variation reaching a number of 32.820 images. The 80% of the images are used to train and validate the model, and the remaining 20% for testing purposes. The result obtained in terms of accuracy is 91.6.

Wang et al. [25] have proposed another approach that faces the grading problem with a different dataset. The authors develop and validate a new histological Grade model named DeepGrade based on digital whole-slide images and deep learning. For the optimization and validation phases, a set of images was derived from 1567 patients, while to verify the model's generalization capability, an external test set, with 1262 patients, was used. The goal was to classify the images according two classes: Grade 1 and Grade 3. The authors divided the WSI images into tiles of 598 × 598, and color normalization was applied. DeepGrade is an ensemble of twenty convolutional neural networks based on InceptionV3.

Jaroensri et al. [12] proposed their approach on a proprietary dataset. Their solution consists of four separate deep-learning models. The first is used to segment the invasive carcinoma within a whole slide image. The model was trained to distinguish between non-tumor, carcinoma in situ, and invasive carcinoma. The other three models were used to predict the score for each of the three tumor features comprising the Nottingham histopathologic Grade.

3   Materials and Methods

3.1   The Used Datasets

The current study is focused on the automatic grading of histopathology images and was tested on two open-access datasets. The first dataset [8] is named *Agios Pavlos* (from the hospital that allowed images collection) and contains 300 images with a resolution of 1280 × 960 from 21 patients with invasive ductal carcinoma. The images are obtained from regions affected by the tumor and were captured using a Nikon digital camera with a magnification factor of 40×. The dataset contains 107 images of the Grade 1 class, 102 images of the Grade2 class, and 91 images of the Grade 3 class; row A of Fig. 1 shows a sample of the images in the dataset. Through a deep investigation of the dataset, we noticed nine duplicated images, i.e., belonging at the same time to the Grade 1 and Grade 3 classes. We omitted these images, reducing the number of images to 282.

The other dataset is named *PathoIDCG* and was proposed in [27]. It contains 3644 images of resolution 1000 × 1000 acquired at two magnification factors 20× and 40×. Table 1 reports the sample distribution of the dataset, while row B of Fig. 1 shows a sample of the images in the dataset.

Table 1. The distribution of the samples in the used datasets, for the Agios Pavlos dataset. We did not consider the duplicated images in Grade 1 and Grade 3 classes.

Dataset	Magn. Factor	Grade 1	Grade 2	Grade 3	# of images
Agios Pavlos	40×	98	102	82	282
PathoIDCG	20×	600	641	1245	2486
	40×	361	480	317	1158

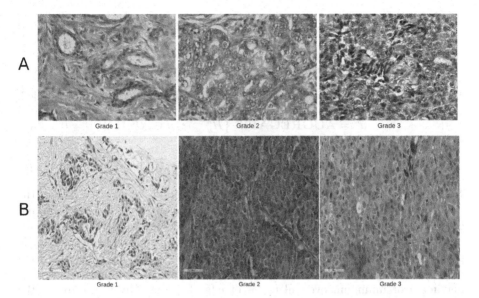

Fig. 1. Some samples from the two datasets: **A**: Agios Pavlos dataset; **B**: PathoIDCG dataset.

3.2   Graph Neural Networks

Graph Neural networks are a class of neural networks for processing data represented as a graph. These models can be used to perform different tasks such as node classification, link prediction, and graph classification. The first two tasks use a single graph as input. In the case of node classification, the graph has a set of labelled nodes, and the goal is to predict the class of unlabelled nodes. Link prediction has, instead, as a goal the prediction of missing or future links in the graph, starting from labelled links. Finally, in graph classification, the input is a set of graphs that belong to different classes, and the goal is to find the class of a given input graph. In this work, we approached histology image grading identification as a graph classification problem. In the following the basic notions of Graph Neural Networks (GNN) are introduced.

Let $\mathcal{G} = (\mathcal{V}, \mathcal{E})$ a graph where $\mathcal{V}$ is the set of nodes of the graph and $\mathcal{E}$ is the set of edges; $A \in \mathbb{R}^{|\mathcal{V}| \times |\mathcal{V}|}$ is the so-called adjacency matrix of the graph $\mathcal{G}$. Associated to each node v there is a vector of attributes $x \in \mathbb{R}^C$ where C is the dimension of input; the set of all the features are represented in a matrix $X \in \mathbb{R}^{|\mathcal{V}| \times C}$. The goal of the GNN is to learn a new matrix representation $H \in \mathbb{R}^{|\mathcal{V}| \times F}$ where F is the number of features for each node, obtained by combining the graph structure information and the node attributes. A graph neural network is made by a set of layers l_k with $k = 1, 2, \ldots, K$. The idea is to iteratively update the node representations by combining them with the node representations of their neighbours [26]. Starting from the initial node representation $H^0 = X$, in each layer of the network l_k, two steps are performed:

- **AGGREGATE**: tries to aggregate the information from the neighbours of each node;
- **COMBINE**: tries to update the node representation by combining the aggregated information from neighbours with the current node representations.

Mathematically, we start with $H^0 = X$, and the following equations are applied:

$$a_v^k = \textbf{AGGREGATE}^k \left\{ H_u^{k-1} \; : \; u \in N(v) \right\} \tag{1}$$

$$H_v^k = \textbf{COMBINE}^k \left\{ H_v^{k-1}, a_v^k \right\} \tag{2}$$

where $N(v)$ is the set of the neighbours of the $v-th$ node and does not include the node v. The node representation $H^K \in \mathbb{R}^{N \times F}$ is the output of the last layer and can be used as the final node representation. This representation can be used for downstream tasks. In the case of graph classification, in order to make a single prediction is necessary to aggregate the node representations. This operation is called **READOUT**. Common readout operations are summation, average, maximum or minimum over all nodes or edge features. Given a graph g, the average node feature readout is defined as:

$$h_{\mathcal{G}} = \frac{1}{|\mathcal{V}|} \sum_{v \in \mathcal{V}} h_v \tag{3}$$

where $|\mathcal{V}|$ is the number of nodes in $\mathcal{G}$, h_v is the feature node of v, and $h_{\mathcal{G}}$ is the calculated representation of the graph $\mathcal{G}$. Once $h_{\mathcal{G}}$ is available, it is the input of an MLP layer for classification output.

Graph Convolutional Networks (GCN) [15] are among the most used graph neural network architectures due to their effectiveness in many applications. GCN are neural networks used to perform the convolution operation on undirected graphs. A single layer within a GCN can be described by the following equation:

$$H^{k+1} = \sigma \left(\tilde{D}^{-\frac{1}{2}} \tilde{A} \tilde{D}^{-\frac{1}{2}} H^k W^k \right) \tag{4}$$

where $\tilde{A} = A + I$ is the adjacency matrix of the graph $\mathcal{G}$ with self-connections for incorporating the node feature itself, and $I \in \mathbb{R}^{|\mathcal{V}| \times |\mathcal{V}|}$ is the identity matrix. $\tilde{D}$ is a diagonal matrix with $\tilde{D}_{ij} = \sum_j \tilde{A}_{ij}$ called degree matrix used to normalize nodes with large degrees and $\sigma(\cdot)$ is an activation function such as ReLu or Tanh. The term $\tilde{D}^{-\frac{1}{2}} \tilde{A} \tilde{D}^{-\frac{1}{2}}$ is a slight modification of the normalized symmetric Laplacian of the graph. $W^k \in \mathbb{R}^{F \times F'}$ (F and F' are the dimension of the node representations in the layer k and $k+1$ respectively) implements a linear transformation matrix which will be trained during the optimization phase. The Eq. 4 projects the node embeddings, computes each new embedding as the average of all node's neighbours and applies an element-wise non-linearity to the new embeddings. To better understand the **AGGREGATE** and **COMBINE** functions used in the GCN, for a node i, the Eq. 4 can be reformulated as:

$$H_i^k = \sigma \left(\sum_{j \in N(i)} \frac{A_{ij}}{\sqrt{\tilde{D}_{ii} \tilde{D}_{jj}}} H_j^{k-1} W^k + \frac{1}{\tilde{D}_i} H_i^{k-1} W^k \right) \tag{5}$$

In Eq. 5, the **AGGREGATE** is the weighted average of the neighbour node representations (first term of the sum). The weight of the neighbour j is the weight of the edge between i and j (i.e. A_{ij} normalized by the degrees of the two nodes). The **COMBINE** function is the sum of the aggregated messages and the node representation itself, normalized by its degree (second term of the sum).

Figure 3 shows the GCN used in this work, the network has seven layers, an average **READOUT** function (as the one in Eq. 3), and a linear layer with Softmax activations.

3.3   Graph Construction

In order to build a graph starting from a histopathological image, we follow an approach similar to the one reported in [13, 20]. We begin segmenting the image using the Simple Linear Iterative Clustering (SLIC) algorithm [2]. The output of this algorithm is a set of so-called superpixels, which are groups of pixels that share common characteristics, such as their colour and proximity in the image plane. The SLIC works by taking as inputs the number of segments, the compactness value used to balance colour and spatial proximities (higher values give more weight to space proximity), and the value of the width of the Gaussian smoothing kernel used to pre-process the image. Figure 2A shows an example of a histopathological input image Fig. 2B its segmentation by SLIC. After the segmentation of the image, we build a Region Adjacency Graph (RAG) [21] using the mean colour. Each node of the RAG is a set of pixels (called superpixels) within the image with the same label (the labels came from the segmentation of the image by SLIC). The weight of two adjacent regions represents the similarity between the two areas. The weight between two adjacent regions is $|c_1 - c_2|$ where c_1 and c_2 are the mean colours of the two regions. After this step, to reduce the complexity of the graph, we merge the regions characterized by weights lower than a prefixed threshold (see Fig. 2C for an example).

The final graph is obtained so that the nearest pseudo centres of the adjacent regions are connected with an edge (see 2D for an example).

To feed the graph to a GCN, initial features are assigned to each node of the graph. For this purpose, we consider the centroid coordinates and extract a patch from the image starting from the centroid. The following equation gives the patch size:

$$ps = \min(w, h) \times \alpha \tag{6}$$

i.e., the minimum between the image's width and height multiplied by a scale factor α. The obtained patch is fed to a ResNet50 [10] pre-trained on the ImageNet dataset [7]; the output is a vector of dimension 2048.

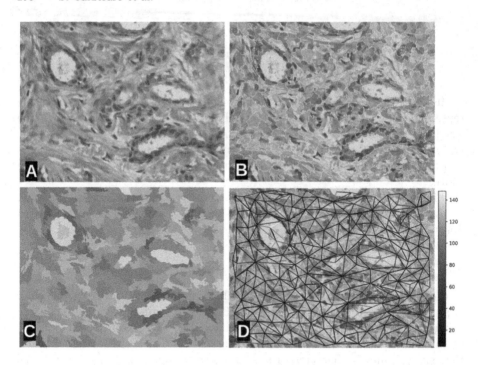

Fig. 2. A representation of the image processing: **A**. The original image; **B**. An example of a segmented image obtained using the SLIC algorithm; **C**. The regions of the image after the thresholding operation; **D**. The final graph, the colour of the edges represents how similar/dissimilar the two connected regions are. Similar areas will have a black edge, while dissimilar ones will have an orange edge. The colour bar on the right represents the variation of these colours based on the similarity value. These weights are not used by the GNN.

3.4   Experimental Setup

Experiments have been performed using a workstation equipped with a 12th Gen Intel®Core™ i9-12900KS CPU and an Nvidia 3090 Ti GPU. The total amount of system memory is 32 Gbyte of DDR5. The GPU is also supplied with 24 Gbyte of DDR5 memory and adopts a CUDA parallel computing platform. The operating system is Ubuntu 22.10. TensorFlow and Python scikit image libraries were used to create the graphs from the images [1,23]; the Deep Graph Library and PyTorch Python libraries [19,24] were used to construct and train the GCN. For the construction of the graph, we chose a number of segments equal to 300 for the Agios Pavlos dataset and 100 for the PathoIDCG dataset, with a compactness value of 20 and a sigma value of 1. We choose a different number of segments for the PathoIDCG dataset because its images have a smaller dimension than the Agios Pavlos one, and the images probably have fewer segments. To reduce the complexity of the graph, we have chosen a threshold value equal to 5.

To obtain the feature for each graph node, we use patches of the image with dimensions given by Eq. 6 with scale factor $\alpha = 0.1$, obtaining a patch size 96 for the Agios Pavlos dataset and 100 for the PathoIDCG dataset. The total time to build the graphs starting from the histopathological images is about 117.5 min for the Agios Pavlos dataset and 1032.47 for the PathoIDCG one. The networks were trained for 60 epochs without any EarlyStopping condition or any check on loss value with a mini-batch size of 32, with the Adam optimisation algorithm [14] and a learning rate of 1×10^{-3}. The training time for the first dataset is about 1 min and 20 s, while for the second dataset about six minutes.

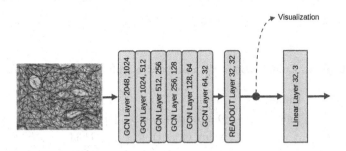

Fig. 3. The architecture of the neural network used for the experiments. The number within the parenthesis represents the number of input and output features. The dotted line indicates the output used for visualisation and clusters analysis.

4 Results

The proposed GCN has been evaluated considering four metrics: accuracy, precision, recall, and f1-score. Regarding the training-test split paradigm, we have adopted a five-fold cross-validation testing protocol. Table 2 reports the values of the adopted metric together with their standard deviation across the five folds for the two considered datasets.

Table 3 reports the comparison between the proposed method and some state-of-the-art. The experiments with the Agios Pavlos dataset are compared with (1) the methodology of the original paper that proposed the dataset [8]; (2) the proposal by Nanni et al. [18]; (3) the approach by Li et al. [16]. The first two methodologies use a five-fold cross-validation testing protocol, while the latter uses five random splits so as a consequence their results cannot be directly comparable with ours. Note that in the Table we omitted the work of Senousy et al. [22] because they use a modification of the accuracy measure that does not account for the images on which the classification system is uncertain. In the same table are reported the results and the comparisons related to the PathoIDCG dataset. The first work of Yan et al. [27] introduced the PathoIDCD dataset and proposed a deep-learning approach, while the second one [28] proposed another deep-learning approach tested on the same dataset. In both works, the authors test their proposed methods' effectiveness using a random split between train and test. Conversely, we use a five-fold cross-validation testing protocol, and for this reason, our results are more reliable.

Table 2. Mean and standard deviation values of the classification metrics on the two datasets.

Dataset	Accuracy	Precision	Recall	F1-Score
Agios Pavlos	96.80 ± 2.06	96.88 ± 2.00	96.80 ± 2.06	96.79 ± 2.07
PathoIDCG	96.67 ± 0.98	96.74 ± 0.92	96.67 ± 0.98	96.67 ± 1.00

Table 3. Comparison of the results with state-of-the-art methods

Dataset	Method	Accuracy
Agios Pavlos	Dimitropoulos et al. [8]	95.80
	Nanni et al. [18]	94.33
	Li et al. [16]	93.01
	Proposed method	**96.80**
PathoIDCG	Yan et al. [27]	93.4
	Yan et al. [28]	91.6
	Proposed method	**96.67**

For further analysis, we have analyzed visually the embeddings produced by the output of the last layer of our GCN (the one shown in Fig. 3 with the dotted arrow).

In particular, the Uniform Manifold Approximation and Projection for Dimension Reduction (UMAP) algorithm [17]. was used to visualise in 2-d the obtained embeddings.

This algorithm use the Riemannian geometry and algebraic topology to find the best representation of the data in the new space. Figure 4 shows this visualisation for train and test data of Fold 1 for both datasets, where we can notice three well-separated clusters and a few misplaced points (the same configuration is visible for the other 4 folds). The axis don't have a particular meaning.

5   Ablation Study

To further demonstrate our methodology's effectiveness, we conducted an ablation study aimed at understanding the result of the graph processing by the GNN. For all the training images, the **READOUT** function was directly applied to the graph representing the images, i.e., bypassing all the GCN layers in Fig. 3. The obtained vector $h_G \in \mathbb{R}^{2048}$ was used to train:

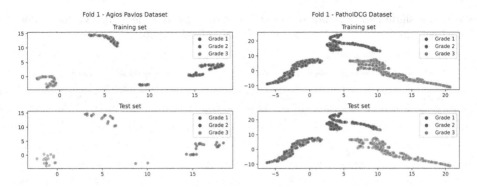

Fig. 4. Bidimensional plot obtained by using UMAP, of the embedding for Agios Pavlos (on the left) and PathoIDCG (on the right) datasets. The upper part is referred to the training set; the lower part to the test set.

- the last linear layer (a matrix of dimension 2048 × 3); we have trained this network for 60 epochs.
- a Support Vector Machine using a linear kernel, in order to check if a classifier is able to separate the three kinds of graphs without the GCN processing. Note that a Gaussian kernel was also tested but the accuracy is significantly lower than the version with the linear kernel (results available upon request).

In Table 4, we have reported the mean and standard deviation of the accuracy metric for both datasets used to perform our experiments. Interestingly, we can notice that a linear SVM can reach results comparable to the GCN network just using the REDOOUT representation, but with a much bigger vector representation (of length 2048) compared to the one proposed by the GCN with its last layer (length 32). This can suggest that the GCN is able to find a reliable and compact representation of the image through the subsequent processing of its layers.

Table 4. Ablation study results.

Dataset	Method	Accuracy
Agios Pavlos dataset	READOUT + Trained Linear	93.60 ± 2.21
	READOUT + Linear SVM	97.50 ± 1.92
PathoIDCG dataset	READOUT + Trained Linear	93.38 ± 0.86
	READOUT + Linear SVM	96.62 ± 0.77

This can be demonstrated by visualizing the embeddings of the **READOUT** operation in 2-d using the UMAP algorithm (Fig. 5). In both datasets, the images of the different grades seem not to be well-separated, showing a visible overlap. This trend is more visible for the PathoIDCG dataset (see Fig. 5 on the right), where a significant overlap between the samples of Grade 1 and Grade 3 and between the samples of Grade 2 and Grade 3 is observable.

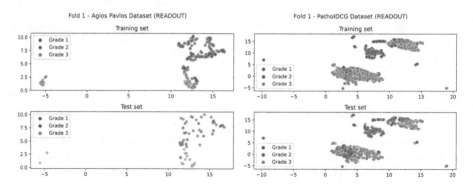

Fig. 5. (Ablation Study) Bidimensional plot obtained using UMAP, of the embedding without any GCN layer, only with READOUT operation. Dataset Agios Pavlos (on the left) and PathoIDCG (on the right). The training set is shown in the upper part and the test set is in the lower part.

6   Conclusions

The adoption of a graph representation for an image gives the possibility to represent in a compact way local features altogether with their spatial relationships, even for large images. Graph neural networks can effectively exploit this representation for the classification of histopathological images. We used this model to assess the grading of breast tumors by processing digitized biopsy images. The experiments, carried out on two independent datasets, show very good results. Moreover, the representation induced by the proposed graph neural network shows a relevant separation capability with respect to a direct representation by GCN. As future work, we plan to use this framework to classify whole slide images (WSI) and to integrate information from multiple patients with the aim to obtain a more general representation, also validating the network's feasibility by testing biopsy images from other datasets or patients.

References

1. Abadi, M., et al.: TensorFlow: large-scale machine learning on heterogeneous distributed systems. CoRR abs/1603.04467 (2016)
2. Achanta, R., Shaji, A., Smith, K., Lucchi, A., Fua, P., Süsstrunk, S.: SLIC superpixels compared to state-of-the-art superpixel methods. IEEE Trans. Pattern Anal. Mach. Intell. **34**(11), 2274–2282 (2012)
3. Arnold, M., et al.: Current and future burden of breast cancer: global statistics for 2020 and 2040. Breast **66**, 15–23 (2022)

4. Calderaro, S., Lo Bosco, G., Rizzo, R., Vella, F.: Fuzzy clustering of histopathological images using deep learning embeddings. In: CEUR Workshop Proceedings, vol. 3074 (2021)
5. Calderaro, S., Lo Bosco, G., Rizzo, R., Vella, F.: Deep metric learning for histopathological image classification. In: 2022 IEEE Eighth International Conference on Multimedia Big Data (BigMM), pp. 57–64 (2022)
6. Calderaro, S., Lo Bosco, G., Rizzo, R., Vella, F.: Deep metric learning for transparent classification of COVID-19 X-Ray images. In: 2022 16TH International Conference On Signal Image Technology & Internet Based Systems (SITIS) (2022)
7. Deng, J., Dong, W., Socher, R., Li, L.J., Li, K., Fei-Fei, L.: ImageNet: a large-scale hierarchical image database. In: 2009 IEEE Conference on Computer Vision and Pattern Recognition, pp. 248–255. IEEE (2009)
8. Dimitropoulos, K., Barmpoutis, P., Zioga, C., Kamas, A., Patsiaoura, K., Grammalidis, N.: Grading of invasive breast carcinoma through Grassmannian VLAD encoding. PLoS ONE **12**(9), e0185110 (2017)
9. Elston, C.W., Ellis, I.O.: Pathological prognostic factors in breast cancer. I. The value of histological grade in breast cancer: experience from a large study with long-term follow-up. Histopathology **19**(5), 403–410 (1991)
10. He, K., Zhang, X., Ren, S., Sun, J.: Deep residual learning for image recognition. In: Proceedings of the IEEE Conference on Computer Vision and Pattern Recognition, pp. 770–778 (2016)
11. Huang, G., Liu, Z., Weinberger, K.Q.: Densely connected convolutional networks. CoRR abs/1608.06993 (2016)
12. Jaroensri, R., et al.: Deep learning models for histologic grading of breast cancer and association with disease prognosis. NPJ Breast Cancer **8**(1), 1–12 (2022)
13. Jaume, G., Pati, P., Anklin, V., Foncubierta, A., Gabrani, M.: HistoCartography: a toolkit for graph analytics in digital pathology. In: MICCAI Workshop on Computational Pathology, pp. 117–128 (2021)
14. Kingma, D.P., Ba, J.: Adam: a method for stochastic optimization. CoRR abs/1412.6980 (2014)
15. Kipf, T.N., Welling, M.: Semi-supervised classification with graph convolutional networks (2016)
16. Li, L., et al.: Multi-task deep learning for fine-grained classification and grading in breast cancer histopathological images. Multimed. Tools Appl. **79**(21), 14509–14528 (2020)
17. McInnes, L., Healy, J., Melville, J.: UMAP: uniform manifold approximation and projection for dimension reduction. CoRR abs/1802.03426 (2018)
18. Nanni, L., Maguolo, G., Lumini, A.: Exploiting Adam-like optimization algorithms to improve the performance of convolutional neural networks. CoRR abs/2103.14689 (2021)
19. Paszke, A., Gross, et al.: Pytorch: An imperative style, high-performance deep learning library. In: Advances in Neural Information Processing Systems, vol. 32 (2019)
20. Pati, P., et al.: Hierarchical graph representations in digital pathology. Med. Image Anal. **75**, 102264 (2022)
21. Potjer, F.K.: Region adjacency graphs and connected morphological operators. In: Mathematical Morphology and its Applications to Image and Signal Processing, pp. 111–118. Computational Imaging and Vision (1996)
22. Senousy, Z., Abdelsamea, M.M., Mohamed, M.M., Gaber, M.M.: 3E-Net: entropy-based elastic ensemble of deep convolutional neural networks for grading of invasive breast carcinoma histopathological microscopic images. Entropy **23**(5), 620 (2021)

23. Van der Walt, S., et al.: scikit-image: image processing in python. PeerJ **2**, e453 (2014)
24. Wang, M., et al.: Deep graph library: a graph-centric, highly-performant package for graph neural networks. arXiv preprint arXiv:1909.01315 (2019)
25. Wang, Y., et al.: Improved breast cancer histological grading using deep learning. Ann. Oncol. **33**(1), 89–98 (2022)
26. Xu, K., Hu, W., Leskovec, J., Jegelka, S.: How powerful are graph neural networks? CoRR abs/1810.00826 (2018)
27. Yan, R., et al.: Nuclei-guided network for breast cancer grading in he-stained pathological images. Sensors **22**(11), 4061 (2022)
28. Yan, R., Yang, Z., Li, J., Zheng, C., Zhang, F.: Divide-and-attention network for he-stained pathological image classification. Biology **11**(7), 982 (2022)

A Pilot Study of Neuroaesthetics Based on the Analysis of Electroencephalographic Connectivity Networks in the Visualization of Different Dance Choreography Styles

Almudena González[1]([✉]) [iD], José Meléndez-Gallardo[2] [iD], and Julian J. Gonzalez[3]

[1] Departamento de Historia del Arte y Filosofía, Facultad de Humanidades, Universidad de La Laguna, San Cristóbal de La Laguna, Spain
agonzabr@ull.edu.es
[2] Departamento de Educación Física y Salud del ISEF, CURE sede Maldonado de la Universidad de la República, Punta del Este, Uruguay
jose.melendez@cure.edu.uy
[3] Departamento de Ciencias Médicas Básicas, Universidad de La Laguna, San Cristóbal de La Laguna, Tenerife, Spain
jugonzal@ull.edu.es

Abstract. Neuroaesthetics allows us to understand how the brain works in different artistic languages and, therefore, to broaden the knowledge of our aesthetic judgments. The present pilot study is an interdisciplinary work that aims to differentiate different aesthetic dance choreography styles and to demonstrate the influence of training, learning, enculturation and familiarization of these styles on their brain perception by means of neurophysiological measurements of EEG signals and neural connectivity network analysis techniques. To this end, EEGs of non-expert dancers are recorded while viewing two fragments (film clips) of classical and modern dance and during other control conditions. Measures of functional connectivity between recorded regions are obtained from phase synchronization measurements between pairs of EEG signals in each EEG frequency band (FB). The responses of each FB are evaluated from indices obtained from models of EEG connectivity networks -graphs and connectomes- constructed from graph theory and network-based statistics (NBS) in a global and local context. Thus, significant alterations -in some of the indices- are observed between different contrasts and conditions in certain areas and specific EEG connections that depend on the EEG frequency band under consideration. These first results, therefore, suggest the usefulness of this neuroaesthetic experimental paradigm. On the other hand, these neuroaesthetic procedures may be of special interest in biomedicine because they provide knowledge about different languages that can be applied in therapies and treatments.

Keywords: Neuroaesthetic · EEG Functional Connectivity · Graph Theory · Connectome · Networks

© The Author(s), under exclusive license to Springer Nature Switzerland AG 2023
I. Rojas et al. (Eds.): IWBBIO 2023, LNBI 13920, pp. 297–310, 2023.
https://doi.org/10.1007/978-3-031-34960-7_21

1 Introduction

Dance in its different choreographic styles has different technical and aesthetic characteristics. This is due to the difference between artistic historical periods, trends and techniques developed among many other factors. The professionals of this art, the dancers, are an interesting paradigm of study that neuroscience has treated because their training begins at an early age and develops throughout life. This fact allows them to obtain improvements in their skills of different kinds such as being able to adapt to musical rhythms and movements more effectively in street dance [1, 2] or in ballroom dancing being able to improve synchronization and sensorimotor plasticity [3]. In addition, due to lifelong training they are known to develop greater proprioceptive skills [4], improve motor efficiency of movements as well as possess also improve memory for complex motor movement sequences [5]. These previous examples contemplate different choreographic styles that despite their technical and stylistic differences show common results suggesting that dance requires high levels of functioning and involvement of different brain areas and different cognitive domains and long-term dance training is associated with brain plasticity.

Next, we will look at the dance visualization research paradigm that we follow in this study. Dance choreography visualization has been studied under the prism of neuroscience [6, 7] with the electroencephalography (EEG) technique where dancers demonstrate greater theta phase synchrony in fronto central brain regions compared to non-dancers when observing an emotionally evocative dance performance and this may be due to a reflection of cognitive or affective skills that would have developed as a result of dance training over time [8]. Also, they show higher alpha synchronization both during the alternative uses task, a creativity task, and during improvisational dance imagination [9].

Using the technique of functional magnetic resonance imaging (fMRI) image analysis, the visualization of dance has also been studied in relation to the mirror neuron system because this is believed to be one that supports the observation and simulation of the actions of others [10] and has been localized in areas of the premotor and parietal cortex [11]. Thus, it is reported that brain activity associated with observation and imagining/simulating movement was found in observation and action simulation networks, which include the premotor cortex and inferior parietal lobe. This network was also related to the amount of experience, skill, and familiarity in dance along with the involvement of other brain areas such as supplementary motor, superior temporal, groove, and primary motor cortex [12].

An interesting concept of analysis in the work on visualization is familiarization/enculturation with the choreographic style. Thus, a higher activity of the action observation network has been observed when watching familiar movements in professional dancers [13, 14]; furthermore in two choreographic styles, ballet and capoeira, in professionals greater bilateral activation was found in motor regions of the action observation network, including premotor cortex, intraparietal sulcus, right superior parietal lobule and left posterior superior temporal sulcus when dancers viewed their own dance style with which they were familiar compared to other styles they found greater activity, particularly in premotor cortex [15]. A subsequent functional MRI study found that expert ballroom dancers showed greater activation in the ventral premotor cortex while watching ballroom dance videos than inexperienced ballroom dancers [16]. From the above studies conducted on professional dancers we can observe the fundamental role of premotor cortex plasticity in dance viewing and its relationship to experience and familiarity with the style of dance being viewed.

In this paper we will focus on visual perception of dance in general population with no training in dance. In this line, visualization studies carried out on non-professionals using the EEG technique measured event-related desynchronization in a sample of expert dancers and non-dancers while observing dance and non-dance movements, with dancers showing greater event-related desynchronization. Events while watching the dance moves, indicating increased activation in the action observation system [17]. On the other hand, in another choreographic style, tango, EEG was used to measure event-related potentials in a sample of expert dancers, novice dancers and non-dancers while watching videos of correctly or incorrectly executed tango steps and found that the activity anticipatory activity generated by the frontal, parietal, and occipital brain regions showed differences between groups and also predicted subsequent activity in motor and temporal regions [18].

In summary, the aforementioned works using signaling or imaging techniques derived from EEG or fMRI report functional brain differences in the action observation network in both dancers and non-dancers. That is why we, using the EEG signal analysis technique, analyze the visualization of different choreographic styles to obtain a map of functional connectivity and differences between styles.

2   Methods of Data Recording and Analysis

2.1   Experimental Paradigm

• Participants: 8 volunteers (4 women and 4 men) between 25–55 years old without experience in the practice of dance, right-handed, without disorders and having signed the informed consent. Experiments were conducted in accordance with the Helsinki Declaration and the experimental protocols approved by our university (CEIBA2021-3088). Procedure: the subjects are taken to the laboratory where the examination is performed in a soundproofed room equipped with a Faraday cage. The recordings are made with the participants seated in front of a television monitor with the EEG electrode cap in place. Using the appropriate software, successive one-minute blocks of visual stimuli (video clips) of the swan dance of death (SW), the dance of a bagatelle (BA) and the blue screen -as a control condition- (BS) are presented randomly on the screen. Each stimulus is run through the screen 5 times. At the beginning and end of the experiment, 2-min recordings are made with eyes open (EO), watching TV black screen and eyes closed (EC) as standard control conditions.

2.2   Processing and Analysis Methods

Most of the concepts and procedures used in this section have been reported by the authors in previous works on brain responses to listening and performance of different musical styles [19–21], so they will be briefly detailed in this paper. First, EEG low-density monopolar recordings were performed following electrode locations of International 10–20 System for EEG. EEG electrodes (19) were:

- 2-frontals Fp1-Fp2
- 2-fronto-dorsals F3-F5
- 2-fronto-ventrals F7-F8,
- One interhemispheric frontal Fz, central Cz and parietal Pz
- 2-temporals T3-T4
- 2-temporo-occipital posteriors T5-T6
- 2-central-parietals C3-C4
- 2-parietal-dorsals P3-P4
- 2-occipitals O1-O2

Additionally, 2-(A1-A2) references electrode for monopolar EEG recordings placed at the mastoids and complementary channels/electrodes for EOG (electro-oculogram), EMG (biceps electromyogram), Respiration monitoring and ECG were used.

Raw EEG recording stored at the computer are then preprocessed for the selection of non-artifactual EEG episodes using ad-hoc techniques.

Measures of functional connectivity between recorded regions are obtained from phase synchronization measurements between pairs of EEG signals in each EEG frequency band (FB), delta, theta, alpha, beta and gamma. For the functional connectivity measurement, we use a Phase Locking Value (PLV) phase synchronization index [see software in HERMES toolbox (http://hermes.ctb.upm.es/)].

Responses for each FB are evaluated from indices obtained from models of EEG connectivity networks -graphs and connectomes- constructed from graph theory and network-based statistics (NBS), in a global and local context.

The main concepts about Graph Theory analysis are described below: The 19 electrodes/channels and the 171 connections/links between them constitute respectively the nodes and edges of the graph representing the EEG neural network in a given FB. Arranging the 19 electrodes in matrix form, we obtain the 19x19 adjacency matrix of the graph (A), in our context, the EEG connectivity matrix, whose elements a(i,j) constitute the functional connectivity values (connectivity weights) between all channel/signal pairs, satisfying: a(i,j) = a(j,i) and diagonal elements a(i,i) = (j,j) = 1 (i,j = 1,...,19).Two central topological indices of each EEG network were obtained for each node and then averaging for all nodes: node degree (DEG) as the number n of effective connections (a(i,j) # 0) reaching a node and the nodal strength or intensity (STR) equal to the average connectivity of these connections ($\sum$ a(i,j)/n,, (i,j = 1,...,19)]). The topological inter-node organization of one EEG network/graph g was assessed through two measures using the A matrix:

a) the global efficiency (GE), a measure of the ability of the graph g to interconnect and transmit information between distant nodes. GE(g) is defined as the average of the inverse of the shortest path length from each node to all other nodes [for N nodes is calculated as GE(g) = 1 / N(N-1) * $\sum$ 1 / d(i,j), (i,j = 1,...,19) where d(i,j) is the shortest path length between node i and node j in the graph g and is calculated as the smallest sum of edge/connections lengths throughout all possible paths from node i and node j [22, 23]. Here the length of a connection was considered as the reciprocal of their connectivity weight $d(i,j) = 1/a(i,j)$, under the assumption that the distance between two nodes is inversely proportional to their connectivity;

b) the local efficiency (LE), which for a node i is defined as the GE of the node and calculated in the subgraph g(i) created by its neighbors. The LE of the graph (g) is the average of the LEs of all nodes. It is calculated by applying the same steps in the subgraph g(i) formed by the neighbors of node i. $LE(g) = 1/N * \sum GE(g(i))$ (i = 1,..,19). From the LE and GE measurements, a regular or lattice network is defined as one with high LE and low GE, a random network as one with low LE and high GE, and a small-world (SW) network would lie somewhere between a regular and a random network, with high LE and GE. To test whether a real network has SW structure, one of the proposed methods is to normalize the LE and GE in the real network, relative to those computed in matched random networks, i.e., by dividing the LE and GE by the corresponding mean [LE(r) or GE(r)] obtained from 100 random networks that retained the same number of nodes, connections, and degree distributions as the real brain networks [24–26]. Thus, a SW network would be characterized by having LE > LE(r), and a comparable GE ~ GE(r). If we considered the normalized versions of NLE = LE/ LE(r) and NGE = GE/ GE(r), a network will have SW structure if NLE > 1 and NGE ~ 1. Young adult brain neural networks structure has been considered as a paradigm of SW-like architecture, with functional connectivity consisting of clusters of brain regions with strong local connections to each other, along with some global connections between these clusters in the form of neural hubs [27, 28].

LE/GE parameters of brain networks structure may be influenced by the magnitude of the degree of the network and thus, in comparisons between groups/conditions, differences that may be found in LE/GE values could be biased by differences in degree. For this reason, it has been suggested to compute those parameters with different node degree thresholds and investigate for possible differences in network topology [29]. In this study, LE and GE have been calculated using the connectivity matrix A resulting from applying the surrogate data test to the PLV (i,j) values between nodes which corrects or thresholds, as we have seen, the degree of the network according to the statistical significance of these PS values. From above, it seems necessary to consider changes in the degree of the graph when comparing different groups/situations as a function of alterations in the topological properties of the LE or GE graph.

Another graph parameter we compute was the betweenness centrality BC which is a measure of the importance of a node in a graph; is defined using the number of the shortest paths passing through the node. The BC of node nx is the summation of the ratio of the number of the shortest paths from a node ns to a node nt which pass through nx, to the number of the shortest paths from ns to nt, over all pairs of ns and nt such that ns $\neq$ nx and n t $\neq$ ns, nx. The MATLAB Brain Connectivity Toolbox (http://www.brain-con nectivity-toolbox.net) was used to calculate the above graph indices from graph theory [30].

Permutation-based statistical procedures were used to analyze the statistical significance of conditions and contrasts in the different recorded EEG nodes/zones. For drawing cortical topographic images of inter-nodes connectivity and for the nodal graph parameters, we use MATLAB function *topoplot* modified by our team mainly to draw the picture of the statistical significance level of the graph parameters. We also use permutation tests for multiple comparisons between the nodal indices of two graphs, and to construct topographical maps of the statistical significance of the indices/parameters between different contrasts. The permutation tests we use can be found at the URL https://www.mathworks.com/matlabcentral/fileexchange/29782-mult_comp_perm_t1-data-n_perm-tail-alpha_level-mu-reports-seed_state).

Connectome of significant subnetworks at each condition and frequency band were obtained from NBS procedures. The Network Based Statistic (NBS) is a non-parametric statistical procedure based on the permutation of the compared groups or conditions (e.g. by t-test) in which the FWER (family wise error rate) correction is used to solve the problem of multiple comparisons in a graph/network in order to select those sub-networks formed by connections whose weights (in our case, the magnitude of the estimated functional connectivity) are significantly different between groups or conditions, regardless of whether they are strong or weak. This method makes use of the original 19x19 matrix of connectivities of each participant (without thresholding) of each group/condition.

The NBS toolbox (https://www.nitrc.org/projects/nbs/) informs -in its manual- about the steps to be carried out for the analysis, which briefly refer to: the preparation of the two matrices to be compared according to whether they correspond to paired or independent conditions/groups; the choice of the t-test (paired or unpaired) and its threshold value; the number of permutations to be performed; the component of the network to be measured (its extension number of connections or, its sum strength T-values) and, to the limit value of p for the FWER correction, in order to limit the p-values obtained from the set of permutations. The NBS application finally provides us with the value of p lower than the selected limit, which will allow us to confirm or reject the existence of a subnetwork whose connections have a greater or lesser functional connectivity in relation to the other group or condition compared.

In the present work we selected the paired t-test, p<0.05 for the FWER correction, 1000 as the permutation number and intensity as the graph component. The results obtained by NBS were plotted using the MATLAB function f_PlotEEG_BrainNetwork (https://www.mathworks.com/matlabcentral/fileexchange/57372-easy-plot-eeg-brain-network-matlab).

3   Results

Figures 1 and 2 show the average functional connectivity (FC) between different electrodes/nodes in the different conditions and EEG frequency bands. Only FC values higher than 0.75 are shown. The three higher FC for each condition in each band were:

- Delta band: SWAN DEATH: C3FZ, P3PZ, C3C4; BAGATELLE: C3FZ, P3PZ, C3C4; BLUE CREEN: C3FZ, P3PZ, C3C4; OPEN EYES: C3FZ, P3PZ, C3C4; CLOSED EYES: C3FZ, P3PZ, C3C4; SWAN DEATH, C3FZ, P4PZ, P3PZ.
- Theta band: BAGATELLE: P4PZ-P3PZ-C3FZ; BLUE CREEN: P4PZ-P3PZ-C3FZ; OPEN EYES: P4PZ-P3PZ-C3FZ; CLOSED EYES: C3FZ-P4PZ-P3PZ.
- Alpha band: SWAN DEATH-P4PZ-Fp1Fp2-F3CZ; Alpha band: BAGATELLE: Fp1Fp2, F3CZ, P4PZ; BLUE CREEN: Fp1Fp2, F3CZ, P4PZ; OPEN EYES: F3CZ, Fp1Fp2, P4PZ; CLOSED EYES: Fp1Fp2, F3CZ, F4CZ.

- Beta band: SWAN DEATH: P4PZ, F3CZ, FZCZ; BAGATELLE: P4PZ, F3CZ, FZCZ; BLUE CREEN: P4PZ, F3CZ, FZCZ; OPEN EYES: P4PZ, F3CZ, FZCZ; CLOSED EYES: F3CZ, P4PZ, FZCZ.
- Gamma band: SWAN DEATH: P4PZ, P3PZ, C3FZ; BAGATELLE: P4PZ, P3PZ, C3FZ; BLUE CREEN: P4PZ, P3PZ, FZPZ; OPEN EYES: P4PZ, P3PZ, FZPZ; CLOSED EYES: P4PZ, P3PZ, C3FZ.

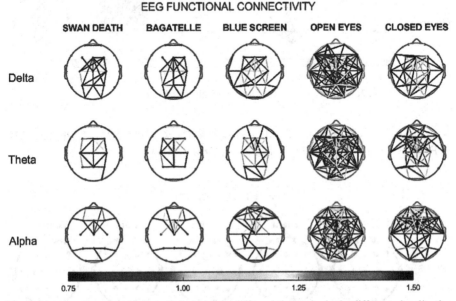

Fig. 1. Average Functional connectivity between different electrodes in the different visualization conditions indicated in the first row -above each head plot- and for the EEG frequency bands consigned at left column.

The results for the four Graph Theory parameters computed were:

a) *Graph node Strength*: In Figure 3 it is shown the electrodes whose statistical significant level was greater than 0.05 in the contrasts SW-BS (Swan Death – Blue Screen) and BA-BS (Bagatelle-Blue Screen). The nodes affected with the probability value in parenthesis were:
- Contrast SW - BS: Delta Band: T6(0.015625).
- Contrast SW - BS: Theta Band: O1(0.0625).
- Contrast SW - BS: Alpha Band: Fp2(0.046875), F8(0.015625), Fz(0.039063).
- Contrast SW - BS: Beta Band: Cz(0.039063).
- Contraste BA - BS: Delta Band: O2(0.039063).
- Contraste BA - BS: Theta Band: O1(0.054688), O2(0.039063), Fz(0.046875).
- Contrast BA - BS: Alpha Band: Fp1(0.046875), F3(0.023438), F7(0.046875), C3(0.046875), Fp2(0.039063), F4(0.023438), F8(0.023438), T4(0.046875), Fz(0.023438), Cz(0.046875).

EEG FUNCTIONAL CONNECTIVITY

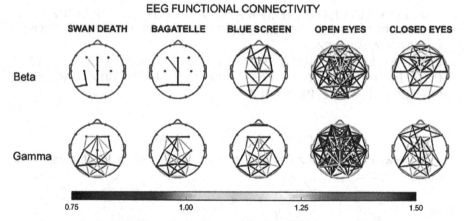

Fig. 2. As in Fig. 1 for the Beta and Gamma EEG frequency bands.

- Contrast BA - BS:, Beta Band: Fp1(0.039063), C3(0.046875), Fp2(0.039063), F4(0.039063), F8(0.046875), C4(0.0625), Cz(0.039063).

GRAPH NODE STRENGTH

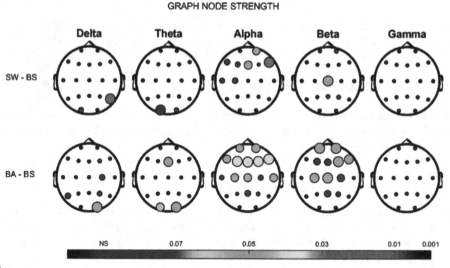

Fig. 3. Nodes that showed significant differences for the graph node strength parameter of the Graph Theory in the contrasts SW-BS (Swan Death – Blue Screen) or BA-BS (Bagatelle-Blue Screen) are shown in different colors according their level of statistical significance indicated in the bottom color bar (NS = no significant).

b) *Graph Node degree*
- Contrast SW - BS, Delta Band: T6(0.054688).
- Contrast BA - BS, Theta Band: O2(0.054688).

- Contrast BA - BS: Alpha Band: F3(0.046875), Fp2(0.023438), T4(0.046875), Cz(0.046875).

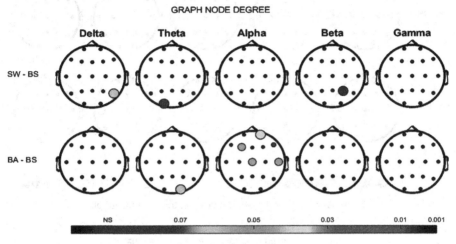

Fig. 4. As Fig. 3 for the node graph degree.

c) *Betweenness Centrality*
- Contrast SW - BS: Gamma Band: T4(0.015625).
- Contrast BA - BS: Delta Band: O2(0.03125).
- Contrast BA - BS: Alpha Band: F8(0.0078125).
- Contrast BA - BS: Beta Band: T4(0.023438).

d) *Graph Normalized Local Efficiency*
- Contrast BA - BS: Theta Band: C3(0.054688), F8(0.015625).
- Contrast BA - BS: Alpha Band: Fp2(0.03125).

Finally, we show (Fig. 7) results of NBS analysis for the comparison between the two contrasts considered (SW-BS versus BA-BS): only delta and theta bands exhibited significant results. In the delta band only Fp2-T3 connection appears very sensitive to the contrast. However, in theta band a larger subnetwork including 21 connections appear to be involved. In this band the right frontal F4 node appears as the node of maximum degree. The results below show node links with T value in parenthesis and node degree with the degree magnitude in parenthesis:

- Delta-band:

a) Node-links, Delta-band (total N = 2), Connections: T3Fp2(4.9), F7Fp2(2.6);
b) Node-degree, Delta-band (total N = 3), nodes with degree k >= 2: Fp2(2);

- Theta-band:

a) Node-links, (total N=21), Connections: Fp1Fz(4.0), F7F4(3.8), O1T4(3.7), C4O2(3.7), C3F4(3.7), Fp1C3(3.6), F4O2(3.1), F4Fz(2.9), F4Pz(2.9), O2Pz(2.8),

GRAPH BETWEENNESS CENTRALITY

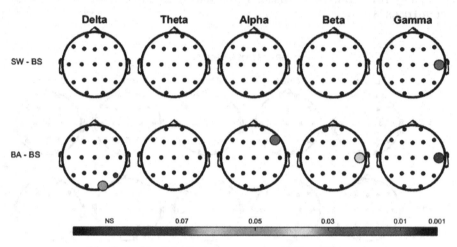

Fig. 5. As Fig. 3 for the betweenness centrality of the graph.

GRAPH NORMALIZED LOCAL EFFICIENCY

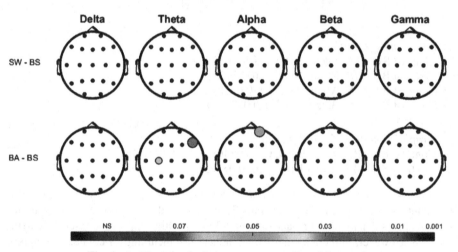

Fig. 6. As Fig. 3 for the betweenness centrality of the graph

F7C3(2.8), F3O2(2.7), T5T6(2.7), P3T5(2.7), O1Cz(2.7), F7Fz(2.7), O1F4(2.6), T5F8(2.6), T5O2(2.6), T3O1(2.6), F7Cz(2.5);

b) Node-degree, (total N=17), nodes with degree k>=2: F4(6), O2(5), O1(4), F7(4), T5(4), C3(3), Fz(3), Fp1(2), Cz(2), Pz(2);

EEG connectome. Contrast between different dance styles (visualization)

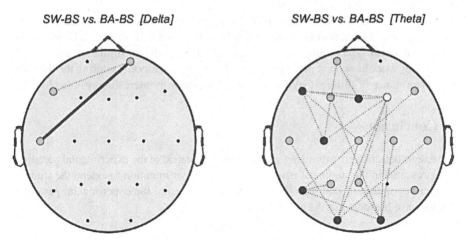

Fig. 7. Electroencephalographic (EEG) connectome subnets that were significant (FDR <0.05) in the visualizations [SW-BS vs. BA-BS] contrasts. Only delta and theta bands show significant results. The contrast conditions are indicated above each head plot. The color range for the degree of the nodes are: green [1, 2], purple [3–5] and yellow [>= 6]. The strength of all connections is in blue dotted lines when the T statistic = <4 and in red thick line when T >4.

4 Discussion

This paper is a preliminary study on the usefulness of EEG functional connectivity network analysis using graph theory and NBS techniques in the context of neuroaesthetics. The results we present, although interesting, correspond to a small population of non-dancer participants and therefore should be taken with caution.

Figures 1 and 2 shows functional connectivity between different electrodes/nodes in the different conditions and EEG frequency bands. From them, it is possible to observe differences between those conditions in the different EEG frequency bands. Note that in the higher alert exposure condition (EO) the connections involved are much larger than during the concentration state (EC) and also differences can be seen between different viewing conditions of the different dance styles (SW and BA) versus the control (BS). Therefore, functional connectivity estimated from phase synchronization indices appears very sensitive to changes in different cognitive visual-acoustic conditions therefore a measure of great interest in for neuroaesthetic quantifications.

Figures 3, 4, 5 and 6 show the statistical significance of the graph theory nodal intensity/strength, degree, betweenness centrality and the normalized local efficiency parameters in the different frequency bands in the SW-BS and BA-BS contrasts. Note that even for the small population size used here, clear differences between the two contrasts appears in most of the frequency bands mainly for the degree and strength parameters. The results from NBS analysis (Fig. 7) show differences between the brain perception (visual and acoustic) of the two dance styles considered although only in the delta and theta EEG connectivity networks. The results obtained could be of interest for future studies related with the musical and dance enculturation and familiarization and also in the therapeutic studies of music and dance.

Due to the absence of this type of EEG connectivity network analysis in the neuroaesthetic literature, we cannot compare our results with others in this context. However, we can state that this type of analysis with experimental paradigms similar to those developed in this work have been performed in studies on brain responses during the listening and performance of different styles of classical music with interesting results [20, 21].

5 Conclusions

The results obtained are promising and confirm the interest of the experimental paradigm here presented in the studies of neuroaesthetics. It is our intention to extend the study to a larger population of both non-dancers and dancers using the experimental paradigm that we have presented in the present work.

References

1. Miura, A., Kudo, K., Nakazawa, K.: Action–perception coordination dynamics of whole-body rhythmic movement in stance: a comparison study of street dancers and non-dancers. Neurosci. Lett. **544**, 157–162 (2013). https://doi.org/10.1016/j.neulet.2013.04.005
2. Miura, A., Fujii, S., Okano, M., Kudo, K., Nakazawa, K.: Finger-to-beat coordination skill of non-dancers, street dancers, and the world champion of a street-dance competition. Frontiers in Psychology **7**, 542 (2016). https://doi.org/10.3389/fpsyg.2016.00542
3. Jin, X., Wang, B., Lv, Y., Lu, Y., Chen, J., Zhou, C.: Does dance training influence beat sensorimotor synchronization? Differences in finger-tapping sensorimotor synchronization between competitive ballroom dancers and nondancers. Exp. Brain Res. **237**(3), 743–753 (2019). https://doi.org/10.1007/s00221-018-5410-4
4. Kiefer, A.W., Riley, M.A., Shockley, K., et al.: Lower-limb proprioceptive awareness in professional ballet dancers. J. Dance Med. Sci. **17**(3), 126–132 (2013). https://doi.org/10.12678/1089-313X.17.3.126
5. Stevens, C.J., Vincs, K., De Lahunta, S., Old, E.: Long-term memory for contemporary dance is distributed and collaborative. Acta Physiol. (Oxf) **194**, 17–27 (2019). https://doi.org/10.1016/j.actpsy.2019.01.002
6. Bläsing, B., Calvo-Merino, B., Cross, E., Jola, C., Honisch, J., Stevens, K.: Neurocognitive control in dance perception and performance. Acta Psychol. **139**, 300–308 (2012). https://doi.org/10.1016/j.actpsy.2011.12.005
7. Bläsing, B., Zimmermann, E.: Dance is more than meets the eye—how can dance performance be made accessible for a non-sighted audience? Front. Psychol. **12**, 643848 (2021). https://doi.org/10.3389/fpsyg.2021.643848
8. Poikonen, H., Toiviainen, P., Tervaniemi, M.: Dance on cortex: enhanced theta synchrony in experts when watching a dance piece. Cogn. Neurosci. **45**, 5 (2018). https://doi.org/10.1111/ejn.13838
9. Fink, A., Graif, B., Neubauer, A.: Brain correlates underlying creative thinking: EEG alpha activity in professional vs. novice dancers. Neuroimage **46**, 854–862 (2009). https://doi.org/10.1016/j.neuroimage.2009.02.036
10. Rizzolatti, G., Craighero, L.: The mirror neuron system. Annu. Rev. Neurosci. **27**, 169–192 (2004)

11. Krüger, B., et al.: Parietal and premotor cortices: activation reflects imitation accuracy during observation, delayed imitation and concurrent imitation. NeuroImage **100** (2014) https://doi.org/10.1016/j.neuroimage.2014.05.074

12. Cross, E.S., Hamilton, A.F., Grafton, S.T.: Building a motor simulation de novo: observation of dance by dancers. Neuroimage **31**, 1257–1267 (2006). https://doi.org/10.1016/j.neuroimage.2006.01.033

13. Gardner, T., Goulden, N., Cross, S.E.: Dynamic modulation of the action observation network by movement familiarity, 2015. J. Neurosci. **35**(4), 1561–1572 (2015)

14. Calvo-Merino, B., Grèzes, J., Glaser, D.E., et al.: Seeing or doing? Influence of visual and motor familiarity in action observation. Curr. Biol. **16**, 1905–1910 (2006)

15. Calvo-Merino, B., Glaser, D.E., Grézes, J., et al.: Action observation and acquired motor skills: an fMRI study with expert dancers. Cereb. Cortex **15**, 1243–1249 (2005)

16. Pilgramm, S., Lorey, B., Stark, R., et al.: Differential activation of the lateral premotor cortex during action observation. BMC Neurosci **11**, 89 (2010). https://doi.org/10.1186/1471-2202-11-89

17. Orgs, G., Dombrowski, J.H., Heil, M., et al.: Expertise in dance modulates alpha/beta event-related desynchronization during action observation. Eur. J. Neurosci. **27**, 3380–3384 (2008)

18. Amoruso, L., Sedeño, L., Huepe, D., et al.: Time to tango: expertise and contextual anticipation during action observation. Neuroimage **98**, 366–385 (2014)

19. González, A., Pérez, O., Santapau, M., González, J.J., Modroño, C.D.: A neuroimaging comparative study of changes in a cellist's brain when playing contemporary and Baroque styles. Brain and Cognition **145**, 105623 (2020) Ihttps://doi.org/10.1016/j.bandc.2020.105623

20. González, A., Santapau, M., Gamundí, A., Pereda, E., González, J.J.: Modifications in the topological structure of EEG functional connectivity networks during listening tonal and atonal concert music in musicians and non-musicians. Brain Sci. **11**(2), 159 (2021). https://doi.org/10.3390/brainsci11020159

21. González, A., Modroño, C., Santapau, M., González, J.J.: Comparing brain responses to different styles of music through their real and imagined interpretation: an analysis based on EEG connectivity networks. Med. Sci. Forum **8**(1), 13 (2022). https://doi.org/10.3390/IECBS2021-10667

22. Latora, V., Marchiori, M.: Efficient behavior of small-world networks. Phys. Rev. Lett. **87**(19), 198701 (2001). https://doi.org/10.1103/PhysRevLett.87.198701

23. Achard, S., Bullmore, E.: Efficiency and cost of economical brain functional networks. PLOS Comput. Biol. **3**(2), e17 (2007). https://doi.org/10.1371/journal.pcbi.0030017

24. Maslov, S., Sneppen, K.: Specificity and stability in topology of protein networks. Science **296**(5569), 910–913 (2002). https://doi.org/10.1126/science.1065103

25. Zhu, J., Zhuo, C., Liu, F., Qin, W., Xu, L., Yu, C.: Distinct disruptions of resting-state functional brain networks in familial and sporadic schizophrenia. Sci. Rep. **6**, 23577 (2016). https://doi.org/10.1038/srep23577

26. Ma, X., Jiang, G., Fu, S., Fang, J., Wu, Y., Liu, M., Xu, G., Wang, T.: Enhanced network efficiency of functional brain networks in primary insomnia patients. Front Psychiatry **9**, 46 (2018). https://doi.org/10.3389/fpsyt.2018.00046

27. Meunier, D., Lambiotte, R., Fornito, A., Ersche, K., Bullmore, E.: Hierarchical modularity in human brain functional networks. Frontiers in Neuroinformatics **3** (2009). https://doi.org/10.3389/neuro.11.037.2009

28. Wu, J., Zhang, J., Ding, X., Li, R., Zhou, C.: The effects of music on brain functional networks: a network analysis. Neuroscience **10**(250), 49–59 (2013). https://doi.org/10.1016/j.neuroscience.2013.06.021

29. Zalesky, A., Fornito, A., Bullmore, E.T.: Network-based statistic: identifying differences in brain networks. Neuroimage **53**(4), 1197–1207 (2010). https://doi.org/10.1016/j.neuroimage.2010.06.041

30. Rubinov, M., Sporns, O.: Complex network measures of brain connectivity: uses and interpretations. Neuroimage **52**(3), 1059–1069 (2010). https://doi.org/10.1016/j.neuroimage.2009.10.003

Machine Learning in Bioinformatics and Biomedicine

Ethical Dilemmas, Mental Health, Artificial Intelligence, and LLM-Based Chatbots

Johana Cabrera(✉) , M. Soledad Loyola , Irene Magaña , and Rodrigo Rojas

University of Santiago de Chile, Santiago, Chile
{johana.cabrera,maria.loyola.f,irene.magana,
rodrigo.rojas.a}@usach.cl

Abstract. The present study analyzes the bioethical dilemmas related to the use of chatbots in the field of mental health. A rapid review of scientific literature and media news was conducted, followed by systematization and analysis of the collected information. A total of 24 moral dilemmas were identified, cutting across the four bioethical principles and responding to the context and populations that create, use, and regulate them. Dilemmas were classified according to specific populations and their functions in mental health. In conclusion, bioethical dilemmas in mental health can be categorized into four areas: quality of care, access and exclusion, responsibility and human supervision, and regulations and policies for LLM-based chatbot use. It is recommended that chatbots be developed specifically for mental health purposes, with tasks complementary to the therapeutic care provided by human professionals, and that their implementation be properly regulated and has a strong ethical framework in the field at a national and international level.

1 Introduction

Over time, many tasks once considered human have been gradually replaced by machines. This phenomenon has been widely observed since the Industrial Revolution [1]. Currently, we are said to be experiencing the Fourth Industrial Revolution, characterized by a series of technological, digital, physical, and even biological convergences, including the development of the internet, automation, robotics, blockchain, cloud computing, 3D printing, and artificial intelligence [2]. The technology that, in the last period, has caused great controversy is Artificial Intelligence (AI), which was defined by John McCarthy as "the combination of science and engineering to create intelligent devices for human welfare" [3].

The current debate is generated by certain types of AI chatbots based on large-scale language models (LLM) [4]. Among these, ChatGPT-3 and ChatGPT-4 [5] have had the greatest impact thus far because of their large capacity for both function and interaction with humans. Such is the evolution of this technology that Bill Gates has called "the most important technological breakthrough since the graphical user interface" [6].

It is important to consider that this scenario takes place in a delicate context for humanity; in recent years, the world has experienced a series of economic, health, and sociopolitical problems at the national and international levels, which will also have

I. Rojas et al. (Eds.): IWBBIO 2023, LNBI 13920, pp. 313–326, 2023.
https://doi.org/10.1007/978-3-031-34960-7_22

a reciprocal impact, affecting people's health and mental health as well as driving the progress of scientific-technological devices [2, 7, 8].

As artificial intelligence becomes more prominent in society, it is crucial to address ethical concerns and enable scientific studies on its development, regulation, and implementation [6, 9, 10].

A topic of relevance to the ethical field can be seen from the perspective of the "ethical dilemma" or "moral dilemma," which are situations in which decisions must be made that can have implications ranging from moral to paradoxical [11–13]. These situations can be challenging because they often involve competing values or interests [14].

These ethical problems and dilemmas are often addressed in the field of "AI ethics," which seeks to establish reflections and guidelines for the development and implementation of AI technology [15].

However, what specific considerations should these analyses include from the fields in which these technologies are applied?

In this sense, international organizations, and donors such as Bill Gates [6, 16] have announced plans for two areas of development and LLM chatbots: health and education in a global scenario. This implies an investment in digital infrastructure to implement both health and education of advanced human capital and improve the population's access to these types of tools, which would constitute part of the main actions to be developed in these areas.

One particularly sensitive field that requires special attention when addressing ethical concerns regarding AI applications is Mental Health. Mental health has been defined as "The ability of individuals to interact with each other and the environment, to promote subjective well-being, the development and optimal use of their psychological, cognitive, emotional, and relational potential, the achievement of their individual and collective goals, in accordance with justice and the common good [32]." In this field, professionals and researchers from various disciplines seek the well-being of people from different perspectives.

In this sense, AI can improve various counseling axes, professional functions, patient care, help with diagnostic accuracy, optimize treatment plans, create intervention design, monitor treatment, optimization, accompaniment, interaction, referral decisions, development of support material, and different potentials for research. For example, it has been shown that these tools achieve a 19% increase in conversational empathy when people use them to generate text [17–19].

In this regard, ethical dilemmas related to mental health and chatbots could include issues surrounding privacy, transparency, accountability, conflicts of interest, cultural differences and biases, involuntary interventions, and balancing patient autonomy with their safety, informed consent, and problems related to power dynamics and inequalities [20–25]. Others may be related to confidentiality, such as deciding whether to violate a patient's privacy to prevent harm to themselves or to others [26, 27].

Some experts argue that, in the meantime, these chatbots will continue to impact human professions, spreading the biases of their creators, advancing persuasive interaction in political and marketing fields, and directly impacting people's thoughts, emotions, and behaviors [8, 28].

Due to the above, immediate actions are required, such as ethical reflections and scientific development, to shed light on developments, uses, and regulations surrounding what is cataloged as "the new era of Artificial Intelligence" [6], which we know will continue to generate changes in our society.

Identifying and addressing ethical dilemmas are related to decision-making and, if done correctly, can contribute to strengthening moral acts and promoting the development of positive behaviors [29].

The objective of this article is to identify and reflect on the bioethical dilemmas (beneficence, non-maleficence, justice, and autonomy) involved in the use of chatbots based on large-scale language models (LLM) in professional practice related to mental health [30]. The results seek to provide relevant information and recommendations that serve as interdisciplinary inputs for the regulation, use, research, and training of these bots in mental health.

To achieve these objectives, it is necessary to conduct interdisciplinary research involving ethics experts, mental health professionals, AI developers, and policymakers. These collaborative efforts can help ensure that ethical concerns are adequately addressed, and effective guidelines for the responsible use of LLM chatbots in the field of mental health are developed.

2 Methodology

To achieve the main objective of this research, a rapid review [31] of scientific literature and news articles from the media was carried out. Subsequently, a systematization and analysis of the collected information were performed.

The type of evidence included in this document considers more than just scientific evidence, as science and technology do not advance at the same pace. Media outlets could offer more recent information and experience; therefore, a review of news that responds to different contexts with multidisciplinary and/or multisectoral approaches is also included.

Search and selection strategy - Scientific literature.

For the literature review, the academic search engine "Web of Science" was used to identify relevant abstracts. The keyword used to refine the search was: "chatbot mental health."

The abstracts were evaluated according to the following inclusion and exclusion criteria:

Inclusion criteria:

- Scientific articles investigating the relationship between chatbots and mental health.
- Articles published up to April 27, 2023.
- Articles focused on any type of chatbot.

Exclusion criteria:

- Documents published before 2020.

After conducting the search on the "Web of Science" engine, a careful review of the results was performed to review the abstracts and apply the inclusion and exclusion criteria. Finally, 33 abstracts were selected for a more detailed analysis.

Table 1 provides a complete list of the articles included in this document, along with relevant information related to each source.

Search and selection strategy for media news.

Using the same rapid review technique, a search was carried out in the media, through the "news segments" using the Microsoft Bing/Edge search engine, using the keyword "chatbots mental health."

This type of search engine was chosen because it incorporates ChatGPT into its new technology.

The inclusion and exclusion criteria were established to evaluate the retrieved articles as follows:

Inclusion criteria:

- News articles and blog entries published between November 2022 (public release milestone) and April 25, 2023
- Focus on any LLM chatbot.
- Inclusion of mental health in the text.
- News articles in English.

Exclusion criteria:

- News published before 2022.
- Websites not directly linked to a specific news entry.
- Documents without free access to the text.
- News that did not provide information about chatbots or mental health.
- Articles without information or mention of LLM-based chatbots.

After the initial search, 140 news articles were retrieved. Following the previous step, the titles and publication dates of each entry were reviewed. In cases where the title was unclear, a full news article was read to determine its relevance. Based on this evaluation, 13 articles were selected for inclusion.

Table 2 provides a complete list of the news articles included in this document, along with relevant information related to each source. These tables facilitate transparency and reproducibility in news search and selection processes.

Table 1. Listing of scientific abstracts included in the review.

Article Title	Source Title	DOI
Perceptions and Opinions of Patients About Mental Health Chatbots: Scoping Review	Journal Of Medical Internet Research	10.2196/17828
Designing Personality-Adaptive Conversational Agents for Mental Health Care	Information Systems Frontiers	10.1007/s10796-022-10254-9
Development of a Positive Body Image Chatbot (KIT) With Young People and Parents/Carers: Qualitative Focus Group Study	Journal Of Medical Internet Research	10.2196/27807
Development process of artificial intelligence based chatbot to support and promote mental wellbeing in sparsely populated areas of five European countries	European Psychiatry	10.1192/j.eurpsy.2022.446
Artificially intelligent chatbots in digital mental health interventions: a review	Expert Review Of Medical Devices	10.1080/17434440.2021.2013200
Identifying Potential Gamification Elements for A New Chatbot for Families with Neurodevelopmental Disorders: User-Centered Design Approach	Jmir Human Factors	10.2196/31991
The Challenges in Designing a Prevention Chatbot for Eating Disorders: Observational Study	Jmir Formative Research	10.2196/28003
A Chatbot for Perinatal Women's and Partners' Obstetric and Mental Health Care: Development and Usability Evaluation Study	Jmir Medical Informatics	10.2196/18607
Implementation of Cognitive Behavioral Therapy in e-Mental Health Apps: Literature Review	Journal Of Medical Internet Research	10.2196/27791
A Mental Health Chatbot for Regulating Emotions (SERMO)-Concept and Usability Test	Ieee Transactions On Emerging Topics In Computing	10.1109/TETC.2020.2974478
Development of a chatbot for depression: adolescent perceptions and recommendations	Child And Adolescent Mental Health	10.1111/camh.12627
AI-based chatbot micro-intervention for parents: Meaningful engagement, learning, and efficacy	Frontiers In Psychiatry	10.3389/fpsyt.2023.1080770
Engagement and Effectiveness of a Healthy-Coping Intervention via Chatbot for University Students During the COVID-19 Pandemic: Mixed Methods Proof-of-Concept Study	Jmir Mhealth And Uhealth	10.2196/27965
Co-developing a Mental Health and Wellbeing Chatbot with and for Young People	Frontiers In Psychiatry	10.3389/fpsyt.2020.606041
Chatbot-Based Assessment of Employees' Mental Health: Design Process and Pilot Implementation	Jmir Formative Research	10.2196/21678
Chatbots to Support Young Adults' Mental Health: An Exploratory Study of Acceptability	Acm Transactions On Interactive Intelligent Systems	10.1145/3485874
A Test Platform for Managing School Stress Using a Virtual Reality Group Chatbot Counseling System	Applied Sciences-Basel	10.3390/app11199071
A Chatbot to Support Young People During the COVID-19 Pandemic in New Zealand: Evaluation of the Real-World Rollout of an Open Trial	Journal Of Medical Internet Research	10.2196/38743
Participatory Development and Pilot Testing of an Adolescent Health Promotion Chatbot	Frontiers In Public Health	10.3389/fpubh.2021.724779
Emotional Reactions and Likelihood of Response to Questions Designed for a Mental Health Chatbot Among Adolescents: Experimental Study	Jmir Human Factors	10.2196/24343
Improving body image at scale among Brazilian adolescents: study protocol for the co-creation and randomized trial evaluation of a chatbot intervention	Bmc Public Health	10.1186/s12889-021-12129-1
Combined Use of Virtual Reality and a Chatbot Reduces Emotional Stress More Than Using Them Separately	Journal Of Universal Computer Science	10.3897/jucs.77237
Assisting Personalized Healthcare of Elderly People: Developing a Rule-Based Virtual Caregiver System Using Mobile Chatbot	Sensors	10.3390/s22103829
Developing, Implementing, and Evaluating an Artificial Intelligence-Guided-Mental Health Resource Navigation Chatbot for Health Care Workers and Their Families During and Following	Jmir Research Protocols	10.2196/33717

Table 2. Listing of the news included in the present review.

Title	Media Communication	Web
ChatGPT Therapy: Why People Ask AI For Mental Health Advice	Inquirer	https://technology.inquirer.net/123143/chatgpt-therapy-why-people-ask-ai-for-mental-health-advice
ChatGPT and health care: Could the AI chatbot change the patient experience?	Fox News	https://www.foxnews.com/health/chatgpt-health-care-could-ai-chatbot-change-patient-experience
Artificial empathy: the dark side of AI chatbot therapy	Cyber News	Artificial empathy: the dark side of AI chatbot therapy | Cybernews
AI bots have been acing medical school exams, but should they become your doctor?	ZDNET	AI bots have been acing medical school exams, but should they become your doctor? | ZDNET
ChatGPT: From Healthcare to Banking, These Sectors Can Benefit Most From AI Chatbots	News 18	https://www.msn.com/en-in/money/news/chatgpt-from-healthcare-to-banking-these-sectors-can-benefit-most-from-ai-chatbots/ar-AA19T4S0
Could Bing Chat replace my Doctor?	Windows Central	https://www.msn.com/en-us/health/other/could-bing-chat-replace-my-doctor/ar-AA1a9WR6
3 ways AI will stand out in healthcare delivery in 2023	Health Care Asia	https://www.msn.com/en-xl/news/other/3-ways-ai-will-stand-out-in-healthcare-delivery-in-2023/ar-AA19NSJX
Conversing with AI: Revolutionizing Mental Health Support through Chatbots	Linkedin	https://www.linkedin.com/pulse/conversing-ai-revolutionizing-mental-health-support-through-pandey/
World Health Day 2023: How AI and ChatGPT Are Revolutionizing Telemedicine and Remote Patient Care	Tecnopedia	World Health Day 2023: How AI and ChatGPT Are Revolutionizing Telemedicine and Remote Patient Care - Techopedia
AI Therapy Is Here. But the Oversight Isn't. (msn.com)	MSN News	https://www.msn.com/en-us/health/other/ai-therapy-is-here-but-the-oversight-isn-t/ar-AA19AJTa
Exploring the Limits of Artificial Intelligence in Mental Health: ChatGPT as a Therapist for a Day (thequint.com)	The Quint	Exploring the Limits of Artificial Intelligence in Mental Health: ChatGPT as a Therapist for a Day (thequint.com)
Conversing with AI: Revolutionizing Mental Health Support through Chatbots	The Strait times	https://www.straitstimes.com/world/united-states/people-are-using-ai-for-therapy-even-though-chatgpt-wasn-t-built-for-it

3   Data Extraction and Analysis Procedure

After selecting the articles and news based on the inclusion and exclusion criteria, manual coding was performed based on the researchers' inference criteria to identify the bioethical dilemmas present in each of the abstracts and selected media texts.

Following the rapid review and using the GPT-4 chatbot, studied in this work, the following prompt was requested from the tool: "Sort, classify, analyze, and explain ethical dilemmas from the bioethical principle of (selection of the ethical principle) based on the following information (list of codes created either from abstracts or news texts)."

This prompt was used for each of the bioethical principles mentioned previously.

4   Results

After using the prompt with Chatgpt 4, the following results were generated for each of the bioethical principles:

- Identification and classification of bioethical dilemmas
- Ethical analysis

To present the key findings, Table 3 shows the results of the sorting and identification of bioethical dilemmas in the review of scientific abstracts, whereas Table 4 provides an overview of the sorting and identification of bioethical dilemmas present in media news, both categorized according to the four bioethical principles.

Table 3. Identification of ethical dilemmas from the primary four bioethical principles present in scientific literature.

Autonomy	Beneficence	No Maleficence	Justice
Limited access to mental health care: Insufficient access to mental health services can raise an ethical dilemma in relation to the principle of autonomy, as patients may not have access to necessary medical care to make informed decisions about their treatment and care.	The lack of access to mental health services and online medical care can be an ethical dilemma, as individuals are being deprived of receiving the treatment, they need to improve their well-being.	The ethical dilemma associated with non-maleficence in the use of chatbots in mental health care can cause harm to patients if not properly addressed.	The need to ensure equity in access to mental health chatbots for all patients.
Evaluation of conversational agents: The lack of standard measures for evaluating conversational agents can raise an ethical dilemma in relation to the principle of autonomy, as patients may not have access to accurate and reliable information about the benefits and risks of these agents.	The use of conversational agents and chatbots to address the shortage of healthcare providers can raise ethical questions about the quality and effectiveness of treatment, as it is unknown whether these agents can provide the same level of care as a healthcare professional.	The ethical dilemma of using experimental therapies in terminally ill patients, as there may be unknown and potentially harmful risks to the patient.	Possible discrimination and bias in the programming of mental health chatbots, which can affect the quality of care and treatment for different groups of people.
Vulnerable populations: The pediatric population and those with schizophrenic or bipolar disorders can raise ethical dilemmas in relation to the principle of autonomy, as these patients may have difficulty making informed decisions about their treatment and care due to their age or medical condition.	The inability of conversational agents and chatbots to capture dynamic human behavior can raise ethical questions about the privacy and confidentiality of users, as it is unknown whether the information shared with these agents will be used appropriately and securely protected.	The ethical dilemma of using assisted reproductive techniques in couples with fertility problems, as there may be risks to the health of the mother and fetus, as well as ethical issues related to embryo selection and genetic manipulation. The ethical dilemma of using gene therapies to treat genetic diseases, as there may be unknown and potentially harmful risks to the patient, as well as ethical issues related to genetic manipulation and the selection of physical and mental characteristics.	The need to ensure that mental health chatbots are culturally appropriate and sensitive to the needs of different groups of people. Lack of access to mental health services for certain groups of people due to a lack of access to the technology needed to use mental health chatbots.
Informed decision-making: The ethical dilemma of the users' ability to make informed decisions about their mental health when interacting with LLM-based chatbots, and whether these programs are designed to provide accurate and complete information.	The need to design adaptive conversational agents and chatbots to personality can raise ethical questions about manipulation and influence on users' decision-making, as it is unknown whether these agents can influence users' decisions in an unethical way.	Confidentiality: The use of a chatbot to monitor a person's health can raise concerns about the confidentiality of the information collected, as it may be shared with third parties without the person's consent. Bias: The chatbot may be programmed with unconscious biases that can affect the accuracy of the results and the healthcare provided to the person.	Possible lack of privacy and security of user data when interacting with mental health chatbots, which can compromise confidentiality and trust in the healthcare system. Lack of regulation and supervision of mental health chatbots, which can lead to the proliferation of ineffective or even harmful chatbots for users' mental health.
Equity in access: The ethical dilemma of equity in access to chatbots for mental health care, as not all users have access to the necessary technology to use them.	The ethical dilemma of using mental health chatbots as a quick and easy solution instead of addressing the underlying causes of mental health problems, which may perpetuate the lack of access to adequate and personalized mental health care.	Stigma: The use of a chatbot to monitor mental health can raise concerns about the stigma associated with mental health and the perception that the person needs constant monitoring. Integrity: It is important to ensure that the chatbot is used ethically and not used for malicious purposes, such as discrimination or manipulation of the collected information.	Lack of equitable access to the technology needed to use mental health chatbots and health games in healthcare. Possible discrimination in the selection of patients who can use mental health chatbots to receive care. Lack of human contact and empathy in mental health care through chatbots, which can negatively affect the quality of care and the therapeutic relationship.
Privacy and confidentiality: The ethical dilemma of privacy and confidentiality of user information when interacting with mental health chatbots, as these	The need to ensure the privacy and confidentiality of users when interacting with mental health chatbots, which can be a technical and legal challenge.	The need to ensure that chatbots are programmed with accurate and up-to-date information about mental health, and that they do not perpetuate stereotypes or	Possible exploitation of participants using mental health chatbots for research and chatbot development without their adequate informed consent. Possible lack of transparency

Table 4. Identification of ethical dilemmas from the primary four bioethical principles present in communication news media.

Autonomy	Beneficence	No Maleficence	Justice
Autonomy in choice of help: Concerns about the quality of information provided by chatbots compared to human experts and whether people have enough information to make autonomous decisions about their mental health care.	Accessibility and speed of access: Concerns about the potential for chatbots to improve accessibility and provide more immediate access to mental health care for those in need.	Quality and accuracy of information: Risks of inadequate, outdated, or biased information provided by chatbots.	Ethical dilemma of quality and accuracy in chatbots: Risk of low-quality, outdated, or biased information affecting users.
Autonomy in emotional interaction: Concerns about social isolation and technological dependence that may result from sharing emotions with chatbots instead of human professionals.	Impact on the doctor-patient relationship: Concerns about the quality and strength of the therapeutic relationship between the patient and the chatbot, and whether this may negatively affect the effectiveness of mental health care.	Human vs. machine interaction: Questions about the ability of chatbots to empathize and genuinely understand complex emotional situations.	Ethical dilemma of human vs. machine: Concerns about the empathy and understanding of chatbots in complex situations.
Autonomy in seeking information: Issues around the quality and accuracy of information provided by chatbots and people's ability to make informed autonomous decisions.	Bias and discrimination: Risks of bias and discrimination in the programming and use of chatbots in mental health care, which may affect the quality of care and treatment of different groups of people.	Cost and accessibility: Possible dehumanization and excessive dependence on technology instead of seeking human professional help.	Ethical dilemma of cost and accessibility: Dehumanization of care and excessive dependence on technology in mental health.
Autonomy in interaction with chatbots and replacement of human professionals: Concerns about whether chatbots can provide the same empathy, understanding, and personalized attention as human professionals, and how this affects the valuation of mental health care.	Capacity for personalized care: Limitations on the ability of chatbots to provide personalized care and tailor treatment to the individual needs of patients.	Substitution of human professionals: Impact on the quality of care and therapeutic relationship between patients and professionals.	Ethical dilemma of substitution of professionals: Impact on the quality of care and therapeutic relationship.
Autonomy and privacy of information: Concerns about the privacy and security of data shared with chatbots, and whether people are making informed decisions about privacy and use of their data.	Technological limitations: Limitations on the ability of chatbots to handle crisis or emergency situations, which may jeopardize patient safety and well-being.	Privacy and data security: Concerns about protection and proper handling of sensitive information and the risk of misuse.	Ethical dilemma of privacy and data security: Adequate protection of sensitive information and risk of misuse.
Autonomy and responsibility in the use of chatbots: Responsibility of developers, healthcare providers, and users in case of errors or harm caused by information provided by chatbots, and whether people are adequately evaluating the quality and reliability of the information provided.	Supervision and monitoring: Need for adequate supervision and monitoring to ensure that patients are receiving appropriate care and that any ethical or safety issues arising from the use of chatbots in mental health care are addressed.	Responsibility and supervision: Challenges in establishing responsibilities and ensuring that chatbots follow ethical guidelines and provide high-quality support.	Ethical dilemma of responsibility and supervision: Establishing responsibilities and ensuring quality of information and support.
Autonomy and equitable access to mental health care: Inequalities in access to chatbots, especially for marginalized or disadvantaged populations, and whether the use of chatbots promotes autonomy and equitable access to care.	Accessibility: Limitations on the accessibility of mental health chatbots for people with visual or hearing impairments, which may negatively impact their ability to receive quality care.	Digital divide: Risk of exacerbating inequalities and excluding populations without access to digital resources.	Ethical dilemma of digital divide: Exclusion of marginalized populations without access to digital technologies.
Autonomy and reliability of chatbots: Issues around the quality and reliability of information provided by	Inequities in access: Inequalities in access to mental health chatbots due to factors such as lack of access to necessary technology or digital literacy.	Excessive dependence on technology: Negative impact on long-term mental health and emotional well-being.	Ethical dilemma of technological dependence: Negative impact on mental health and long-term emotional well-being.
	Ethics and responsibility of developers: Ethical and legal responsibility of mental	Research and validation: Need for rigorous research and validation of the efficacy and safety of chatbots in mental health.	Ethical dilemma of research and validation: Need for rigorous studies to evaluate the effectiveness and safety of chatbots.
		Regulation and policies: Importance of policies and regulations to ensure quality, safety, and user data protection.	Ethical dilemma of regulation and policies: Ensuring quality, safety, and data protection, preventing inadequate liability and protection.
		Balance between innovation and caution: Adapting the promotion of innovation and caution to address ethical concerns and protect user	Ethical dilemma of balancing innovation and caution: Balance between fostering innovation and protecting the well-being of users.
			Ethical dilemma of education and public awareness:

5 Ethical Analysis

With the results obtained, we proceeded to unify the content and conduct an ethical analysis of the review results at both the scientific and news levels. This process was carried out by ChatGPT 4 and the research group.

In this sense, a total of 24 moral dilemmas were identified, most of which are transversal to the four bioethical principles. Similarly, these bioethical dilemmas respond to the context and populations that create, use, and regulate them. To make sense of the collected information, bioethical dilemmas have been classified according to specific populations and their functions in mental health. These are as follows:

- Mental health technology developers
- Mental health beneficiaries of mental health services
- Mental health professionals
- Mental health researchers and developers
- Mental health regulators

Table 5 presents the synthesis of these results and the main bioethical dilemmas classified according to the user population.

Table 5. Main bioethical dilemmas classified according to the user population.

Mental health technology developers
Chatbots as conversational agents and the quality of care:
1 Uncertainty about whether conversational agents can provide the same level of attention as healthcare professionals.
2 Designing adaptive and personalized chatbots raises ethical dilemmas about manipulation and influences user decision-making.
3 Possibility of unconscious biases in chatbot programming, affecting the accuracy of results and the medical care provided.
4 The risk that chatbots are not equipped to handle crises or emergency situations endangers the safety and well-being of the user.
5 Need to ensure that chatbots are culturally appropriate and sensitive to the needs of different groups of people.
6 How can the use of chatbots complement and improve mental health care provided by human professionals, rather than replacing them?
7 Privacy and confidentiality of user information when interacting with mental health chatbots.
Mental health professionals
Human responsibility and supervision:
8 Proper accountability and oversight in the use of chatbots in mental health care. Challenges in establishing accountability and ensuring chatbots follow ethical guidelines and offer high-quality support.
9 Proper accountability and oversight in the use of chatbots in mental health care. Challenges in establishing accountability and ensuring chatbots follow ethical guidelines and offer high-quality support.
Mental health beneficiaries of mental health services
Access and exclusión:
10 Not all users may have access to the necessary technology to use Chat LLM.
11 Gaps in vulnerable populations such as the pediatric population, those with schizophrenic or bipolar disorders, or with intellectual or physical disabilities.
12 Limited access to medical care and accurate, reliable information on the benefits and risks of chatbots.
13 Users' ability to make informed decisions about their mental health when interacting with LLM chatbots.
Interacción emocional y dependencia tecnológica
15 Limitations of chatbots in addressing complex mental health issues that require empathetic understanding.
16 Negative impact on long-term mental health and emotional well-being due to excessive dependence on technology.
Mental health researchers and developers:
17 Need for solid and ethical research to determine safety, effectiveness, and potential risks of chatbots.
18 Ensuring that chatbots are programmed with accurate and up-to-date mental health information and do not perpetuate stereotypes or biases.
19 Risks of inadequate, outdated, or biased information provided by chatbots due to their design.
20 Possible exploitation of participants using mental health chatbots for research and chatbot development without their proper informed consent.
Mental health regulators
21 Importance of policies and regulations to ensure quality, safety, and data protection for users.
22 Balancing innovation and caution: Adjusting the promotion of innovation and caution to address ethical concerns and protect user well-being.
23 How the use of chatbots can complement and improve mental health care provided by human professionals, rather than replacing them.
24 Education and public awareness: The need to inform about the benefits and risks of chatbots in mental health, and understand their role as a complementary resource.

6 Conclusions

Considering the results and analysis carried out, it can be interpreted that there are various bioethical dilemmas present in the relationship between the use of chatbots in mental health, which are mostly transversal and changing, depending on the creating, benefiting, and regulating population, as well as the context in which they are developed.

In conclusion, there are 4 major areas in which bioethical dilemmas are categorized, requiring further reflection and analysis for the development of the creation and use of LLM chatbots in the field of mental health:

7 LLM Chatbots, Quality of Care, Responsible Research, and Development in Mental Health

To improve the quality of care, research, and technological development in mental health provided by chatbots, it is important to consider the different actors involved in this process, such as creators, technologists, professionals, users, researchers, and decision-makers in mental health.

To reduce risks and increase benefits, it is necessary to incorporate clarity in both the scope and risks of the technology itself, as well as the inclusion of various theories and practical knowledge of mental health with which chatbots are trained.

Therefore, it is recommended to create chatbots specifically designed to address mental health issues, with purposes and tasks that do not replace therapeutic care provided by human professionals. For example, technology could be used to provide companionship and emotional support, but therapeutic care should be centered or guided by human professionals, who in turn must have clear protocols to proceed in cases of emergencies detected or created by chatbots in mental health settings. This is because LLM chatbots can complement and improve the mental health care provided by human professionals rather than replace them. Therefore, it is essential to establish appropriate regulatory and ethical frameworks that allow various users to have autonomy, justice, and well-being.

8 Access, Exclusion, and User Dependence on Chatbots

In this regard, it is important to expand access to vulnerable populations and reduce elitism in mental healthcare, taking into account gender, social class, age, and other biases. Public and private institutions should focus their efforts with special emphasis on identifying and supporting marginalized and/or displaced populations to facilitate and educate the population about the use of chatbots in mental health topics.

It is also important to consider the interaction and dependence that these types of tools can induce in users, either in the public seeking care or professionals using chatbots as auxiliary tools. It is necessary to anticipate problems related to the preference that some people may have for chatbots, which can be a risk factor for social isolation, contributing to greater mental health problems and associated social issues.

9 Responsibility and Human Supervision of Chatbots

As it is already known, chatbots are imperfect tools, which can be misused for manipulation or influence based on misinformation. In this sense, the research group proposed to enhance critical thinking and verify the results provided by AI to ensure effectiveness and safety in the care provided to the public.

Likewise, it is important to consider the need for continuous supervision of AI in mental healthcare. Thus, it is necessary to question the future role of professionals and researchers in the field of mental health and LLM chatbots. Some of these roles could incorporate "the supervision of AI, both from the technological and mental health fields," who should be highly trained to detect and address problems in the services provided by these chatbots.

There are still questions about the responsibility that each actor must have in this process to ensure that AI is used effectively and ethically in mental healthcare.

10 Regulation and Chatbot Usage Policies

To achieve greater benefits in mental health, it is essential to establish clear regulations at the institutional and governmental levels in national and international settings.

In this sense, it is important to define who will be responsible for regulating and solving problems that may arise in the mental health care provided by chatbots. Furthermore, it is necessary to consider the liability for damage caused by chatbots and the biopolitical control that can be exerted through them.

Public and private institutions responsible for mental health care and those areas of political decision-making should promote the idea of human supervisors, who must ultimately be responsible for ensuring the effectiveness and safety of the care provided.

In conclusion, it is necessary to address the issue of mental health care through chatbots from a broad and conscious perspective, considering the different actors involved, the risks and benefits associated with their use, and the importance of establishing appropriate regulatory and ethical frameworks. All this, with the aim of improving the quality of mental health care and ensuring the widest and most equitable possible access, provides safe, effective, and inclusive care for all.

11 Limitations of the Present Review

Several limitations were identified, which could have affected the results and conclusions. The limitations of this study are as follows.

Dependence on prompts: LLM-based chatbots operate using prompts provided by users, meaning that the quality and relevance of the generated responses are directly related to the accuracy and clarity of the input prompts. This dependence can generate inaccurate or incomplete responses if prompts are inadequate.

User influence: Because LLM-based chatbots learn from interactions with their users, it is likely that the information provided will be influenced by the opinions, knowledge, and biases of these users.

Technical issues: On some occasions, the chatbot did not function properly, forcing researchers to use other versions of the LLM chatbots. These technical issues may have affected the efficiency and effectiveness of the research process and the quality of the data collected.

Peer control: Following the rapid review technique, it is reported that the inferences were not passed through controls by other peers, which could have influenced the results of the present investigation.

References

1. Sutz, J.: Engenharia e preocupação social: rumo a novas práticas, vol. 14 (2019)
2. CEPAL: Repercusiones en América Latina y el Caribe de la guerra en Ucrania: ¿cómo enfrentar esta nueva crisis? (2022)
3. Rupali, M., Amit, P.: A review paper on general concepts of 'artificial intelligence and machine learning.' Int. Adv. Res. J. Sci. Eng. Technol. 4(4), 79–82 (2017). https://doi.org/10.17148/IARJSET/NCIARCSE.2017.22
4. Eloundou, T., Manning, S., Mishkin, P., Rock, D.: GPTs are GPTs: an early look at the labor market impact potential of large language models. arXiv, Mar. 21, 2023. Accessed: Mar. 25, 2023. http://arxiv.org/abs/2303.10130
5. Open AI: GPT-4 (2023). https://openai.com/research/gpt-4
6. Gates, B.: A new era, the age of AI has begun. Gates Notes (2023). https://www.gatesnotes.com/The-Age-of-AI-Has-Begun
7. Sadasivan, V.S., Kumar, A., Balasubramanian, S., Wang, W., Feizi, S.: Can AI- generated text be reliably detected? arXiv Mar. 17, 2023. Accessed: Mar. 25, 2023. http://arxiv.org/abs/2303.11156
8. Burtell, M., Woodside, T.: Artificial influence: an analysis of AI-driven persuasion. arXiv, Mar. 15, 2023. Accessed: Mar. 25, 2023. http://arxiv.org/abs/2303.08721
9. Sandu, I., Wiersma, M., Manichand, D.: Time to audit your AI algorithms. Maandblad voor Accountancy en Bedrijfseconomie 96(7/8), 253–265 (2022). https://doi.org/10.5117/mab.96.90108
10. Pereira, G.V., et al.: South American expert roundtable: increasing adaptive governance capacity for coping with unintended side effects of digital transformation. Sustainability 12(2), 718 (2020). https://doi.org/10.3390/su12020718
11. Schofield, G., Dittborn, M., Selman, L.E., Huxtable, R.: Defining ethical challenge(s) in healthcare research: a rapid review. BMC Med. Ethics 22(1), 135 (2021). https://doi.org/10.1186/s12910-021-00700-9
12. Gotowiec, S., Cantor-Graae, E.: The burden of choice: a qualitative study of healthcare professionals' reactions to ethical challenges in humanitarian crises. J. Int. Humanitarian Action 2(1), 2 (2017). https://doi.org/10.1186/s41018-017-0019-y
13. Molyneux, S., et al.: Model for developing context-sensitive responses to vulnerability in research: managing ethical dilemmas faced by frontline research staff in Kenya. BMJ Glob. Health 6(7), e004937 (2021). https://doi.org/10.1136/bmjgh-2021-004937
14. Van Bavel, J.J., Packer, D.J., Haas, I.J., Cunningham, W.A.: The importance of moral construal: moral versus non-moral construal elicits faster, more extreme, universal evaluations of the same actions. PLoS ONE 7(11), e48693 (2012). https://doi.org/10.1371/journal.pone.0048693
15. Hagendorff, T.: The ethics of AI ethics: an evaluation of guidelines. Mind. Mach. 30(1), 99–120 (2020). https://doi.org/10.1007/s11023-020-09517-8
16. Gates Foundation Announces $1.27B in Health and Development Commitments to Advance Progress Toward the Global Goals. Bill & Melinda Gates Foundation. https://www.gatesfoundation.org/ideas/media-center/press-releases/2022/09/gates-foundation-unga-global-fund-replenishment-commitment (accessed May 01, 2023)
17. Fiske, A., Henningsen, P., Buyx, A.: Your robot therapist will see you now: ethical implications of embodied artificial intelligence in psychiatry, psychology, and psychotherapy. J. Med. Internet Res. 21(5), e13216 (2019). https://doi.org/10.2196/13216
18. Philip, A., Samuel, B., Bhatia, S., Khalifa, S., El-Seedi, H.: Artificial intelligence and precision medicine: a new frontier for the treatment of brain tumors. Life 13(1), 24 (2022). https://doi.org/10.3390/life13010024

19. Goisauf, M., Cano Abadía, M.: Ethics of AI in radiology: a review of ethical and societal implications. Frontiers in Big Data **5** (2022). Accessed: May 01, 2023. https://www.fronti ersin.org/articles/https://doi.org/10.3389/fdata.2022.850383

20. Renier, L., Mast, M., Dael, N., Kleinlogel, E.: Nonverbal social sensing: what social sensing can and cannot do for the study of nonverbal behavior from video. Front. Psychol. **12**, 2874 (2021). https://doi.org/10.3389/fpsyg.2021.606548

21. Sollini, M., Bartoli, F., Marciano, A., Zanca, R., Slart, R.H.J.A., Erba, P.A.: Artificial intelligence and hybrid imaging: the best match for personalized medicine in oncology. Eur. J. Hybrid Imaging **4**(1), 1–22 (2020). https://doi.org/10.1186/s41824-020-00094-8

22. Robillard, J.M., et al.: Scientific and ethical features of English-language online tests for Alzheimer's disease. Alzheimers Dement (Amst) **1**(3), 281–288 (2015). https://doi.org/10.1016/j.dadm.2015.03.004

23. Mallakin, M., Dery, C., Vaillancourt, S., Gupta, S., Sellen, K.: Web-based co-design in health care: considerations for renewed participation. Interact. J. Med. Res. **12**(1), e36765 (2023). https://doi.org/10.2196/36765

24. Reamer, F.G.: The evolution of social work ethics: bearing witness. Adv. Soc. Work **15**(1), 163–181 (2013). https://doi.org/10.18060/14637

25. Kacetl, J., Maresova, P.: Legislative and ethical aspects of introducing new technologies in medical care for senior citizens in developed countries. CIA **11**, 977–984 (2016). https://doi.org/10.2147/CIA.S104433

26. Pourvakhshoori, N., Norouzi, K., Ahmadi, F., Hosseini, M., Khankeh, H.: Nurse in limbo: a qualitative study of nursing in disasters in Iranian context. PLoS ONE **12**(7), e0181314 (2017). https://doi.org/10.1371/journal.pone.0181314

27. Khatiban, M., Falahan, S.N., Soltanian, A.R.: Professional moral courage and moral reasoning among nurses in clinical environments: a multivariate model. JMEHM **14** (2022). https://doi.org/10.18502/jmehm.v14i20.8180

28. Park, S.: Heterogeneity of AI-Induced Societal Harms and the Failure of Omnibus AI Laws. arXiv, Mar. 15

29. Taufiq, A., Saripah, I., Herdi, H.: The role of education and supervision toward the candidates of group counselor competencies. Presented at the 3rd Asian Education Symposium (AES 2018), pp. 118–122. Atlantis Press (2019). https://doi.org/10.2991/aes-18.2019.28

30. Córdoba, A., Mejía, L.F., Mannis, M.J., Navas, A., Madrigal-Bustamante, J.A., Graue-Hernandez, E.O.: Current global bioethical dilemmas in corneal transplantation. Cornea **39**(4), 529–533 (2020). https://doi.org/10.1097/ICO.0000000000002246

31. Goris, G., Adolf, S.J.: Utilidad y tipos de revisión de literatura. Ene **9**(2) (2015). https://doi.org/10.4321/S1988-348X2015000200002

32. Ministry of Health of Chile, Mental Health National Plan 2027 – 2025. Gobierno de Chile (2017). https://www.minsal.cl/wp-content/uploads/2017/12/PDF-PLAN-NACIONAL-SALUD-MENTAL-2017-A-2025.-7-dic-2017.pdf

Cyclical Learning Rates (CLR'S) for Improving Training Accuracies and Lowering Computational Cost

Rushikesh Chopade[1], Aditya Stanam[2], Anand Narayanan[3], and Shrikant Pawar[3,4(✉)]

[1] Department of Geology and Geophysics, Indian Institute of Technology, Kharagpur, India
[2] Department of Toxicology, University of Iowa, Iowa City, IA 52242-5000, USA
aditya-stanam@uiowa.edu
[3] Yale Center for Genomic Analysis, Yale School of Medicine, Yale University, New Haven, CT 30303, USA
{anand.narayanan,shrikant.pawar}@yale.edu
[4] Department of Computer Science & Biology, Claflin University, Orangeburg, SC 29115, USA

Abstract. Prediction of different lung pathologies using chest X-ray images is a challenging task requiring robust training and testing accuracies. In this article, one-class classifier (OCC) and binary classification algorithms have been tested to classify 14 different diseases (atelectasis, cardiomegaly, consolidation, effusion, edema, emphysema, fibrosis, hernia, infiltration, mass, nodule, pneumonia, pneumothorax and pleural-thickening). We have utilized 3 different neural network architectures (MobileNetV1, Alexnet, and DenseNet-121) with four different optimizers (SGD, Adam, and RMSProp) for comparing best possible accuracies. Cyclical learning rate (CLR), a tuning hyperparameters technique was found to have a faster convergence of the cost towards the minima of cost function. Here, we present a unique approach of utilizing previously trained binary classification models with a learning rate decay technique for re-training models using CLR's. Doing so, we found significant improvement in training accuracies for each of the selected conditions. Thus, utilizing CLR's in callback functions seems a promising strategy for image classification problems.

Keywords: One-Class Classifier · Optimizer · Cyclical Learning Rates

1 Introduction

Speech recognition, computer vision and text analysis are major fields in which deep learning is prominently used for image classification [1–3]. Cyclical learning rates (CLR's) allow the learning rates to vary between a range of boundary values. Selecting learning rate manually is a time consuming and computationally costly task [4]. Optimal learning rate is important as the model can converge slowly if the learning rate is too slow or the model can diverge from the minima of the cost function if the learning rate is too high [5]. Even if an optimal learning rate for the model is achieved, the model can take many epochs to reach the minima of the loss function. The model doesn't have a

© The Author(s), under exclusive license to Springer Nature Switzerland AG 2023
I. Rojas et al. (Eds.): IWBBIO 2023, LNBI 13920, pp. 327–342, 2023.
https://doi.org/10.1007/978-3-031-34960-7_23

regular cost function, moreover, the gradient of the cost function is different in different parts of the cost function curve [6]. To overcome this issue, instead of using constant single learning rate, a learning rate decay policy can be used to obtain better results. However, the learning rate decay also has several drawbacks including getting stuck in a local minimum or plateau of cost function due to very small learning rates in later epochs [7]. CLR's can be an effective technique to make the model converge faster in minimal number of epochs and to decrease the efforts of finding optimal learning rates.

2 Experimental Results and Analysis

2.1 Data Collection, Preprocessing, Model Architecture, and Learning Rates

2.1.1 Data Collection

The data used for binary and one-class classification has been made available by National Institutes of Health (NIH), USA [8]. This dataset consists of 112,120 chest X-ray images, each with a 1024 * 1024-pixel resolution. Images belong to 15 classes, 14 classes of diseased individuals and 1 class of healthy individuals ('No Finding'). The disease classes contain 'Atelectasis', 'Cardiomegaly', 'Consolidation', 'Effusion'; 'Emphysema', 'Edema', 'Fibrosis', 'Infiltration', 'Mass', 'Nodule', 'Pneumonia', 'Pneumothorax', 'Pleural Thickening' and 'Hernia'. A metadata associated with the image dataset consists of patient's age, gender, unique patient id, and the view position (anterior-posterior and posterior-anterior) of the X-ray image.

2.1.2 Exploratory Data Analysis

From the total set, 60,361 images have the label 'No Finding' (healthy), while others have multiple labels with combinations of 14 classes. Overall, the unique constitutes to around 836 labels. Unique can be any of the 14 primary classes ('No Finding' label excluded) or any combination of these 14 primary classes. Figure 1 depicts the distribution of these 15 unique labels.

A one-hot encoding was applied to convert 836 unique labels to 15 primary class labels [9]. Comparison of the number of images in 15 primary classes before and after performing one-hot encoding is shown in Table 1. A plot for the number of images after performing one-hot encoding is shown in Fig. 2.

Binary classifiers have been developed on each disease and the 'No Finding' class. The 'No Finding' class has approximately 3 times more images than the 'Infiltration' class, this type of unbalanced dataset can raise a state where the algorithm will overfit the class having more images. To avoid this, the number of images in the 'No Finding' class has been taken approximately the same as the number of images in the class for which the binary classifier was developed.

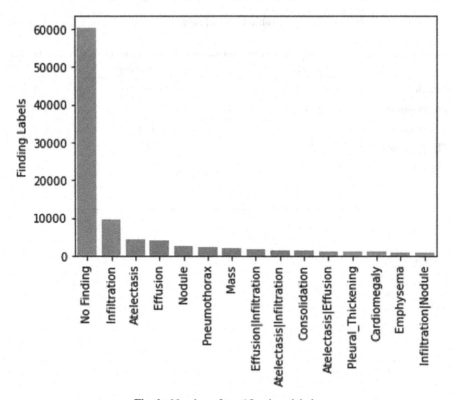

Fig. 1. Number of top 15 unique labels.

Table 1. Counts per class for primary labels before and after one-hot encoding.

Image Label	No. of Images before One Hot Encoding	No. of Image before One Hot Encoding
No Finding	60361	60361
Atelectasis	4215	11559
Cardiomegaly	1093	2776
Consolidation	1310	4667
Edema	628	2303
Emphysema	892	2516
Effusion	3955	13317
Fibrosis	727	1686
Infiltration	9547	19894
Mass	2139	5782

(continued)

Table 1. (*continued*)

Image Label	No. of Images before One Hot Encoding	No. of Image before One Hot Encoding
Nodule	2705	6331
Pneumothorax	2194	5302
Pneumonia	322	1431
Pleural Thickening	1126	3385
Hernia	110	227

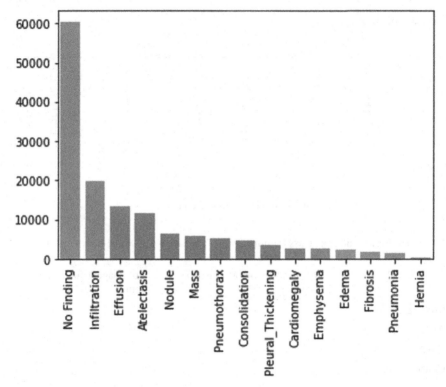

Fig. 2. Counts per class for primary labels after one-hot encoding.

2.1.3   Pre-processing of Data

Binary Classifier A 1:4-fold split of test to training set was performed for 14 binary classifiers (Table 2). To save overhead memory and making model more robust, we passed all the images through a ImageDataGenerator class of Keras [10] (shear range of 0.05, zoom range of 0.1, rotation range of 7°, width, and height shift range of 0.1, brightness range of 0.4 to 1.5 with a horizontal flip), while subsequently applying image augmentation technique. These techniques helped the model to generalize and reduce the overfitting state.

Table 2. List of binary classifiers and the number of images in their training and test sets.

Binary Classifier	No. of images containing respective disease label	No. of Images with 'No Finding' Label	Total Images	No. of training images (80% of total images)	No. of test images (20% of total images)
Atelectasis	11559	12000	23599	18847	4712
Cardiomegaly	2776	2800	5576	4460	1116
Consolidation	4667	4700	9367	7493	1874
Edema	2303	2300	4603	3682	921
Emphysema	2516	2600	5116	4092	1024
Effusion	13317	13500	26817	21453	5364
Fibrosis	1686	1700	3386	2708	678
Infiltration	19894	20000	39894	31915	7979
Mass	5782	6000	11782	9425	2357
Nodule	6331	6500	12831	10264	2567
Pneumothorax	5302	5500	10802	8641	2161
Pneumonia	1431	1500	2931	2344	587
Pleural Thickening	3385	3500	6885	5508	1377
Hernia	227	250	447	381	96

A dynamic batch training was utilized to decrease computational time and memory. Based on optimal performance, an iterative loop of 32 images/batch was used for training till all the images in batch were exhausted. Apart from utilizing less memory, this method helps to save fewer errors in the memory for updating hyperparameters through backpropagation which increases the training speed drastically. The high-resolution X-ray images for training have higher fractional improvements in area under curve (AUC) [11], and also can help localize a disease pattern (Table 3).

One Class Classifier. With the idea of choosing a balanced data, the dataset for one-class classifier contains 2,800 images of "No Finding" class and 200 images from each disease class. We again choose 1:4-fold split of test to training set to be consistent with binary classifiers. Further, the preprocessing through ImageDataGenerator class with same parameters as binary classifiers was performed for this split. Dynamic training with an optimal batch of 16 images/batch was performed.

Table 3. Number of training and testing batches with respective batch sizes for all the binary classifiers.

Binary Classifier	Total Images	Batch Size	No. of training batches	No. of test batches
Atelectasis	23599	16	1178	295
Cardiomegaly	5576	32	140	35
Consolidation	9367	32	235	59
Edema	4603	32	116	29
Emphysema	5116	32	128	32
Effusion	26817	32	671	168
Fibrosis	3386	32	85	22
Infiltration	39894	16	1995	499
Mass	11782	32	295	74
Nodule	12831	32	321	81
Pneumothorax	10802	32	271	68
Pneumonia	2931	16	147	37
Pleural Thickening	6885	32	173	44
Hernia	447	4	96	24

2.1.4   Model Architectures for Binary & One-Class Classifiers

Binary Classifier. A 2D convolutional neural network is applied using an MobileNetV1 network architecture [12]. The model parameters of MobileNet previously trained on ImageNet have been utilized using transfer learning (Fig. 3).

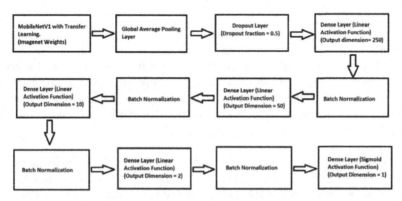

Fig. 3. Model architecture used for all the binary classifiers.

For MobileNetV1 previously trained ImageNet weights are passed through a global average pooling layer considering averages of each feature map instead of adding fully connected layers. This technique helps to easily interpret feature maps as categories confidence maps, to reduce overfitting, and is more robust to spatial translations of the input as it sums out the spatial information [13]. To further reduce overfitting, a dropout regularization layer to drop ~50% of the input units for variance reduction has been applied after the global average pooling layer. The model is then passed through 4 dense layers of output nodes 250, 50, 10, and 2 with linear activation functions in them. In each dense layer, L1 and/or L2 regularization is applied to the layer's kernel, bias, and activity. Kernel regularizer with both L1 and L2 penalties of 0.001 and 0.01 respectively are applied on the kernel's layer. A bias regularizer with an L2 penalty of 0.01 is applied on the layer's bias. Activity regularizer with an L2 penalty of 0.001 is applied on the layer's output. After each dense layer, batch normalization is used to stabilize the learning process and dramatically reduce the number of training epochs required to train a deep neural network. Finally, the model architecture is complete with application of a dense layer comprising of sigmoid activation function and 1 output node. The stochastic gradient descent (SGD) optimizer with learning rate decay has been used to train the model as it gave a superior performance compared to RMSProp and adam optimizer for all the classifiers except "Hernia". Adam optimizer with a learning rate of 0.01 has been found to perform better in case of "Hernia". A momentum parameter has been used to help accelerate gradient vectors in right directions (Table 4).

Table 4. Chart showing optimizer, its momentum, learning rates, and the decay constants used with SGD optimizer for all binary classifiers (Except Hernia which has Adam optimizer).

Binary Classifier	Optimizer Used	Learning Rate	Decay constant	Momentum
Atelectasis	SGD	0.01	0.001	0.9
Cardiomegaly	SGD	0.1	0.0005	0.9
Consolidation	SGD	0.05	0.0005	0.9
Edema	SGD	0.01	0.0005	0.9
Emphysema	SGD	0.01	0.0005	0.9
Effusion	SGD	0.01	0.001	0.9
Fibrosis	SGD	0.001	0.00005	0.9
Infiltration	SGD	0.01	0.001	0.9
Mass	SGD	0.01	0.001	0.9
Nodule	SGD	0.01	0.001	0.9
Pneumothorax	SGD	0.01	0.0005	0.9
Pneumonia	SGD	0.01	0.0001	0.9
Pleural Thickening	SGD	0.05	0.0005	0.9
Hernia	Adam	0.01	0.0001	0.9

One Class Classifier. A false-positive predictions arise when the algorithm is unable to identify the "No Finding" class, a problem falling under the category of "Anomaly

Detection". One-class classifier is an unsupervised learning algorithm focusing on the problem of anomaly detection [14]. The model contains a negative class (inlier or normal class) and a positive class (outlier or anomaly class). In our case, the normal class or inlier class is the "No Finding" class. The anomaly class is formed by combining 200 images of each disease class. The benefit of this approach is that if the prediction/test image fed to the algorithm is not from any of the 14 disease classes, it will still categorize it as an "Anomaly" simply because the algorithm could not classify it as an image with "No Finding" class. If the algorithm classifies the image with a disease other than these 14 diseases as a "No Finding" class, it will give rise to a problem of false negative prediction. One-class classifier serves the purpose of solving the problem of both false positives and false negative predictions. The model architecture for one-class classifier is same as the binary classifier.

2.1.5 Cyclical Learning Rates

The first step in applying CLR's is to define a maximum learning rate and a base learning rate [4]. The learning rate can then be allowed to vary between maximum learning rate and base learning rate. We have utilized learning rate finder technique (described in section A6) to decide maximum learning and the base learning rates. For one condition, "Pneumothorax" binary classifier, maximum and the base learning rates of 0.03 and 0.0075 respectively were obtained using learning rate finder. A step size is an important parameter which simply is the number of batches in which the learning rate will become equal to the maximum learning rate starting from the base learning rate or vice-versa. It is the number of training batches to reach half cycle. Typically, the step size of 2–8 times the number of training batches in 1 epoch is ideal [4]. For "Pneumothorax", the total number of training batches in 1 epoch is equal to 541. Therefore, a step size of 1082 was used for learning rate finder. Finally, a mode policy needs to be defined for calculating learning rates. Mode is the pattern in which the learning rate will vary within the bounds of maximum and minimum learning rates. The "triangular" policy for "Pneumothorax" binary classifier is shown in Fig. 4. The learning rate monotonically increases to maximum learning rate from base learning rate in two epochs and decreases back to base learning rate in the next two epochs. Since the "Pneumothorax" model with CLR technique and "triangular" policy is trained for 36 epochs, a total of 9 full cycles can be observed in Fig. 4.

We also have parallelly utilized a more complex policy called a "modified triangular2" policy. In this policy, the maximum learning rate is not taken to be the average of previous maximum learning rate unlike "triangular2" policy. After 3 complete cycles of the "triangular2" policy, the training is continued with "triangular2 policy" with original maximum learning rate obtained from the learning rate finder technique. This process is carried out until whole training is exhausted. In the "Pneumothorax" binary classifier, the maximum learning rate in the first cycle is 0.03 from the first learning rate finder cycle, followed by second cycle with maximum learning rate of 0.01875, followed by third cycle with maximum learning rate of 0.013125 (Fig. 5), etc.

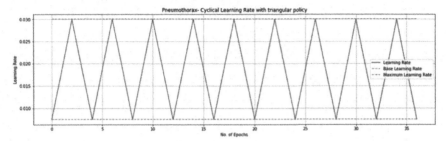

Fig. 4. Plot showing the "Triangular" policy for "Pneumothorax" binary classifier trained for 36 epochs.

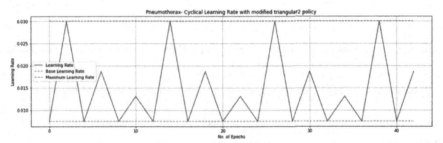

Fig. 5. Plot for the "modified triangular2" policy of "Pneumothorax" binary classifier trained for 42 epochs.

2.1.6 Learning Rate Finder

The upper and lower bounds of the CLR have been determined by learning rate finder technique where the cost function is minimum. Training the model with a learning rate finder as a callback for 1-5 epochs was enough to get the learning rate with minimum cost function. In case of the "Pneumonia" binary classifier, the minimum and maximum values for the learning rates were $1e{-}7$ as minimum and 1 as maximum (Fig. 6). The training increases exponentially after each batch on minimum learning rate. The "Pneumothorax" model loss vs. learning rate curve trained for 10 epochs is found to have a learning rate of $3e{-}2$ with minimum loss (Fig. 6). This loss increased as the learning rate approached to 1. The base learning rate for CLR can be accounted to one-fourth of the maximum learning rate [4].

2.1.7 With Binary Classifiers CLR'S Out-Perform Normal Training with a Learning Rate Decay Policy

We have run 3 model architectures (MobileNetV1, AlexNet, and DenseNet121) for comparing the performance (computational cost & accuracy) of classifiers [15, 16]. MobileNetV1 with an SGD optimizer was found to be most efficient, while DenseNet121 had good accuracy but significantly more computational cost, AlexNet had significantly lower accuracies when trained for the same number of epochs (Table 5).

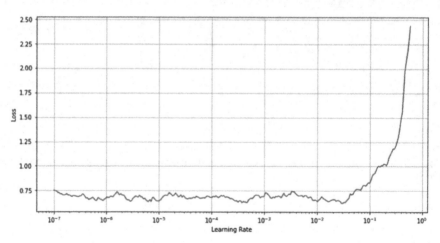

Fig. 6. Loss vs. learning rate plot for "Pneumothorax" binary classifier trained for 10 epochs.

Table 5. Accuracies of all the binary classifiers after training for given number of epochs.

Binary Classifier	No. of Epochs	Accuracy (in %)
Atelectasis	10	75.10
Cardiomegaly	12	75.78
Consolidation	10	73.32
Edema	12	93.37
Emphysema	10	85.60
Effusion	10	86.53
Fibrosis	10	66.58
Infiltration	10	64.60
Mass	10	70.11
Nodule	10	68.23
Pneumothorax	10	70.12
Pneumonia (with CLR)	30	88.43
Pleural Thickening	10	71.67
Hernia	30	90.81

The problem of false-positive predictions was addressed using one-class classifiers. For the models of "Infiltration", "Atelectasis", "Fibrosis" & "Pneumothorax" the accuracies have been consistently low after training for the selected number of epochs. So, we chose these conditions to test CLR's on (Table 6).

The problem of false-positive and false-negative predictions was resolved with one class classifiers. After which, a selected model trained for 32 epochs using CLR's with

Table 6. Comparison of the network architectures for "Atelectasis" binary classifier.

Name of Model Architecture	Approx. training time per epoch in hours	Training Epochs	Accuracy (in %)
MobileNetV1	2.5	10	75.10
DenseNet121	5	10	78.07
AlexNet	1.5	30	75.50

a maximum learning rate of 0.1, a base learning rate of 0.025, a step size of 2, and with a "triangular" policy provided a final training accuracy of 83.01%. CLR's showed improved accuracy and a lower computational cost compared to training a network with constant learning rates (Tables 7 and 8).

Table 7. Classifier accuracies after application of CLR's.

Binary Classifier	Accuracy before CLR application (in %)	Epochs taken to achieve the accuracy before CLR application	Accuracy after CLR application (in %)	Epochs taken to achieve the accuracy after CLR application	Policy Used
Atelectasis	75.10	10	79.59	32	Triangular
Infiltration	64.6	10	76.15	10	Modified Triangular2
Fibrosis	66.58	10	88.96	32	Modified Triangular2
Pneumothorax	70.12	10	79.83	36	Triangular
Pneumonia	–	–	88.43	30	Triangular

Table 8. Parameters and specifications of the CLR's.

Binary Classifier	Policy Used	Step Size	Epochs	Maximum Learning Rate	Base Learning Rate
Atelectasis	Triangular	2	32	0.1	0.025
Infiltration	Modified Triangular2	2	10	0.02	0.005
Fibrosis	Modified Triangular2	2	32	0.002	0.0005
Pneumothorax	Triangular	2	36	0.03	0.0075
Pneumonia	Triangular	2	30	0.01	0.0001

The "Pneumothorax" model is found to perform best when the CLR's is used with a "triangular" policy. As shown in Fig. 7, it took 47 epochs for the model with a constant learning rate to reach an accuracy of 79.26%. With CLR using "modified triangular2" policy crossed the accuracy level of 79.26% at 38 epoch and reached the accuracy of 80.92% in 41 epochs. While, the "Pneumothorax" model with CLR using a "triangular" policy crossed the accuracy level of 79.26% in just 36 epochs to achieve final accuracy of 79.83%.

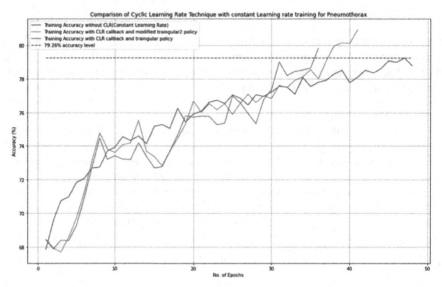

Fig. 7. Accuracy plot for "Pneumothorax" binary classifier with constant learning rate, CLR with "triangular" and CLR with "modified triangular2" policies.

The loss compared to the number of epochs was seen to be decreased with CLR's in both "triangular" and "modified triangular2" policies (Fig. 8). The loss of the "Pneumothorax" model with CLR reduced quicker than the "Pneumothorax" model with a constant learning rate.

The "Fibrosis" model was found to give better results in the case of the CLR technique with a "modified triangular2" policy. A comparison of "fibrosis" model trained for 32 epochs is shown in Fig. 9. The model reached an accuracy of 85.04% in 32 epochs when trained with a constant learning rate policy. The model reached an accuracy of 86.96% in 32 epochs when trained with CLR using a "triangular" policy. It crossed the 85% accuracy level in 30 epochs. The model reached an accuracy of 88.15% in 32 epochs when trained with CLR using a "modified triangular2" policy. It crossed the accuracy level of 85% in just 25 epochs. The loss was observed to be always less in CLR's with a "modified triangular2" (Fig. 10).

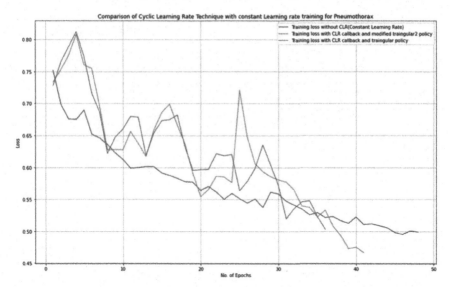

Fig. 8. Loss for "Pneumothorax" model with constant learning rate, CLR with "triangular" policy and CLR with "modified triangular2" policy.

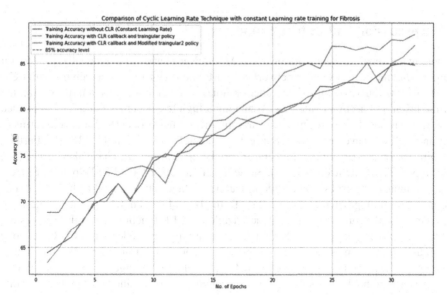

Fig. 9. Accuracy plot comparing "Fibrosis" binary classifier with constant learning rate, CLR with "triangular" policy and CLR with "modified triangular2" policy.

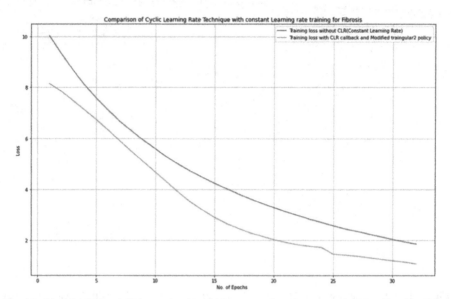

Fig. 10. Loss for "Fibrosis" binary classifier with constant learning rate and CLR with a "modified Triangular2" policy.

3 Discussion and Future Scope

Depthwise separable convolutions like MobileNets have been gradually pruned for improving the speed of dense network [17]. MobileNetV1 Imagenet weights with SGD optimizer is found to outperform other optimizers and architectures in terms of training time taken and accuracy attained. Achieving a high test accuracy is directly depended on learning rate hyper-parameter for training neural networks [18–21]. Three forms of triangle CLR's have been stated to accelerate neural network training [18, 19]. Further, tuning the batch size hyper-parameter for adjusting learning rates have also been shown to improve learning accuracy [22]. Some hyperparameter tools like Hyperopt, SMAC, and Optuna, using grid search, random search and Bayesian optimization have been seen efficient in tuning batch sizes [23, 24]. To the best of our knowledge, our work is the first to present a comprehensive characterization of CLR function on training and testing accuracy of dense network models. In general, training any model with a CLR technique is found to perform better than training with a constant learning rate. For the "Pneumothorax" binary classifier, the CLR technique with the "triangular" policy is found to outperform both CLR with the "modified triangular2" policy and constant learning rate training. For the "Fibrosis" binary classifier, the CLR with the "modified triangular2" policy was found to give better results than the rest two policies. Primarily, we found that there are two main advantages of training with CLR's over constant learning rates, with decay learning rates the model can get stuck into the saddle points or local minima due to low learning rates, and secondly CLR's reduces the effort of choosing an optimal learning rate by hit and trial method. Poor choice of initial learning rate can make the model circle infinitely. In setting a learning rate, there is a trade-off between the rate of convergence and overshooting, a high learning rate will make the learning jump over minima but a too low learning rate will either take too long to converge or get stuck in an undesirable local minimum [25]. The CLR's cyclically provided higher learning rates too, which helped the model to jump out of the local minima of the cost function. With these findings, implementing CLR's for improving prediction accuracies seems a promising strategy for object detection and machine translation.

Author Contributions. SP, AS, AN, and RC conceived the concepts, planned, and designed the article. SP, AS, AN, and RC primarily wrote and edited the manuscript.

Funding. This work was primarily supported by the National Science Foundation EPSCoR Program under NSF Award # OIA-2242812.

Data Availability Statement. The datasets generated during and/or analyzed during the current study are available in the repository: https://bitbucket.org/chestai/chestai_rushikes_code/src/mas ter/ https://www.kaggle.com/nih-chest-xrays/data.

Conflict of Interest for All Authors. None.

References

1. Girshick, R., Donahue, J., Darrell, T., Malik, J.: Rich feature hierarchies for accurate object detection and semantic segmentation. In: 2014 IEEE Conference on Computer Vision and Pattern Recognition (CVPR), pp. 580–587 (2014)
2. Graves, A., Jaitly, N.: Towards end-to-end speech recognition with recurrent neural networks. In Proceedings of the 31st International Conference on Machine Learning (ICML14), pp. 1764–1772 (2014)
3. Taigman, Y., Yang, M., Ranzato, M., Wolf, L.: Deepface: closing the gap to human-level performance in face verification. In: 2014 IEEE Conference on Computer Vision and Pattern Recognition (CVPR), pp. 1701–1708. IEEE (2014)
4. Smith, L.N.: Cyclical Learning Rates for Training Neural Networks (2017). arXiv:1506.011 86v6
5. Wilson, R.C., Shenhav, A., Straccia, M., et al.: The Eighty Five Percent Rule for optimal learning. Nat. Commun. **10**, 4646 (2019). https://doi.org/10.1038/s41467-019-12552-4
6. Pattanayak, S.: A mathematical approach to advanced artificial intelligence in Python. In: Pro Deep Learning with TensorFlow (2017). https://doi.org/10.1007/978-1-4842-3096-1
7. Bukhari, S.T., Mohy-Ud-Din, H.: A systematic evaluation of learning rate policies in training CNNs for brain tumor segmentation. Phys. Med. Biol. **66**(10) (2021). https://doi.org/10.1088/ 1361-6560/abe3d3
8. Wang, X., Peng, Y., Lu, L., Lu, Z., Bagheri, M., Summers, R.M.: ChestX-ray8: Hospital-scale Chest X-ray Database and Benchmarks on Weakly-Supervised Classification and Localization of Common Thorax Diseases. IEEE CVPR (2017)
9. Zhang, S.W., Zhang, X.X., Fan, X.N., Li, W.N.: LPI-CNNCP: prediction of lncRNA-protein interactions by using convolutional neural network with the copy-padding trick. Anal. Biochem. **601**, 113767 (2020). https://doi.org/10.1016/j.ab.2020.113767
10. Chollet, F., et al.: Keras. GitHub (2015). https://github.com/fchollet/keras
11. Sabottke, C.F., Spieler, B.M.: The effect of image resolution on deep learning in radiography. Radiology Artif. Intell. **2**(1), e190015 (2020). https://doi.org/10.1148/ryai.2019190015

12. Pang, S., Wang, S., Rodríguez-Patón, A., Li, P., Wang, X.: An artificial intelligent diagnostic system on mobile Android terminals for cholelithiasis by lightweight convolutional neural network. PLoS ONE **14**(9), e0221720 (2019). https://doi.org/10.1371/journal.pone.0221720

13. Lin, M., Chen, Q., Yan, S.: Network in Network (2014). https://arxiv.org/pdf/1312.4400v3.pdf

14. Dai, H., Cao, J., Wang, T., Deng, M., Yang, Z.: Multilayer one-class extreme learning machine. Neural Netw. **115**, 11–22 (2019). https://doi.org/10.1016/j.neunet.2019.03.004

15. Chen, J., et al.: Medical image segmentation and reconstruction of prostate tumor based on 3D AlexNet. Comput. Methods Programs Biomed. **200**, 105878 (2021). https://doi.org/10.1016/j.cmpb.2020.105878

16. Urinbayev, K., Orazbek, Y., Nurambek, Y., Mirzakhmetov, A., Varol, H.A.: End-to-end deep diagnosis of X-ray images. In: Annual International Conference of the IEEE Engineering in Medicine and Biology Society. IEEE Engineering in Medicine and Biology Society. Annual International Conference, pp. 2182–2185 (2020). https://doi.org/10.1109/EMBC44109.2020.9175208

17. Tu, C.-H., Chan, Y.-M., Lee, J.-H., Chen, C.-S.: Pruning depthwise separable convolutions for MobileNet compression. In: IEEE WCCI.https://doi.org/10.1109/IJCNN48605.2020.9207259

18. Smith, L.N., Topin, N.: Super-Convergence: Very Fast Training of Neural Networks Using Large Learning Rates (2017). arXiv e-prints: arXiv:1708.07120

19. Smith, L.N.: Cyclical Learning Rates for Training Neural Networks (2015). arXiv e-prints: arXiv:1506.01186

20. Goyal, P., et al.: Accurate, large minibatch SGD: training imagenet in 1 hour. CoRR (2017). http://arxiv.org/abs/1706.02677

21. Zulkifli, H.: Understanding learning rates and how it improves performance in deep learning (2018). https://towardsdatascience.com/understanding-learning-rates-and-how-it-improves-performance-indeep-learning-d0d4059c1c10. Accessed 23 Sep 2018

22. Hutter, F., Hoos, H.H., Leyton-Brown, K.: Sequential model-based optimization for general algorithm configuration. In: Coello, C.A.C. (ed.) Learning and Intelligent Optimization, pp. 507–523. Springer Berlin Heidelberg, Berlin, Heidelberg (2011). https://doi.org/10.1007/978-3-642-25566-3_40

23. Hyperopt Developers. hyperopt – distributed asynchronous hyperparameter optimization in python (2019). http://hyperopt.github.io/hyperopt/. Accessed 13 Aug 2019

24. Akiba, T., Sano, S., Yanase, T., Ohta, T., Koyama, M.: Optuna: a next generation hyperparameter optimization framework. In: Proceedings of the 25th ACM SIGKDD International Conference on Knowledge Discovery & Data Mining, ser. KDD 2019, pp. 2623–2631 (2019). ACM, New York, NY, USA

25. Buduma, N., Locascio, N.: Fundamentals of Deep Learning: Designing Next-Generation Machine Intelligence Algorithms, p. 21 (2017). O'Reilly. ISBN: 978-1-4919-2558-4

Relation Predictions in Comorbid Disease Centric Knowledge Graph Using Heterogeneous GNN Models

Saikat Biswas[1]($\boxtimes$)(iD), Koushiki Dasgupta Chaudhuri[2], Pabitra Mitra[3], and Krothapalli Sreenivasa Rao[3]

[1] Advanced Technology Development Centre, Indian Institute of Technology, Kharagpur, India
saikatbiswas17@iitkgp.ac.in
[2] Mathematics and Computing, Indian Institute of Technology, Kharagpur, India
koushiki@iitkgp.ac.in
[3] Department of Computer Science and Engineering, Indian Institute of Technology, Kharagpur, India
pabitra@cse.iitkgp.ac.in, pabitra@gmail.com

Abstract. Disease comorbidity has been an important topic of research for the last decade. This topic has become more popular due to the recent outbreak of COVID-19 disease. A comorbid condition due to multiple concurrent diseases is more fatal than a single disease. These comorbid conditions can be caused due to different genetic as well as drug-related side effects on an individual. There are already successful methods for predicting comorbid disease associations. This disease-associated genetic or drug-invasive information can help infer more target factors that cause common diseases. This may further help find out effective drugs for treating a pair of concurrent diseases. In addition to that, the common drug side-effects causing a disease phenotype and the gene associated with that can be helpful in finding important biomarkers for further prognosis of the comorbid disease. In this paper, we use the knowledge graph (KG) from our previous study to find out target-specific relations apart from sole disease-disease associations. We use four different heterogeneous graph neural network models to perform link prediction among different entities in the knowledge graph and we perform a comparative analysis among them. It is found that our best heterogeneous GNN model outperforms existing state-of-the-art models on a few target-specific relationships. Further, we also predict a few novel drug-disease, drug-phenotype, disease-phenotype, and gene-phenotype associations. These interrelated associations are further used to find out the common phenotypes associated with a comorbid disease as well as caused by the direct side effects of a treating drug. In this regard, our methodology also predicts some novel biomarkers and therapeutics for different fatal prevalent diseases.

Keywords: Disease comorbidity · knowledge graph · graph neural network

I. Rojas et al. (Eds.): IWBBIO 2023, LNBI 13920, pp. 343–356, 2023.
https://doi.org/10.1007/978-3-031-34960-7_24

1   Introduction

Comorbid disease condition refers to the concurrent presence of one or more diseases along with a primary disease. An individual suffering from a comorbid disease is more likely to be in high mortality risk zone than if he was suffering from a single disease. For instance, a patient suffering from Type II Diabetes is highly susceptible to getting affected by different neurodegenerative disorders like Parkinson's or Alzheimer's disease [8,14,37]. Past years of massive COVID-19 outbreaks have already shown the disastrous effect of disease comorbidity in large populations [7,24,47]. The disease and its associated genes may also be related to some fatal phenotypic conditions. Similarly, disease-curing drugs may also be related to those phenotypic conditions, which may be directly related to certain other diseases and their associated comorbid conditions. So these disease, gene, phenotype, and drug inter-related associations can be visualized as a complex graph structure. Recently, there have been rapid advancements in the field of knowledge graph (KG) building and inferring new relations from KGs. A KG is a triple representation, composed of *head* entity, *relation*, and *tail* entity. Due to the logical and robust structure of KGs, this graph representation has been successfully applied to solve many biological and biomedical problems [39,40]. In their recent work, Liu et al. [41] have proposed a novel multitasking healthcare management recommendation system leveraging knowledge graphs.

Recently, task-specific learning from heterogeneous graph structure is being explored in various literature studies [28,30,48,51]. The graph neural network model can effectively infuse the graph entity and corresponding relational information into the node embeddings of the graph for better learning which leads to better logical inference. So, in our proposed methodology, we used different heterogeneous graph neural network models to predict different relationships among knowledge graph entities and also predict the novel links among comorbid diseases, genes, phenotypes, and drugs. In this work, we used our earlier published comorbid disease-centric KG [10] to perform link prediction between different pairs of entities. We produced comparative analysis results for each of the fifteen different relation types using 4 different heterogeneous graph neural network models. We further analyzed the link prediction results obtained from the best-performing model for our KG and used them to predict some novel links between these entities. These novel findings can shed more light on the study of hidden biological relationships in comorbid disease associations.

2   Related Work

There are three major ways to perform link prediction in knowledge graphs, namely, weight rule learning, graph random walk, and tensor factorization. Weight rule learning methods combine probability [21] and relational logic to assign weights in KG data structure. Link prediction in these methods is performed by learning algorithms based on the conjugate gradient algorithm, pseudo-likelihood, and inductive logic programming [22]. Graph random walk models define similarity measures between nodes of a network based on a random walk on a graph. Curado et al. [19] proposed a new method namely, return random walk, for link prediction to infer new intra-class edges while minimizing the amount of inter-class noise. Tensor factorization approaches [2,23] predict the relationship between different nodes by learning the embeddings for each node and their corresponding relationships. The tensor factorization methods can be grouped into three main approaches - translation-based, deep learning based, and multiplicative-based. Here, we are mainly concerned with deep learning-based approaches for link prediction.

Liu et al. [38] proposed three different deep-learning methods to predict links in online social network services. Zhang et al. [53] integrated topological features and features based on social patterns into a deep learning framework to predict links between authors in bibliographic networks. In recent times, research has shifted to predicting link prediction in a dynamic setting, that is, in networks that evolve over time. Weak estimators were added to traditional similarity metrics to build an effective feature vector for a deep neural network [15], which resulted in increased prediction accuracy on several real-world dynamic networks. A novel encoder-LSTM-decoder (E-LSTM-D) deep learning model was proposed in 2019 [13] to predict dynamic links end to end. Zhang et al. [54] described a new method to learn heuristics from local subgraphs using a graph neural network (GNN). Neighborhood Overlap-aware Graph Neural Networks (Neo-GNNs) were proposed to perform link prediction in 2021 [52]. Cai et al. [12] solved link prediction problems by performing node classification in the corresponding line graphs.

Due to the robustness and better expressibility of GNN models for graph-based structures, these GNN-based models are being used to solve biological and biomedical data-centric tasks. Crichton et al. [18] investigated how inputs from four node representation algorithms affect the performance of a neural link predictor on random and time-sliced biomedical graphs of real-world sizes. Breit et al. [11] introduced a large-scale, high-quality, and highly challenging benchmarking framework for biomedical link prediction. In another work [46], two models - graph convolutional network (GCN) and Node2Vec were used to predict links in a cancer dataset and a plant genetics dataset. In 2021, Coskun et al. [17] used node similarity-based convolution matrices in GCNs to compute node embeddings for link prediction. A novel Pre-Training Graph Neural Networks-based framework named PT-GNN to integrate different data sources for link prediction in biomedical networks was proposed in 2022 [42].

3   Dataset and Method

3.1   Brief Description of the Biological KG and Its Associated Datasets

In this work, a biological KG [10] from our previously published work is being used for the comorbid disease-centric entity pair link analysis task. This KG is composed of different ontologies, namely, Gene Ontology (GO) [5], Human Phenotype Ontology (HPO) [34], and Disease Ontology (DO) [32]. Genes and their corresponding functions are obtained from the GO database. The HPO database gives information about genes and their corresponding phenotypes. Human GO annotations are taken from SwissProt [16] and human protein-protein interactions (PPIs) are obtained from STRING database [49]. Drug side-effects and interactions were obtained from the SIDER database [35]. Disease-gene associations are obtained from the disease databases, namely, Online Mendelian Inheritance in Man (OMIM) [26] and Genetic Association Database (GAD) [9]. Disease-pathway relations are taken from the Comparative Toxicogenomics Database [20]. The Reactome database [25] is used to download gene Entrez identifiers and their corresponding pathway ids.

The final constructed knowledge graph has 6 different types of nodes, namely, disease, drug, pathway, gene, phenotype, and GO function. This KG contains 15 different types of edges or relations among these 6 different entities. The complete schematic diagram of KG is shown in Fig. 1.

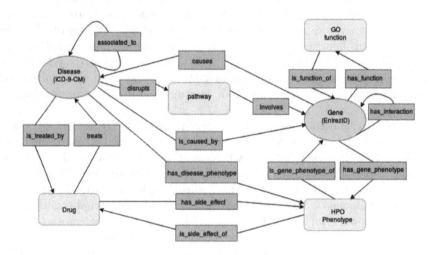

Fig. 1. Schematic Representation of Knowledge Graph [10]

The total numbers of each node entity type and their corresponding edge instances, present in the KG are given in Table 1.

3.2   BioBERT Embedding for Entity Concepts and Data Pre-processing

We first collectively find out all the entity instances of each of the 6 different types of entities in the KG. Then for each unique entity, an entity concept is formed by appending the associated biological or biomedical information regarding the entity. There is a total of 52,382 unique entities present in the KG. We then

Table 1. Detailed description of our knowledge graph

Entity Type	Name of Entity	Number of entities of that type
Node	'GO'	17582
	'disease'	2464
	'drug'	1460
	'gene'	20849
	'pathway'	2172
	'phenotype'	7855
Edge	'GO_is_function_of_gene'	267510
	'gene_has_function_GO'	267510
	'disease_associated_to_disease'	207460
	'disease_is_treated_by_drug'	7074
	'drug_treats_disease'	7074
	'disease_disrupts_pathway'	91101
	'disease_is_caused_by_gene'	168937
	'gene_causes_disease'	168937
	'pathway_involves_gene'	107352
	'gene_has_interaction_gene'	332658
	'gene_has_gene_phenotype_phenotype'	146838
	'phenotype_is_gene_phenotype_of_gene'	146838
	'drug_has_side_effects_phenotype'	47904
	'phenotype_side_effect_of_drug'	47904
	'disease_has_disease_phenotype_phenotype'	72639

use the BioBERT (Bidirectional Encoder Representations from Transformers for Biomedical Text Mining) [36] model to convert these entity-associated concepts to vector embeddings. Thus we get a 768-dimensional vector embedding for each of the unique entities present in the KG.

After getting all the entity vector embeddings, we convert the KG into DGL [50] compatible heterogeneous graph. To build this heterogeneous graph, we first divide the node data file and edge data file on the basis of different node types and different edge types respectively. Then we build a meta file to encode each node type and edge type-specific IDs. Each of the node type-specific and edge type-specific instances are being stored in a set of comma-separated files and tagged with their corresponding node and edge type labels. Finally, the customized data parser is being built up to process the whole graph data with their pre-trained BioBERT node embeddings to build a DGL-compatible heterogeneous network. In the final heterogeneous graph, the node and edge attributes are stored as ndata and edata respectively.

3.3   Brief Overview of GNN

Graph neural networks (GNNs) are neural networks that operate on graph data. They are based on a simple differentiable message-passing framework :

$$h_i^{l+1} = \sigma(\sum_{m \in M_i} g_m(h_i^{(l)}, h_j^{(l)})), \tag{1}$$

where $h_i^{(l)} \in \mathbb{R}^{d^{(l)}}$ is the hidden state of the v_i-th node in the $l-th$ layer of the neural network and $d^{(l)}$ is the dimensionality of the $l-th$ layer. Incoming messages of the form $g_m(.,.)$ are passed through an element-wise activation function (σ) like ReLU. M_i denotes the set of incoming messages for node v_i and is usually chosen to be identical to the set of incoming edges. $g_m(.,.)$ is typically chosen to be a message-specific neural network-like function or simply a linear transformation $g_m(h_i, h_j) = W h_j$ with a weight matrix W.

The Heterogenous GNN Models Used for Our Analysis

– **RGCN**: Relational Graph Convolutional Networks (R-GCNs) [48] is an extension of GCNs (Graph Convolutional Networks) [33] that operate on local graph neighborhoods. The forward pass update of a node v_i in a relational multigraph in the RGCN model is given by:

$$h_i^{l+1} = \sigma(\sum_{r \in \mathbb{R}} \sum_{j \in N_i^r} \frac{1}{c_{i,r}} W_r^{(l)} h_j^{(l)} + W_0^{(l)} h_i^{(l)}), \tag{2}$$

where N_i^r denotes the set of neighbor indices of node i under relation $r \in \mathbb{R}$. $c_{i,r}$ is a problem-specific normalization constant that can either be learned or chosen in advance.
– **HGT:** Heterogeneous Graph Transformer (HGT) [28] is an architecture for modeling web-scale heterogeneous graphs. Given a sampled heterogeneous sub-graph, HGT extracts all linked node pairs, where target node t is linked by source node s via edge e. The goal of HGT is to aggregate information from source nodes to get a contextualized representation for the target node t. Such a process can be decomposed into three components: Heterogeneous Mutual Attention, Heterogeneous Message Passing, and Target-Specific Aggregation. HGT uses the meta relations of heterogeneous graphs to parameterize weight matrices for the heterogeneous mutual attention, message passing, and propagation steps. A relative temporal encoding mechanism is introduced into the model to further incorporate network dynamics.
– **HAN:** Heterogeneous Graph Attention Network (HAN) [51] is a semi-supervised graph neural network for heterogeneous graphs. It follows a hierarchical attention structure from node-level attention to semantic-level attention. Given the node features as input, a type-specific transformation matrix is used to project different types of node features into the same space. The node-level attention is able to learn the attention values between the nodes and their meta-path-based neighbors, while the semantic-level attention aims

to learn the attention values of different meta-paths for the specific task in the heterogeneous graph. Based on the learned attention values in terms of the two levels, HAN gets the optimal combination of neighbors and multiple meta-paths in a hierarchical manner, which enables the learned node embeddings to better capture the complex structure and rich semantic information in a heterogeneous graph.

- **HPN**: Heterogeneous Graph Propagation Network (HPN) [30] is a type of a heterogeneous graph neural network (HeteGNN) that is able to alleviate the semantic confusion phenomenon in node-level based on theoretical analysis. HPN consists of a semantic propagation mechanism and a semantic fusion mechanism. Inspired by meta-path-based random walks with restart, the semantic propagation mechanism emphasizes a node's local semantics in a node-level aggregating process, alleviating the semantic confusion at the node level. The semantic fusion mechanism is also able to learn the importance of meta-paths and get the optimally weighted combination of semantic-specific node embeddings for the specific task.

4   Experiment

The dataset is split into train, test, and validation parts in a 7:2:1 ratio. All the models were trained for 200 epochs to maintain uniformity. A learning rate of 0.01 was chosen for each model with the Adam optimizer.

The dataset is trained with RGCN which had 3 hidden layers, 40 bases, a hidden dimension of 64, a weight decay of 0.0001, and a dropout rate of 0.2. The HGT model used on our dataset has 2 hidden layers, a hidden dimension of 64, a weight decay of 0.0001, 2 heads, and a dropout rate of 0.4. The HAN model used on the dataset has a hidden dimension of 8, 8 heads, and a dropout rate of 0.6. An HPN model was also used on our dataset. It has a weight decay of 0.001, a dropout rate of 0.6, a hidden dimension of 64, and an alpha value of 0.1. All models are implemented using the Deep Graph Library (DGL) [50], which is a Python package built for easy implementation of graph neural network models on top of existing deep learning frameworks like Pytorch. We also take the help of OpenHGNN [27], which is an open-source toolkit for Heterogeneous Graph Neural Networks based on DGL and Pytorch. All computer operations are performed in a Python environment in the CPU.

5   Results and Discussion

The results from all 4 models are summarized in Table 2 and shown in Fig. 2. It is evident that HGT performed better than the rest of the 3 models in predicting 14 out of the 15 different target edge types. This result is not surprising since HGT uses node and edge-type dependent parameters to characterize the heterogeneous attention over each edge, allowing it to maintain dedicated representations for different types of nodes and edges. Further, a relative temporal encoding technique is present in HGT, which allows it to capture the dynamic

Table 2. Comparison of performance of different GNN models on test data in the prediction of different target links in our biological KG (accuracy scores in ROC_AUC metric)

Predicted Edge Type	Models Used							
	RGCN		HGT		HAN		HPN	
	accuaracy	loss	accuracy	loss	accuracy	loss	accuracy	loss
GO_is_function_of_gene	0.7695	0.6524	**0.7883**	0.6575	0.6844	0.6854	0.607	0.6962
gene_has_function_GO	0.8641	0.6141	**0.8881**	0.6057	0.8341	0.6456	0.8596	0.6081
disease_associated_to_disease	0.8904	0.5979	**0.8968**	0.5895	0.622	0.6948	0.5049	0.7241
disease_is_treated_by_drug	0.7703	0.6618	**0.8589**	0.6143	0.7389	0.6689	0.7864	0.6484
drug_treats_disease	0.9259	0.5814	**0.9432**	0.5661	0.8717	0.5827	0.942	0.5777
disease_disrupts_pathway	**0.7693**	0.6638	0.7683	0.6686	0.6341	0.7801	0.7432	0.6733
disease_is_caused_by_gene	0.8532	0.6299	**0.8781**	0.6115	0.8514	0.6412	0.8494	0.6363
gene_causes_disease	0.911	0.6012	0.9094	0.6069	0.9281	0.5986	**0.9303**	0.5938
pathway_involves_gene	0.8574	0.6302	**0.8635**	0.6154	0.7564	0.833	0.8378	0.6419
gene_has_interaction_gene	0.773	0.6605	**0.8142**	0.6458	0.6557	0.693	0.6576	0.7004
gene_has_gene_phenotype_phenotype	0.8457	0.6259	**0.9049**	0.601	0.8588	0.6427	0.8827	0.6191
phenotype_is_gene_phenotype_of_gene	0.9102	0.5589	**0.951**	0.5578	0.9087	0.5620	0.9412	0.5715
drug_has_side_effects_phenotype	0.9662	0.5269	0.9822	0.5213	**0.9893**	0.5261	0.9873	0.5244
phenotype_side_effect_of_drug	0.4276	0.6937	0.5793	0.6983	**0.6176**	0.6917	0.6117	0.7004
disease_has_disease_phenotype_phenotype	0.7535	0.6681	**0.8618**	0.6181	0.6784	0.6833	0.717	0.6719

structural dependency of heterogeneous graphs with arbitrary durations. HGT model has consistently outperformed all state-of-the-art GNN baselines by 9%-21% on various tasks like node classification and link prediction [28]. Among the fifteen different target edge types, all models seem to have performed the best while predicting the edges of type 'drug_treats_disease'. This can be attributed to how dense or comprehensive our graph is in relation to specific edge types. Further investigation into the predicted links can yield previously unknown biologically relevant relations. That can help us identify new drugs that target a specific pathway or new co-morbid disease pairs.

We also compare our results to the results obtained from state-of-the-art models used for link prediction in biomedical knowledge graphs. For the edge type 'drug_treats_disease', it is found that the HGT model performed better than other commonly used models. In our proposed framework, the HGT model yields an AUC-ROC score of 0.9432 while a logistic regression with neuro-symbolic feature learning model [4] gives only 0.83. REDIRECTION architecture [6] has a ROC-AUC score of 0.93, artificial neural network [3] has a score of 0.884, random forest [3] scored 0.895 and the average commute time (ACT) [1] model gives a score of 0.8134. For the edge type 'gene_has_gene_phenotype_phenotype', the HGT model gives a score of 0.9049 while the neuro-symbolic feature learning model scores 0.90, XGBoost [45] scores 0.81 and LightGBM [45] scores 0.90. For the target edge 'drug_has_side_effects_phenotype', our HGT model scored 0.9822 compared to a score of 0.94 obtained by the neuro-symbolic feature learning model. The HGT model also scores 0.8618 while predicting the edge 'disease_has_disease_phenotype_phenotype' whereas the

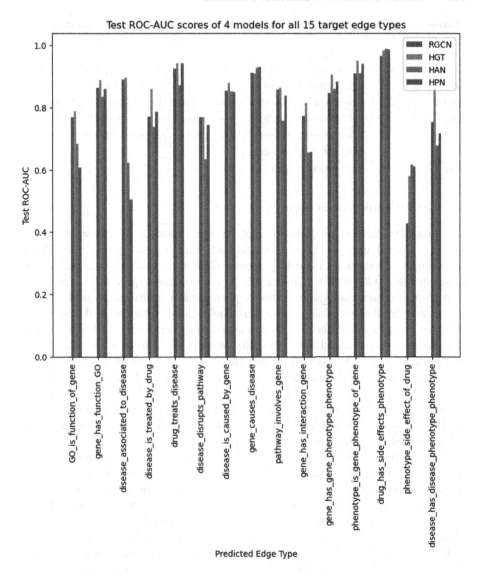

Fig. 2. Performance results of all 4 models in predicting 15 target edge types

neuro-symbolic feature learning model scored only 0.77. For the edge 'disease_is_caused_by_gene', the HGT scores 0.8532 while the ACT model [1] scores 0.8134. Our proposed model also outperforms the neuro-symbolic feature learning method on the other relations, namely, 'gene_has_gene_phenotype_phenotype' (ROC_AUC = 0.90), 'drug_has_side_effects_phenotype' (ROC_AUC = 0.94), and 'disease_has_disease_phenotype_phenotype' (ROC_AUC = 0.77).

Table 3. Few of the predicted disease-phenotype associations

Disease (ICD-CM-9)	Phenotype	PMID
702	HP:0000789	28554943
614.9	HP:0025134	36765791
531	HP:0000966	32772616
382.9	HP:0000966	24080322
684	HP:0001944	34833375
410.9	HP:0003040	20836862

Table 4. Few of the predicted drug side-effects phenotype associations

Drug	Phenotype	PMID
Clomiphene	HP: 0000989	28319428
Rosuvastatin	HP: 0000966	24063830
Metolazone	HP: 0003418	27134890
Benzathine	HP: 0010307	31297163

5.1 Novel Link Predictions

Our proposed method predicts some of the novel associations between disease and phenotypes and also the drug-related side effect causing phenotypes. A few of our novel predictions regarding disease-phenotype and drug-phenotype are given in the tables Table 3 and Table 4. Among these predictions, we show that some diseases may cause complex phenotypic conditions, as well as some phenotypic conditions, which may occur due to a few drug-induced effects. For instance, our proposed method predicts that actinic keratosis a flaky skin-related disorder is also associated with infertility phenotypic condition and also biologically validated [31]. It has been shown that the SPINK family protein's altered expression can lead to cause atopic dermatitis or actinic keratosis-related skin disorders. This SPINK family protein deficient individual can also be affected by infertility phenotypic condition. Similarly, the predicted link between the increased estrogen phenotypic condition and pelvic inflammatory disease is also being verified in another recent literature [29].

The study has shown that pelvic inflammatory disease can directly stimulate estrogen production and activates eutopic endometrium. Similarly, our proposed method also predicts a few drug-induced phenotypic conditions as well as validated through available literature. For instance, clomiphene is predicted to cause the phenotypic condition pruritus and also verified through the study by Mutlu et al. [44]. This study has shown the effect of clomiphene causing pruritus in intrahepatic cholestasis of pregnancy-disordered condition in a patient. The drug-induced side effect of benzathine can cause respiratory-related disorder stridor in an individual suffering from rheumatic heart disease prophylaxis [43]. Our prediction also has shown this association with a high probability linking score. Thus our prediction model identifies a few of the important disease-phenotype and drug-phenotype associations with proper literature validation.

6   Conclusion

Link prediction of comorbid disease pairs and inferring novel hidden patterns among them are highly important topics of study due to their combined fatal effect on a patient in the post-COVID era. In this work, we propose a new framework to perform link analysis tasks on a comorbid disease-centric KG. Our contributions can be reflected through four of the salient components of our framework. First, to induce the domain-specific biological and biomedical entity information, we use BioBERT node embedding in the network. Second, we use four different types of heterogeneous graph neural network models and perform a comparative analysis of the target edge-type link prediction tasks. Third, for a few of the target edge-type link predictions, our proposed method outperforms the existing state-of-the-art models for similar tasks. Finally, through our model, we predict a few significant new *disease-phenotype* and *drug-phenotype* associations and further validated them with the available literature. Our link analysis result shows that HGT performs the best, giving the highest test ROC_AUC scores for fourteen out of fifteen different target edge types. This outcome undoubtedly justifies the efficacy of the HGT model for heterogeneous network structures.

Instead of the proven efficacy of our proposed method, there is a scope for further modifications of our model in the future. The work can be modified by hybridizing the models together and building a more intuitive negative sample set to improve the learning model. Finally, our proposed work can be used to predict novel biomarkers and therapeutics for complex comorbid diseases while inferring new latent genomic and phenotypic patterns.

References

1. Abbas, K., et al.: Application of network link prediction in drug discovery. BMC Bioinf. **22**, 1–21 (2021)
2. Acar, E., Dunlavy, D.M., Kolda, T.G.: Link prediction on evolving data using matrix and tensor factorizations. In: 2009 IEEE International Conference on Data Mining Workshops, pp. 262–269. IEEE (2009)
3. Alshahrani, M., Hoehndorf, R.: Drug repurposing through joint learning on knowledge graphs and literature. Biorxiv, p. 385617 (2018)
4. Alshahrani, M., Khan, M.A., Maddouri, O., Kinjo, A.R., Queralt-Rosinach, N., Hoehndorf, R.: Neuro-symbolic representation learning on biological knowledge graphs. Bioinformatics **33**(17), 2723–2730 (2017)
5. Ashburner, M., et al.: Gene ontology: tool for the unification of biology. Nat. Genet. **25**(1), 25–29 (2000)
6. Ayuso Muñoz, A., et al.: Redirection: generating drug repurposing hypotheses using link prediction with DISNET data. bioRxiv, pp. 2022–07 (2022)
7. Baradaran, A., Ebrahimzadeh, M.H., Baradaran, A., Kachooei, A.R.: Prevalence of comorbidities in COVID-19 patients: a systematic review and meta-analysis. Arch. Bone Joint Surg. **8**(Suppl 1), 247 (2020)
8. Barbagallo, M., Dominguez, L.J.: Type 2 diabetes mellitus and Alzheimer's disease. World J. Diabetes **5**(6), 889 (2014)
9. Becker, K.G., Barnes, K.C., Bright, T.J., Wang, S.A.: The genetic association database. Nat. Genet. **36**(5), 431–432 (2004)
10. Biswas, S., Mitra, P., Rao, K.S.: Relation prediction of co-morbid diseases using knowledge graph completion. IEEE/ACM Trans. Comput. Biol. Bioinf. **18**(2), 708–717 (2019)

11. Breit, A., Ott, S., Agibetov, A., Samwald, M.: Openbiolink: a benchmarking framework for large-scale biomedical link prediction. Bioinformatics **36**(13), 4097–4098 (2020)
12. Cai, L., Li, J., Wang, J., Ji, S.: Line graph neural networks for link prediction. IEEE Trans. Pattern Anal. Mach. Intell. (2021)
13. Chen, J., et al.: E-LSTM-D: A deep learning framework for dynamic network link prediction. IEEE Trans. Syst. Man Cybern.: Syst. **51**(6), 3699–3712 (2019)
14. Cheong, J.L., de Pablo-Fernandez, E., Foltynie, T., Noyce, A.J.: The association between type 2 diabetes mellitus and Parkinson's disease. J. Parkinson's Dis. **10**(3), 775–789 (2020)
15. Chiu, C., Zhan, J.: Deep learning for link prediction in dynamic networks using weak estimators. IEEE Access **6**, 35937–35945 (2018)
16. Consortium, U.: Uniprot: a hub for protein information. Nucleic Acids Res. **43**(D1), D204–D212 (2015)
17. Coşkun, M., Koyutürk, M.: Node similarity-based graph convolution for link prediction in biological networks. Bioinformatics **37**(23), 4501–4508 (2021)
18. Crichton, G., Guo, Y., Pyysalo, S., Korhonen, A.: Neural networks for link prediction in realistic biomedical graphs: a multi-dimensional evaluation of graph embedding-based approaches. BMC Bioinf. **19**(1), 1–11 (2018)
19. Curado, M.: Return random walks for link prediction. Inf. Sci. **510**, 99–107 (2020)
20. Davis, A.P., et al.: A CTD-Pfizer collaboration: manual curation of 88 000 scientific articles text mined for drug-disease and drug-phenotype interactions. Database 2013 (2013)
21. De Raedt, L., Kimmig, A., Toivonen, H.: Problog: A probabilistic prolog and its application in link discovery. In: IJCAI, vol. 7, pp. 2462–2467. Hyderabad (2007)
22. Domingos, P., Kok, S., Lowd, D., Poon, H., Richardson, M., Singla, P.: Markov logic. In: De Raedt, L., Frasconi, P., Kersting, K., Muggleton, S. (eds.) Probabilistic Inductive Logic Programming. LNCS (LNAI), vol. 4911, pp. 92–117. Springer, Heidelberg (2008). https://doi.org/10.1007/978-3-540-78652-8_4
23. Dunlavy, D.M., Kolda, T.G., Acar, E.: Temporal link prediction using matrix and tensor factorizations. ACM Trans. Knowl. Disc. Data (TKDD) **5**(2), 1–27 (2011)
24. Ejaz, H., et al.: Covid-19 and comorbidities: deleterious impact on infected patients. J. Infect. Pub. Health **13**(12), 1833–1839 (2020)
25. Fabregat, A., et al.: Reactome graph database: efficient access to complex pathway data. PLoS Comput. Biol. **14**(1), e1005968 (2018)
26. Hamosh, A., Scott, A.F., Amberger, J., Valle, D., McKusick, V.A.: Online mendelian inheritance in man (OMIM). Hum. Mutat. **15**(1), 57–61 (2000)
27. Han, H., et al.: Openhgnn: an open source toolkit for heterogeneous graph neural network. In: Proceedings of the 31st ACM International Conference on Information & Knowledge Management, pp. 3993–3997 (2022)
28. Hu, Z., Dong, Y., Wang, K., Sun, Y.: Heterogeneous graph transformer. In: Proceedings of The Web Conference 2020, pp. 2704–2710 (2020)
29. Huang, J.Y., et al.: The risk of endometrial cancer and uterine sarcoma following endometriosis or pelvic inflammatory disease. Cancers **15**(3), 833 (2023)
30. Ji, H., Wang, X., Shi, C., Wang, B., Yu, P.: Heterogeneous graph propagation network. IEEE Trans. Knowl. Data Eng. (2021)
31. Kherraf, Z.E., et al.: Spink 2 deficiency causes infertility by inducing sperm defects in heterozygotes and azoospermia in homozygotes. EMBO Mol. Med. **9**(8), 1132–1149 (2017)

32. Kibbe, W.A., et al.: Disease ontology 2015 update: an expanded and updated database of human diseases for linking biomedical knowledge through disease data. Nucleic Acids Res. **43**(D1), D1071–D1078 (2015)
33. Kipf, T.N., Welling, M.: Variational graph auto-encoders. arXiv preprint arXiv:1611.07308 (2016)
34. Köhler, S., et al.: The human phenotype ontology project: linking molecular biology and disease through phenotype data. Nucleic Acids Res. **42**(D1), D966–D974 (2014)
35. Kuhn, M., Campillos, M., Letunic, I., Jensen, L.J., Bork, P.: A side effect resource to capture phenotypic effects of drugs. Mol. Syst. Biol. **6**(1), 343 (2010)
36. Lee, J., et al.: Biobert: a pre-trained biomedical language representation model for biomedical text mining. Bioinformatics **36**(4), 1234–1240 (2020)
37. Li, X., Song, D., Leng, S.X.: Link between type 2 diabetes and Alzheimer's disease: from epidemiology to mechanism and treatment. Clin. Intervent. Aging, 549–560 (2015)
38. Liu, F., Liu, B., Sun, C., Liu, M., Wang, X.: Deep learning approaches for link prediction in social network services. In: Lee, M., Hirose, A., Hou, Z.-G., Kil, R.M. (eds.) ICONIP 2013. LNCS, vol. 8227, pp. 425–432. Springer, Heidelberg (2013). https://doi.org/10.1007/978-3-642-42042-9_53
39. Liu, T., Pan, X., Wang, X., Feenstra, K.A., Heringa, J., Huang, Z.: Exploring the microbiota-gut-brain axis for mental disorders with knowledge graphs. J. Artif. Intell. Med. Sci. **1**(3–4), 30–42 (2021)
40. Liu, T., Pan, X., Wang, X., Feenstra, K.A., Heringa, J., Huang, Z.: Predicting the relationships between gut microbiota and mental disorders with knowledge graphs. Health Inf. Sci. Syst. **9**, 1–9 (2021)
41. Liu, W., Yin, L., Wang, C., Liu, F., Ni, Z., et al.: Multitask healthcare management recommendation system leveraging knowledge graph. J. Healthc. Eng. 2021 (2021)
42. Long, Y.: Pre-training graph neural networks for link prediction in biomedical networks. Bioinformatics **38**(8), 2254–2262 (2022)
43. Marantelli, S., Hand, R., Carapetis, J., Beaton, A., Wyber, R.: Severe adverse events following benzathine penicillin g injection for rheumatic heart disease prophylaxis: cardiac compromise more likely than anaphylaxis. Heart Asia **11**(2) (2019)
44. Mutlu, M.F., et al.: Two cases of first onset intrahepatic cholestasis of pregnancy associated with moderate ovarian hyperstimulation syndrome after IVF treatment and review of the literature. J. Obstet. Gynaecol. **37**(5), 547–549 (2017)
45. Patel, R., Guo, Y., Alhudhaif, A., Alenezi, F., Althubiti, S.A., Polat, K.: Graph-based link prediction between human phenotypes and genes. Math. Prob. Eng. 2022 (2021)
46. Pham, C., Dang, T.: Link prediction for biomedical network. In: The 12th International Conference on Advances in Information Technology, pp. 1–5 (2021)
47. Sanyaolu, A., et al.: Comorbidity and its impact on patients with Covid-19. SN Compr. Clin. Med. **2**, 1069–1076 (2020)
48. Schlichtkrull, M., Kipf, T.N., Bloem, P., van den Berg, R., Titov, I., Welling, M.: Modeling relational data with graph convolutional networks. In: Gangemi, A., et al. (eds.) ESWC 2018. LNCS, vol. 10843, pp. 593–607. Springer, Cham (2018). https://doi.org/10.1007/978-3-319-93417-4_38
49. Szklarczyk, D., et al.: The string database in 2011: functional interaction networks of proteins, globally integrated and scored. Nucleic Acids Res. **39**(suppl_1), D561–D568 (2010)

50. Wang, M.Y.: Deep graph library: towards efficient and scalable deep learning on graphs. In: ICLR Workshop on Representation Learning on Graphs and Manifolds (2019)

51. Wang, X., et al.: Heterogeneous graph attention network. In: The World Wide Web Conference, pp. 2022–2032 (2019)

52. Yun, S., Kim, S., Lee, J., Kang, J., Kim, H.J.: Neo-GNNs: neighborhood overlap-aware graph neural networks for link prediction. Adv. Neural. Inf. Process. Syst. **34**, 13683–13694 (2021)

53. Zhang, C., Zhang, H., Yuan, D., Zhang, M.: Deep learning based link prediction with social pattern and external attribute knowledge in bibliographic networks. In: 2016 IEEE International Conference on Internet of Things (iThings) and IEEE Green Computing and Communications (GreenCom) and IEEE Cyber, Physical and Social Computing (CPSCom) and IEEE Smart Data (SmartData), pp. 815–821. IEEE (2016)

54. Zhang, M., Chen, Y.: Link prediction based on graph neural networks. In: Advances in Neural Information Processing Systems, vol. 31 (2018)

Inter-helical Residue Contact Prediction in α-Helical Transmembrane Proteins Using Structural Features

Aman Sawhney[(✉)] [iD], Jiefu Li [iD], and Li Liao [iD]

University of Delaware, Newark, DE 19716, USA
{asawhney,lijiefu,liliao}@udel.edu
https://www.cis.udel.edu/

Abstract. Residue contact maps offer a 2-d, reduced representation of 3-d protein structures and constitute a structural constraint and scaffold in structural modeling. Precise residue contact maps are not only helpful as an intermediate step towards generating effective 3-d protein models, but also useful in their own right in identifying binding sites and hence providing insights about a protein's functions. Indeed, many computational methods have been developed to predict residue contacts using a variety of features based on sequence, physio-chemical properties, and co-evolutionary information. In this work, we set to explore the use of structural information for predicting inter-helical residue contact in transmembrane proteins. Specifically, we extract structural information from a neighborhood around a residue pair of interest and train a classifier to determine whether the residue pair is a contact point or not. To make the task practical, we avoid using the 3-d coordinates directly, instead we extract features such as relative distances and angles. Further, we exclude any structural information of the residue pair of interest from the input feature set in training and testing of the classifier. We compare our method to a state-of-the-art method that uses non-structural information on a benchmark data set. The results from experiments on held out datasets show that the our method achieves above 90% precision for top $L/2$ and L inter-helical contacts, significantly outperforming the state-of-the-art method and may serve as an upper bound on the performance when using non-structural information. Further, we evaluate the robustness of our method by injecting Gaussian normal noise into PDB coordinates and hence into our derived features. We find that our model's performance is robust to high noise levels.

Keywords: PDB database · PDBTM database · Protein structure · Contact map prediction · Transmembrane proteins · Protein Structure modeling · Alpha helix · Machine learning

1 Introduction

Transmembrane (TM) proteins are involved in several critical cell processes such as receptor and signaling transduction pathways, transport of ions & molecules

© The Author(s), under exclusive license to Springer Nature Switzerland AG 2023
I. Rojas et al. (Eds.): IWBBIO 2023, LNBI 13920, pp. 357–371, 2023.
https://doi.org/10.1007/978-3-031-34960-7_25

across membranes and protein targeting [7,28]. About 30 percent of the proteins encoded by the human genome are membrane proteins and a similar proportion is observed in other organisms [4,6,18]. Further, TM proteins are predominantly α-helical [5]. There exists a dramatic gap between protein sequences and experimentally determined structures, especially in the case of TM proteins [39] as crystallizing membrane proteins can be particularly challenging [25].

The 3-d structure of TM proteins is fundamental in understanding their function and to aid drug design [25]. In the absence of 3-d structures, residue contact map offers a reduced 2-d representation of 3-d protein structure [1] which is translation and rotation invariant and, is easily ingested by learning models. Generating a 3-d protein model using 2-d contact maps is an actively researched problem with several promising models. Residue contact predictions have also been used for protein-protein interaction prediction [43,47], accelerating molecular dynamics simulations [31,34] and in predicting binding affinity in docking simulations [13]. Further, a 2-d contact map can be an effective tool in itself, to identify binding sites (especially for inter-helical contact points) and hence provides insights about a protein's functions.

A variety of features based on sequence, physio-chemical properties and co-evolutionary information [39] have been used in literature to predict residue contacts. Evolutionary coupling approaches such as direct coupling analysis [8], EVFold [38] , FreeContact [22] input multiple sequence alignments (MSAs) to learn the constraints between residues and have proven more effective than other approaches. Further, several methods exist that employ supervised learning methods and use predictions from different evolutionary coupling approaches as input features. These include deep learning approaches such as DeepMetaP-SICOV [23], Wang et al. [44] and DeepHelicon [39]. Recent approaches such as Alphafold2 [21] and trRossetta [15] input MSAs directly and employ a deep learning pipeline to predict residue 3-d coordinates, providing an increasingly effective end to end solution.

Recently, it has been shown that topological patterns in the neighborhood of a residue pair in the contact map, such as contact propensity of the surrounding positions, can further enhance the prediction accuracy [30]. Inspired by this development, we set to explore the use of structural information. While the number of available structures at the atomic level remains relatively low, existing structures should be leveraged to provide structural insights and help with extracting relevant features, with the hope that the gained knowledge can then be used to predict contacting residues for proteins with unknown structures. Specifically, from the PDB coordinate data [9], we extract structural features in the neighborhood around a residue pair of interest and train a classifier on these features to determine whether the residue pair is a contact point or not. Since the contact map is an intermediate step towards finding the 3-d structure, in order to make the task more practical and to avoid being circular in logic, we do not use the 3-d coordinates directly, instead we extract derivative features such as relative distances and angles. Also, we exclude any structural information of the residue pair of interest from the input feature set in training and testing of the classifier. We compare our method on a benchmark data set to a state-of-the-art method that uses non-structural information, including several co-evolutionary models as input features to a Residual neural network based deep learning pipeline [39].

The results of our experiments on held out datasets show that the our method achieved above 90% precision for top $L/2$ and L inter-helical contacts, significantly outperforming the state-of-the-art method. We emphasize that these results provide evidence that neighborhood structural information is information rich and may serve as an upper bound on a classifier's performance when using non-structural information. Further, we evaluate the robustness of our method by injecting Gaussian normal noise into the 3-d coordinates from the PDB files, analogous to noisy crystallography output, which percolates into our derived features. We find that our model's performance is robust to high noise levels, and is pessimistically reliable up to $2 A°$ of coordinate noise. Since, both experimentally determined and predicted 3-d structures can be noisy, our robust method trained using quality experimental structures could be used to improve them.

2   Materials and Methods

2.1   Dataset

In this work, we use a dataset of α-helical TM proteins created by Sun et al. [39]. It was constructed using 5606 α-helical TM proteins chains from the PDBTM database [27,41,42], each with resolution better than 3.5 A°. It was made non-redundant at 23% sequence identity level to ensure that protein chains didn't share structural similarity. The sequences were filtered for a maximum TM score [46] of 0.4. There are a total of 222 protein chains in the dataset, with the number of TM helices varying between 2 to 17. The dataset is divided into three sub-datasets a) TRAIN - a training set of 165 sequences b) TEST - a held out set of 57 sequences and c) PREVIOUS - a held out set of 44 sequences [19,45]. For each protein chain, the dataset includes the protein sequence, annotations of which residue pair positions are contacting, which residue positions are in the TM zone, and the 3-d coordinates for heavy atoms of each residue in PDB atomic coordinate format. [9] The model predictions for the TEST and PREVIOUS datasets are included in the dataset.

Based on the atomic structures, two residues are deemed to be in contact if the distance between their heavy atoms is below a certain threshold. Limiting the distance between C_α atoms or C_β atoms for the residue pair to be less 8 A° are some of the contact definitions that have been used in literature [19]. Our dataset defines two residues to be in contact (contact point), if the least distance between any pair of their heavy atoms is less than 5.5 A° and if they are sequence separated by a minimum of 5 residues [39].

Using the aforementioned criteria, several sequences were removed from the the original dataset. Specifically, sequences '5yi2B', '5lkiA', '5bw8D' in the TRAIN dataset have no positive inter-helical contacts, which would have led to Recall score being undefined, consequently we removed them from the

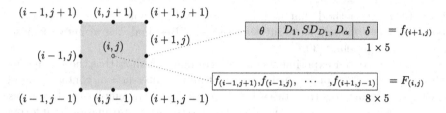

Fig. 1. Neighborhood feature vector - Features extracted from a 3×3 neighborhood window (excluding the center) around a residue pair of interest (i, j) for a feature vector of length 40.

dataset. Further, for sequences '4p79A', '4qtnA', '4f35B' in the TEST dataset and '2rh1A', '3ukmA', '3m73A', '3m7lA in the PREVIOUS dataset, some of the residue positions annotated to be in TM zone don't match with the positions that Sun et. al [39] predicted on, hence we removed these sequences from the dataset as well. With a final total of 162 sequences in the TRAIN dataset, 40 sequences in the PREVIOUS dataset and 54 sequences in the TEST dataset. These changes and contact ratio (CR) (for residue pair positions within TM zones) for all datasets are summarized in table 1, where CR is defined as Eq. 1.

$$CR = \frac{\#contact\ points}{\#residue\ pair\ positions} \tag{1}$$

Table 1. Dataset statistics - Total number of sequences and contact ratio for TRAIN, TEST and PREVIOUS datasets.

Dataset	#Sequences	#Filtered Sequences	$CR \times 100$
TRAIN	165	162	2.10
TEST	57	54	2.07
PREVIOUS	44	40	1.95

2.2 Feature Extraction

We explore the use of structural features derived from coordinate data for residue contact prediction. For a residue pair position (i, j), where i, j are amino acid sequence positions, s.t. $|i - j| > 5$ and i and j are on separate helices (inter-helical), we select a neighborhood window of size 3×3 around it. For each of the eight positions around (i, j) (excluding the center (i, j)), we construct a feature vector of length 5 - consisting of the inter-helical tilt angle, relative residue distances and relative residue angle. By concatenating the five features for each of the eight neighboring positions, we obtain a feature vector of length 40 (8×5) for any inter-helical residue pair position of interest. This process is illustrated in Fig. 1. The definition of residue contact (hence, positive label) is

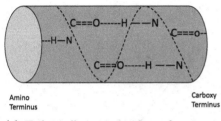

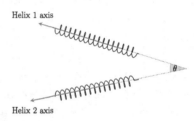

Helix 1 axis

Helix 2 axis

(a) Toilet roll representation of main chain hydrogen bonding in alpha-helix, adapted from [12]

(b) Inter-helical tilt angle θ between the two helical axes

Fig. 2. Inter helical tilt angle

based on minimum distance between the residue pair's heavy atoms, while we use residue pair's neighborhood structural information (distance functions used are different as well) to predict whether it is in contact. In the following sections, we describe the extracted features in some detail.

Inter-helical Tilt Angle (θ) - For any pair of residues, inter-helical tilt angle is the angle between the helices the residues reside on [29]. In an α-helix, each main-chain $C = O$ and $N - H$ group is hydrogen bonded to a peptide bond four residues away i.e. $O(i)$ to $N(i+4)$ (where i is the ith residue), giving it a stable arrangement [12].

Further, the peptide planes are roughly parallel with the helical axis and the dipoles within the helix are aligned, i.e. all $C = O$ point in the same direction and all $N - H$ point in the other direction, while the side chains point outward from the helical axis (generally oriented towards the amino-terminal) [12]. This bond pattern is depicted in Fig. 2a.

The helical axis orientation is computed by averaging the direction of $C(i) = O(i) - N(i+4)$ for all residues in the helix. The angle between the axes of two helices is the inter-helical tilt angle. Figure 2b shows the inter-helical tilt angle between two helical axes. We use the Pymol package for these computations [35–37].

Relative Residue Distance - We define three relative residue distance features:-

1. **D_1 distance (mean relative residue distance)** [24, 32] - For any pair of residues, we compute the mean euclidean distance between all pairs of their heavy atoms. Let $\{A_x^1, \dots A_x^M\}$ and $\{A_y^1, \dots A_y^N\}$ be the euclidean (3-d) coordinates of the atoms of residues R_x and R_y respectively. Further, let $dist(i, j)$ denote the euclidean distance between two 3-d coordinates i and j then mean residue distance between R_x and R_y is

$$D_1(R_x, R_y) = \frac{1}{MN} \sum_{i=1}^{M} \sum_{j=1}^{N} dist(A_x^i, A_y^j) \tag{2}$$

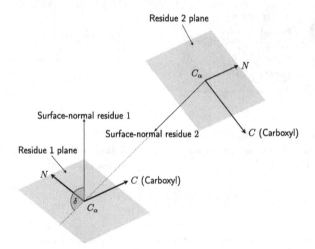

Fig. 3. Relative residue angle (δ) - Angle between the Surface-normals to the residue planes

2. D_1 deviation (relative residue distance deviation) [24,32] - For any pair residues, we compute the standard deviation of the euclidean distances between all pairs of their heavy atoms. So relative residue distance deviation between residues R_x and R_y is

$$SD_{D_1}(R_x, R_y) = \sqrt{\left\{\frac{1}{MN}\sum_{i=1}^{M}\sum_{j=1}^{N}[(dist(A_x^i, A_y^j) - D_1(R_x, R_y))^2]\right\}} \tag{3}$$

3. D_α (C_α distance) [24,32] - For any pair of residues, we compute the euclidean distance between their alpha carbons. Let $atom(R, k)$ be a function that returns the kth atom for residue R. Further, let $atom(R_x, i) = C_\alpha$ and $atom(R_y, j) = C_\alpha$ then

$$D_\alpha(R_x, R_y) = dist(A_x^i, A_y^j) \tag{4}$$

Relative residue angle (δ) - For any residue, we define its plane as formed by the two vectors: one is between C_α and N atom and the other is between C_α and C atom of the carboxyl group. We define the relative residue angle as the absolute angle between the surface-normals of the residue planes for these residues [32]. The angle is represented as δ in Fig. 3.

2.3   Classification Experiment

The prediction of whether a pair of inter-helical TM residue positions are in contact or not is handled as a supervised binary classification problem. As stated previously, we only consider residue positions that are sequence separated by at least 5 residue positions. Our training set consists of 162 sequences from the

TRAIN dataset, while we test on PREVIOUS (40 sequences) and TEST (54 sequences) datasets. For each inter-helical TM residue pair position, we construct a neighborhood feature vector of length 40 (described in Sect. 2.2).

As is common practice, these features are first normalized to a $[0,1]$ scale before being used for classification, such that $f^t_{i_{scaled}} = \frac{f^t_i - min(f_i)}{max(f_i) - min(f_i)}$ where f^t_i is the tth sample for the feature f_i, $max(.)$ and $min(.)$ compute the maximum and minimum observed value for the feature f_i and $f^t_{i_{scaled}}$ represents the scaled value of tth sample for the feature f_i.

We use a random forest classifier (100 trees) [10,33] to learn using the training set and assess training performance with 5 fold cross validation [17,20,26]. In each fold, 80% of randomly selected training sequences are used for training and 20% are held out for validation. Finally, we retrain on the full TRAIN dataset and assess the performance of this trained classifier on the PREVIOUS and TEST datasets.

2.4   Performance Metrics

We evaluate the performance of our classifier at all thresholds using :-

1. AUC-ROC - The area under the Receiver operating characteristic curve computed using the trapezoidal rule. [3,16,33]
2. AUPR - The Area under the precision-recall curve computed using the trapezoidal rule. [14,33]
3. Average precision - Average precision summarizes the precision recall curve as weighted mean of precision at each threshold, wherein the increase in recall from the previous threshold is used as the weight. [2,33]

$$Average\ Precision = \sum_n (R_n - R_{n-1})P_n \tag{5}$$

where P_n and R_n are precision and recall at the nth threshold.

Since the dataset is skewed – with a much larger proportion of negative data, it has been reported [14] that in such cases Average precision and AUPR present a more realistic picture as compared to AUC-ROC which tends to be more optimistic .

Further, we also evaluate our performance using

$$Precision = \frac{TP}{TP + FP} \quad \& \quad Recall = \frac{TP}{TP + FN} \tag{6}$$

where TP is the number of true positives, FP is the number of false positives and FN is the number of false negatives at a particular threshold. Precision and recall were computed for the top L, $L/2$, $L/5$, $L/10$ residue pair predictions where L denoted the total concatenated length of the TM helices for a sequence. For all metrics, we report the mean value across all sequences.

2.5   Noise Injection

In order to evaluate the robustness of the classifier we measure its prediction performance when noisy features are used i.e. when noisy experimental structures or predicted structures are available during testing. For any atom in a protein chain, we assume that the injected Gaussian normal noise deviates its resolved position by $dA°$. Further, we assume that the noise is isotropic in orientation, i.e. equal for x,y and z coordinates.

$$(d_x)^2 + (d_y)^2 + (d_z)^2 = d^2$$

$$\implies d_x = d_y = d_z = \frac{d}{\sqrt{3}} \tag{7}$$

The tuple (d_x, d_y, d_z) is point-wise added to the coordinate of an atom to generate its new noisy coordinates. This is done for all atomic coordinates in our datasets. We vary the deviation as $d = 0.5 \times x$, s.t. $x \in [1, 7]$ & $x \in \mathcal{I}$.

Since the definition of a contact point is dependent on the distance between the pair of residues, understandably, addition of noise into the coordinates can alter the distances between the residues, and hence may affect how a pair of residues are annotated i.e. as a contact point or as a non-contact point. We refer to a label of contact that was generated based on the noisy data as noisy label, in contrast to the original label based on PDB data which we refer to as pristine label. So, we monitor and assess the impact of noise injection on the contacts, using the following metrics.

Mismatch proportion (MP) - For any dataset, we compute the proportion of annotated contacts that flips as a result of noise injection i.e., contact point to non-contact point or vice versa. Then MP is defined in Eq. 8. We report the effect of noise injection measured in terms of MP for TRAIN, TEST and PREVIOUS datasets at all noise deviations levels in Table 2. It can be seen that, as expected, there is a monotonic increase in MP with an increase in noise deviation for all datasets, from around 0.3% at noise deviation of 0.5 A° to around 5% at a noise deviation of 3.5 A°. The MP values are generally highest for TRAIN dataset, followed by TEST and PREVIOUS datasets; in accordance with the dataset sizes.

$$MP = \frac{\#contact\ points\ flipped+\ \#non-contact\ points\ flipped}{\#residue\ pair\ positions} \tag{8}$$

Per sequence mismatch proportion ($Avg\ MP$) - For each dataset, we compute the average mismatch proportion across all sequences. Let MP_i be the mismatch proportion of a sequence in the dataset and S be the number of sequences in the dataset, then $Avg\ MP$ is defined in Eq. 9. We report the effect of noise injection measured in terms of $Avg\ MP$ for TRAIN, TEST and PREVIOUS datasets at all noise deviations levels in table 2. Again, there is a monotonic increase in $Avg\ MP$ with an increase in noise deviation for all datasets. TRAIN dataset experiences the highest change, followed by TEST and PREVIOUS datasets; in accordance with the dataset sizes.

$$Avg\ MP = \frac{1}{S} \sum_{i=1}^{S} MP_i \tag{9}$$

Change in contact ratio (ΔCR) - In addition to MP, we also monitor how injected noise affects the number of contact points in particular. In Eq. 1, we defined CR, which is the proportion of relevant residue pair positions of a dataset that are in contact. Now, ΔCR is defined as the change in CR caused as a result of noise injection and is computed using Eq. 10, where $CR_{original}$ is the CR for the non-noisy dataset and CR_{noisy} is the CR after noise is injected. We report the effect of noise injection measured in terms of ΔCR for TRAIN, TEST and PREVIOUS datasets at all noise deviations levels in Fig. 4. Like MP and $AvgMP$, ΔCR also monotonically increases with an increase noise deviation for all datasets, from around 0.12% at a noise deviation of 0.5 $A°$ to about 5% at a noise deviation level of 3.5 $A°$. The change in positive contact ratio is generally highest for TRAIN dataset, followed by TEST and PREVIOUS datasets; in accordance with the dataset sizes. At first glance, though, this increase of the positive contact ratio seems to be strange, as one would expect a symmetric (such as Gaussian) random noise to have the equal chance to flip a positive to negative as it flips a negative to a positive, therefore leaves the CR unchanged. That is only true when the data is balanced number of positive and negative examples. However, our datasets are highly skewed; there are far fewer contact points, i.e., residue pairs at distances less than 5.5 $A°$ from each other. Since the injected noise is independent of the distance between a residue pair, more non-contact points would flip simply because they are more in number. We illustrate this in Table 3, which lists the percentage of total contacts that would flip when the distance is altered by $\sigma, 2\sigma$ & 3σ at various noise deviation levels. The number of flips observed from contact point to non-contact point maxes out as we go down the table, but the number of non-contact point to contact point flips increase.

$$\Delta CR = CR_{noisy} - CR_{original} \tag{10}$$

Similar to the classification experiment for data without noise, a random forest classifier (100 trees) is trained using features (described in Sect. 2.2) derived from noisy datasets. However, we now have access to both the *pristine labels* i.e., labels generated from dataset with no noise and the *noisy labels* i.e., the labels generated from dataset with injected noise. To properly assess the robustness of our method and at the same time to make the situation realistic, we device the following scheme. First, we train our classifier on the TRAIN dataset assuming no noise and therefore using the pristine labels. Second, for testing (on both TEST and PREVIOUS datasets), we use features generated using datasets with noise injected and measure the classifier's performance using both pristine and noisy labels. Pristine labels help us assess the true performance of the classifier, whereas the noisy labels can tell us how high the noise level can go when classification (using pristine labels) is no longer tenable - when classification with pristine labels is seemingly outperformed by the classification with noisy labels.

Table 2. Effect of noise injection on contact labels, measured in terms of MP and Avg MP for TRAIN, TEST and PREVIOUS datasets.

Noise deviation (in $A°$)	Dataset	MP	Avg MP
0.5	TRAIN	0.0029	0.0045 ± 0.004
	TEST	0.0030	0.0039 ± 0.003
	PREVIOUS	0.0028	0.0034 ± 0.002
1.0	TRAIN	0.0072	0.0106 ± 0.007
	TEST	0.0073	0.0095 ± 0.005
	PREVIOUS	0.0066	0.0083 ± 0.004
1.5	TRAIN	0.0145	0.0206 ± 0.013
	TEST	0.0131	0.0171 ± 0.009
	PREVIOUS	0.0123	0.0155 ± 0.007
2.0	TRAIN	0.0212	0.0322 ± 0.021
	TEST	0.0211	0.0272 ± 0.119
	PREVIOUS	0.0194	0.0241 ± 0.010

Noise deviation (in $A°$)	Dataset	MP	Avg MP
2.5	TRAIN	0.0308	0.0469 ± 0.028
	TEST	0.0305	0.0396 ± 0.018
	PREVIOUS	0.0283	0.0348 ± 0.139
3.0	TRAIN	0.0420	0.0646 ± 0.039
	TEST	0.0412	0.0539 ± 0.024
	PREVIOUS	0.0383	0.0471 ± 0.019
3.5	TRAIN	0.0538	0.0808 ± 0.045
	TEST	0.0530	0.0680 ± 0.029
	PREVIOUS	0.0493	0.0607 ± 0.023

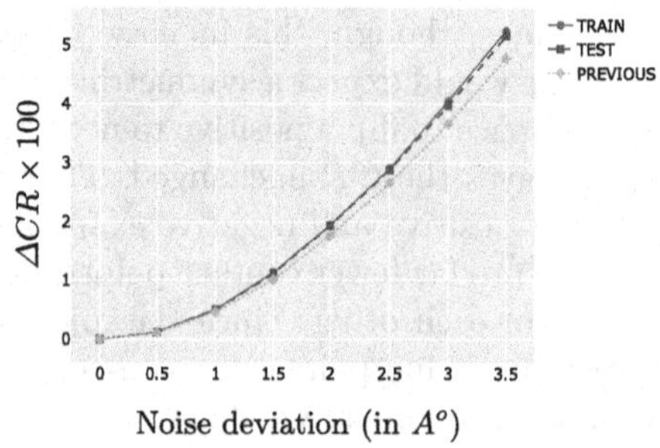

Noise deviation	$\Delta CR \times 100$		
	TRAIN	TEST	PREVIOUS
0.5	0.12	0.12	0.11
1.0	0.50	0.50	0.46
1.5	1.13	1.11	1.01
2.0	1.94	1.93	1.76
2.5	2.90	2.88	2.66
3.0	4.05	3.97	3.66
3.5	5.25	5.16	4.78

(a)   (b)

Fig. 4. Change in contact ratio with variation in noise deviation for TRAIN, TEST, PREVIOUS datasets.

3   Results and Discussion

3.1   Classification

We report the classification performance in Table 4. AUPR, Average precision and AUC-ROC scores for TEST, PREVIOUS datasets and average 5 fold cross validated scores for TRAIN dataset are reported in Table 4a. In terms of AUPR, there is a significant gain of 38.81% and 39.36% over DeepHelicon for the TEST and PREVIOUS datasets; in terms of AUC-ROC, the performance is improved by 7.12% and 7.45% for TEST and PREVIOUS datasets respectively.

The performance in terms of precision and recall scores for top $L, L/2, L/5, L/10$, where L is the combined helical length of a sequence, are computed for all sequences. We report the precision and recall scores averaged across all sequences in Table 4b and Table 4c respectively. The classifier achieves near perfect precision for top $L/10$ and $L/5$ predictions for both the TEST and

Table 3. Percentage proportion of Max possible contact flips (contact to non-contact or non-contact to contact) computed till 3 noise standard deviations from contacting definition distance ($D = 5.5$ A°)

Noise deviation σ	Max contact to non-contact flips (%)			Max non-contact to contact flips(%)			ΔCR
	$D-3\sigma$	$D-2\sigma$	$D-\sigma$	$D+\sigma$	$D+2\sigma$	$D+3\sigma$	
0.5	1.034	0.642	0.356	0.561	1.411	2.481	0.12
1.0	2.074	1.697	0.642	1.411	3.658	6.144	0.50
1.5	2.099	2.074	1.034	2.481	6.144	10.424	1.13
2.0	2.099	2.099	1.697	3.658	8.893	15.635	1.94
2.5	2.099	2.099	1.997	4.876	12.064	21.362	2.90
3.0	2.099	2.099	2.074	6.144	15.635	27.22	4.05
3.5	2.099	2.099	2.097	7.458	19.459	33.278	5.25

(a) TRAIN dataset

Noise deviation σ	Max contact to non-contact flips (%)			Max non-contact to contact flips(%)			ΔCR
	$D-3\sigma$	$D-2\sigma$	$D-\sigma$	$D+\sigma$	$D+2\sigma$	$D+3\sigma$	
0.5	1.016	0.623	0.350	0.577	1.425	2.491	0.12
1.0	2.056	1.673	0.623	1.425	3.661	6.141	0.50
1.5	2.069	2.056	1.016	2.491	6.141	10.334	1.11
2.0	2.069	2.069	1.673	3.661	8.841	15.564	1.93
2.5	2.069	2.069	1.98	4.895	11.986	21.218	2.88
3.0	2.069	2.069	2.056	6.141	13.74	27.096	3.97
3.5	2.069	2.069	2.068	7.449	19.309	33.307	5.16

(b) TEST dataset

Noise deviation σ	Max contact to non-contact flips (%)			Max non-contact to contact flips (%)			ΔCR
	$D-3\sigma$	$D-2\sigma$	$D-\sigma$	$D+\sigma$	$D+2\sigma$	$D+3\sigma$	
0.5	0.981	0.586	0.319	0.530	1.320	2.332	0.11
1.0	1.948	1.613	0.586	1.320	3.454	5.779	0.46
1.5	1.948	1.944	0.981	2.332	5.779	9.725	1.01
2.0	1.948	1.948	1.613	3.454	8.315	14.610	1.76
2.5	1.948	1.948	1.884	4.591	11.253	19.939	2.66
3.0	1.948	1.948	1.944	5.779	14.610	25.497	3.66
3.5	1.948	1.948	1.948	7.005	18.152	31.347	4.78

(c) PREVIOUS dataset

Table 4. Classification performance of constructed features compared with Deephelicon

Classifier	Dataset	AUPR	Average precision	AUC-ROC
RF-100	TRAIN (Avg. 5 folds)	0.9289 ± 0.0050	0.9278 ± 0.0048	0.9968 ± 0.0005
RF-100	TEST	0.9383	0.9369	0.9975
Deephelicon	TEST	0.5502	0.5523	0.9263
RF-100	PREVIOUS	0.9395	0.9382	0.9979
Deephelicon	PREVIOUS	0.5459	0.5480	0.9234

(a) Classification performance measured at all thresholds.

Classifier	Dataset	$L/10$	$L/5$	$L/2$	L
RF-100	TRAIN (Avg. 5 folds)	0.9842 ± 0.0206	0.9713 ± 0.0325	0.9292 ± 0.0359	0.8144 ± 0.0308
RF-100	TEST	1	1	0.9834	0.9078
Deephelicon	TEST	0.8784	0.8403	0.7526	0.6200
RF-100	PREVIOUS	1	0.9922	0.9673	0.9177
Deephelicon	PREVIOUS	0.9049	0.8693	0.7805	0.6397

(b) Classification performance measured in terms of Precision at L thresholds

Classifier	Dataset	$L/10$	$L/5$	$L/2$	L
RF-100	TRAIN (Avg. 5 folds)	0.1366 ± 0.0342	0.2334 ± 0.0277	0.4820 ± 0.0212	0.7454 ± 0.0128
RF-100	TEST	0.0865	0.1686	0.4000	0.6931
Deephelicon	TEST	0.0718	0.1322	0.2858	0.4621
RF-100	PREVIOUS	0.0929	0.1700	0.3649	0.6382
Deephelicon	PREVIOUS	0.0686	0.120	0.2770	0.4341

(c) Classification performance measured in terms of Recall at L thresholds

PREVIOUS datasets. In terms of precision, for the TEST dataset we achieve an increment of 23.08% for top $L/2$ and 28.78% for top L predictions. While, for the PREVIOUS dataset we gain 18.68% and 27.80% for top $L/2$ and top L predictions respectively.

3.2   Robustness

In Fig. 5, we report the classification performance on the TEST and PREVIOUS datasets when noise is injected at various levels. As it can be seen from the performance results, even at the highest noise level of 3.5 $A°$, the trained classifier still achieves an AUPR of about 0.65 assessed with the pristine labels, comfortably outperforming DeepHelicon [39]. In a real world situation, when the data are noisy we would generally not have access to the pristine labels, rather the performance of a classifier would typically be measured using the noisy labels. As shown in Fig. 5, below a noise level of 2 $A°$, the performance of our classifier is underestimated with the noisy labels. In other words, we know the classifier actually performs better than what is measured with the noisy labels. When the noise is a level of 2 $A°$, the classifier performance measured using noisy labels becomes an overestimate, which is likely attributed to a higher ΔCR, as explained in Sect. 2.5. So the crossing point at the noise level 2 $A°$ of the plots for the AUPR of using the pristine labels vs the noisy labels suggests a level of noise tolerance of the classifier. In our setup, at σ noise deviation level, we expect the majority of the atoms' coordinates to shift at most by a distance of $\sigma A°$ (since noise is assumed to be Gaussian normal). These noise levels are a proxy for X-ray resolution, however the relationship between resolution and coordinate error is not straightforward and true coordinate error is likely much lower as reported in [11,40]. This then bolsters our case, since then for reasonable resolutions our classifier's performance estimated using noisy or pristine labels is very close and not overestimated.

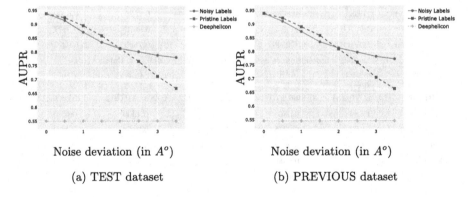

(a) TEST dataset   (b) PREVIOUS dataset

Fig. 5. Classifier performance in terms of AUPR using noisy and pristine labels

4   Conclusion

Since experimentally determined atomic structures are still scarce, a plethora of literature is devoted to developing methods that can predict residue contact based on protein primary structure. In this work, we explored how features can be derived from the atomic structures in the neighborhood of a residue pair and used for predicting inter-helical residue contact for α-helical TM proteins. We demonstrated that the trained classifier achieved near perfect precision for both TEST and PREVIOUS datasets for top $L/10$ and $L/5$ predictions, significantly outperforming the state-of-the-art [39] which relies on co-evolutionary information based features as input. The robustness experiments using injected Gaussian noise into the coordinates showed that, up to a practical level of 2 A°, the classifier's performance is only marginally affected and is underestimated. This work affirms that the structure of the residue' s sequence neighborhood is very informative and that the derived features are generalizable across sequences. It is conceivable that the performance of the classifier from these structurally derived features provides an useful upper bound for what a classifier can achieve without using structure based information and that noisy and predicted structures can be improved using these features.

Acknowledgment. Support from the University of Delaware CBCB Bioinformatics Core Facility and use of the BIOMIX compute cluster was made possible through funding from Delaware INBRE (NIH NIGMS P20 GM103446), the State of Delaware, and the Delaware Biotechnology Institute. The authors would also like to thank the National Science Foundation (NSF-MCB1820103), which partly supported this research.

References

1. Contact maps (molecular biology). https://what-when-how.com/molecular-biology/contact-maps-molecular-biology/. Accessed 26 Jan 2022
2. Information retrieval - wikipedia. https://en.wikipedia.org/w/index.php?title=Information_retrieval&oldid=793358396#Average_precision. Accessed 26 Jan 2022
3. Receiver operating characteristic - Wikipedia. https://en.wikipedia.org/wiki/Receiver_operating_characteristic. Accessed 26 Jan 2022
4. Scientists alter membrane proteins to make them easier to study - sciencedaily. https://www.sciencedaily.com/releases/2018/08/180828104043.htm. Accessed 26 Jan 2022
5. Albers, R.W.W.: Cell membrane structures and functions. In: Basic Neurochemistry, pp. 26–39. Elsevier (2012)
6. Almén, M.S., Nordström, K.J., Fredriksson, R., Schiöth, H.B.: Mapping the human membrane proteome: a majority of the human membrane proteins can be classified according to function and evolutionary origin. BMC Biol. **7**(1), 1–14 (2009)
7. Attwood, M.M., Schiöth, H.B.: Characterization of five transmembrane proteins: with focus on the tweety, sideroflexin, and YIP1 domain families. Front. Cell Dev. Biol. **9**, 1950 (2021)

8. Baldassi, C., et al.: Fast and accurate multivariate gaussian modeling of protein families: predicting residue contacts and protein-interaction partners. PLoS One **9**(3), e92721 (2014)

9. Berman, H.M., Battistuz, T., Bhat, T.N., Bluhm, W.F., Bourne, P.E., Burkhardt, K., Feng, Z., Gilliland, G.L., Iype, L., Jain, S., et al.: The protein data bank. Acta Crystallogr. D Biol. Crystallogr. **58**(6), 899–907 (2002)

10. Breiman, L.: Random forests. Mach. Learn. **45**(1), 5–32 (2001)

11. Brünger, A.T.: X-ray crystallography and NMR reveal complementary views of structure and dynamics. Nat. Struct. Biol. **4**, 862–865 (1997)

12. Cooper, J.: Alpha-Helix geometry part. 2 – cryst.bbk.ac.uk (1995). https://www.cryst.bbk.ac.uk/PPS95/course/3_geometry/helix2.html. Accessed 25 Jan 2022

13. Dago, A.E., Schug, A., Procaccini, A., Hoch, J.A., Weigt, M., Szurmant, H.: Structural basis of histidine kinase autophosphorylation deduced by integrating genomics, molecular dynamics, and mutagenesis. Proc. Natl. Acad. Sci. **109**(26), E1733–E1742 (2012)

14. Davis, J., Goadrich, M.: The relationship between precision-recall and roc curves. In: Proceedings of the 23rd International Conference on Machine Learning, pp. 233–240 (2006)

15. Du, Z., et al.: The trRosetta server for fast and accurate protein structure prediction. Nat. Protoc. **16**(12), 5634–5651 (2021)

16. Fawcett, T.: An introduction to ROC analysis. Pattern Recogn. Lett. **27**(8), 861–874 (2006)

17. Friedman, J., Hastie, T., Tibshirani, R., et al.: The Elements of Statistical Learning. Springer Series in Statistics, vol. 1. Springer, New York (2001). https://doi.org/10.1007/978-0-387-84858-7

18. Frishman, D., Mewes, H.W.: Protein structural classes in five complete genomes. Nat. Struct. Biol. **4**(8), 626–628 (1997)

19. Hönigschmid, P., Frishman, D.: Accurate prediction of helix interactions and residue contacts in membrane proteins. J. Struct. Biol. **194**(1), 112–123 (2016)

20. James, G., Witten, D., Hastie, T., Tibshirani, R.: An Introduction to Statistical Learning, vol. 112. Springer, Heidelberg (2013)

21. Jumper, J., Evans, R., Pritzel, A., Green, T., Figurnov, M., Ronneberger, O., Tunyasuvunakool, K., Bates, R., Žídek, A., Potapenko, A., et al.: Highly accurate protein structure prediction with alphafold. Nature **596**(7873), 583–589 (2021)

22. Kaján, L., Hopf, T.A., Kalaš, M., Marks, D.S., Rost, B.: FreeContact: fast and free software for protein contact prediction from residue co-evolution. BMC Bioinform. **15**(1), 1–6 (2014)

23. Kandathil, S.M., Greener, J.G., Jones, D.T.: Prediction of interresidue contacts with DeepMetaPSICOV in CASP13. Proteins Struct. Funct. Bioinform. **87**(12), 1092–1099 (2019)

24. Karlin, S., Zuker, M., Brocchieri, L.: Measuring residue association in protein structures possible implications for protein folding. J. Mol. Biol. **239**(2), 227–248 (1994)

25. Kermani, A.A.: A guide to membrane protein X-ray crystallography. FEBS J. **288**(20), 5788–5804 (2021)

26. Kohavi, R., et al.: A study of cross-validation and bootstrap for accuracy estimation and model selection. In: Ijcai, Montreal, Canada, vol. 14, pp. 1137–1145 (1995)

27. Kozma, D., Simon, I., Tusnady, G.E.: PDBTM: protein data bank of transmembrane proteins after 8 years. Nucleic Acids Res. **41**(D1), D524–D529 (2012)

28. Lagerström, M.C., Schiöth, H.B.: Structural diversity of G protein-coupled receptors and significance for drug discovery. Nat. Rev. Drug Discovery **7**(4), 339–357 (2008)

29. Lee, H.S., Choi, J., Yoon, S.: QHELIX: a computational tool for the improved measurement of inter-helical angles in proteins. Protein. J. **26**(8), 556–561 (2007)
30. Li, J., Sawhney, A., Lee, J.Y., Liao, L.: Improving inter-helix contact prediction with local 2D topological information (2023)
31. Lubecka, E.A., Liwo, A.: Introduction of a bounded penalty function in contact-assisted simulations of protein structures to omit false restraints. J. Comput. Chem. **40**(25), 2164–2178 (2019)
32. Mahbub, S., Bayzid, M.S.: EGRET: edge aggregated graph attention networks and transfer learning improve protein-protein interaction site prediction. bioRxiv, pp. 2020–11 (2021)
33. Pedregosa, F., et al.: Scikit-learn: machine learning in Python. J. Mach. Learn. Res. **12**, 2825–2830 (2011)
34. Raval, A., Piana, S., Eastwood, M.P., Shaw, D.E.: Assessment of the utility of contact-based restraints in accelerating the prediction of protein structure using molecular dynamics simulations. Protein Sci. **25**(1), 19–29 (2016)
35. Schrödinger, LLC: The AxPyMOL molecular graphics plugin for Microsoft PowerPoint, version 1.8 (2015)
36. Schrödinger, LLC: The JyMOL molecular graphics development component, version 1.8 (2015)
37. Schrödinger, LLC: The PyMOL molecular graphics system, version 1.8 (2015)
38. Sheridan, R., et al.: EVfold. org: evolutionary couplings and protein 3D structure prediction. biorxiv, p. 021022 (2015)
39. Sun, J., Frishman, D.: DeepHelicon: accurate prediction of inter-helical residue contacts in transmembrane proteins by residual neural networks. J. Struct. Biol. **212**(1), 107574 (2020)
40. Torda, A.: Powerpoint presentation. https://www.zbh.uni-hamburg.de/forschung/bm/lehre/downloads/ws1718/67-104/1-genauigkeit.pdf. Accessed 07 Apr 2022
41. Tusnády, G.E., Dosztányi, Z., Simon, I.: Transmembrane proteins in the protein data bank: identification and classification. Bioinformatics **20**(17), 2964–2972 (2004)
42. Tusnády, G.E., Dosztányi, Z., Simon, I.: PDB_TM: selection and membrane localization of transmembrane proteins in the protein data bank. Nucleic Acids Res. **33**(suppl_1), D275–D278 (2005)
43. Vangone, A., Bonvin, A.M.: Contacts-based prediction of binding affinity in protein-protein complexes. Elife **4**, e07454 (2015)
44. Wang, S., Sun, S., Li, Z., Zhang, R., Xu, J.: Accurate de novo prediction of protein contact map by ultra-deep learning model. PLoS Comput. Biol. **13**(1), e1005324 (2017)
45. Wang, X.F., Chen, Z., Wang, C., Yan, R.X., Zhang, Z., Song, J.: Predicting residue-residue contacts and helix-helix interactions in transmembrane proteins using an integrative feature-based random forest approach. PLoS One **6**(10), e26767 (2011)
46. Xu, J., Zhang, Y.: How significant is a protein structure similarity with TM-score = 0.5? Bioinformatics **26**(7), 889–895 (2010)
47. Zhang, H., et al.: Evaluation of residue-residue contact prediction methods: from retrospective to prospective. PLoS Comput. Biol. **17**(5), e1009027 (2021)

Degree-Normalization Improves Random-Walk-Based Embedding Accuracy in PPI Graphs

Luca Cappelletti[1] , Stefano Taverni[1], Tommaso Fontana[1] ,
Marcin P. Joachimiak[2] , Justin Reese[2] , Peter Robinson[3] ,
Elena Casiraghi[1,2,4] , and Giorgio Valentini[1,4,5(✉)]

[1] AnacletoLab, Department of Computer Science "Giovanni degli Antoni",
Universitá degli Studi di Milano, 20133 Milan, Italy
valentini@di.unimi.it
[2] Environmental Genomics and Systems Biology Division,
Lawrence Berkeley National Laboratory, Berkeley, CA, USA
[3] The Jackson Laboratory for Genomic Medicine, Farmington, USA
[4] CINI, Infolife National Laboratory, Rome, Italy
[5] ELLIS, European Laboratory for Learning and Intelligent Systems,
Tuebingen, Germany

Abstract. Among the many proposed solutions in graph embedding,
traditional random walk-based embedding methods have shown their
promise in several fields. However, when the graph contains high-degree
nodes, random walks often neglect low- or middle-degree nodes and
tend to prefer stepping through high-degree ones instead. This results
in random-walk samples providing a very accurate topological representation of neighbourhoods surrounding high-degree nodes, which contrasts
with a coarse-grained representation of neighbourhoods surrounding middle and low-degree nodes. This in turn affects the performance of the
subsequent predictive models, which tend to overfit high-degree nodes
and/or edges having high-degree nodes as one of the vertices. We propose
a solution to this problem, which relies on a degree normalization approach. Experiments with popular RW-based embedding methods applied
to edge prediction problems involving eight protein-protein interaction
(PPI) graphs from the STRING database show the effectiveness of the
proposed approach: degree normalization not only improves predictions
but also provides more stable results, suggesting that our proposal has
a regularization effect leading to a more robust convergence.

1 Introduction

State of the art, graph representation learning and random-walk (RW)-based
embedding methods [6,10,11] for edge-prediction have been successfully applied
in several areas of computational biology and network medicine [7,16].

Given a graph $\mathcal{G} = (V, E)$, where V is the set of nodes ($|V| = N$) and E is
the set of (directed) edges, RW-based edge prediction models start by computing

I. Rojas et al. (Eds.): IWBBIO 2023, LNBI 13920, pp. 372–383, 2023.
https://doi.org/10.1007/978-3-031-34960-7_26

vectorial representations (embeddings) of each graph-node $v \in V$ by starting several RWs from each node in the graph. The neighborhoods sampled by the RWs are then input to supervised neural network models [8] that are trained to compute embeddings that optimally maintain the topological neighborhoods in G, as represented by the RWs. In simpler words, the embeddings of nodes that are neighbors in G should represent points that are also neighbors in the embedding space.

Given the embeddings of the two vertices of a specific edge, $e = (s, d)$, its embedding is simply computed by an element-wise operator (e.g. the Hadamard product) applied to the embeddings of the vertices, s and d [6]. Once the edges have been represented as points in an embedding space, any supervised prediction model can be trained to recognize existent/non-existent edges.

PPI networks, such as STRING [14], are widely used to model known interactions between proteins, improving knowledge about the mechanisms underlying various biological processes. Henceforth, the prediction of novel PPIs is of paramount importance for increasing knowledge in the field. This problem can be naturally formulated as an RW-based edge prediction problem, as witnessed by recent works proposed in literature [15].

However, when the graph contains high-degree nodes, the RWs tend to prefer them and neglect low- or middle-degree nodes. As a result, low- or middle-degree nodes are rarely sampled, and their embeddings do not faithfully represent the topological structure in their neighborhoods. This leads to poor edge-prediction results, often overfitting edges having one high-degree node as one of the vertices.

In this work, we aim to address the issue of the bias towards high-degree nodes in edge prediction methods. To this aim, we propose a simple Degree-Normalization (DN) approach to counter-bias the walks and improve the representation of low-degree nodes (see Sect. 3 and Fig. 1). The degree normalization is inspired by solutions for handling unbalanced classes [5,12], and it is general enough to be applied to other graph-based methods, different from those exploiting random-walks.

We evaluate the effectiveness of the proposed method by performing edge prediction tasks, on a set of weighted PPI graphs, with classic random-walk-based embedding models that use the degree-normalization technique.

The obtained performance measures are compared to those obtained by using the same methods without normalization. We aim to demonstrate that the degree-normalization approach we are proposing can significantly improve the representation of low- and middle-degree nodes and leads to better prediction performance.

2   Background

A (weighted) directed graph G with N nodes is generally represented by a tuple $\mathcal{G} = (V, E)$, where V is the set of nodes ($|V| = N$) and E is the set of (directed) edges, composed of elements $e = (s, d, w_{sd})$, where $s \in V$ is the source (outbound) node, $d \in V$ is the destination (inbound) node, and the optional element

w_{sd} represents the strength of the relationships between s and d. The degree of an outbound node s is the cardinality of the set of its inbound neighbors $\mathcal{N}(s)$. Alternatively, a (weighted) directed graph G may also be represented by a (weighted) adjacency matrix $A \in \Re^{N \times N}$, where the element $A(i, j)$ contains the weight of the edge connecting the ith node to the jth node (if any exists).

Given the graph G, a (first-order) RW of length l that begins at some node v is a random process that moves from one node to the other for l time steps. At each time step, when the graph is unweighted, the destination node the RW moves to, u, is uniformly chosen among the destination neighbors of the present source node, v. In other words, the RW will step to one of the inbound neighbors $u \in \mathcal{N}(v)$ according to a (transition) probability $\pi_{uv} = P(u|v) = \frac{1}{|\mathcal{N}(v)|}$. The edge weights, when available, can be used to weigh the transition probability.

DeepWalk [10] computes node embeddings by generating several first-order RW samples starting from all the nodes in the graph. Next, the node embeddings are computed by a SkipGram algorithm [8], a language model that is trained on the RW samples to maximize the probability of co-occurrence of nodes in the same RW.

Node2Vec [6] leverages DeepWalk by substituting the first-order RW with a second-order RW whose peculiarity relies on the fact that the probability of stepping from one source node v to any of its direct neighbors considers the preceding step of the walk. More precisely, when the walk resides at v, where it has arrived transitioning from t to v, Node2Vec defines the un-normalized transition probability π_{uv} of moving from v to any direct neighbor u, specifically considering that the previous step started from node t. In detail, π_{uv} is a function of the weight w_{vu} on the edge (v, u) connecting vertices v and u, and a search bias $\alpha_{pq}(t, u)$:

$$\pi_{uv} = \alpha_{pq}\left(t, u\right) w_{vu}$$

The search bias $\alpha_{pq}(t, u)$ is defined as a function of the distance $d(t, u)$ between t and u, and two parameters p and q, called, respectively, the *return* and *in-out* parameters [6]:

$$\alpha_{pq}(t, u) = \begin{cases} \frac{1}{p} & \text{if } d(t, u) = 0 \\ 1 & \text{if } d(t, u) = 1 \\ \frac{1}{q} & \text{if } d(t, u) = 2 \end{cases} \tag{1}$$

If the return parameter p is small, the walk will be enforced to return to the preceding node; if p is large, the walk will otherwise be encouraged to visit new nodes. The in-out parameter q allows to vary smoothly between Breadth First Search (BFS) and Depth First Search (DFS) behavior. Indeed, when q is small the walk will prefer outward nodes, thus mimicking DFS; it will otherwise prefer inward nodes emulating in this case BFS.

The Walklets method [11] leverages DeepWalk by performing first-order RWs at different scales, where the specific scale is defined by the length of the walk. We observe that Walklets creates **multiple** node embeddings (one per scale).

3   Methods: Degree Normalization (DN)

As shown in Fig. 1, when applying both first-order and second-order RWs, higher-degree nodes have a higher probability to be sampled by an RW compared to other nodes.

Henceforth, they have the largest influence on the optimization process used by following embedding models, which essentially overfits them while neglecting other nodes.

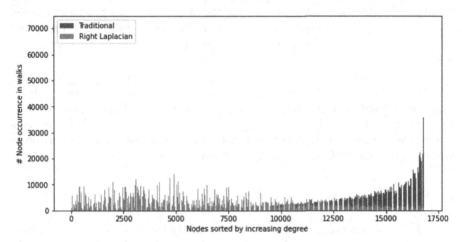

Fig. 1. Frequency of node sampling in a random walk: Comparison of traditional (blue) and degree-normalized (orange) methods. The traditional method is biased towards high-degree nodes sampled at a higher frequency. The degree-normalized method removes this bias and samples low-degree nodes with a higher frequency. (Color figure online)

To alleviate this effect, at each step of the walk, when the walk resides at node v and must decide where to move next, we propose normalizing each of the transition probabilities computed by either first- or second-order RWs by the inverse of the degree of the respective destination node. In practice, for each inbound neighbor u, we obtain a new transition probability such that:

$$\bar{\pi}_{uv} = \bar{P}(u|v) = \begin{cases} \frac{\pi_{uv}}{|\mathcal{N}(u)|} * c(u,v)^{-1} & \text{if } \mathcal{N}(u) \neq \emptyset \\ \pi_{uv} & \text{otherwise} \end{cases}$$

where $c(u,v) = \sum_{u \in \mathcal{N}(v)} \frac{\pi_{uv}}{|\mathcal{N}(u)|}$ is a normalization constant, depending both on v and on v's outbound neighbors, ensuring that

$$\sum_{u \in \mathcal{N}(v)} \bar{P}(u|v) = 1$$

This DN approach is a simple and computationally inexpensive approach for attenuating the oversampling of high-degree nodes in graph node sampling schemas, such as random walks.

DN can be practically computed by weighting the adjacency matrix $A = w_{ij}$ by the Moore-Penrose inverse degree diagonal matrix D^+ [1,2,13], to obtain a matrix $\hat{A}$ where high-degree nodes are less likely to be sampled.

$$\hat{A} = AD^+ \qquad D^+_{i,j} = \begin{cases} D^+_{i,j} = D_{i,j}{}^{-1} & \text{if } i = j \wedge D_{i,j} > 0 \\ 0 & \text{otherwise} \end{cases}$$

Besides reducing the oversampling of high-degree nodes, the normalization techniques have the side-effect of reducing the likelihood of sampling a sink node, i.e. a node in a directed graph with no destination nodes. Indeed, a sink node u will only and exclusively be sampled when the previous node v has exclusively only sink node successors.

Further, we note that the degree-normalization approach may be equivalently implemented by modifying the weights in the (weighted) adjacency matrix; therefore, even though in this paper we tested it on RW-based embedding models, it can be as well applied to any graph-embedding/processing method that uses such weights, e.g., factorization-based embedding techniques, any other node or edge sampling methods, or even graph-neural network models.

4   Datasets

To evaluate our proposal we employ eight datasets from the STRING database (version 11.5) [14]. They provide a comprehensive representation of PPI networks for eight organisms (see Table 1), from mammals to fungi and even reptiles, therefore allowing us to evaluate the generalizability of the proposed method across different biological species.

All edges in the PPI graphs are weighted with a score ranging from 100 to 900, which represents the likelihood of the existence of a specific interaction, stemming from computational prediction, from knowledge transfer between organisms, and from interactions aggregated from other (primary) databases. The curators of the STRING database suggest filtering "unreliable" edges by keeping only edges with scores greater than or equal to 700. However, such a filtering process has the effect of creating multiple disconnected components, some of which are singletons. Unfortunately, the presence of different graph components might bias the obtained evaluations, leading to overoptimistic results. Indeed, the embedding of nodes and edges in different components might be very different from each other, especially when the components' diameters differ a lot. This eases the classification task because the following predictive model will have no problem in finding disconnected nodes (i.e. unexistent edges). To avoid biases, and considering that the largest components obtained after filtering have considerable sizes, we decided to evaluate the considered algorithms on the largest component of each filtered graph. Table 1 details the PPI graph characteristics before and after the filtering operation.

Table 1. Summary statistics of the datasets used in the experiments, including the number of nodes, edges, and maximum degree of the original and filtered graphs.

Graph	Nodes	Edges	Max degree
Alligator Sinensis	18348	5725813	7014
Alligator Sinensis (filtered)	14002	163885	470
Amanita Muscaria Koide	17947	1102956	1817
Amanita Muscaria Koide (filtered)	4229	78507	302
Canis Lupus	19854	5677845	5430
Canis Lupus (filtered)	13501	173028	692
Drosophila Melanogaster	13932	2171899	3487
Drosophila Melanogaster (filtered)	10473	140619	366
Homo Sapiens	19566	5969249	7507
Homo Sapiens (filtered)	16584	252833	747
Mus Musculus	22048	7248179	7669
Mus Musculus (filtered)	16180	233340	684
Saccharomyces Cerevisiae	6691	994296	2729
Saccharomyces Cerevisiae (filtered)	5936	120357	385
Sus Scrofa	21597	6890582	5995
Sus Scrofa (filtered)	15173	170918	536

5 Experimental Results

We compared edge prediction results computed using RW-based embeddings with and without our proposed DN approach. More precisely we applied Deep-Walk, Node2Vec, and Walklets embedding algorithms and a random forest classifier to predict existing/nonexisting edges.

Table 2 lists the hyperparameter values set for the embedding and the RF models[1].

Since the primary goal of DN is to enhance the representation of low-degree nodes in the embeddings, we focused on predictions of edges characterized by low-degree nodes, ranging from a minimum degree of 5 to a maximum degree of 100. To this end, we used a 10 connected Monte Carlo holdout strategy with 80% of samples for training and 20% for test. To create holdouts that maintain a closed world assumption, we execute connected holdouts, i.e. holdouts where the training graph has the same connected components of the original graph. To ensure this property, the training graph is built starting from a skeleton graph based on a random spanning tree, which ensures the training graph has the same connected components as the original graph.

As mentioned above, we intended to evaluate the impact of the models on low to medium-degree nodes, and we did so by sampling the test set edges having

---

[1] The bias-correction schema proposed here is made available in Rust with Python bindings as part of the GRAPE library for graph machine learning [3]. Besides novel implementations of DeepWalk, Node2Vec, and Walklets, GRAPE integrates the random forest implementation from sklearn [9].

Table 2. Table of DeepWalk, Node2Vec, and Walklets parameters (left) and Random Forest parameters (right) used in the edge prediction task.

Parameter	Value	Parameter	Value
embedding size	100	edge embedding method	Concatenate
epochs	30	use scale-free distribution	True
number of negative samples	10	n estimators	1000
walk length	128	criterion	gini
iterations per node	10	max-depth	10
return weight (Node2Vec)	0.25	min samples split	2
explore weight (Node2Vec)	4.0	min samples leaf	1
window size (DeepWalk, Node2Vec)	5	min weight fraction leaf	0.0
window size (Walklets)	4	max features	sqrt
learning rate	0.01	max-leaf nodes	None
learning rate decay	0.9	min impurity decrease	0.0
use scale-free distribution	True	bootstrap	True

source and destination nodes with node degree in the range $[5, 50]$. In Fig. 2 we show the difference in node-degree distribution between the original graph and the train and test graphs from one of the holdouts, with a noticeable "dent" in the training test node degree distribution corresponding to the edges moved into the test graph.

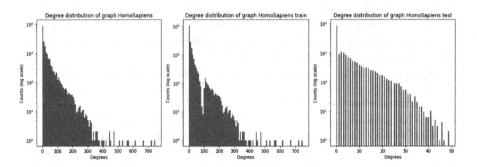

Fig. 2. Node degree distributions of STRING Homo Sapiens: (left) the degree distribution of the original graph; (center) the degree distribution of the training set, where the gap is due to the removal of edges moved to the test set; (right) degree distribution of the test graph.

For each holdout, we sampled a set of negative examples with cardinality equal to that of the positive examples. To avoid bias in sampling, we sampled negatives according to the node degree distribution rather than using a uniform distribution. Moreover, the negative edges' source and destination nodes have

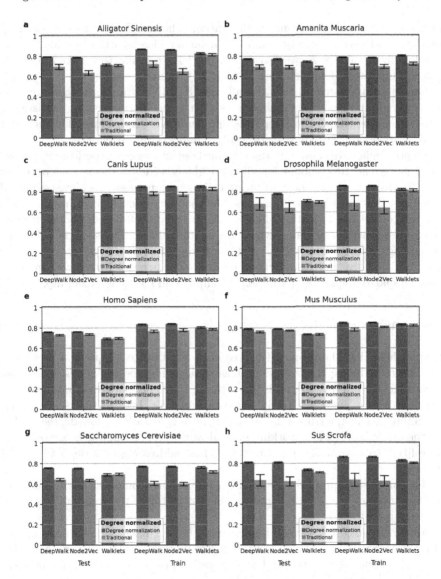

Fig. 3. Average (and standard deviation) of the MCC scores of classic (orange bars) versus degree-normalized (blue bars) DeepWalk, Node2Vec, and Walklets across the 10 test (left) and train (right) holdouts. (Color figure online)

been sampled from the same connected component to maintain a closed-world assumption.

We evaluated the performance of the tested models on the PPI prediction task by using the tailed Wilcoxon rank-sum test (p-value < 0.05) to compare the distributions of the Matthews correlation coefficients (MCC) (Fig. 3), obtained by all the models across the 10 holdouts. We present results obtained with the

MCC metric since recently in [4] the authors showed the advantages of using MCC instead of F1 score or accuracy in binary classification problems. Nevertheless in Appendix A we also report the F1 score and the Area Under the Receiver Operating Characteristics (AUROC) results.

Our analysis evidenced that, for all eight STRING graphs, the computation of DeepWalk and Node2Vec embeddings with degree normalization consistently outperforms the traditional DeepWalk and Node2Vec embeddings. On the other hand, Walklets did not show statistically significant improvements.

Of note, when we compared the standard deviations of the performance measures across the 10 holdouts for all the eight considered species, we observed that the proposed degree-normalization technique consistently achieves standard deviations that are smaller than those obtained by classic embedding methods. This suggests a regularization effect on the random walk-based models, leading to a more robust convergence, which is essential when training large-scale node embeddings, as it may help to prevent overfitting and increase the model robustness. Additionally, this regularization effect may be beneficial when working with noisy or sparse graphs, as it can help to improve model generalizability.

6   Conclusion

Results show that the proposed degree normalization technique is a promising solution for reducing the overfitting bias towards high-degree nodes. This is an essential feature in several graph-based tasks, where high-degree nodes are often those for which there is already much information. Instead, users are generally interested in inferring information regarding middle or low-degree nodes. On the other hand, when working on real-world graphs, high-degree nodes are more likely to have noisy data or errors; thus, addressing high degree biases is important.

The proposed approach is simple to implement, computationally efficient, and does not add any high computational cost on top of the computation of the embedding.

Moreover, since DN can be implemented by weighing the adjacency matrix with the Moore-Penrose inverse of the degree diagonal matrix, it can be applied to weigh graph-representation learning techniques other than random-walk-based models (e.g., GNNs or matrix-factorization-based techniques).

Despite the promising results, the lack of change in performance for the Walklets method is a result that warrants further investigation.

Acknowledgment. This research was supported by the "National Center for Gene Therapy and Drugs based on RNA Technology", PNRR-NextGenerationEU program [G43C22001320007].

A   Additional Results

(See Figs. 4 and 5).

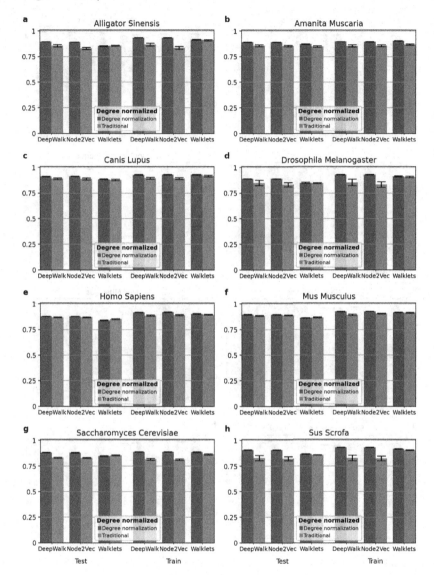

Fig. 4. Average (and standard deviation) of the F1 scores of classic (orange bars) versus degree-normalized (blue bars) DeepWalk, Node2Vec, and Walklets across the 10 test (left) and train (right) holdouts (Color figure online)

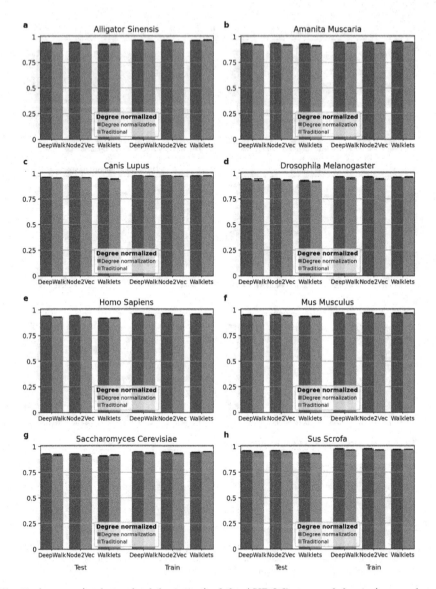

Fig. 5. Average (and standard deviation) of the AUROC scores of classic (orange bars) versus degree-normalized (blue bars) DeepWalk, Node2Vec, and Walklets across the 10 test (left) and train (right) holdouts. (Color figure online)

References

1. Ben-Israel, A., Greville, T.N.E.: Generalized Inverses: Theory and Applications, vol. 15. Springer, Heidelberg (2003). https://doi.org/10.1007/b97366
2. Campbell, S.L., Meyer, C.D.: Generalized Inverses of Linear Transformations. SIAM (2009)
3. Cappelletti, L., et al.: GRAPE: fast and scalable graph processing and embedding. arXiv preprint arXiv:2110.06196 (2022)
4. Chicco, D., Jurman, G.: The advantages of the Matthews correlation coefficient (MCC) over f1 score and accuracy in binary classification evaluation. BMC Genom. **21**(1), 1–13 (2020)
5. Cuzzocrea, A., Cappelletti, L., Valentini, G.: A neural model for the prediction of pathogenic genomic variants in mendelian diseases. In: Proceedings of the 1st International Conference on Advances in Signal Processing and Artificial Intelligence (ASPAI 2019), Barcelona, Spain, pp. 34–38 (2019)
6. Grover, A., Leskovec, J.: node2vec: Scalable feature learning for networks. In: Proceedings of the 22nd ACM SIGKDD International Conference on Knowledge Discovery and Data Mining, pp. 855–864 (2016)
7. Li, M.M., Huang, K., Zitnik, M.: Graph representation learning in biomedicine and healthcare. Nat. Biomed. Eng. **6**(12), 1353–1369 (2022)
8. Mikolov, T., Chen, K., Corrado, G., Dean, J.: Efficient estimation of word representations in vector space. arXiv preprint arXiv:1301.3781 (2013)
9. Pedregosa, F., et al.: Scikit-learn: machine learning in python. J. Mach. Learn. Res. **12**, 2825–2830 (2011)
10. Perozzi, B., Al-Rfou, R., Skiena, S.: DeepWalk: online learning of social representations. In: Proceedings of the 20th ACM SIGKDD International Conference on Knowledge Discovery and Data Mining, pp. 701–710 (2014)
11. Perozzi, B., Kulkarni, V., Chen, H., Skiena, S.: Don't walk, skip! Online learning of multi-scale network embeddings. In: Proceedings of the 2017 IEEE/ACM International Conference on Advances in Social Networks Analysis and Mining 2017, pp. 258–265 (2017)
12. Petrini, A., et al.: parSMURF, a high-performance computing tool for the genome-wide detection of pathogenic variants. GigaScience **9**(5), giaa052 (2020)
13. Radhakrishna Rao, C., Mitra, S.K., et al.: Generalized inverse of a matrix and its applications. In: Proceedings of the Sixth Berkeley Symposium on Mathematical Statistics and Probability, vol. 1, pp. 601–620. University of California Press, Oakland (1972)
14. Szklarczyk, D., et al.: The STRING database in 2021: customizable protein-protein networks, and functional characterization of user-uploaded gene/measurement sets. Nucleic Acids Res. **49**(D1), D605–D612 (2021)
15. Yi, H.-C., You, Z.-H., Huang, D.-S., Kwoh, C.K.: Graph representation learning in bioinformatics: trends, methods and applications. Brief. Bioinform. **23**(1), bbab340 (2021)
16. Yue, X., et al.: Graph embedding on biomedical networks: methods, applications and evaluations. Bioinformatics **36**(4), 1241–1251 (2019)

Medical Image Processing

Role of Parallel Processing in Brain Magnetic Resonance Imaging

Ayca Kirimtat[1] and Ondrej Krejcar[1,2(✉)]

[1] Faculty of Informatics and Management, Center for Basic and Applied Research, University of Hradec Kralove, Rokitanskeho 62, Hradec Kralove 500 03, Czech Republic
{ayca.kirimtat,ondrej.krejcar}@uhk.cz

[2] Malaysia Japan International Institute of Technology (MJIIT), Universiti Teknologi Malaysia, Kuala Lumpur, Malaysia

Abstract. Parallel processing is a procedure for making computation of more than a processor to overcome the difficulty of separate parts of an overall task. It is really crucial for some medicine-related tasks since the method provide time-efficient computation by a program, thus several calculations could be made simultaneously. Whereas, magnetic resonance imaging (MRI) is one of the medical imaging methods to show form of an anatomy and biological progressions of a human body. Parallel processing methods could be useful for being implemented in MRI with the aim of getting real-time, interventional and time-efficient acquisition of images. Given the need of faster computation on brain MRI to get early and real-time feedbacks in medicine, this paper presents a systematic review of the literature related to brain MRIs focusing on the emerging applications of parallel processing methods for the analysis of brain MRIs. We investigate the articles consisting of these kernels with literature matrices including their, materials, methods, journal types between 2013 and 2023. We distill the most prominent key concepts of parallel processing methods.

Keywords: parallel processing · MRI · brain · Web of Science · review

1 Introduction

Among various medical image modalities such as computed tomography (CT), positron emission tomography (PET) and ultrasound; magnetic resonance imaging (MRI) is the most employed technique for brain imaging [1–3]. Basically, MRI has been extensively used for the analysis of the whole brain structure [4]. In order to do a healthy treatment for brain area, MRI is generally used as a medical assessment, and MRI is performed through powerful and expensive equipment and computers. In addition to these, the data acquisition and image rendering in MRI spend waste of time. In order to speed up the process, accurate and fast brain MRI processing is needed for a large variety of brain manipulations [5–7] such as analysis, diagnostics and examination of most of the brain diseases in medicine [8, 9]. However, due to sudden increase in the medical image data sizes, the conventional computational Central Processing Unit (CPU) became insufficient

© The Author(s), under exclusive license to Springer Nature Switzerland AG 2023
I. Rojas et al. (Eds.): IWBBIO 2023, LNBI 13920, pp. 387–397, 2023.
https://doi.org/10.1007/978-3-031-34960-7_27

for accurate and fast brain MRI processing. Instead, Graphics Processing Unit (GPU) started to become a novel technology for finding solutions to complex computational problems in medicine [10]. A GPU is defined as a multiple core processor which can be used for efficient parallelization [11].

In this review, we analyze the combination of brain MRI methods, parallel processing and the use of GPU architectures. We investigate the various applications of brain MRI that have been previously carried out in the recent literature. In addition, recent image processing techniques including image reconstruction, denoising, visualization and segmentation algorithms are included in this review. We also include several types of implementations, solutions and parallel approaches used in the field of parallel processing of brain MRI. This review is an extended version of our earlier work by [12], building on a detailed literature search performed in the Web of Science (WoS). Since GPU could be used to solve a wide variety of problems in the field of brain MRI, several experiments with various parameter settings and algorithms are possible. Moreover, in general, a single GPU core is not faster than a CPU core, but it is slower in terms of frequency and FLOPs (floating point operations per second). However, given the number of cores in GPU, GPU overcomes CPU in parallelized performance. This review will give important remarks and gentle introduction to the readers that are eager to learn the details of GPU programming and brain MRI analysis highlighting important developments within the technology, which enables high-level of programming interfaces.

GPU is a modern terminology having a complicated architecture, which varies from one model or manufacturer to another. Parallel computing could also be accelerated by resorting to the computational power of GPU, since it is quite affordable and energy-efficient. Actually, the main reason for the use of GPU is that it becomes highly powerful for realistic computer games. Historically, the computational performance of CPUs was the highest, however today's GPUs' computational performance outperforms traditional CPUs. Furthermore, computer rendering requires complex computer graphics for complex scenes where a scene should be rendered within less than 25 ms. The rendering process is normally performed in parallel on individual pixels exploiting the data level parallelism of an image, which is called Single Instruction Multiple Data (SIMD). In addition to these, today's GPU runs a very suitable SIMD algorithm, which progresses on over thousands of threads simultaneously [13].

As per today, the computational power of GPU has overcome traditional CPU. Furthermore, in modern GPU, multi-processors have their processor cores and memory chips separated. For instance, NVIDIA GTX 680 has 8 multi-processors and 192 processor cores for each one, thus computations are done by conforming a great number of threads, and these threads are divided into different thread blocks that can be controlled by each multi-processor. Therefore, the performance of modern GPU implementations significantly depends on the parallelization possibility of the underlying algorithm being run. Since image processing algorithms are generally parallel by nature, adapting them to run on GPUs is usually a simple task. For instance, image registration algorithms are quite suitable for GPU parallelism, since GPU could be used for calculating any selected similarity degree in parallel, on the other hand CPU could only make a serial optimization. However, not all algorithms are suitable for GPU parallelism. Given the

GPU's massive parallelism, GPU programming models and tools have already evolved themselves in order to yield the best performance [14].

The CUDA computational model runs thousands of threads, which are also conforming a virtual architecture. A piece of code that is performed on the GPU model is called kernel, which conforms the limits of the architecture. Therefore, when the kernel is performed, thread blocks are assigned to parallel multiprocessors [15]. Figure 1 illustrates the difference between CPU and GPU architecture. Based on Fig. 1 we observe that the optimization of graphics performance for GPU vs CPU highly differs from each other. For instance, there are too many vertices in CPU for processing rather than GPU, however, rendering is not a big problem in modern GPU [16].

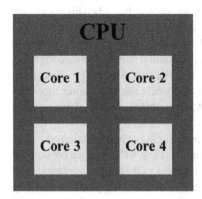 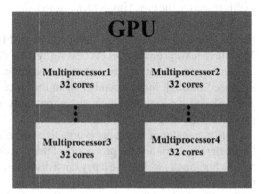

Fig. 1. Typical CPU vs GPU architecture.

The significant examples on GPU performance, from our literature survey, are summarized as in the following sentences. For instance, in the study of [17], an application of the Fast Fourier Transform in a GPU was presented by carrying out image reconstruction in MRI. The implementation focused on the balance between vertex and fragment processor, and the rasterizer using a high-performance NVIDIA Quadro NV4x. In this research, it is concluded that GPU has a higher performance in comparison with traditional CPU for reconstructing the images. The authors in [18] proposed a distributed parallel hybrid architecture with the aim of solving computation time problem and filters' coefficient computation. The authors basically focused on an implementation of filters' parameters with the neighborhood of the current voxel. The main contribution of this study is the reduction in the number of shared memories. In addition to these, the brain-web database was used by performing the proposed method in order to diagnose various levels of noise.

According to [19], SENSE (Sensitivity Encoding) reconstruction for brain MRI was proposed to solve large-scale constrained optimization problems. In the paper, firstly, regularized SENSE reconstruction was formulated as an unconstrained optimization task, and secondly, the authors converted it into a set of constrained problems. The proposed method was quite applicable to a general class of regularization. As a conclusion, the augmented Lagrangian (AL) approach was also highly efficient for solving constrained optimization problems for SENSE reconstruction. Another important study by [10] presented the performance evaluation of some algorithms that were used for analyzing MRI dataset on the GPU. According to the authors, since CPUs have limited computational power, GPUs are generally used for solving huge medical data. In [10], K-means segmentation for feature extraction was used for MRI volumes using GPU computing. Based on the reported experiments, GPU-based implementation achieved faster computing than traditional CPU processors in the proposed model.

The study by [20] proposed a parallel Hidden Markov Model algorithm for the segmentation of 3D brain MRI image through two different approaches. In the first one, multilevel parallel approach was used to reach higher performance for processing of the algorithm, on the other hand the second approach was orthogonal to the first one and tried to minimize the errors for 3D MRI brain segmentation by multiple processors. As a result of this research, the novel implementation of the authors' approach reached high accuracy and low error, which has the level of 98%. In the study of [21], the authors investigated the detection of the edges of brain tumor in MRI images using Wavelet Transform Modulus Maxima on Parallel Processing system. As evaluation metrics, processing time, efficiency and velocity of the processes were reported. For instance, the efficiency was increased through higher number of multi-processors, while the processing time was also reduced by less complicated computation. Also, within the optimal time of the processes, speedup was reached.

The remainder of this paper is organized as follows. Section 2 basically introduces to the reader the scope and the procedure of this systematic review. In Sect. 3, we present our review findings as well as we carry out a discussion regarding GPU implementations and the case studies by including detailed literature matrix summarizing prominent legacy studies. The future trends, prospects and concluding remarks are given in the last section of this review study.

2   Methodology

This systematic review addresses the main question that creates a framework for the existing studies on parallel processing methods for brain MRI manipulations in the WoS: What are the parallel processing methods that provide faster computation time with suitable algorithms while analyzing brain MRIs? This question mainly canalizes us to explore specific methods and procedures that could be implemented in the analysis stage of brain MRIs. The apical targets are to investigate and analyze the most advanced computer algorithms and computational methods that enable time-efficient computation for medicine related works, especially for brain MRI studies. In addition to these, the awareness on the proposed topic is crucial for making the readers of this review comprehend the latest advances and drawbacks.

Within the previously conducted works on parallel processing for brain MRI [22], there are several studies that possess a higher priority among other studies when performing a keyword-based search from the WoS. In this review, we address these papers using different search strategy rather than the existing literature surveys. MRI on the parallel processing has become very popular recently because of the technological developments that enable faster computation with advanced algorithms in terms of clinical aspects. However, research papers on brain MRI and parallel processing are quite rare, even this review focuses on very prominent examples from the existing literature (Fig. 2).

On the other hand, Fig. 3 shows the conceptual framework of this review study according to the relevant references from the WoS. Even though there are some other database searches such as Google Scholar (GS) and Scopus, WoS is specifically chosen since it is mandatory in the west countries across the world. If we explain the framework for this systematic review, firstly we start with making keyword-based filtering among thousands of articles related with medicine and parallel processing. The filtering is basically depending upon 3 keywords namely as "parallel processing", "brain" and "MRI". Furthermore, we have an access to the articles through the WoS and the topics of these articles basically are related with conceptual themes, different methods and brain implementations. On the other hand, the categorization of the previously-conducted studies is made based on four different criteria that are shown in Fig. 3 as methods, materials, publication and case studies.

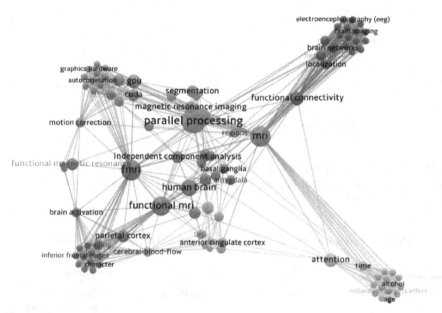

Fig. 2. The network visualization of the keywords based on WoS data using VOSviewer software.

3   Results and Discussion

This section aims to present the implementations and case studies from the WoS in the form of a literature matrix. Based on each previous study, which are relevant to parallel processing of brain MRI, a comprehensive literature matrix is generated by using some specific criteria as seen in Table 1, which are methods, brain MRI materials, publication types, and contributions of the studies. The articles published between the years 2013 and 2023 are gathered in Table 1 with the aim of presenting a brief, but comprehensive matrix. Some articles are specifically not related with the topics of "parallel processing" and

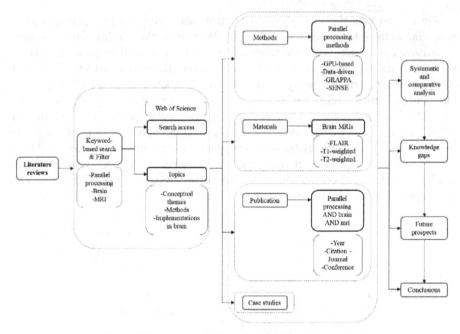

Fig. 3. The conceptual framework of this review study.

"MRI", therefore we remove them from Table 1. Given the technological advancements within the last years, a great number of studies on parallel processing for brain MRI images has emerged.

Table 1. The previous articles on parallel processing and brain MRI image analysis.

Ref.	Method	MRI material	Publication	Contribution
[23]	Parallel processing	T1w and T2w	Medical Physics	Suffering from structure-dependent artifacts
[14]	Parallel processing	T1w	Frontiers in Neuroinformatics	Reduce in processing time
[24]	Parallel processing	N/A	Nature Communications	Double dissociation by parallel processing
[25]	N/A	T1w and T2w	Neuroscience	Demanding of parallel processing
[26]	Parallel processing	T1w	Experimental Brain Research	Good decision-making performance

(*continued*)

Table 1. (*continued*)

Ref.	Method	MRI material	Publication	Contribution
[27]	Parallel processing	T1w	Brain and Language	A challenge to single-mechanism models
[28]	N/A	N/A	Statistical Methods in Medical Research	Balanced and symmetrically distributed data
[29]	Serial and parallel processing	T1w	BMC Neuroscience	Successful tactile signal processing
[30]	Parallel processing	N/A	Neurocomputing	Successful segmentation
[31]	Parallel processing	N/A	Frontiers in Neuroscience	Faster and robust platforms
[32]	Parallel processing	N/A	Medical Engineering & Physics	Smoother distributions
[33]	Parallel processing	T1w	Medical Engineering & Physics	Basis for future research on real-time connectivity-based brain–computer
[34]	Parallel processing	N/A	PLoS One	Speed proposed scheme
[35]	Parallel processing	N/A	Turkish Journal of Electrical Engineering & Computer Sciences	Computation time is markedly decreased
[36]	Parallel processing	N/A	NeuroImage	Fast parallel processing and easy to use
[37]	Parallel processing	T1w	Frontiers in Neuroscience	Computational time saving
[38]	Parallel processing	N/A	International Journal of Computers Communications & Control	Answering successfully to computational challenge
[39]	Parallel processing	N/A	Concurrency and Computation-Practice & Experience	Reduce in processing time

(*continued*)

Table 1. (*continued*)

Ref.	Method	MRI material	Publication	Contribution
[40]	Parallel processing	T1w	Journal of Supercomputing	Reducing analysis time by 70%
[41]	Cooperative parallel processing	N/A	Journal of Medical Imaging and Health Informatics	Higher error minimization

T1w: T1-weighted
T2w: T2-weighted
N/A: Not Available

4 Conclusions

Given the sound breaking spread of parallel processing applications in medicine, parallel computers should be used more extensively in brain MRI. Therefore, the significance of parallel processing tools for brain MRI detection is highlighted in this review study. Furthermore, recent developments are gathered and presented in a systematic manner. Taking new perspectives into account, here we summarize the practical future opportunities as listed below:

- Grid computing and modern web technologies could be used for post-processing of brain MRI in terms of Alzheimer diseases since they are quite useful for execution time of the post-processing.
- Additional research should be conducted in order to build technologically more faster supercomputers in laboratories across the world, even the existing ones are quite powerful.
- Even supercomputers play significant role in computer science area such as quantum mechanics, weather forecasting, molecular modeling etc., the implementation area should be shifted to medicine much more.
- Special-purpose supercomputers could be conceived with the aim of exclusive usage in medicine applications, even there are some dedicated to astrophysics, even playing chess.
- Supercomputers should be connected to grid computing in order to carry out larger computing tasks in medicine, even in brain MRI detection.
- The performances of the existing supercomputers should be compared and benchmarked in case of emergency situations in medicine related studies.
- Among the existing NVIDIA solutions, the most promising and the cost-efficient ones such as Turing and Game CPU cards could be optimal alternative because of low price on one hand and high computational efficiency on the other hand reaching 20 TFLOPS in single precision.
- Reliable, quick, powerful, parallel, user-friendly, cheap and smart mobile phone applications should be designed for physicians in order to help them solve neurodegeneration, which is a heavy burden and geometrically escalate in terms of number of patients, cost on care and drug therapy.

- Thanks to advancements in high-supercomputing (HPC) technologies and artificial intelligence (AI), digital twin technology could be used by medical companies to build digital twins of organs.
- More supercomputers should be created for designing new medicines in order to solve complicated molecular processes and interactions of new drugs with human tissues.

We investigate how brain MRI image processing can benefit from the development in GPU programming and parallel processing. We present the use of GPU programming for analysis of brain MRI images. We also highlight the general ideas on differences between CPU and GPU programming to the readers and show the main differences between them. The related works on parallel processing on brain MRI analysis are carefully chosen from the Web of Science, considering their novel application areas and implementation on human brain MRI. After analyzing the search results, we list the selected articles in a detailed literature matrix. By taking these results into consideration, we conduct a comprehensive discussion on our findings, and present the future trends and prospects of this field by highlighting the future aspects on parallel processing for brain MRI. We believe this review is able to present a comprehensive view on this topic, and can provide an important advancement by facilitating easier understanding of the current literature.

Acknowledgments. The work and the contribution were also supported by the SPEV project, University of Hradec Kralove, Faculty of Informatics and Management, Czech Republic (ID: 2102–2023), "Smart Solutions in Ubiquitous Computing Environments". We are also grateful for the support of student Michal Dobrovolny in consultations regarding application aspects.

References

1. Pizarro, R., et al.: Using deep learning algorithms to automatically identify the brain MRI contrast: implications for managing large databases. Neuroinformatics **17**(1), 115–130 (2018). https://doi.org/10.1007/s12021-018-9387-8
2. Xu, Z., Wang, S., Li, Y., Zhu, F., Huang, J.: PRIM: an efficient preconditioning iterative reweighted least squares method for parallel brain MRI reconstruction. Neuroinformatics **16**(3–4), 425–430 (2018). https://doi.org/10.1007/s12021-017-9354-9
3. Liu, Y., Unsal, H.S., Tao, Y., Zhang, N.: Automatic brain extraction for rodent MRI images. Neuroinformatics **18**(3), 395–406 (2020). https://doi.org/10.1007/s12021-020-09453-z
4. Kontos, D., Megalooikonomou, V., Gee, J.C.: Morphometric analysis of brain images with reduced number of statistical tests: a study on the gender-related differentiation of the corpus callosum. Artif. Intell. Med. **47**, 75–86 (2009). https://doi.org/10.1016/j.artmed.2009.05.007
5. Sun, L., Zu, C., Shao, W., Guang, J., Zhang, D., Liu, M.: Reliability-based robust multi-atlas label fusion for brain MRI segmentation. Artif. Intell. Med. **96**, 12–24 (2019). https://doi.org/10.1016/j.artmed.2019.03.004
6. Richard, N., Dojat, M., Garbay, C.: Automated segmentation of human brain MR images using a multi-agent approach. Artif. Intell. Med. **30**, 153–176 (2004). https://doi.org/10.1016/j.artmed.2003.11.006
7. González-Villà, S., Oliver, A., Valverde, S., Wang, L., Zwiggelaar, R., Lladó, X.: A review on brain structures segmentation in magnetic resonance imaging. Artif. Intell. Med. **73**, 45–69 (2016). https://doi.org/10.1016/j.artmed.2016.09.001

8. Youssfi, M., Bouattane, O., Bensalah, M.O., Cherradi, B.: A fast massively parallel fuzzy C-means algorithm for brain MRI segmentation **22**, 19 (2015)

9. Murugavalli, S., Rajamani, V.: A high speed parallel fuzzy c-mean algorithm for brain tumor segmentation. BIME J. **06**, 29–34 (2006)

10. Kalaiselvi, T., Sriramakrishnan, P., Somasundaram, K.: Survey of using GPU CUDA programming model in medical image analysis. Inform. Med. Unlock. **9**, 133–144 (2017). https://doi.org/10.1016/j.imu.2017.08.001

11. Cuda, C.: Programming Guide (2017)

12. Kirimtat, A., Krejcar, O., Dolezal, R., Selamat, A.: A mini review on parallel processing of brain magnetic resonance imaging. In: Rojas, I., Valenzuela, O., Rojas, F., Herrera, L.J., Ortuño, F. (eds.) IWBBIO 2020. LNCS, vol. 12108, pp. 482–493. Springer, Cham (2020). https://doi.org/10.1007/978-3-030-45385-5_43

13. Eklund, A., Dufort, P., Forsberg, D., LaConte, S.M.: Medical image processing on the GPU – past, present and future. Med. Image Anal. **17**, 1073–1094 (2013). https://doi.org/10.1016/j.media.2013.05.008

14. Eklund, A., Dufort, P., Villani, M., LaConte, S.: BROCCOLI: software for fast fMRI analysis on many-core CPUs and GPUs. Front. Neuroinform. **8** (2014). https://doi.org/10.3389/fninf.2014.00024

15. Lamas-Rodríguez, J., Heras, D.B., Argüello, F., Kainmueller, D., Zachow, S., Bóo, M.: GPU-accelerated level-set segmentation. J. Real-Time Image Proc. **12**(1), 15–29 (2013). https://doi.org/10.1007/s11554-013-0378-6

16. Wawrzonowski, M., Szajerman, D., Daszuta, M., Napieralski, P.: Mobile devices' GPUs in cloth dynamics simulation. In: Presented at the 2017 Federated Conference on Computer Science and Information Systems September 24 (2017). https://doi.org/10.15439/2017F191

17. Sumanaweera, T., Liu, D.: Medical image reconstruction with the FFT **5** (2005)

18. Nguyen, T.-A., Nakib, A., Nguyen, H.-N.: Medical image denoising via optimal implementation of non-local means on hybrid parallel architecture. Comput. Methods Programs Biomed. **129**, 29–39 (2016). https://doi.org/10.1016/j.cmpb.2016.02.002

19. Ramani, S., Fessler, J.A.: Parallel MR image reconstruction using augmented lagrangian methods. IEEE Trans. Med. Imaging. **30**, 694–706 (2011). https://doi.org/10.1109/TMI.2010.2093536

20. El-Moursy, A.A., ElAzhary, H., Younis, A.: High-accuracy hierarchical parallel technique for hidden Markov model-based 3D magnetic resonance image brain segmentation: high-accuracy hierarchical parallel HMM for 3D MRI brain segmentation. Concurr. Comput. Pract. Expert. **26**, 194–216 (2014). https://doi.org/10.1002/cpe.2959

21. Sulaiman, H., Said, N.M., Ibrahim, A., Alias, N.: High performance visualization of human tumor detection using WTMM on parallel computing system. In: 2013 IEEE 9th International Colloquium on Signal Processing and its Applications, pp. 205–208. IEEE, Kuala Lumpur (2013). https://doi.org/10.1109/CSPA.2013.6530042

22. Bernal, J., et al.: Deep convolutional neural networks for brain image analysis on magnetic resonance imaging: a review. Artif. Intell. Med. **95**, 64–81 (2019). https://doi.org/10.1016/j.artmed.2018.08.008

23. Kwon, K., Kim, D., Park, H.: A parallel MR imaging method using multilayer perceptron. Med. Phys. **44**, 6209–6224 (2017). https://doi.org/10.1002/mp.12600

24. Ahveninen, J., et al.: Evidence for distinct human auditory cortex regions for sound location versus identity processing. Nat. Commun. **4**, 2585 (2013). https://doi.org/10.1038/ncomms3585

25. Gilat, M., et al.: Dysfunctional limbic circuitry underlying freezing of gait in Parkinson's disease. Neuroscience **374**, 119–132 (2018). https://doi.org/10.1016/j.neuroscience.2018.01.044

26. Gathmann, B., et al.: Stress and decision making: neural correlates of the interaction between stress, executive functions, and decision making under risk. Exp. Brain Res. **232**(3), 957–973 (2014). https://doi.org/10.1007/s00221-013-3808-6

27. Cummine, J., et al.: Manipulating instructions strategically affects reliance on the ventral-lexical reading stream: converging evidence from neuroimaging and reaction time. Brain Lang. **125**, 203–214 (2013). https://doi.org/10.1016/j.bandl.2012.04.009

28. Khondoker, M., Dobson, R., Skirrow, C., Simmons, A., Stahl, D.: A comparison of machine learning methods for classification using simulation with multiple real data examples from mental health studies. Stat. Methods Med. Res. **25**, 1804–1823 (2016). https://doi.org/10.1177/0962280213502437

29. Chung, Y., et al.: Intra- and inter-hemispheric effective connectivity in the human somatosensory cortex during pressure stimulation. BMC Neurosci. **15**, 43 (2014). https://doi.org/10.1186/1471-2202-15-43

30. De, A., Zhang, Y., Guo, C.: A parallel adaptive segmentation method based on SOM and GPU with application to MRI image processing. Neurocomputing **198**, 180–189 (2016). https://doi.org/10.1016/j.neucom.2015.10.129

31. Mano, M., Lécuyer, A., Bannier, E., Perronnet, L., Noorzadeh, S., Barillot, C.: How to build a hybrid neurofeedback platform combining EEG and fMRI. Front. Neurosci. **11** (2017). https://doi.org/10.3389/fnins.2017.00140

32. Minati, L.: Fast computation of voxel-level brain connectivity maps from resting-state functional MRI using l1-norm as approximation of Pearson's temporal correlation: proof-of-concept and example vector hardware implementation. Med. Eng. **6** (2014)

33. Minati, L., Nigri, A., Cercignani, M., Chan, D.: Detection of scale-freeness in brain connectivity by functional MRI: signal processing aspects and implementation of an open hardware co-processor. Med. Eng. Phys. **35**, 1525–1531 (2013). https://doi.org/10.1016/j.medengphy.2013.04.013

34. Wang, W.-J., Hsieh, I.-F., Chen, C.-C.: Accelerating computation of DCM for ERP in MATLAB by external function calls to the GPU. PLoS ONE **8**, e66599 (2013). https://doi.org/10.1371/journal.pone.0066599

35. Saran, A.N., Nar, F., Saran, M.: Vessel segmentation in MRI using a variational image subtraction approach. 18 (2014)

36. Meunier, D., et al.: NeuroPycon: an open-source python toolbox for fast multi-modal and reproducible brain connectivity pipelines. Neuroimage **219**, 117020 (2020). https://doi.org/10.1016/j.neuroimage.2020.117020

37. Cui, Z., Zhao, C., Gong, G.: Parallel workflow tools to facilitate human brain MRI post-processing. Front. Neurosci. **9** (2015). https://doi.org/10.3389/fnins.2015.00171

38. Ţugui, A.: GLM analysis for fMRI using Connex array. Int. J. Comput. Commun. **9**, 768 (2014). https://doi.org/10.15837/ijccc.2014.6.1482

39. Thiruvenkadam, K., Nagarajan, K., Padmanaban, S.: An automatic self-initialized clustering method for brain tissue segmentation and pathology detection from magnetic resonance human head scans with graphics processing unit machine. Concurr. Comput. Pract. Expert. **33** (2021). https://doi.org/10.1002/cpe.6084

40. Pantoja, M., Weyrich, M., Fernández-Escribano, G.: Acceleration of MRI analysis using multicore and manycore paradigms. J. Supercomput. **76**(11), 8679–8690 (2020). https://doi.org/10.1007/s11227-020-03154-9

41. ElAzhary, H., Younis, A.: Cooperative parallel processing with error curve sensing: a novel technique for enhanced hidden Markov model training for 3D medical image segmentation. J. Med. Imaging Health Inform. **6**, 1605–1611 (2016)

Deep Learning Systems
for the Classification of Cardiac
Pathologies Using ECG Signals

Ignacio Rojas-Valenzuela[1](✉) (iD), Fernando Rojas[2] (iD), Juan Carlos de la Cruz[3],
Peter Gloesekoetter[4] (iD), and Olga Valenzuela[5](✉) (iD)

[1] Information Technology and Telecommunications Engineering,
University of Granada, Granada, Spain
e.ignaciorojas@go.ugr.es
[2] Department of Computer Engineering, Automatics and Robotics,
University of Granada. C.I.T.I.C., University of Granada, Granada, Spain
[3] Department of Physical Education and Sport, University of Granada,
Granada, Spain
[4] FH Muenster University of Applied Sciences, 48329 Steinfurt, Germany
[5] Department of Applied Mathematics, University of Granada, Granada, Spain
olgavc@ugr.es

Abstract. In this paper, several deep learning models are analyzed for
the construction of the automated helping system to ECG classification.
The methodology presented in this article begins with a study of the
different alternatives for performing the discrete wavelet transform-based
scalogram for an ECG. Then, several Deep Learning architectures are
analysed. Due to the large number of architectures in the literature,
seven have been selected as they have a high degree of acceptance in
the scientific community. The influence of the number of epochs used
for training will also be analysed. In addition to the development of
a classifier able to accurately solve the multi-class problem of, given
an ECG signal, deciding which pathology the subject is suffering from
(main interest for a medical expert), we also want to rigorously analyze,
through the use of a statistical tool (ANOVA), the impact the main
functional blocks of our system have on its behaviour. As a novel result
of this article, different homogeneous groups of deep learning systems
are analysed (from a statistical point of view, they have the same impact
on the accuracy of the system). As can be seen in the results, there are
four homogeneous groups, with the group with the lowest accuracy index
obtaining an average value of 76,48% in the classification and the group
with the best results, with an average accuracy of 83,83%.

Keywords: Deep Learning architectures · ECG classification ·
statistical tool · ANOVA · Scalogram · Wavelet

1 Introduction

Cardiovascular diseases are the main cause of death worldwide, way over any
kind of cancer, neurodegenerative diseases or traffic accidents. The World Health

I. Rojas et al. (Eds.): IWBBIO 2023, LNBI 13920, pp. 398–412, 2023.
https://doi.org/10.1007/978-3-031-34960-7_28

Organization (WHO) estimates 17.7 million people died in 2015 because of cardiovascular diseases.

There exist plenty of factors related to the appearance of arrhythmias, which can result in the alteration of the heart's performance leading to other kinds of heart diseases and, in some cases, may cause death. Depending on the type of arrhythmia, the heart will be affected in different ways, meaning a tachycardia (accelerated functioning of the heart) won't have the same consequences as an atrial fibrillation, since in this last one heart rate may increase to 220 beats per minute and the person may suffer from thrombosis, embolisms, or even death if the treatment is not adequate or other complications appear due to arterial hypertension. Atrial fibrillation is one of the most common causes of external consultation and the demand of emergency services in hospitals, mainly associated to old age. Once a person suffers from an arrhythmia, they must remain under cardiological surveillance and receive the appropriate medical care.

The procurement of an automated diagnosis model based on electrocardiograms generates a real impact in medical service, which makes the search for mechanisms that improve the classification accuracy an indispensable factor to collaboratively and objectively help clinical decision making.

Nowadays, a methodology widely used in the scientific community are Deep Learning Systems. These systems have great potential and accuracy in a variety of tasks, including the task which is the target of this project, multi-class classification. Deep learning are based on multi-level neural network architectures (may be several hundred levels). A huge number of neurons, synapses and levels in Deep Learning systems require a training process quite costly in regards to it's computation time. Because of this, the rise of GPUs and advance computation platforms (parallelism) has positively impacted the ability of this DL systems to be trained in a manageable amount of time.

During the development of a Deep Learning system, there are, according to us, two matters which may have a great repercussion over the system. The first one would be the Deep Learning network architecture or model used [20,23]. The second is the number of epochs used for training [16]. There are other factors (MiniBatchSize, Learning rate, ADAM/SGDM optimization method, etc.), but we focused on the first two since the analysis of multiple factors exponentially increases the amount of simulations needed to obtain a relevant result [17,25].

In this article we propose to carry out an exhaustive and rigorous study by means of a powerful statistical analysis tool (ANOVA), of the influence of different deep learning models (seven models widely used in the literature) and different values for the number of epochs, in the classification of cardiac diseases from the ECG. For this purpose, a first step of pre-processing of the ECG signal (using the Wavelet transform) and obtaining the Scalogram will be carried out. This Scalogram figure will be the input to the different Deep Learning models.

This article is structured as follows. After this introductory section, the electrocardiogram is briefly analysed in Sect. 2, and in Sect. 3 the concept of arrhythmia is defined. Section 4 summarises the Discreet Wavelet transform and the obtaining of the scalogram. Section 5 presents the public database used for

ECG acquisition: Physionet, together with a brief summary of the three ECG datasets used in this study. Section 6 discusses the different deep learning models frequently used in the literature (a total of 7 different models have been used, with different number of layers, structure, number of weights and complexity). Section 7 presents the results of the statistical analysis performed on the different deep learning models and on the number of epochs used for training. The main objective is the clustering of deep learning models with statistically similar behaviour. Finally, Sect. 8 presents the main conclusions of this work.

Figure 1 shows a complete diagram of the methodology used in this article to perform the statistical analysis of the influence of different deep learning models in the classification of cardiac pathologies using the ECG signal scalogram.

2   Electrocardiogram

In an electrocardiogram (ECG) we can find, represented by a graph, the heart's voltage variations captured by electrodes on the surface of our body, in relation to time. Therefore, the average voltage is represented on the vertical axis and time on the horizontal one. Said voltage variations result from the cardiac muscle's depolarization and repolarization, leading to electrical changes which reach our body's surface. An average cardiac pulse is represented streamlined in the upper part of the Fig. 1.

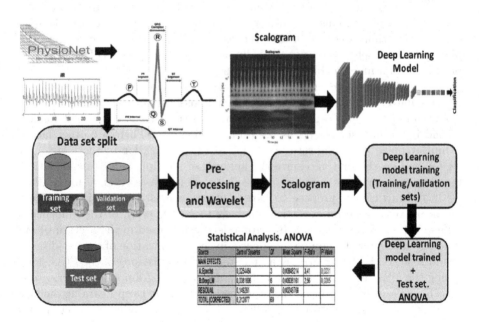

Fig. 1. Block diagram of the methodology used in this paper to perform the ANOVA for the different Deep Learning model used for ECG classification

The nomenclature used in the electrocardiogram will be determined by the diastolic phase, represented by the isoelectric line containing two positive potentials over it and two negative ones beneath. On the other side, the systolic phase is formed by two opposing processes, myocardial activation and recovery, designated by the letters P, Q, R, S, T and U.

3 Arrhythmias

It is considered an arrhythmia any abnormal heart rhythm due to alterations in the creation or transmission of the impulses. In pathological settings, the depolarization can be initiated outside the SA node and several kinds of extra-systoles or ectopic beats may be present. Electrical impulses may also be blocked, accelerated, deviated by alternative trajectories and can change their origin from one beat to the next, resulting in several kind of blockages and anomalous transmissions. In both cases the effects they have on the ECG can be changes to the signal's morphology and durations of it's waves and intervals, as well as the absence of some of it's waves [7,21,24]. From a clinical standpoint, arrhythmia classification is based on its origin and the way they are represented. Depending on it's origin they are divided into supraventricular (sinus node, atria and AV junction) and ventricular; based on their presentation they can be either paroxysmal or permanent. Under regular circumstances, the electrical impulse originates in the sinus node, travels through the atria depolarizing them (P wave) and reaches the atrioventricular node (AV node). From there, through the Bundle of His, it will attempt to reach the ventricles traveling simultaneously by the two branches of the Bundle, depolarizing. When both ventricles are depolarized at the same time we observe a narrow QRS complex, meanwhile if one is depolarized before the other, as happens during branch blockages or in the presence of an ectopic ventricular focus, the QRS complex is wide [2,14].

4 Discrete Wavelet Transform (DWT)

The Wavelet Transform can be described in several ways. An elegant way is Mallats Multi-resolution Analysis. It describes a Wavelet Transform as a process to represent a function with different approximation levels [22]. The first level estimates the function very generally, projecting it onto a space generated by two large-scale base functions. The second level is a bit less general (using four base functions) and so on. The last level will completely represent the signal, and this representation's inverse transform will recreate the original signal. Thus the DWT may be used to partition the characteristic vector of a sub-space sequence, each of them representing a different scale level. The DWT algorithm operates by transforming the initial characteristic vector into a new factor sequentially filled with the different scales' wavelet coefficients. Each scale corresponds with a different dilation of the Mother Wavelet. The lesser coefficients represent the largest scale characteristics (low frequency) in the original characteristic vector, and the larger ones represent the smallest scale characteristics. The discretization enables the signal to be represented in terms of elemental functions accompanied by coefficients

$$f(t) = \sum_{\lambda} c_\lambda \varphi_\lambda \tag{1}$$

In Wavelet systems, the Mother Wavelets $\psi(t)$ have several scale functions $\phi(t)$ associated, the first ones are in charge of representing the function's small details, while the scale functions perform the approximation. It is thus possible to represent a signal $f(t)$ as a summation of wavelet functions and scale functions:

$$f(t) = \sum_{k}\sum_{j} c_{j,k}\phi(t) + \sum_{k}\sum_{j} d_{j,k}\psi(t) \tag{2}$$

4.1   Scale Functions and Wavelet Functions

A way of discretizing the scale and frequency parameters is through exponential sampling, to guarantee a more accurate approximation, by which we may redefine the parameters to discreet values as follows:

$$s = a^{-j} \quad u = kna^{-j} \tag{3}$$

we obtain the discretized function family, which form Wavelet orthonormal bases in $\mathbb{L}^2(\mathbb{R})$.

$$\psi_{j,k}(t) = \frac{1}{\sqrt{a^{-j}}}\psi\left(\frac{t - kna^{-j}}{a^{-j}}\right) = a^{\frac{j}{2}}\psi\left(a^j t - kn\right) \tag{4}$$

To obtain a better approximation of the signal on very fine resolution levels, it is required the Wavelets be dilated by a factor of 2^{-j}, allowing for a resolution of 2^j, this functions are called Dyadic Wavelets.

$$\psi_{j,k}(t) = 2^{\frac{j}{2}}\psi\left(2^j t - kn\right) \quad j,k \in \mathbb{Z} \tag{5}$$

The DWT takes the form:

$$DW\,\mathrm{Wf}(j,k) = \langle f, \psi_{j,k}\rangle = \int_{-\infty}^{\infty} f(t)\psi_{j,k}(t)dt$$
$$DW\,\mathrm{Tf}(j,k) = \int_{-\infty}^{\infty} f(t)2^{\frac{j}{2}}\psi\left(2^j t - kn\right) dt \tag{6}$$

Given the preceding procedure, it is possible to generate a family of scale functions defined:

$$\phi_{j,k}(t) = 2^{\frac{j}{2}}\phi\left(2^j t - kn\right) \quad j,k \in \mathbb{Z} \tag{7}$$

The general representation of signal $f(t)$ will be in the form

$$f(t) = \sum_{k}\sum_{j} c_{j,k}2^{\frac{j}{2}}\phi\left(2^j t - kn\right) + \cdots$$
$$+ \sum_{k}\sum_{j} d_{j,k}2^{\frac{j}{2}}\psi\left(2^j t - kn\right) \tag{8}$$

4.2 Scale Coefficients ($c_{j,k}$) and Wavelet Coefficients ($d_{j,k}$)

In order to represent a signal $f(t)$ and taking into account Eq. 8, we must find the values of the coefficients $(c_{j,k})$ and $(d_{j,k})$, which finally allow for the approximation of the signal. They are the result of a vector multiplication between function $f(t)$ and scale function (ϕ) or wavelet function (ψ). For the scale coefficients we have

$$
\begin{aligned}
c_{j,k} &= <f(t), \phi_{j,k}(t)> = \int_{-\infty}^{\infty} |f(t)\phi_{j,k}(t)|\, dt \\
<f(t), \phi_{j,k}> &= c_{j,-\infty} <\phi_{j,-\infty}(t), \phi_{j,k)(t)}> + \dots \\
&\quad + c_{j,k} <\phi_{j,k}(t), \phi_{j,k)(t)}> + \dots \\
&\quad + c_{j,-\infty} <\phi_{j,\infty}(t), \phi_{j,k)(t)}>
\end{aligned}
\tag{9}
$$

Since the wavelet and scale functions satisfy the orthonormality property, we can be sure one of the vector products will be different from zero, $(<\phi_{,k}(t), \phi_{,m}(t) = \delta(k-m))$ or $(<\psi_{,k}(t), \phi_{,m}(t) = \delta(k-m))$ thus

$$
\begin{aligned}
c_{j,k} &= <f(t), \phi_{j,k}(t)> \\
&= \int_{t_1}^{t_2} f(t)\phi_{j,k}\left(2^j t - k\right) dt
\end{aligned}
\tag{10}
$$

Equally for the Wavelet coefficients

$$
\begin{aligned}
d_{j,k} &= <f(t), \psi_{j,k}(t)> \\
&= \int_{t_1}^{t_2} f(t)\psi_{j,k}\left(2^j t - k\right) dt
\end{aligned}
\tag{11}
$$

The Discreet Wavelet Transform has been frequently used in ECG study and characterization [1, 3, 4, 13, 15, 26].

4.3 Scalogram Based on Discrete Wavelet Transform

The revolution of wavelet theory comes exactly from this fact: the two parameters (time u and scale s) of the CWT (or DWT) enable the study of a signal in both domains (time and frequency) simultaneously, with a resolution depending on the scale of interest. According to these considerations, the CWT provides a time-frequency decomposition of f in the so called time-frequency plane. This approach, as it is discussed in [9], is more accurate and efficient than other procedures such as the windowed Fourier transform (WFT).

The scalogram of f is defined by the function [8]:

$$
\mathcal{S}(s) := \|Wf(s, u)\| = \left(\int_{-\infty}^{+\infty} |Wf(s, u)|^2\, du\right)^{\frac{1}{2}},
\tag{12}
$$

representing the energy of Wf at a scale s. Obviously, $\mathcal{S}(s) \geq 0$ for all scale s, and if $\mathcal{S}(s) > 0$ we will say that the signal f has details at scale s. Thus, the scalogram allows the detection of the most representative scales (or frequencies) of a signal, that is, the scales that contribute the most to the total energy of the signal [5, 8].

If we are only interested in a set time interval $[t_0, t_1]$, we can define the associated windowed scalogram by

$$S_{[t_0,t_1]}(s) := \|Wf(s,u)\|_{[t_0,t_1]} = \left(\int_{t_0}^{t_1} |Wf(s,u)|^2 \, du \right)^{\frac{1}{2}}. \tag{13}$$

Therefore, the scalogram of f captures the "energy" of the CWT of the time series f at this particular scale [5]. It allows for the identification of the most representative scales of a time series, that is, the scales that contribute most to its total energy. The reason behind the use of this measure is that if two time series (in our study these will be ECG, already analyzed in the bibliography, for example in [3, 6, 19, 26]) display a similar pattern, then their scalograms should be very similar. In this regard, it is vital to point out certain requisites for when two time series have the same scalogram.

5   Data Set. Physionet

In this paper, we will focus on three patient groups of the Physionet portal (https://physionet.org/): people with cardiac arrhythmia (ARR), people with congestive heart failure (CHF), and people with normal sinus rhythms (NSR). Despite the possibility of expanding it, the computational need would be greatly increased, and having three categories already we consider to be before an multi-class problem.

Thus we will use the following three databases:

- **MIT-BIH Arrhythmia Database.** The MIT-BIH Arrhythmia Database encompasses 48 half-hour excerpts of two-channel ambulatory ECG recordings, collected from 47 subjects studied by the BIH Arrhythmia Laboratory between 1975 and 1979. Twenty-three recordings were chosen at random from a collection of 4000 24-h ambulatory ECG recordings obtained from a mixed population of inpatients (about 60%) and outpatients (about 40%) at Boston's Beth Israel Hospital; the other 25 recordings were chosen from the same set to include less frequent but still clinically significant arrhythmias that would not be well-represented in a small random sample. About half (25 of 48 complete records, and reference annotation files for all 48 records) of this database has been freely available here since PhysioNet's birth in September 1999. For our study specifically, 96 recordings from persons with arrhythmia will be used.
- **MIT-BIH Normal Sinus Rhythm Database.** This database contains 18 long-term ECG recordings of subjects referred to the Arrhythmia Laboratory at Boston's Beth Israel Hospital (now the Beth Israel Deaconess Medical Center). Subjects presented in this database were found to have had no significant arrhythmias; they include 5 men, aged 26 to 45, and 13 women, aged 20 to 50 (more details in https://www.physionet.org/content/nsrdb/1.0.0/). For our

study specifically, 36 recordings from persons with normal sinus rhythms will be used.

– **BIDMC Congestive Heart Failure Database.** This database contains long-term ECG recordings from 15 subjects (11 men, aged 22 to 71, and 4 women, aged 54 to 63) suffering from severe congestive heart failure (NYHA class 3–4). This group of subjects was part of a larger study group receiving conventional medical therapy prior to receiving an oral inotropic agent, milrinone. A number of other studies have made use of these recordings. The individual recordings are about 20 h in duration each, and include two ECG signals sampled at 250 samples per second with 12-bit resolution over a ±10 millivolt range. The original analog recordings were made at Boston's Beth Israel Hospital (now the Beth Israel Deaconess Medical Center) using ambulatory ECG recorders with a typical recording bandwidth of approximately 0.1 Hz to 40 Hz. Annotation files (with the suffix .ecg) were prepared using an automated detector and have not been corrected manually.

5.1   ECG an Scalogram Examples from the Database

This section attempts to display some of the patients' ECG signals, for all three groups, together with their Scalogram.

6   Different Deep Learning Models for ECG Classification

In the last few years, thanks to the advancements made to powerful computing platforms (specially GPU), the field dedicated to Deep Learning has had a large impact on problems related to biomedical information treatment (signals as well as images). Succinctly, this Deep Learning methodology is based on the implementation of multi leveled deep neural networks imitating the human brain's own neural networks.

The models have been selected in order to use architectures that are diverse in size and performance. We have not tried to make an exhaustive analysis of which models to select, but rather, we have taken models that normally show good performance in image classification tasks and whose architectures are varied in number of parameters and layers. Thus, the models used were *VGG16 and VGG19* [18], *MobileNetv2* [11], *ResNet50 and ResNet101* [10] and *DenseNet201* [12].

Table 1 summarizes different deep learning models that were used in this paper alongside pre-trained weights (using information from Keras Applications). Depth refers to the topological depth of the network (including activation layers, batch normalization layers, etc.).

7   Results

In this section, we will proceed to present the results obtained from the ANOVA statistical analysis, regarding the repercussion of different functional blocks

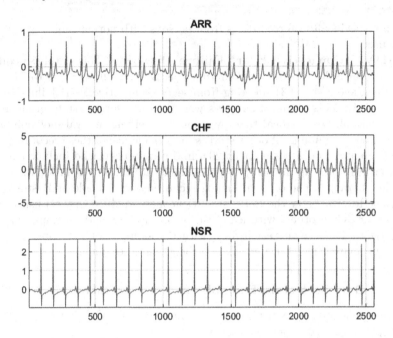

Fig. 2. ECG example for a different set of three patients.

inside the deep learning structure, based on the Scalogram (Fig. 3) of the ECG (Fig. 2) on the classification system (from an accuracy point of view).

The factors considered in the analysis are listed in the Table 2. By analyzing the different levels of each of these factors, it is possible to determine their influence on the performance of the analyzed system (accuracy) when different deep learning models and epochs for training are used.

Table 3 gives the two-way variance analysis for the whole set of processing examples of the deep learning system analysed in this contribution. The ANOVA table containing the sum of squares, degrees of freedom, mean square, test statistics, etc., represents the initial analysis in a compact form. This kind of tabular representation is customarily used to set out the results of the ANOVA calculations.

The p-value is often set at 0.05. Any value lower than this leads to significant effects, while any value higher than this leads to non-significant effects.

With respect to the experiment above, this would mean that the main effects have a significant effect on the accuracy of the system. As can be seen from Table 3, the factor number of epochs has the bigger F-Ratio.

The Table 4 shows the mean Accuracy for each level of the factors. It also shows the standard error of each mean, which is a measure of its sampling variability. The rightmost two columns show 95,0% confidence intervals for each of the means.

Thus, a detailed analysis will be performed now for each of the factors examined, using the Multiple Range Test (MRT). the MRT table applies a multiple

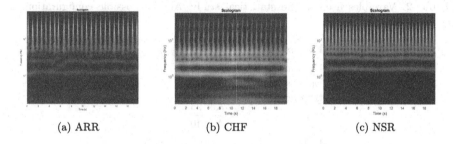

(a) ARR   (b) CHF   (c) NSR

Fig. 3. Scalogram for different ECG signals.

Table 1. Summary of Different Neural Network Architectures

Model	Size (MB)	Parameters	Depth	Main Features
VGG16	528	138.4M	16	VGG16 and VGG19 use multiple convolutional layers with a small kernel size (3 × 3) to replace a single convolutional layer with a large kernel size
VGG19	549	143.7M	19	VGG19 has a higher number of CNN
ResNet50	98	25.6M	107	It provides an innovative way to add more convolutional layers to a CNN
ResNet101	171	44.7M	209	
MobileNet	16	4.3M	55	It is a simple, yet efficient and computationally inexpensive convolutional neural network for mobile vision applications
MobileNetv2	14	3.5M	105	
DenseNet121	33	8.1M	242	DenseNets simplify the connectivity pattern between layers
DenseNet201	80	20.2M	402	

comparison procedure to determine which means are significantly different from which others. This is presented in Table 5 and Table 6.

From Table 5, the bottom half of the output shows the estimated difference between each pair of means. An asterisk has been placed next to 3 pairs, indicating that these pairs show statistically significant differences at the 95,0% confidence level. At the top of the page, 2 homogenous groups are identified using columns of letters. Within each column, the levels containing the same letter form a group of means within which there are no statistically significant differences. The method currently being used to discriminate among the means is Fisher's least significant difference (LSD) procedure. With this method, there is a 5,0% risk of calling each pair of means significantly different when the actual difference equals 0.

Table 2. Variables used in the statistical study. All the possible configurations of factors levels.

Factors	Levels
D: Number of Epochs for each training DL	A1: 10
	A2: 17
	A3: 25
	A4: 40
E: Deep Learning Model	B1: GoogleNet
	B2: Resnet-50
	B3: Resnet-101
	B4: VGG16
	B5: VGG19
	B6: Densenet-201
	B7: Mobilenet-v2

Table 3. Analysis of Variance for Accuracy - Type III Sums of Squares

Source	Sum of Squares	Df	Mean Square	F-Ratio	P-Value
MAIN EFFECTS					
A:Epochs	0,0254464	3	0,00848214	3,41	0,0231
B:DeepLM	0,0381696	6	0,00636161	2,56	0,0285
RESIDUAL	0,149261	60	0,00248768		
TOTAL (CORRECTED)	0,212877	69			

Table 4. Table of least squares means for accuracy with 95% confidence intervals

Level	Count	Mean	Stnd. Error	Lower Limit	Upper Limit
GRAND MEAN	70	0,802827			
Epochs					
A1	21	0,77753	0,010884	0,755759	0,799301
A2	21	0,789435	0,010884	0,767663	0,811206
A3	21	0,798363	0,010884	0,776592	0,820134
A4	7	0,845982	0,0188516	0,808273	0,883691
DeepLM					
B1	10	0,813318	0,0160515	0,781211	0,845426
B2	10	0,785193	0,0160515	0,753086	0,817301
B3	10	0,789881	0,0160515	0,757773	0,821989
B4	10	0,838318	0,0160515	0,806211	0,870426
B5	10	0,825818	0,0160515	0,793711	0,857926
B6	10	0,802381	0,0160515	0,770273	0,834489
B7	10	0,764881	0,0160515	0,732773	0,796989

Similar analysis can be performed for Table 6, in which an asterisk has been placed next to 5 pairs, indicating that these pairs show statistically significant differences at the 95,0% confidence level. At the top of the page, 3 homogenous groups are identified using columns of letters.

Table 5. Multiple Range Tests for Accuracy by Epochs

Method: 95,0 percent LSD

Epochs	Count	LS Mean	LS Sigma	Homogeneous Groups
A1	21	0,77753	0,010884	A
A2	21	0,789435	0,010884	A
A3	21	0,798363	0,010884	A
A4	7	0,845982	0,018851	B

Contrast	Sig.	Difference	+/- Limits
A1 - A2		-0,0119048	0,0307892
A1 - A3		-0,0208333	0,0307892
A1 - A4	*	-0,0684524	0,0435424
A2 - A3		-0,00892857	0,0307892
A2 - A4	*	-0,0565476	0,0435424
A3 - A4	*	-0,047619	0,0435424

* denotes a statistically significant difference.

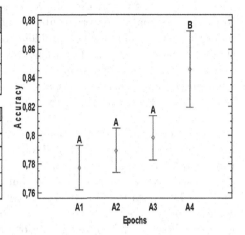

Table 6. Multiple Range Tests for Accuracy by Deep Learning Models (DeepLM)

Method: 95,0 percent LSD

DeepLM	Count	LS Mean	LS Sigma	Homogeneous Groups
B7	10	0,764881	0,0160515	A
B2	10	0,785193	0,0160515	AB
B3	10	0,789881	0,0160515	AB
B6	10	0,802381	0,0160515	ABC
B1	10	0,813318	0,0160515	BC
B5	10	0,825818	0,0160515	BC
B4	10	0,838318	0,0160515	C

Contrast	Sig.	Difference	+/- Limits
B1 - B2		0,028125	0,0446177
B1 - B3		0,0234375	0,0446177
B1 - B4		-0,025	0,0446177
B1 - B5		-0,0125	0,0446177
B1 - B6		0,0109375	0,0446177
B1 - B7	*	0,0484375	0,0446177
B2 - B3		-0,0046875	0,0446177
B2 - B4	*	-0,053125	0,0446177
B2 - B5		-0,040625	0,0446177
B2 - B6		-0,0171875	0,0446177
B2 - B7		0,0203125	0,0446177
B3 - B4	*	-0,0484375	0,0446177
B3 - B5		-0,0359375	0,0446177
B3 - B6		-0,0125	0,0446177
B3 - B7		0,025	0,0446177
B4 - B5		0,0125	0,0446177
B4 - B6		0,0359375	0,0446177
B4 - B7	*	0,0734375	0,0446177
B5 - B6		0,0234375	0,0446177
B5 - B7	*	0,0609375	0,0446177
B6 - B7		0,0375	0,0446177

* denotes a statistically significant difference.

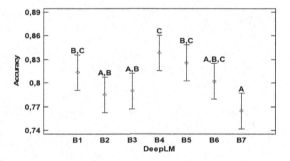

8   Conclusion

It is frequent in the literature using a certain machine learning system without
an in depth and exhaustive analysis of the impact that different alternatives of
that model has on the behaviour of the system. In this paper we have carried out

an exhaustive analysis of how a certain Deep Learning model (analysing seven models widely used in the literature) influences an important classification problem: the automatic determination of a cardiac pathology from ECG information. The first phase of the methodology consists of applying the Wavelet transform to the ECG signal, in order to subsequently obtain the Scalogram image of the different ECGs available. The problem to be solved is multi-class, as there are three different types of ECG patients. We are going to use MIT-BIH Arrhythmia Database (i.e. patients suffering from arrhythmia, MIT-BIH Normal Sinus Rhythm Database (i.e. healthy patients without arrhythmia) and BIDMC Congestive Heart Failure Databas (patients suffering from severe congestive heart failure).

These scalograms are used as input to the different Deep Learning models, in which the influence of having different numbers of epochs for training is also investigated. To examine the behaviour of the different alternatives, a statistical study based on ANOVA was carried out. The conclusion of this article is that both the Deep Learning model and the number of epochs have an impact on the accuracy of the system. For the number of epochs, there are two homogeneous groups, a first group consisting of a low number of epochs (A1: 10 epochs ,A2: 17 epochs and A3: 25 epochs) and a fourth group consisting of a high number of epochs (A4: 40 epochs). The range of accuracy varies from 77.75% on average for 10 epochs, to 84.59% for 40 epochs. The different Deep Learning models are also examined, there are four homogeneous groups, with the model with the lowest accuracy index obtaining an average value of 76.48% in the classification (Mobilenet-v2 Deep Learning model) and the model with the best results, with an average accuracy of 83.83% (VGG16 Deep Learning model).

Acknowledgements. This work was funded by the Spanish Ministry of Sciences, Innovation and Universities under Project PID2021-128317OB-I00 and the projects from Junta de Andalucia P20-00163 and A-TIC-530-UGR20.

References

1. Ahmed, S.M., Al-Ajlouni, A.F., Abo-Zahhad, M., Harb, B.: ECG signal compression using combined modified discrete cosine and discrete wavelet transforms. J. Med. Eng. Technol. **33**(1), 1–8 (2009)
2. Ali, M., Haji, A.Q., Kichloo, A., Grubb, B., Kanjwal, K.: Inappropriate sinus tachycardia: a review. Rev. Cardiovasc. Med. **22**(4), 1331 (2021). https://doi.org/10.31083/j.rcm2204139
3. Arefnezhad, S., Eichberger, A., Frühwirth, M., Kaufmann, C., Moser, M., Koglbauer, I.V.: Driver monitoring of automated vehicles by classification of driver drowsiness using a deep convolutional neural network trained by scalograms of ECG signals. Energies **15**(2), 480 (2022)
4. Asaduzzaman, K., Reaz, M.B.I., Mohd-Yasin, F., Sim, K.S., Hussain, M.S.: A study on discrete wavelet-based noise removal from EEG signals. In: Arabnia, H. (ed.) Advances in Computational Biology. Advances in Experimental Medicine and Biology, vol. 680, pp. 593–599. Springer, New York (2010). https://doi.org/10.1007/978-1-4419-5913-3_65

5. Bolós, V.J., Benítez, R.: The wavelet scalogram in the study of time series. In: Casas, F., Martínez, V. (eds.) Advances in Differential Equations and Applications. SSSS, vol. 4, pp. 147–154. Springer, Cham (2014). https://doi.org/10.1007/978-3-319-06953-1_15

6. Chourasia, V.S., Tiwari, A.K., Gangopadhyay, R.: Time-frequency characterization of fetal phonocardiographic signals using wavelet scalogram. J. Mech. Med. Biol. **11**(02), 391–406 (2011)

7. Gacek, A.: An introduction to ECG signal processing and analysis. In: Gacek, A., Pedrycz, W. (eds.) ECG Signal Processing, Classification and Interpretation, pp. 21–46. Springer, London (2011). https://doi.org/10.1007/978-0-85729-868-3_2

8. Ghobber, S.: Some results on wavelet scalograms. Int. J. Wavelets Multiresolut. Inf. Process. **15**(03), 1750019 (2017)

9. Gomes, J., Velho, L.: Orthogonal wavelets. In: Gomes, J., Velho, L. (eds.) From Fourier Analysis to Wavelets. IM, vol. 3, pp. 143–155. Springer, Cham (2015). https://doi.org/10.1007/978-3-319-22075-8_11

10. He, K., Zhang, X., Ren, S., Sun, J.: Deep residual learning for image recognition (2015). https://doi.org/10.48550/ARXIV.1512.03385, https://arxiv.org/abs/1512.03385

11. Howard, A.G., et al.: MobileNets: efficient convolutional neural networks for mobile vision applications. arXiv preprint arXiv:1704.04861 (2017)

12. Huang, G., Liu, Z., Van Der Maaten, L., Weinberger, K.Q.: Densely connected convolutional networks. In: 2017 IEEE Conference on Computer Vision and Pattern Recognition (CVPR), pp. 2261–2269 (2017). https://doi.org/10.1109/CVPR.2017.243

13. Jang, S.W., Lee, S.H.: Detection of ventricular fibrillation using wavelet transform and phase space reconstruction from ECG signals. J. Mech. Med. Biol. **21**(09) (2021). https://doi.org/10.1142/s0219519421400364

14. Leguizamón, G., Coello, C.A.: An introduction to the use of evolutionary computation techniques for dealing with ECG signals. In: Gacek, A., Pedrycz, W. (eds.) ECG Signal Processing, Classification and Interpretation, pp. 135–153. Springer, London (2011). https://doi.org/10.1007/978-0-85729-868-3_6

15. Minhas, F.U.A.A., Arif, M.: Robust electrocardiogram (ECG) beat classification using discrete wavelet transform. Physiol. Meas. **29**(5), 555–570 (2008)

16. Nandhini, S., Ashokkumar, K.: An automatic plant leaf disease identification using DenseNet-121 architecture with a mutation-based henry gas solubility optimization algorithm. Neural Comput. Appl. **34**(7), 5513–5534 (2022)

17. Panda, M.K., Sharma, A., Bajpai, V., Subudhi, B.N., Thangaraj, V., Jakhetiya, V.: Encoder and decoder network with ResNet-50 and global average feature pooling for local change detection. Comput. Vis. Image Underst. **222**, 103501 (2022)

18. Simonyan, K., Zisserman, A.: Very deep convolutional networks for large-scale image recognition. arXiv preprint arXiv:1409.1556 (2014)

19. Singh, S.A., Majumder, S.: A novel approach OSA detection using single-lead ECG scalogram based on deep neural network. J. Mech. Med. Biol. **19**(04), 1950026 (2019)

20. Srinivasu, P.N., SivaSai, J.G., Ijaz, M.F., Bhoi, A.K., Kim, W., Kang, J.J.: Classification of skin disease using deep learning neural networks with MobileNet v2 and LSTM. Sensors **21**(8), 2852 (2021)

21. Varma, N.: Bidirectional atrial tachycardia ablated from an aortic sinus. JACC Clin. Electrophysiol. **7**(10), 1326–1327 (2021). https://doi.org/10.1016/j.jacep.2021.06.013

22. Vyas, A., Yu, S., Paik, J.: Wavelets and wavelet transform. In: Vyas, A., Yu, S., Paik, J. (eds.) Multiscale Transforms with Application to Image Processing. Signals and Communication Technology, pp. 45–92. Springer, Singapore (2018). https://doi.org/10.1007/978-981-10-7272-7_3

23. Wang, W., Li, Y., Zou, T., Wang, X., You, J., Luo, Y.: A novel image classification approach via dense-MobileNet models. Mob. Inf. Syst. **2020**, 1–8 (2020)

24. Wasilewski, J., Poloński, L.: An introduction to ECG interpretation. In: Gacek, A., Pedrycz, W. (eds.) ECG Signal Processing, Classification and Interpretation, pp. 1–20. Springer, London (2011). https://doi.org/10.1007/978-0-85729-868-3_1

25. Wen, L., Li, X., Gao, L.: A transfer convolutional neural network for fault diagnosis based on ResNet-50. Neural Comput. Appl. **32**(10), 6111–6124 (2019). https://doi.org/10.1007/s00521-019-04097-w

26. Yanık, H., Değirmenci, E., Büyükakıllı, B., Karpuz, D., Kılınç, O.H., Gürgül, S.: Electrocardiography (ECG) analysis and a new feature extraction method using wavelet transform with scalogram analysis. Biomed. Tech. (Berl.) **65**(5), 543–556 (2020)

Transparent Machine Learning Algorithms for Explainable AI on Motor fMRI Data

José Diogo Marques dos Santos[1,2(✉)] ⓘ, David Machado[1,2] ⓘ,
and Manuel Fortunato[1,2] ⓘ

[1] Faculty of Engineering, University of Porto,
R. Dr Roberto Frias, 4200-465 Porto, Portugal
up201908014@up.pt
[2] Abel Salazar Biomedical Sciences Institute, University of Porto,
R. Jorge de Viterbo Ferreira, 4050-313 Porto, Portugal

Abstract. With the emergence of explainable artificial intelligence (xAI), two main approaches for tackling model explainability have been put forward. Firstly, the use of inherently simple and transparent models that with easily understandable inner-workings (interpretability) and can readily provide useful knowledge about the model's decision making process (explainability). The second approach is the development of interpretation and explanation algorithms that may shed light upon black-box models. This is particularly interesting to apply on fMRI data as either approach can provide pertinent information about the brain's underlying processes. This study aims to explore the capability of transparent machine learning algorithms to correctly classify motor fMRI data, if more complex models inherently lead to a better prediction of the motor stimulus, and the capability of the Integrated Gradients method to explain a fully connected artificial neural network (FCANN) used to model motor fMRI data. The transparent machine learning models tested are Linear Regression, Logistic Regression, Naive Bayes, K-Neighbors, Support Vector Machine, and Decision Tree, while the Integrated Gradients method is tested on a FCANN with 3 hidden layers. It is concluded that the transparent models may accurately classify the motor fMRI data, with accuracies ranging from 66.75% to 85.0%. The best transparent model, multinomial logistic regression, outperformed the most complex model, FCANN. Lastly, it is possible to extract pertinent information about the underlying brain processes via the Integrated Gradients method applied to the FCANN by analyzing the spatial expression of the most relevant Independent Components for the FCANN's decisions.

Keywords: Human Connectome Project · HCP · fMRI · machine learning · logistic regression · linear regression · decision tree · naive bayes · k-neighbors · SVM · artificial neural network

ⓒ The Author(s), under exclusive license to Springer Nature Switzerland AG 2023
I. Rojas et al. (Eds.): IWBBIO 2023, LNBI 13920, pp. 413–427, 2023.
https://doi.org/10.1007/978-3-031-34960-7_29

1   Introduction

Functional magnetic resonance imaging (fMRI) is a non-invasive imaging technique that is used to measure brain activity. An increase in blood flow is detected and the resulting images are used to create maps of brain activity [1].

FMRI data is particularly difficult for most machine learning (ML) classifiers as it has a characteristic unbalance between number of inputs and possible training epochs, which makes the data dimensionality reduction an adequate solution, amongst other challenges, such as a generally low signal to noise ratio and correlated data.

The Human Connectome Project (HCP) [2] is a large-scale research effort aimed at mapping the structural and functional connections within the human brain.

ML classifiers are algorithms that automatically learn patterns on training data and make predictions on new data [3,4]. Here, several supervised learning models are tested on motor fMRI data. The ML algorithms tested are Linear Regression, Logistic Regression, Naive Bayes, K-Neighbors, Support Vector Machine [5,6] and Decision Tree, with the use of an external library (*scikit-learn*).

In the field of explainable artificial intelligence (xAI), an increased importance has been given to the need for easily explainable models that allow for the extraction of useful knowledge from the model's decision making process. Therefore, although deep learning approaches may achieve higher levels of accuracy, these models often have complicated inner-workings, which hinder the explanation process. To overcome such shortcomings two different approaches have been put forward, the use of more transparent models and the creation of model interpretation and explanation methods. This study focus mainly on the first approach, as different simple machine learning models are compared on their capability of analysing motor fMRI data. However, it also partially covers the second approach by including a fully connected aritificial neural network (FCANN) and applying an integrated gradient methods to compute feature importance.

Thus, the present study aims to answer to the questions:

- Are simple transparent machine learning algorithms capable of accurately classifying motor fMRI data?
- Do more complex models necessarily lead to better prediction accuracy?
- Is it possible to extract relevant information about the pertinent underlying brain processes captured in the motor fMRI via the integrated gradients method of computing an ANN's feature importance?

2   Methods

2.1   Dataset

As a dataset, 80 subjects of the HCP Young Adult database motor paradigm subset are used. The paradigm is adapted from [7,8], and consists of two runs per subject. Each run encompasses a sequence of five possible stimuli: left foot (lf), left hand (lh), right foot (rf), right hand (rh) and tongue (t). For the feet stimuli the subjects are asked to squeeze the prompted foot, for the hand the subjects are asked to tap their fingers and for the tongue the subjects are asked to move it. Each stimuli has a 12 s stimulus response time that is preceded by a 3 s cue. For each run there are also three 15 s long fixation crosses, at the beginning, the end and one within the sequence. The runs differ between themselves by sequence of stimuli and by the scanner's phase encoding, one is left to right (LR) while the other is right to left (RL) phase encoding. The stimuli sequences are:

– LR sequence: FIX-RH-LF-T-RF-LH-FIX-T-LF-RH-FIX-LH-RF-FIX;
– RL sequence: FIX-LH-RF-FIX-T-LF-RH-FIX-LH-T-RF-RH-LF-FIX.

The sequences do not differ between subjects. The subjects make no response in this paradigm. Each run has a duration of 3 min and 34 s and the repetition time (TR) is 0.72 s. Each subjects originates two fMRI files with 284 volumes each, a volume is the 3 dimensional image correspondent to a time-point. The first fixation cross is cut to synchronise all the files between runs and subjects. Data in the files is already subject to brain extraction and registered to a standard brain image.

2.2 Data Processing

The pre-processing and processing of fMRI data is adapted from [9] and [10]. To reduce data dimensionality, an Independent Component Analysis (ICA) is run for 20 of the subjects (test set subjects), yielding 46 Independent Components (IC). The ICA is implemented in MELODIC ((Multivariate Exploratory Linear Optimized Decomposition into Independent Components) v. 3.15 [11], part of FSL (FMRIB Software Library). It is important to note that, in this version of MELODIC, the MIGP (MELODIC's Incremental Group-PCA) default value is true, this imposes limitations as MIGP precludes the output of ICs' complete time-courses. Therefore, this step is run in line command with the option –disableMigp. The input data underwent several pre-processing steps before analysis. These steps included masking out non-brain voxels, removing the mean value from each voxel, and normalizing the variance at each voxel.

The pre-processed data is then whitened and reduced to 46 dimensions using probabilistic Principal Component Analysis, an example of one of these dimensions (IC7) is represented in Fig. 1. The number of dimensions is determined using the Laplace approximation to the Bayesian estimation method [11,12]. Next, the whitened observations are separated into different types of variations: those that occur over time (time-courses), across participants or sessions (subject/session domain), and across different spatial locations (maps), using a fixed-point iteration method that optimized for non-Gaussian spatial source distributions [13].

In order to extract relevant information from the 46 ICs, that are obtained from the pre-processed data, the following feature extraction strategy is implemented: taking the average of the seventh, eighth, and ninth signals that occurred

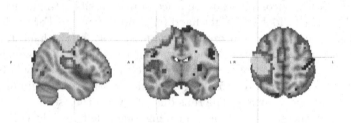

Fig. 1. Sagittal, coronal and axial views of IC7 in the plans x = 46, y = −10 and z = 56).

after the onset of a stimulus. The time difference between the stimulus onset and this feature extraction point is chosen to be 5.285 s, which is close to the peak of the typical hemodynamic response in the brain [14], maximizing the difference between task activation and the baseline state. For the other 60 subjects (train set subjects), the spatial masks of the ICs are extracted and applied to the fMRI data files. This averaged the voxels selected by the mask which yielded 46 timecourses. Their exclusion from the ICA is done to guarantee independence between train, evaluation and test sets as well as due to running other ICAs for the different sets would offer no guarantee of also obtaining 46 ICs as output.

The feature extraction process used is the same as the one used for the ICA output, through the average of the seventh, eighth and ninth signals after the onset of each stimulus. The obtained values are then standardized. These samples of 46-attribute vectors served as inputs for the models trained to classify each movement associated with each sample.

2.3 Machine Learning Models and Algorithms

Linear Regression. Statistical method that models a linear equation between a dependent variable and independent variables. It assumes that the relationship is linear and that there is no high correlations among the independent variables.

Logistic Regression. Uses a logistic function to model the probability of a binary outcome. The goal is to learn a function that maps the input features to the probability that a given example belongs to a certain class. The predicted probability is then converted into a binary class label using a threshold [4].

The logistic function, which is used to model the probability, is defined as:

$$p(y = 1|x) = \frac{1}{1 + e^{-z}} \wedge z = wx + b \tag{1}$$

where z is the linear combination of the input features and the weight.

In a multi-class classification problem, there are two common strategies:

- The Multinomial Logistic Regression classification algorithm is used to directly predict a categorical label that has more than two classes. The probability of each class is modeled by a softmax function and the prediction is based on the highest probability.
- The one-vs-rest (OvR) strategy involves training a separate binary logistic regression model for each class, where each model is responsible for predicting the probability of a given epoch belonging to the model's class. The prediction of new examples is then made by computing a probability for each class and selecting the one with the highest probability. The one-vs-rest strategy has the benefit of being easy to develop and adaptable to any binary classifier. However, because the decision boundary between the classes may not be as accurate as if the classes are modeled jointly, it can be less accurate than other approaches.

Decision Tree. Operates by recursively splitting the input data based on feature values and represent decisions and their potential outcomes in a tree-like structure. Decision Trees are easily interpretable, handle both categorical and numerical data, but can overfit if the tree becomes too complex.

Naive Bayes. Classification algorithm based on the Bayes theorem, which states that the probability of an event occurring is equal to the prior probability of the event multiplied by the likelihood of the event given the evidence. There are different types of Naive Bayes algorithms (Gaussian, Multinomial, Bernoulli), for a multi-class prediction problem the Multinomial approach is used. The training algorithm consists on the estimation of the probability of each class and each feature and the highest probability is then chosen as the predicted class. Also, Multinomial Naive Bayes assumes that features have multinomial distribution which is a generalization of the binomial distribution. Neither binomial nor multinomial distributions can contain negative values, requiring an extra pre-processing step.

K-Nearest Neighbors. Makes predictions based on the features of the data by finding the K (user-specified parameter) nearest examples in the training data and using the class labels or values of those examples to make a prediction for the new example.

Support Vector Machine. The main idea behind SVM is to find hyperplane that maximally separates the different classes in the dataset. The hyperplane is chosen in such a way that it maximizes the distance between the hyperplane and the closest data points from each class (margin). It can be adapted to multi-class classification using the one-vs-rest approach.

Logistic Regression from Scratch. The logistic regression is re-built from scratch in the two modalities (multinomial and one-vs-rest) with the intent of achieving a more transparent model with further control over its characteristics. During the implementation of these two algorithms, the studied hyperparameters are the learning rate (lr), number of epochs (epochs) and regularization term (lambda).

2.4 Artificial Neural Network Models

Fully Connected Artificial Neural Network (FCANN). Considering the input data type, a FCANN is chosen has an adequate artificial neural network to process the features and classify the inputs. A FCANN is an artificial neural network in which every neuron in one layer is connected to every neuron in the subsequent layer. To implement this artificial neural network we made use of the *Python* library *Pytorch*, that provides all the methods and objects needed to implement such a model.

The architecture of the implemented FCANN is arbitrary and consists of 4 layers, with 3 of them being hidden layers. Each hidden layer has a number of neurons that is tuned to approximate the optimal result. The architecture of the model is implemented using the *Pytorch* module *torch.nn* in a class called *SmallNetwork* which inherits from the module *torch.nn.Module*. In Fig. 2, it is possible to view the different layers implemented in the model.

```
SmallNetwork(
    (hiddenLay1): Linear(in_features=46, out_features=100, bias=True)
    (hiddenLay2): Linear(in_features=100, out_features=50, bias=True)
    (hiddenLay3): Linear(in_features=50, out_features=20, bias=True)
    (output): Linear(in_features=20, out_features=5, bias=True)
    (tanHAct): Tanh()
    (softmax): Softmax(dim=1)
    (dropout): Dropout(p=0.1, inplace=False)
    (batchnorm1): BatchNorm1d(100, eps=1e-05, momentum=0.1, affine=True, track_running_stats=True)
    (batchnorm2): BatchNorm1d(50, eps=1e-05, momentum=0.1, affine=True, track_running_stats=True)
    (batchnorm3): BatchNorm1d(20, eps=1e-05, momentum=0.1, affine=True, track_running_stats=True)
)
```

Fig. 2. Layers of the Model (non-sequencial) (printed using *Google Collab*).

The activation function used between the hidden layers of the FCANN is the hyperbolic tangent function, *tanh*. The *tanh* function maps the input to a value between −1 and 1 and is commonly used in artificial neural networks to enable non-linear relations between the input and output (Eq. 2). Due to the fact that its outputs are centered around zero, it ensures that the gradients are of similar magnitudes, which can help with model optimization.

$$\tanh(x) = \frac{e^x - e^{-x}}{e^x + e^{-x}} \tag{2}$$

Since the *tanH* function can saturate for large values, a batch normalization is implemented between the hidden layers and their respective activation function, which helps to stabilize the distribution of the activations. This aids in preventing an internal covariate shift and stabilizes the training process. Therefore, the input of each activation function is normalized to have a mean of 0 and a standard deviation of 1. It is important to denote that batch normalization does not take place in the validation and test phases.

To improve the model's performance by preventing overfitting, a dropout layer is introduced between the first two hidden layers with a dropout rate that is tuned between 0.00 and 0.25. Dropout consists in a regularization technique that drops out a certain proportion of neurons (sets their value to zero) during training. This can help reduce overfitting because it prevents complex co-adaptations on training data.

The chosen optimizer is Adam, an optimizer algorithm that combines gradient descent with the RMSprop algorithm.

To guide the optimization process, the cross-entropy loss function is implemented as the model's criterion, which measures the difference from the predicted probability distribution to the real class probability (Eq. 3). It is adequate for a multiclass classification model when the real class probability is a one-hot vector.

$$\ell(x, y) = -w_y \log \frac{\exp(x_y)}{\sum_{c=1}^{C} \exp(x_c)} \tag{3}$$

(In Eq. 3, C is the number of classes, y is the true label, x_y is the predicted probability of the true label and x_c is the predicted probability for class c.)

To prevent slow convergence or non-convergence of the model, all of linear layers' weights are initialized through a standard normal distribution and its bias set to zero. This is done through the _init_weights method of the *SmallNetwork*.

Several models are tested with different hyperparameters (their architecture, dropout rate, initial learning rate, Optimizer).

2.5   Evaluation Metrics

For comparison and evaluation purposes, the models performance (their capability to distinguish motor stimuli classes) are computed according to the following metrics:

– Accuracy: describes the global performance of model in labeling classes, a useful metric for a balanced data set. It is calculated by the ratio between the correct predictions (true positives) and the total number of predictions.
– Confusion Matrix: summarizes the performance of a classification algorithm, comparing the actual label with the predicted one. Complements the accuracy metric, providing extra information (each row shows the number of occurrence of the predicted class and each column has the occurrence of the actual class).
– ROC-AUC (Area Under the Curve): measures the capability of a classifier to distinguish each class, used as a score of the ROC curve (Receiver Operating Characteristic). It displays the relation between the true positive and false positive rates, in several thresholds. In the context of multi-class classification models, we applied the one-vs-rest approach for each class. The closer

the AUC value is to unit value, greater is the model's capacity of distinguish between classes. A random classification model has an AUC-Score of approximately 0.5.

2.6   Feature Importance

On the FCANN, explain the model's decision making process, a feature importance analysis is implemented. To accomplish this, we made use of the *Captum* library, a *PyTorch* library that provides state-of-the-art interpretability algorithms for artificial neural networks.

One such state-of-the-art algorithm is the *IntegratedGradients* method, a popular technique for computing feature importance. By calculating the gradients of the model's output with respect to the input features, this approach approximates the impact of each feature on the model's classification output. This is done for each class, by creating an instance of the *IntegratedGradients* method in *Captum*, passing the input and target to it, and then computing the feature importance. The feature importance values are averaged over the entire test set.

3   Results

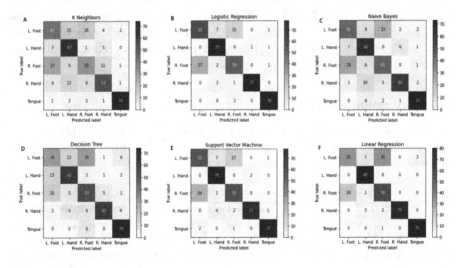

Fig. 3. Confusion matrix of each ML model. A) K-Nearest Neighbors (K = 20). B) Logistic Regression (multinomial). C) Naive Bayes (multinomial). D) Decision Tree. E) Support Vector Machine (kernel = 'linear', C = 1). F) Linear Regression.

The ML algorithm with the highest accuracy is the Logistic Regression (85%), followed by the Support Vector Machine (84.5%), the Linear Regression

Table 1. Test accuracy for each model.

Model	Accuracy (%)
Linear Regression	81.75
Logistic Regression	85.00
Multinomial Naive Bayes	72.25
Decision Tree	71.00
K - Nearest Neighbors	66.75
Support Vector Machine	84.50
Logistic Regression (Multinomial)	85.00
Logistic Regression (One-vs-Rest)	84.50
Artificial Neural Network	84.75

(81.75%). The Multinomial Naive Bayes, the Decision Tree and the K-nearest Neighbors have lower accuracies, around 70%. Analysing the diagonal elements of the confusion matrix of Fig. 3 the Logistic Regression and the Support Vector Machine are the best at correctly classifying the classes (Table 1).

The Logistic Regression algorithm has the best performance from the ML classifiers.

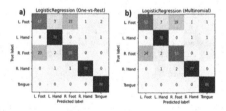

Fig. 4. Confusion matrix for the Logistic Regression model with a) **one-vs-rest** approach (lr = 0.01, epochs = 20000, lambda = 0.0001) and with b) **multinomial** approach (lr = 0.1, epochs = 5000, lambda = 0.0001).

For the logistic regression from scratch, the accuracies of the Multinomial and One-vs-rest approaches have very similar results (85% and 84.5%, respectively), confusion matrices in Fig. 4, and matched the previously obtained accuracy for the logistic model implemented with sk-learn. Figure 5 and Fig. 6 show the ROC curves for each class vs other classes, as well as the probabilities assigned by the model to each input observation. Tongue, left hand and right hand are classes that the models successfully differentiated from the rest, while left foot and right foot classes have worst ROC-AUC scores, in both approaches.

The learning rate that works best for this dataset is 0.1 for the multinomial and 0.01 for the one-vs-rest approach. A small penalization on the weights (lambda of 1E−4, close to zero) and a large enough number of epochs (above 5000) seems to works best.

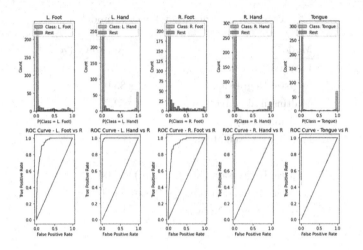

Fig. 5. One-vs-rest probability and ROC curve for each class in the Logistic Regression model with the **multinomial** approach.

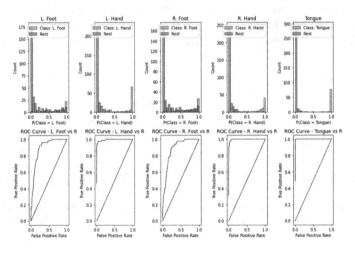

Fig. 6. One-vs-rest probability and ROC curve for each class in the Logistic Regression model with the **one-vs-rest** approach.

3.1   FCANN

For the FCANN, the best model achieved an accuracy of 84.75%. When training with the same specifications, to evaluate the model's reproducibility, we obtained an average accuracy of $(81.80 \pm 3.85)\%$.

The class probabilities distributions and ROC curves for a one-vs-rest analysis of the FCANN is presented in Fig. 7 and the confusion matrix in Fig. 8. Table 2 shows the ROC-AUC scores for the FCANN and the multinomial and one-vs-rest approaches to the logistic regression for each class.

Table 2. ROC-AUC Scores (One-vs-Rest) for the Artificial Neural Network and the two approaches of the Logistic Regression.

Class	LF	LH	RF	RH	T
Multinomial LogR.	0.935	0.998	0.933	0.998	0.999
One-vs-rest LogR.	0.900	0.994	0.880	0.996	0.999
Artificial Neural Network	0.896	0.981	0.924	0.992	0.995

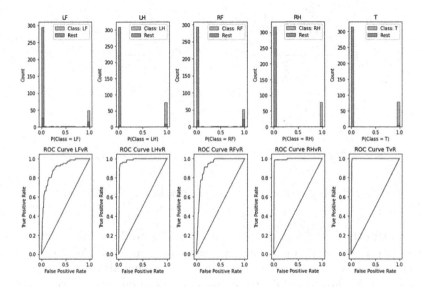

Fig. 7. Class probabilities distributions and ROC Curves for OvR analysis.

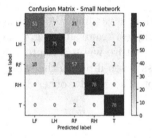

Fig. 8. Confusion Matrix of the FCANN.

3.2   Feature Importance

On Fig. 9, it is possible to view the feature importance values through a bar plot that shows the importance of each feature for each class.

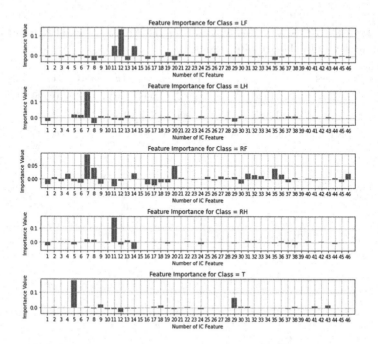

Fig. 9. Feature Importance for the Artificial Neural Network.

Through the analysis of the plots, it is possible to denote that different features have a stronger impact on the classification of each class. In class LF, features 11, 12 and 14 have relatively high positive importance values, whereas in class RH, feature 14 has relatively low negative importance values. Additionally, some features may have similar importance values across multiple classes, such as feature 7 which has high importance values for the LH and RF classes. This suggests that feature 7 is a key feature in the classification of both of those classes.

For classes LH, RH and T, regarding Fig. 8, obtained a near perfect classification throughout their samples. Analysing Fig. 9, we can observe that all three of these classes obtain one feature whose feature value is over 0.1. However the same occurs for the LF class (even if not in the same magnitude), and its results are worse than the ones obtained for the other 3 classes previously mentioned.

4   Discussion

This study aims to answer three main questions:

- Are simple transparent machine learning algorithms capable of accurately classifying motor fMRI data?
- Do more complex models necessarily lead to better prediction accuracy?
- Is it possible to extract relevant information about the pertinent underlying brain processes captured in the motor fMRI via the integrated gradients method of computing an ANN's feature importance?

To answer the first question, every model has an accuracy significantly above 20%, which is the expected accuracy for a completely random classification of a balanced five class problem.

For the ML classifiers, there are two classes in which the models particularly failed the classification task: right foot and left foot. Left foot is miss-classified as right foot motor stimuli and vice-versa in every model. On the other hand, the motor stimuli class that is easier to distinguish is the tongue. According to established neuroscientific knowledge, there are two most likely reasons for this behaviour of the models, the areas of brain activity related to foot movement are along the longitudinal fissure and, due to size of each voxel, it is possible that the acquisition does not have enough spatial resolution to differentiated between an activation on the right or left hemisphere. Another possible reason is related to an unintentional movement of the opposing hand with the foot, as this is a subconscious process associated with helping to retain balance while walking.

The choice between using one-vs-all or multinomial logistic regression can be associated with the computational resources. The multinomial approach requires more computational resources than the one-vs-all approach because the latter trains for each class a single binary classifier, which requires less data, less computation and storage, and the optimization problem is simpler. However, if the data is abundant and computational resources are not a concern, the multinomial approach might be a better choice as it can directly model the relation between the classes.

To answer the second question, through the visualization of Fig. 8, it is apparent that the FCANN struggles to distinguish between the right foot and left foot, most likely due to the same reasons previously mentioned for the ML models. While the models' ability to classify these samples from these classes is not so optimal, its capability to correctly classify elements from the other three classes is highly satisfactory, displaying a higher level of accuracy and proficiency.

The introduction of the dropout layer, as expected, induced a regularization aspect on the models that slowed down the evolution of the training accuracy and allowed for better generalizations in the test set. However the difference is not explored sufficiently to obtain a proper diagnosis on its effect which would be something to take into account to further augment the FCANN's classification ability.

Regarding the logistic regression achieving a higher accuracy than the FCANN, this can possibly be explained by two different approaches. Firstly, the introduction of an ICA step for feature extraction from the fMRI data introduces a linear independency between features that may not accurately model the underlying brain processes. It is also possible that with more training data the FCANN would be able to attain higher prediction accuracy, i.e. with more data from other subjects the FCANN could converge even further during the training process, yielding a model that more accurately predicts the motor stimuli. Therefore, further research is needed to explore the limitations imposed by the ICA by using model-agnostic methods for feature extraction.

To answer the third question, the analysis of the most important features reveals that the most important features are ICs 5, 7, 11 and 12, which is in accordance with the results obtained by [9]. Analysing the spatial dimension of each of these ICs leads us to conclude that, overall, the areas encompassed by them are in accordance with established neuroscientific knowledge. ICs 5 and 12 contain symmetrical activations in the motor cortex while ICs 7 and 11 contain lateralised activations in the right and left hemispheres, respectively. This means we can associate IC 7 with movement on the left hand, although it is present in left hand and right foot stimuli for the reason discussed above. IC 11 can be associated with the right hand stimulus. IC 5 can be associated with movement of the tongue. Lastly, due to its spatial dimension IC 12 would expectedly be associated with left foot and right foot stimuli, however, according to feature importance, it is only associated with the left foot.

In conclusion, the best classification model is the Logistic Regression with the multinomial approach. Therefore, it is concluded that not only can simpler and more transparent methods still be able to accurately classify motor fMRI data, but, as it performed better than the FCANN, the model's complexity does not inherently mean a higher prediction accuracy. Furthermore, the information attained via the computation of feature importance with the integrated gradients method is in accordance with established neuroscientific knowledge.

References

1. Uludağ, K., Dubowitz, D.J., Yoder, E.J., Restom, K., Liu, T.T., Buxton, R.B.: Coupling of cerebral blood flow and oxygen consumption during physiological activation and deactivation measured with fMRI. NeuroImage **23**(1), 148–155 (2004). https://doi.org/10.1016/j.neuroimage.2004.05.013. ISSN 1053-8119
2. Van Essen, D.C., Smith, S.M., Barch, D.M., Behrens, T.E.J., Yacoub, E., Ugurbil, K.: The WU-Minn human connectome project: an overview. Neuroimage **80**, 62–79 (2013). https://doi.org/10.1016/j.neuroimage.2013.05.041
3. Bi, Q., Goodman, K.E., Kaminsky, J., Lessler, J.: What is machine learning? A primer for the epidemiologist. Am. J. Epidemiol. **188**(12), 2222–2239 (2019). https://doi.org/10.1093/aje/kwz189
4. Nasteski, V.: An overview of the supervised machine learning methods. Horizons. b **4**, 51–62 (2017)
5. Pereira, F., Mitchell, T., Botvinick, M.: Machine learning classifiers and fMRI: a tutorial overview. Neuroimage **45**(1 Suppl.), S199–S209 (2009). https://doi.org/10.1016/j.neuroimage.2008.11.007
6. Misaki, M., Kim, Y., Bandettini, P.A., Kriegeskorte, N.: Comparison of multivariate classifiers and response normalizations for pattern-information fMRI. Neuroimage **53**(1), 103–118 (2010). https://doi.org/10.1016/j.neuroimage.2010.05.051

7. Buckner, R.L., Krienen, F.M., Castellanos, A., Diaz, J.C., Yeo, B.T.T.: The organization of the human cerebellum estimated by intrinsic functional connectivity. J. Neurophysiol. **106**, 2322–2345 (2011). https://doi.org/10.1152/jn.00339.2011
8. Yeo, B.T.T., et al.: The organization of the human cerebral cortex estimated by intrinsic functional connectivity. J. Neurophysiol. **106**, 1125–1165 (2011). https://doi.org/10.1152/jn.00338.2011
9. Marques dos Santos, J.D., Marques dos Santos, J.P.: Towards XAI: interpretable shallow neural network used to model HCP's fMRI motor paradigm data. In: Rojas, I., Valenzuela, O., Rojas, F., Herrera, L.J., Ortuño, F. (eds.) IWBBIO 2022. LNCS, vol. 13347, pp. 260–274. Springer, Cham (2022). https://doi.org/10.1007/978-3-031-07802-6_22
10. Marques dos Santos, J.D., Marques dos Santos, J.P.: Path weights analyses in a shallow neural network to reach explainable artificial intelligence (XAI) of fMRI data. In: Nicosia, G., et al. (eds.) LOD 2022. LNCS, vol. 13811, pp. 417–431. Springer, Cham (2023). https://doi.org/10.1007/978-3-031-25891-6_31
11. Beckmann, C.F., Smith, S.M.: Probabilistic independent component analysis for functional magnetic resonance imaging. IEEE Trans. Med. Imaging **23**, 137–152 (2004). https://doi.org/10.1109/TMI.2003.822821
12. Minka, T.P.: Automatic choice of dimensionality for PCA. Technical report 514. MIT Media Lab Vision and Modeling Group, MIT (2000)
13. Hyvärinen, A.: Fast and robust fixed-point algorithms for independent component analysis. IEEE Trans. Neural Netw. **10**, 626–634 (1999). https://doi.org/10.1109/72.761722
14. Buckner, R.L.: Event-related fMRI and the hemodynamic response. Hum. Brain Mapp. **6**, 373–377 (1998). https://doi.org/10.1002/(SICI)1097-0193(1998)6:5/6⟨373::AID-HBM8⟩3.0.CO;2-P

A Guide and Mini-Review on the Performance Evaluation Metrics in Binary Segmentation of Magnetic Resonance Images

Ayca Kirimtat[1] and Ondrej Krejcar[1,2(✉)]

[1] Faculty of Informatics and Management, Center for Basic and Applied Research, University of Hradec Kralove, Rokitanskeho 62, Hradec Kralove 500 03, Czech Republic
{ayca.kirimtat,ondrej.krejcar}@uhk.cz
[2] Malaysia Japan International Institute of Technology (MJIIT), Universiti Teknologi Malaysia, Kuala Lumpur, Malaysia

Abstract. Eight previously proposed segmentation evaluation metrics for brain magnetic resonance images (MRI), which are sensitivity (SE), specificity (SP), false-positive rate (FPR), false-negative rate (FNR), positive predicted value (PPV), accuracy (ACC), Jaccard index (JAC) and dice score (DSC) are presented and discussed in this paper. These evaluation metrics could be classified into two groups namely pixel-wise metrics and area-wise metrics. We, also, distill the most prominent previously published papers on brain MRI segmentation evaluation metrics between 2021 and 2023 in a detailed literature matrix. The identification of illness or tumor areas using brain MRI image segmentation is a large area of research. However, there is no single segmentation evaluation metric when evaluating the results of brain MRI segmentation in the current literature. Also, the pixel-wise metrics should be supported with the area-wise metrics such as DSC while evaluating the image segmentation results and each metric should be compared with other metrics for better evaluation.

Keywords: MRI · brain · image segmentation · metrics

1 Introduction

Image segmentation is the technique for identifying object boundaries. Any objects with various colors and intricate patterns, as well as geometric shapes, might be included in this category such as people and cars. Several computer vision procedures for diagnosis, typically employ detection by image segmentation as an a priori work package. On the other hand, magnetic resonance imaging (MRI) is a crucial imaging method for the early detection of lesions in many areas of the brain. As compared to computed tomography (CT), MRI offers better contrast; thus, most of the studies in the existing literature employ MRI, especially the FLAIR sequence for better contrast. To evaluate the results of brain MRI segmentation, several metrics are proposed and used, therefore, we, in this study, compare and discuss these metrics with their mathematical foundations. We, also, compare different relevant studies on the use of those metrics for the evaluation of

I. Rojas et al. (Eds.): IWBBIO 2023, LNBI 13920, pp. 428–440, 2023.
https://doi.org/10.1007/978-3-031-34960-7_30

brain MRI segmentation between 2021 and 2023 in a detailed literature matrix. These metrics are sensitivity (SE), specificity (SP), false-positive rate (FPR), false-negative rate (FNR), positive predicted value (PPV), accuracy (ACC), Jaccard index (JAC), and dice score (DSC) as seen in Table 1. Since this is a mini-review paper, we, only, show the last three years' search of the whole literature.

Table 1. The previously published papers on different evaluation metrics.

Paper	Year	Pixel-wise						Area-wise	
		SE	SP	FPR	FNR	PPV	ACC	JAC	DSC
[1]	2023		✔					✔	✔
[2]	2023	✔	✔			✔	✔	✔	✔
[3]	2023							✔	✔
[4]	2023	✔	✔			✔	✔	✔	✔
[5]	2023	✔	✔						✔
[6]	2023								✔
[7]	2023							✔	✔
[8]	2023						✔		✔
[9]	2023								✔
[10]	2022								✔
[11]	2022	✔	✔	✔	✔	✔	✔	✔	✔
[12]	2022						✔	✔	
[13]	2022	✔							✔
[14]	2022								✔
[15]	2021	✔	✔			✔			✔
[16]	2021	✔	✔				✔		✔
[17]	2021								✔
[18]	2021						✔		
[19]	2021	✔	✔					✔	✔
[20]	2021	✔	✔			✔			✔

For instance, Alpar et al. [11] used all of these pixel-wise and area-wise metrics in their study to evaluate segmented and Nakagami-applied binary images. Another study by Alpar et al. [15] dealt with the MS identification and segmentation from FLAIR MRI images and SE, SP, pixel-wise, and DSC as area-wise metrics were used for the evaluation of the results. Another relevant study by Alpar et al. [10] presented an automated framework for the intelligent segmentation of brain lesions and based on the preliminary results, the authors used only DSC for the evaluation process. Alpar [1], also, proposed a mathematical fuzzy inference based on the framework for increasing the DSC values of the segmented FLAIR images.

Aghalari et al. [20] made automatic segmentation of brain MRIs using improved UNet-based architectures and used SE, SP, PPV, and DSC to evaluate the segmentation results. Guven and Talu [3] proposed a different architecture to make brain image segmentation and JAC and DSC were used to make analyses of the segmented images. Cao et al. [5] used SE, SP, and DSC to show analyses of the experiments on the segmented images. Moreover, being evaluated by DSC, JAC, SE, and SP, Khosravanian et al. [19] proposed a novel region-based level set method for segmenting the brain images. Kumar et al. [8] proposed a fuzzy k-plane clustering method to provide accurate segmentation by removing noise. In this study, DSC and ACC were two metrics to evaluate the results of the segmentation. We also show the usage of these metrics in all of these studies in Fig. 1 and according to this figure, DSC is the most used metric as one of the area-wise metrics and SE is the second most used metric among others as one of the pixel-wise metrics.

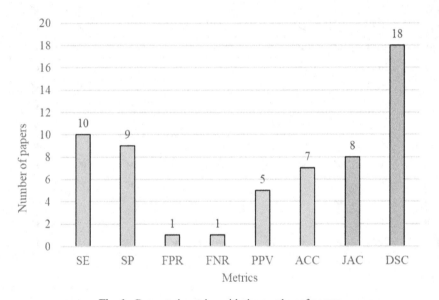

Fig. 1. Presented metrics with the number of papers

As one of the most important aspects considered in medical image segmentation, the choice of the metrics mentioned above should make good correlations with the visual assessment of the segmentation and the grading of these metrics should be well identified. Maybe, the selected metric should also be compared with other metrics for the chosen segmentation study.

2   Prerequisites of Binary Segmentation

All GT (ground truth) images, created by experts in the corresponding medical field, are a binary representation of the lesions found in MRI slices. The plotting methods vary among the experts; while the resulting images are strictly binary, where the lesions are white (1) and the surroundings are black (0). According to the binary structure of the GT images, any segmentation protocol should also generate binary images for performance evaluation. The synthetic images created for GT and segmented classes, which will be used throughout this research paper, are presented in Fig. 2.

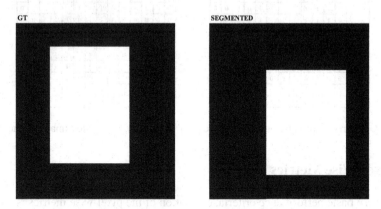

Fig. 2. Raw images, GT on the left, Segmented image on the right

The images in Fig. 2 are 14×15 synthetic images (width $= 14$, height $= 15$) consisting of one large synthetic lesion per each for better comprehension of the performance evaluation step. We used rectangular synthetic images instead of real binary segmented image and corresponding GT for better presentation and ease of comprehension. In addition the images we used are 14×15 pixel synthetic matrices such that it is not possible find this much small real images. On the other hand, the segmented area is deliberately chosen to not obey the GT. We chose rectangular synthetic images for the simplicity to reveal the mismatching pixels better. Although these images seem like black & white photos; they certainly are two matrices with 210 cells of 0 or 1 values as presented in Fig. 3.

Given that, a pixel $g_{x,y}$ on a GT image G of a raw genuine MRI is mathematically notated as:

$$g_{x,y} \in G(x = [1 : w], y = [1 : h]) \tag{1}$$

where $g_{x,y} \in \mathbb{Z}^+ = [0, 1] [0, 1]$; x is the coordinate in the x-axis, y is the coordinate in the y-axis; w is the width and h is the height of the image. A pixel $s_{x,y}$ in the segmented image S has the same specifications, namely:

$$s_{x,y} \in S(x = [1 : w], y = [1 : h]) \tag{2}$$

These notations will be used in the following section as the base of the performance calculations.

GT

SEGMENTED

Fig. 3. Matrix representation of the images, GT on the left, Segmented image on the right

3 Pixel-Wise Metrics

As the very basic performance criteria, evaluation of the pixel-wise metrics is the most common method, especially with the confusion matrix. The fundamentals of the performance evaluation step lie behind the number of the matched pixel when comparing the segmented image with the GT. What we expect in a precise segmentation is a high matching rate, where white pixels must match the white pixels; black pixels must match the black pixels, which is easily represented by a confusion matrix in Fig. 4.

		SEGMENTED	
		Negative (0)	Positive (1)
GT	Negative (0)	True Negative (TN)	False Positive (FP)
	Positive (1)	False Negative (FN)	True Positive (TP)

Fig. 4. Basic Confusion Matrix in Segmentation

Despite the simplicity, the automatic counting of these pixels by a code is rather heavy and needs fundamental matrix operations. On the other hand, it is necessary to consider the superposition image, which is the two-dimensional representation of both images as one, which is presented in Fig. 5, where the yellow highlighted pixels are taken from GT; blue ones from segmented so that the green ones are found in both images.

Although all pixels could be counted and classified into four different categories, it is very possible to reach the same numbers with matrix operations.

3.1 True-Positives and True-Negatives

True positives (TP) are the easiest to calculate in general; since these pixels are also forming the intersection area of two images. Any pixel would be classified as TP if and

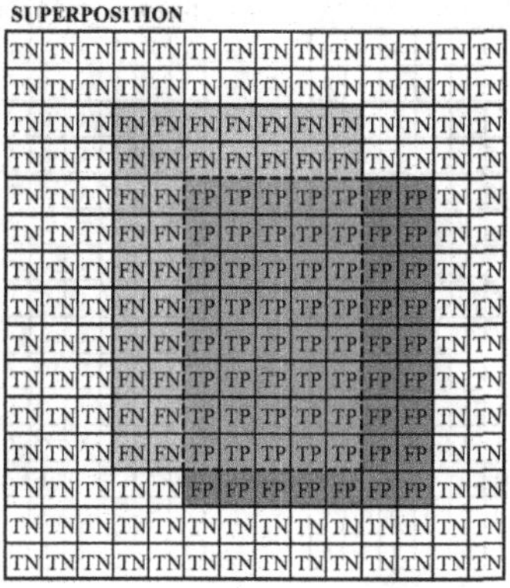

SUPERPOSITION

Fig. 5. Superposition image

only if, it is found in the intersection area, which is easily found by the dot product of the matrices by multiplying the corresponding cells by the dot product, namely:

$$TP = \sum_{a=1}^{w} \sum_{b=1}^{h} s_{a,b} \cdot g_{a,b} \qquad (3)$$

In other words, the TP is the number of pixels if they have the same coordinates and the same value of 1. Therefore, if all corresponding cells have the value 1, these cells will remain; while all other cells will turn to zero. True negative (TN) is the opposite of TP which is calculated by the zeros in each image if they occupy the corresponding cells. Without counting the mutual zeros, it is possible the find the TN by mathematical computations, namely:

$$TN = \sum_{a=1}^{w} \sum_{b=1}^{h} 1 - \left[\left(s_{a,b} + g_{a,b} \right) - \left(s_{a,b} \cdot g_{a,b} \right) \right] \qquad (4)$$

which indeed is 1-(sum of the matrices-dotproduct of the matrices). Given this equation, it is safe to state that the TN is 1-Union of the images, which will be held in the area-wise metrics section. The TN and TP calculations are illustrated by the synthetic images in Fig. 6.

Two more basic metrics represent the yellow and blue residual areas in Fig. 6, which are false positives and false negatives.

3.2 False-Positives and False-Negatives

False Positive (FP) and false negative (FN) values are the unwanted outputs in performance evaluation, which should be as lowest as possible. FP is the number of cells with the value 1 in the segmented image, which should have been 0; while FN is the opposite. The residual yellow areas in Fig. 6 are the FNs found by:

$$FN = \sum_{a=1}^{w} \sum_{b=1}^{h} g_{a,b} - \left(s_{a,b} \cdot g_{a,b} \right) \qquad (5)$$

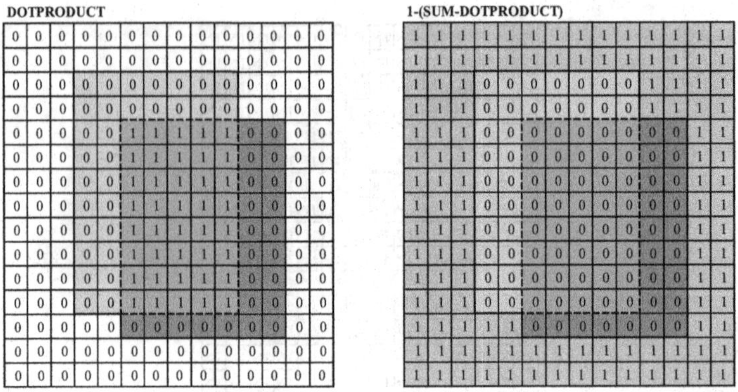

DOTPRODUCT

0	0	0	0	0	0	0	0	0	0	0	0	0	0
0	0	0	0	0	0	0	0	0	0	0	0	0	0
0	0	0	0	0	0	0	0	0	0	0	0	0	0
0	0	0	0	0	0	0	0	0	0	0	0	0	0
0	0	0	0	0	1	1	1	1	1	0	0	0	0
0	0	0	0	0	1	1	1	1	1	0	0	0	0
0	0	0	0	0	1	1	1	1	1	0	0	0	0
0	0	0	0	0	1	1	1	1	1	0	0	0	0
0	0	0	0	0	1	1	1	1	1	0	0	0	0
0	0	0	0	0	1	1	1	1	1	0	0	0	0
0	0	0	0	0	1	1	1	1	1	0	0	0	0
0	0	0	0	0	0	0	0	0	0	0	0	0	0
0	0	0	0	0	0	0	0	0	0	0	0	0	0
0	0	0	0	0	0	0	0	0	0	0	0	0	0
0	0	0	0	0	0	0	0	0	0	0	0	0	0

1-(SUM-DOTPRODUCT)

1	1	1	1	1	1	1	1	1	1	1	1	1	1
1	1	1	1	1	1	1	1	1	1	1	1	1	1
1	1	1	0	0	0	0	0	0	0	1	1	1	1
1	1	1	0	0	0	0	0	0	0	1	1	1	1
1	1	1	0	0	0	0	0	0	0	0	1	1	1
1	1	1	0	0	0	0	0	0	0	0	1	1	1
1	1	1	0	0	0	0	0	0	0	0	1	1	1
1	1	1	0	0	0	0	0	0	0	0	1	1	1
1	1	1	0	0	0	0	0	0	0	0	1	1	1
1	1	1	0	0	0	0	0	0	0	0	1	1	1
1	1	1	0	0	0	0	0	0	0	0	1	1	1
1	1	1	1	1	1	0	0	0	0	0	0	1	1
1	1	1	1	1	1	1	1	1	1	1	1	1	1
1	1	1	1	1	1	1	1	1	1	1	1	1	1

Fig. 6. True Positives are in green on the left and True negatives are in red on the right (color figure online)

which is the GT-dot product of the matrices. Likewise, the residual blue areas are found and counted by:

$$FP = \sum_{a=1}^{w} \sum_{b=1}^{h} s_{a,b} - \left(s_{a,b} \cdot g_{a,b}\right) \tag{6}$$

which is a segmented image-dot product of the matrices as the FP. However, these values don't mean anything without inserting them into the rate functions, which are the beginning of performance metrics.

3.3 False-Positive and False-Negative Rates

Until now, all values are only the numbers that don't provide sufficient information on the performance, alone. Therefore, the performance evaluation step starts with the rates, and the specific value concerning all values in the same class. For instance. False positive rate (FPR) is the rate of FP compared to the negatives; while the false negative rate is the rate of FN compared to all positives in GT. However, there are many inter-relations between the metrics, for instance: all positives in GT are TP + FN and all negatives in GT are TN + FP by definition and according to the superposition image in Fig. 5. The common equation which could be found everywhere are:

$$FNR = \frac{FN}{TP + FN} \tag{7}$$

$$FPR = \frac{FP}{TN + FP} \tag{8}$$

In contrast to the common overappraisal of the rates, their rates could easily be misleading in segmentation. The rates need to be minimum; while the minimization of any division is very easy by turning the dividend to zero, as presented in Fig. 7.

The idea is simple if the segmented image covers the GT as in Fig. 7 right, there will be no FN; since all zeros in the segmented image are also zero in GT. Likewise, if the segmented image is smaller than GT, there would be no FP, as in Fig. 7 left.

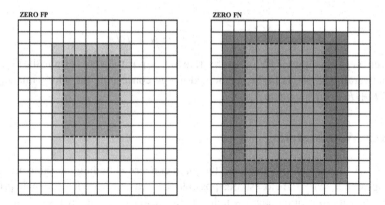

Fig. 7. Zero false positives on the left and zero false negatives on the right

3.4 Sensitivity and Specificity

Sensitivity (SE) is the true positive rate (TPR), which is sometimes referred as "recall" and the specificity (SP) is the true negative rate (TNR), which have their unique names somehow. Very similar to FNR and FPR, these rates are used in performance evaluation tables, which have the basic equations of:

$$SE = TPR = \frac{TP}{TP + FN} = 1 - FNR \tag{9}$$

$$SP = TNR = \frac{TN}{TN + FP} = 1 - FPR \tag{10}$$

where we try to maximize these rates. Despite the common usage and presentation in the final tables, their rates also don't provide any significant information. As presented in Fig. 7 left, zero FP means 100% specificity, and in Fig. 7 right, zero FN means 100% sensitivity; while the segmentation performance is far from being decent. These rates cannot be used in performance evaluation without the area-wise metrics; while still giving some information on the segmentation tendency.

3.5 Precision, Accuracy

Positive predictive value (PPV), or precision, represents the true positive rate over all positives in GT. Accuracy (ACC) is the rate of true positives and true negatives concerning all pixels. Despite the names, precision, and, accuracy, neither of them represents true accuracy nor the real precision of the segmentation. These rates are borrowed from statistics; while the adaptation to segmentation is very limited. While there is also a negative predictive value (NPV), it is not possible to come across these metrics in the papers. The basic equations of the precision and accuracy metrics are below:

$$PPV = \frac{TP}{TP + FP} \tag{11}$$

$$ACC = \frac{TP + TN}{FP + TP + FN + TN} \tag{12}$$

Given these equalities, PPV means the rate of true positives concerning the GT; while ACC is overall true pixels in the segmented image concerning the whole image; since $FP + TP + FN + TN = w.h.$

4 Area-Wise Metrics

In the binary segmentation viewpoint, the main concern is finding the same area in the GT images; or at least, the closest area possible. In area-wise metrics, the main problem arises in the superposition images which contain two-layered data in a two-dimensional plane as presented in Fig. 8 with separated layers.

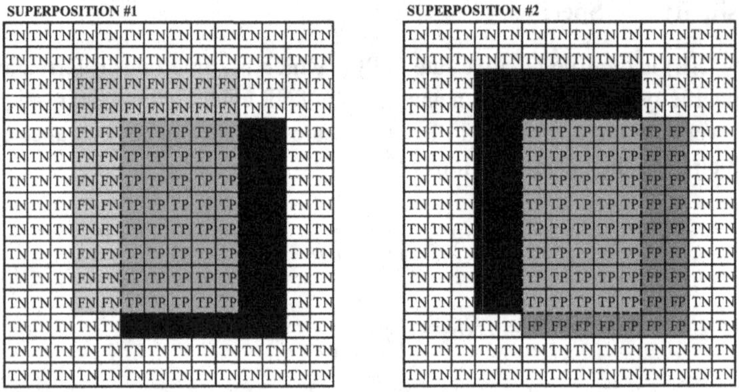

Fig. 8. Separated layers of superposition image

When combined, the resulting image has 1 FN and 1 FP area; but 2 TP areas on each other. To overcome this issue, two major performance metrics are used as the area-wise evaluation.

4.1 Jaccard Index (IoU)

The Jaccard index (JAC), or intersection over union (IoU) is one of the two area-wise metrics which is the ratio between the intersection area and the union area, as presented in Fig. 9.

As seen in Fig. 9, the intersection area is divided by the union area; since the Jaccard index accepts the union and the intersection areas as standalone layers; therefore, the equation of the Jaccard index with intersection and union operators is:

$$JAC = \frac{G \cap S}{G \cup S} \tag{13}$$

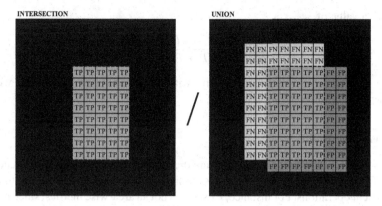

Fig. 9. Areas of intersection as dividend and union as a divider in the Jaccard index

which has a value of

$$JAC = \frac{TP}{TP + FN + FP} \tag{14}$$

However, if we consider the superposition image as two-layered, we have to use the dice score coefficient (DSC) as the main performance indicator.

4.2 Dice Score Coefficient (F1 Score)

Dice score coefficient (DSC), or F1 score, is the most common method to evaluate the overall performance of all segmentation protocols in the literature. Just like the Jaccard index, the TN values are neglected; since finding black pixels as black is not so important. As mentioned above, the DSC accepts the superposition image as two layers and separated them for computations of the intersection and union areas, as presented in Fig. 10.

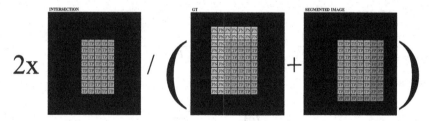

Fig. 10. Areas of intersection as dividend and union as a divider in DSC

As seen in Fig. 10, there are two regions of TP in the dividend and divider. The mathematical notation is:

$$DSC = 2\frac{G \cap S}{G + S} \tag{15}$$

which has a value of

$$DSC = \frac{2TP}{2TP + FP + FN} \tag{16}$$

F1 score has the same value; but is found with a different approach, which will be mentioned in the following subsection.

4.3 Interconvertibility

Although all metrics seem different, they are dictated by only four basic pixel matching metrics, TP, TN, FP and FN. Therefore, it is possible to interchange these metrics through mathematical operations. For instance, F1 score is not an area-wise metrics; since it deals with the harmonic mean of precision (PPV) and sensitivity (SE), namely:

$$F1 = 2\frac{PPV.SE}{PPV + SE} \tag{17}$$

However, it interestingly provides the DSC, when we substitute the metrics with their equations:

$$F1 = 2\frac{\left(\frac{TP}{TP+FP}\right) \cdot \left(\frac{TP}{TP+FN}\right)}{\left(\frac{TP}{TP+FP}\right) + \left(\frac{TP}{TP+FN}\right)} = 2\frac{\frac{TP^2}{(TP+FP)(TP+FN)}}{TP\frac{2TP+FN+FP}{(TP+FP)(TP+FN)}} \tag{18}$$

$$F1 = \frac{2TP}{2TP + FN + FP} = DSC \tag{19}$$

On the other hand, calculation of JAC and DSC together is redundant since they are mathematically interconvertible, namely:

$$2\left(\frac{1}{DSC} - \frac{1}{2}\right) = 2\left(\frac{2TP + FN + FP}{2TP} - \frac{1}{2}\right) = \frac{TP + FN + FP}{TP} \tag{20}$$

so that:

$$\left[2\left(\frac{1}{DSC} - \frac{1}{2}\right)\right]^{-1} = \frac{TP}{TP + FN + FP} = JAC \tag{21}$$

which always gives:

$$JAC = \frac{DSC}{2 - DSC} \tag{22}$$

Given that $0 \leq JAC \leq 1$ and $0 \leq DSC \leq 1$; JAC is always smaller than DSC, for except the boundary values 0 and 1, where they are equal.

5 Discussion and Conclusion

In this study, eight previously proposed segmentation evaluation metrics are compared and discussed with their mathematical foundations. These segmentation evaluation metrics are compared to understand which better fulfills the requirements of a medical image segmentation evaluation for clinical experts. As also seen from our literature matrix on those metrics, some of the authors used only DSC to evaluate their results, and some of them used all of these metrics for the evaluation of the image segmentation.

We can clearly understand that there is no single segmentation evaluation method available in the literature, there are several and some of them can be converted to other mathematically as could be seen in the previous sections. For instance, if an author uses DSC for evaluating the segmentation results, the results could be easily converted to JAC metrics, thus both of them could be presented in a study. Moreover, DSC as an area-wise metric is the most used metric for the evaluation of the segmentation results in most of the papers according to our literature matrix presented in the first section of this study. In addition to these, the pixel-wise metrics are not enough for being used in the evaluation process, the results should be supported with area-wise metrics such as DSC and JAC.

Acknowledgments. The work and the contribution were also supported by the SPEV project, University of Hradec Kralove, Faculty of Informatics and Management, Czech Republic (ID: 2102–2023), "Smart Solutions in Ubiquitous Computing Environments". We are also grateful for the support of student Michal Dobrovolny in consultations regarding application aspects.

References

1. Alpar, O.: A mathematical fuzzy fusion framework for whole tumor segmentation in multimodal MRI using Nakagami imaging. Expert Syst. Appl. **216**, 119462 (2023). https://doi.org/10.1016/j.eswa.2022.119462
2. Sangui, S., Iqbal, T., Chandra, P.C., Ghosh, S.K., Ghosh, A.: 3D MRI segmentation using U-Net architecture for the detection of brain tumor. Procedia Comput. Sci. **218**, 542–553 (2023). https://doi.org/10.1016/j.procs.2023.01.036
3. Altun Güven, S., Talu, M.F.: Brain MRI high-resolution image creation and segmentation with the new GAN method. Biomed. Signal Process. Control **80**, 104246 (2023). https://doi.org/10.1016/j.bspc.2022.104246
4. Gab Allah, A.M., Sarhan, A.M., Elshennawy, N.M.: Edge U-Net: brain tumor segmentation using MRI based on deep U-Net model with boundary information. Expert Syst. Appl. **213**, 118833 (2023). https://doi.org/10.1016/j.eswa.2022.118833
5. Cao, Y., Zhou, W., Zang, M., An, D., Feng, Y., Yu, B.: MBANet: a 3D convolutional neural network with multi-branch attention for brain tumor segmentation from MRI images. Biomed. Signal Process. Control **80**, 104296 (2023). https://doi.org/10.1016/j.bspc.2022.104296
6. Zhu, Z., He, X., Qi, G., Li, Y., Cong, B., Liu, Y.: Brain tumor segmentation based on the fusion of deep semantics and edge information in multimodal MRI. Inf. Fusion **91**, 376–387 (2023). https://doi.org/10.1016/j.inffus.2022.10.022
7. Santosh Kumar, P., Sakthivel, V.P., Raju, M., Sathya, P.D.: Brain tumor segmentation of the FLAIR MRI images using novel ResUnet. Biomed. Signal Process. Control **82**, 104586 (2023). https://doi.org/10.1016/j.bspc.2023.104586

8. Kumar, P., Agrawal, R.K., Kumar, D.: Fast and robust spatial fuzzy bounded k-plane clustering method for human brain MRI image segmentation. Appl. Soft Comput. **133**, 109939 (2023). https://doi.org/10.1016/j.asoc.2022.109939

9. Li, Z., et al.: CAN: context-assisted full attention network for brain tissue segmentation. Med. Image Anal. **85**, 102710 (2023). https://doi.org/10.1016/j.media.2022.102710

10. Alpar, O., Dolezal, R., Ryska, P., Krejcar, O.: Low-contrast lesion segmentation in advanced MRI experiments by time-domain Ricker-type wavelets and fuzzy 2-means. Appl. Intell. **52**, 15237–15258 (2022). https://doi.org/10.1007/s10489-022-03184-1

11. Alpar, O., Dolezal, R., Ryska, P., Krejcar, O.: Nakagami-Fuzzy imaging framework for precise lesion segmentation in MRI. Pattern Recogn. **128**, 108675 (2022). https://doi.org/10.1016/j.patcog.2022.108675

12. Walsh, J., Othmani, A., Jain, M., Dev, S.: Using U-Net network for efficient brain tumor segmentation in MRI images. Healthc. Analyt. **2**, 100098 (2022). https://doi.org/10.1016/j.health.2022.100098

13. Wang, Y., Ji, Y., Xiao, H.: A data augmentation method for fully automatic brain tumor segmentation. Comput. Biol. Med. **149**, 106039 (2022). https://doi.org/10.1016/j.compbiomed.2022.106039

14. Zhao, J., et al.: Automatic macaque brain segmentation based on 7T MRI. Magn. Reson. Imaging **92**, 232–242 (2022). https://doi.org/10.1016/j.mri.2022.07.001

15. Alpar, O., Krejcar, O., Dolezal, R.: Distribution-based imaging for multiple sclerosis lesion segmentation using specialized fuzzy 2-means powered by Nakagami transmutations. Appl. Soft Comput. **108**, 107481 (2021). https://doi.org/10.1016/j.asoc.2021.107481

16. Zhang, Z., Li, J., Tian, C., Zhong, Z., Jiao, Z., Gao, X.: Quality-driven deep active learning method for 3D brain MRI segmentation. Neurocomputing **446**, 106–117 (2021). https://doi.org/10.1016/j.neucom.2021.03.050

17. Weiss, D.A., et al.: Automated multiclass tissue segmentation of clinical brain MRIs with lesions. Neuroimage Clin. **31**, 102769 (2021). https://doi.org/10.1016/j.nicl.2021.102769

18. Zhang, F., et al.: Deep learning based segmentation of brain tissue from diffusion MRI. Neuroimage **233**, 117934 (2021). https://doi.org/10.1016/j.neuroimage.2021.117934

19. Khosravanian, A., Rahmanimanesh, M., Keshavarzi, P., Mozaffari, S.: A level set method based on domain transformation and bias correction for MRI brain tumor segmentation. J. Neurosci. Methods **352**, 109091 (2021). https://doi.org/10.1016/j.jneumeth.2021.109091

20. Aghalari, M., Aghagolzadeh, A., Ezoji, M.: Brain tumor image segmentation via asymmetric/symmetric UNet based on two-pathway-residual blocks. Biomed. Signal Process. Control **69**, 102841 (2021). https://doi.org/10.1016/j.bspc.2021.102841

Next Generation Sequencing
and Sequence Analysis

The Pathogenetic Significance of *miR-143* in Atherosclerosis Development

Mikhail Lopatin[1](✉) ⓘ, Maria Vulf[1] ⓘ, Maria Bograya[1] ⓘ, Anastasia Tynterova[2] ⓘ, and Larisa Litvinova[1] ⓘ

[1] Center for Immunology and Cellular Biotechnology, Immanuel Kant Baltic Federal University, Nevskogo ul. 14 A, 236016 Kaliningrad, Russia
`mikhaillopatin28@gmail.com`
[2] Educational and Scientific Cluster "Institute of Medicine and Life Sciences", Immanuel Kant Baltic Federal University, Nevskogo ul. 14 A, 236016 Kaliningrad, Russia

Abstract. Our pilot studies of blood plasma in patients with comorbidities allow *miR-143* to be regarded as a potential biomarker of atherosclerosis expression, but the literature data on the dynamics of *miR-143* expression are too controversial. The continuation of the study to verify the results by "wet lab" methods is costly and therefore it is advisable to first determine whether the putative biomarker has pathogenic relevance. The aim of the study was to establish and assess the role of *miR-143* in atherosclerosis using a comprehensive bioinformatics analysis, to identify possible pathways of its inclusion in the pathology mechanism. Using open sources, two sets of gene expression data in atherosclerotic plaques were selected, then only differentially expressed genes (DEGs) shared by both datasets were identified, from which a protein-protein interaction (PPI) extended network was then constructed. The next step was network analysis (identification of clusters and hub genes within them) and construction of regulatory networks of miRNAs-hub genes. The analysis revealed that *miR-143* is one of the central miRNAs whose action may be associated with suppression of atherosclerosis formation through its targeting of several hub genes: TLR2, TNF and LYN. However, another target is ITGB1, whose reduction increases autophagy and activates the inflammatory response. Based on established topological and functional characteristics, *miR-143* is of interest for further verification as a biomarker and for possible therapeutic applications in atherosclerosis. We also cut through the perspective of the already known effects of *miR-143* on the NK-kB pathway, but in the context of atherosclerosis (via TNF and TLR2).

Keywords: Atherosclerosis · Comorbidity · *miR-143* · MicroRNA · Bioinformatics

I. Rojas et al. (Eds.): IWBBIO 2023, LNBI 13920, pp. 443–455, 2023.
https://doi.org/10.1007/978-3-031-34960-7_31

Graphical Abstract

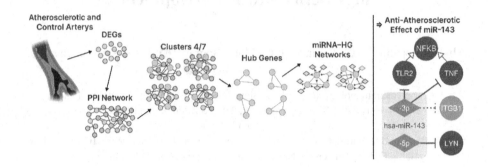

1 Introduction

Cardiovascular diseases (CVDs) remain the leading cause of death worldwide [1]. The presence of obesity, type 2 diabetes mellitus (T2DM) contribute to the formation and progression of atherosclerosis (AS) and significantly increase the risk of ischemic stroke (IS). Thus, our pilot blood plasma studies of patients with comorbidities showed that circulating *miR-143* levels increased in patients with and without IS, reaching maximum values at recurrent stroke (2.87 ± 0.40 vs. 4.40 ± 0.78 conventional units) relative to conditionally healthy donors. It should be noted that the distribution of pathologies in the study groups was as follows: Group 1 conditionally healthy donors (no pathologies); Group 2 – 20% T2DM, 30% obesity, 77% AS; Group 3 – 22% T2DM, 28% obesity, 59% AS, 18% re-IS.

These results allow *miR-143* to be considered as a potential predictor or biomarker of cerebrovascular ischemia in patients with comorbidities. However, the literature data describing the dynamics of *miR-143* expression level changes in atherosclerotic lesions are confusing and controversial.

We used the following search query in PubMed to search the literature: "circulating AND *miR-143* AND atherosclerosis AND *Homo sapiens*". The papers we found reported decreased circulating *miR-143* in AS [2], including that its level was reduced in acute coronary syndrome and inversely correlated with the severity of coronary [3] and cerebral artery stenosis [4]. Meanwhile, in a comprehensive study [5], increased plasma *miR-143* levels were found in patients hospitalized with IS, as demonstrated in three independent samples: detection, validation and replication. The results did not depend on the etiology of IS (especially patients with large artery atherosclerosis, cardioembolic stroke and stroke of unspecified etiology by TOAST classification [6], whereas sample sizes for small vessel stroke and stroke of other established etiology were too small), and did not correlate significantly with infarct volume. The first blood sampling was done on average ~5.0 h from the onset of symptoms – not on an empty stomach and before taking any medication. A longitudinal analysis lasting up to 90 days showed that expression levels decreased from the time of admission first blood draw, and from the second day onwards were not significantly different from the control group. In the cerebral atherosclerosis study [Gao 2019], with overall decreased plasma *miR-143* levels, values in the IS group (represented by only one TOAST subtype – large artery atherosclerosis) were slightly higher than in the AS group without stroke. However, the difference was not statistically significant; we can speculate that this was due to the later sampling (from the onset of symptoms): blood samples were taken early in the morning after admission.

In searching for the source of the increased secretion of *miR-143* in the circulation, the authors [5] found that its level was dependent on the number of platelets in the up-regulated experiment, but was independent of hypoxia in Neuro2a cells and was not increased in experimental models of stroke with middle cerebral artery occlusion. Data describing the release of *miR-143*-enriched microvesicles from endothelial cells have been reported in the literature. The activation of release has been linked to hemodynamic properties, namely changes in shear stress; it has been suggested that expression and release occurs in response to overexpression of the mechanosensitive transcription factor KLF2 [7, 8].

Given the inevitable limitations of any empirical (experimental) data, which may have a significant impact on their reliability, especially considering the mentioned ambiguity (lack of consensus) on *miR-143*, further studies are needed to verify this microRNA as a potential biomarker or predictor. The study of the pathogenetic significance of *miR-143* in atherosclerosis by "dry lab" methods allows to determine the feasibility of spending time and resources on "wet lab" studies, to help develop an optimal study design.

The aim of this study was to define and evaluate the role of *miR-143* in atherosclerosis using a comprehensive bioinformatics analysis and to define the possible ways of its involvement in the mechanism of pathology.

2   Materials and Methods

The design of the research is shown in the diagram below (Fig. 1).

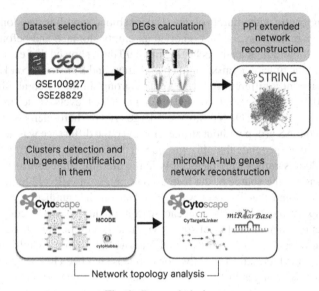

Fig. 1. Research design.

2.1 Dataset Selection

Human gene expression data in atherosclerotic plaques of various stages and in intact arterial sections were downloaded from NCBI Gene Expression Omnibus (GEO) [9]. The search strategy included the following parameters: (1) gene expression profiles were obtained by expression profiling by array; (2) samples were obtained from atherosclerotic plaques; (3) data series contained control samples; (4) samples were obtained from Homo sapiens; (5) the data series contained at least 20 samples. As a result, the GSE100927 and GSE28829 series were selected, comprising 104 and 29 samples respectively. The data we analyzed were publicly available, so no ethics committee approval or informed consent was required.

2.2 Differentially Expressed Genes Calculation

For each data series, statistical analysis of differential expression genes (DEGs) was performed using the web-based GEO2R the web-based GEO2R tool available directly from NCBI GEO, using GEOquery 2.66.0, limma 3.54.0 and DESeq2 1.38.3 R packages. The p-values were corrected using the Benjamini & Hochberg false discovery rate method. Adjusted p-values < 0.05 and |log2 (fold change)| > 1 were set as thresholds for screening DEGs. Finally, the resulting lists of up-regulated and down-regulated DEGs were intersected by separately, i. e. shared DEGs were defined as genes that were simultaneously activated or simultaneously repressed in both the GSE100927 and GSE28829 data series. The web-based application EVenn (http://www.ehbio.com/test/venn/#/) [10] was used to construct Venn/Euler diagrams for overlapping DEGs.

2.3 PPI Extended Network Reconstruction

The protein-protein interaction network (PPI) was reconstructed using STRING v. 11.5. Both physical and functional interactions were included. To increase the coverage of the PPI network, the first envelope of interactome proteins (i. e. interacting only directly) was also added with the restriction: no more than one per DEG.

2.4 Clusters Detection and Hub Genes Identification in Them

For further work (including topological analysis) the PPI extended network was loaded into Cytoscape v. 3.9.1, where:

- Using the MCODE (graph theoretic clustering algorithm "Molecular Complex Detection") plugin v.2.0.2, clusters were extracted in it. For the PPI network, these clusters often represent protein complexes and parts of pathways [11–14]. (Only significant clusters were considered below.)
- Using the cytoHubba plugin, hub genes were identified in each of the significant clusters based on topological parameters using the Maximal Clique Centrality algorithm.

2.5 MicroRNA-Hub Genes Network Reconstruction

Identifying miRNA target genes has proven to be more challenging than initially thought. Nowadays, the best way to do this is usually through bioinformatic prediction followed by experimental validation. Numerous computational approaches have been proposed to predict miRNA-mRNA interactions. And their number continues to increase, as well as the number of tools using/implementing these approaches or a combination of them [15]. As none of them has proved to be good enough. A combination of prediction algorithms based on different approaches can reduce the number of potential targets and at the same time improve prediction accuracy, more precisely obtaining some necessary compromise between sensitivity and specificity [16]. Nevertheless, whichever strategy is used to predict miRNA-mRNA interactions, only targets validated experimentally can be considered fully legitimate (see [15] for an example of criteria for experimental validation). Based on this situation, we used only experimentally validated data from miRTarBase v.9 [17]. This updated database is collected by manually reviewing the relevant literature after a systematic (computer based) analysis of the text data. The choice of miRTarBase is due to the fact that it contains the largest number of validated miRNA-target interactions (MTIs) and at the same time provides the most up-to-date collection compared to other similar, previously developed databases. The fact is that with each new release, not only additions are made, but also hundreds of data checks and system updates are carried out to integrate the MTIs.

The expansion (enrichment) of hub genes networks with complementary microRNAs was performed with the CyTargetLinker v.4.1.0 [18, 19], which requires a network file with linkset information. The linkset obtained for the obsolete miRTarBase v.8. We prepared linkset network files for current miRTarBase v.9 (update 2022) using scripts Linkset-Creator v.2.0 (https://github.com/CyTargetLinker/linksetCreator).

Thus, miRNAs-hub genes regulatory networks were obtained, followed by evaluation of their topological characteristics and clarification of the role of the identified *miR-143* targets.

3 Results

According to established thresholds, 574 DEGs were identified for the GSE100927 series and 386 DEGs for GSE2882. Crossing over individually identified 145 upregulated and 20 downregulated DEGs (a total of 165 DEGs) were identified in common for the two series (Fig. 2).

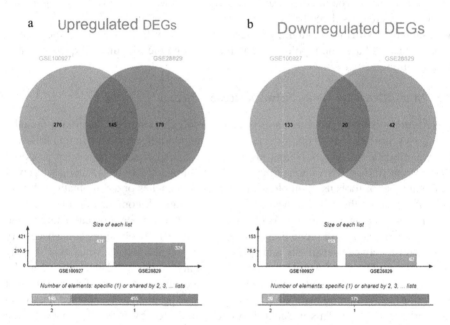

Fig. 2. Veen diagrams for (a) upregulated and (b) downregulated DEGs for GSE100927 and GSE28829.

The reconstructed PPI network, including both physical and functional interactions, based on 165 common DEGs and extended by the first layer of integrators contained 328 nodes (Fig. 3).

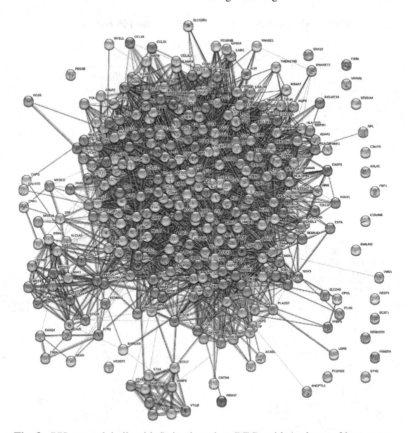

Fig. 3. PPI network built with String based on DEGs with 1st layer of interactors.

The resulting network (of 328 nodes) was loaded into Cytoscape v. 3.9.1, where 7 clusters were identified using the MCODE plugin (see Table 1). Interesting for further work was one of the most significant cluster (computed score ~ 40), as well as 3 other clusters that were moderately significant (7 < computed score < 10).

Then, using the cytoHubba plug-in, hub genes were identified in each of the first four clusters based on topological parameters using the Maximal Clique Centrality algorithm (Fig. 4).

The networks of hub genes were then expanded (enriched) with complementary microRNAs, as experimentally confirmed by miRTarBase v.9.

Analysis of miRNAs-hub genes regulatory networks revealed that the direct targets of *miR-143-3p* are TLR2, TNF, ITGB1 genes, which have high network centrality ranks and relatively few other targeting microRNAs, *miR-143-5p* is targeted to LYN. The summation diagram of the regulation is shown in Fig. 5.

Table 1. Characteristics of the clusters identified in the PPI network by MCODE.

Rank	Nodes	Edges	Score
1	61	1223	40.767
2	33	147	9.188
3	42	188	9.188
4	54	206	7.774
5	4	6	4.000
6	6	3	3.000
7	7	8	2.667

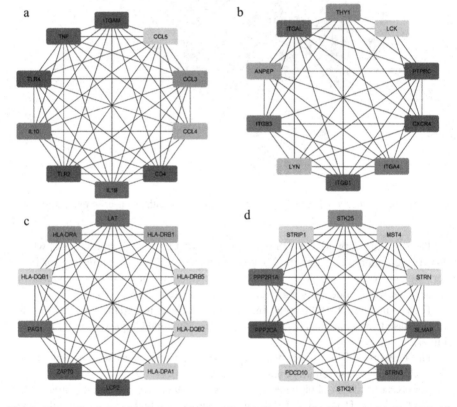

Fig. 4. Networks of top 10 hub genes obtained for (a) Cluster 1; (b) Cluster 2; (c) Cluster 3; (d) Cluster 4. The color of the node indicates its rank (red is the highest) (Color figure online).

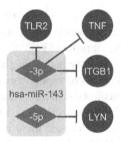

Fig. 5. *miR-143* and its high-ranking hub genes targets.

4 Discussion

All of the above genes have been noted in the works in connection with the development of atherosclerosis. Thus, Toll-like receptors (TLRs) are dominant components of the innate immune system and underlie inflammatory pathologies, including CVDs. TLR2 has been shown to play a key role in inflammation, contributing to the onset and development of AS [20–22]. TLR2 may be involved in the development of AS by combining with MyD88 to activate the NF-κB signaling pathway [23], and insufficient TLR2 expression may have a protective effect on the heart [24–26]. Activation of NF-κB triggers the secretion of tumor necrosis factor α (TNF) [27], which is one of the most potent pro-inflammatory cytokines. Its involvement in the development of atherosclerosis is well known and has been confirmed both in animal models and in humans [28, 29]. TNF and TLR2, both direct targets of *miR-143*, activate the NF-κB signaling pathway [30]. The nuclear transcription factor NF-κB plays a critical role in AS by modulating the expression of several inflammatory mediators. Genes transcribed by NF-κB mediate all stages of AS: atherogenesis, plaque formation and plaque rupture [31] Targeting NF-κB is therefore an attractive approach for anti-inflammatory therapy in AS. At the same time, it should be considered that NF-κB activation is essential for maintaining the immune response and cell survival [32]. There are data in the literature on the effects of *miR-143* on the NF-κB pathway; however, these works have focused on oncological issues [33, 34]. Our study, on the other hand, offers the prospect of exploring a new facet of this interaction – in relation to atherosclerosis. The next target for *miR-143* is Lyn, a protein tyrosine kinase of the Src family. Its activation promotes the accumulation of foamy cells in the arterial intima, leading to the progression of atherosclerosis [35, 36]. *miR-143* can target integrin subunit beta 1 (ITGB1), the integrin β1 encoded by it forms receptors for extracellular matrix proteins and plays a major role in intercellular adhesion in endothelial cells [37], in addition, it is involved in various processes, including tissue repair and immune response [38]. Deletion of β1-integrin impairs blood vessel wall stability [39], which promotes infiltration of lymphocytes and monocytes into subendothelial area, leads to decreased vascular contractile response, impaired in vivo arteriolar narrowing and dilatation, progressive systemic fibrosis and smooth muscle cell death [38]. Reduced ITGB1 levels are characteristic of progressive carotid atherosclerosis, coronary heart disease and stroke. Decreased ITGB1 may play an anti-apoptotic role in progressive carotid atherosclerosis by enhancing autophagy, activating the inflammatory response and signal transduction. Initially, autophagy may be an adaptive mechanism occurring in atherosclerotic plaques to recycle damaged cellular components for cell survival. However, when autophagic activity is continuously over-activated, the plaque becomes unstable [37].

When speaking about evaluation of the pathogenetic significance of microRNAs, it is necessary to consider the ambiguity of correspondences in the regulatory mechanism of RNA interference. The translation of one mRNA can be inhibited by dozens or hundreds of microRNAs; and vice versa – one microRNA can be (partially) complementary to dozens or hundreds of different mRNAs (i.e., there is "many to many" relation). And even if we limit ourselves to experimentally confirmed miRNA-mRNA interactions, as we did in our work, the number is very high per microRNA or gene. Many genes are also involved in the pathogenesis of atherosclerosis, so it is not at all surprising, and even highly likely, that many miRNAs will target some of them. To adequately determine the degree of importance of microRNAs in this complex regulatory network, one can turn to its topological characteristics. Following this approach, we showed that *miR-143* targets very significant genes both topologically (i.e., being the most concentrating nodes in the biological network) and functionally, which in addition have relatively fewer other microRNAs that tread on them. Therefore, we can call *miR-143* one of the central microRNAs with an antiatherosclerotic function – based on the role of the genes it represses. An important finding of the work is also that *miR-143* can suppress NF-kB signaling pathway activity (via targeting TLR2 and TNF), which has been little reported in publications on *miR-143* in the context of atherosclerosis. Moreover, the identified characteristics allow to identify *miR-143* as a possible candidate for pathogenetic therapy of atherosclerotic lesions, and the design of studies on this topic would also need to take into account, for example, risks with possible downregulation of ITGB1 expression and immune response.

5   Conclusion

The release or elevation of circulating microRNA can result from passive release or active secretion. The data presented (in the introduction) suggest, on the one hand, that the spike in IS-related plasma *miR-143* levels is due to its release from damaged brain cells and/or circulating cells. On the other hand, that circulating *miR-143* is mainly released by normal cells and decreases when they are damaged or degenerated by atherosclerosis. However, this contradicts the results of our pilot study, where *miR-143* levels were elevated in comorbid patients without stroke. Considering the above, *miR-143* plays an important role in normal cerebrovascular development.

The functions of *miR-143* target genes and the topological characteristics of miRNAs-hub genes regulatory network suggest that *miR-143* is a central microRNA in suppressing atherosclerotic processes due to its ability to dysregulate proatherosclerotic TLR2, TNF and LYN hub genes at posttranscriptional level. However, another target is ITGB1, the reduction of which enhances autophagy and activates the inflammatory response. Thus, *miR-143* is of interest for further verification of its biomarker potential and the possibility of its therapeutic application to suppress atherogenesis. The effects of *miR-143* (via TNF and TLR2) on the NK-kB pathway in relation to atherosclerosis are also of interest to study.

Acknowledgments. This work was supported by the Ministry of Science and Higher Education of the Russian Federation, Program "Priority-2030".

References

1. Word Health Organization: Cardiovascular Diseases. Fact Sheet (2021). https://www.who.int/news-room/fact-sheets/detail/cardiovascular-diseases-(cvds)
2. Liu, K., et al.: Expression levels of atherosclerosis-associated *miR-143* and miR-145 in the plasma of patients with hyperhomocysteinaemia. BMC Cardiovasc. Disord. **17**, 163 (2017). https://doi.org/10.1186/s12872-017-0596-0
3. Meng, L., et al.: Circulating *miR-143* and miR-145 as promising biomarkers for evaluating severity of coronary artery stenosis in patients with acute coronary syndrome. Clin. Biochem. **111**, 32–40 (2023). https://doi.org/10.1016/j.clinbiochem.2022.10.004
4. Gao, J., Yang, S., Wang, K., Zhong, Q., Ma, A., Pan, X.: Plasma miR-126 and *miR-143* as potential novel biomarkers for cerebral atherosclerosis. J. Stroke Cerebrovasc. Dis. **28**, 38–43 (2019). https://doi.org/10.1016/j.jstrokecerebrovasdis.2018.09.008
5. Tiedt, S., et al.: RNA-Seq identifies circulating miR-125a-5p, miR-125b-5p, and *miR-143*-3p as potential biomarkers for acute ischemic stroke. Circ. Res. **121**, 970–980 (2017). https://doi.org/10.1161/CIRCRESAHA.117.311572
6. Adams, H.P., et al.: Classification of subtype of acute ischemic stroke. Definitions for use in a multicenter clinical trial. TOAST. Trial of Org 10172 in Acute Stroke Treatment. Stroke. **24**, 35–41 (1993). https://doi.org/10.1161/01.STR.24.1.35
7. Neth, P., Nazari-Jahantigh, M., Schober, A., Weber, C.: MicroRNAs in flow-dependent vascular remodelling. Cardiovasc. Res. **99**, 294–303 (2013). https://doi.org/10.1093/cvr/cvt096
8. Kumar, S., Kim, C.W., Simmons, R.D., Jo, H.: Role of flow-sensitive microRNAs in endothelial dysfunction and atherosclerosis: mechanosensitive athero-miRs. Arterioscler. Thromb. Vasc. Biol. **34**, 2206–2216 (2014). https://doi.org/10.1161/ATVBAHA.114.303425
9. Edgar, R.: Gene expression omnibus: NCBI gene expression and hybridization array data repository. Nucleic Acids Res. **30**, 207–210 (2002). https://doi.org/10.1093/nar/30.1.207
10. Chen, T., Zhang, H., Liu, Y., Liu, Y.-X., Huang, L.: EVenn: easy to create repeatable and editable Venn diagrams and Venn networks online. J. Genet. Genomics **48**, 863–866 (2021). https://doi.org/10.1016/j.jgg.2021.07.007
11. Bader, G.D., Hogue, C.W.: An automated method for finding molecular complexes in large protein interaction networks. BMC Bioinform. **4**, 2 (2003). https://doi.org/10.1186/1471-2105-4-2
12. Eisen, M.B., Spellman, P.T., Brown, P.O., Botstein, D.: Cluster analysis and display of genome-wide expression patterns. Proc. Natl. Acad. Sci. U.S.A. **95**, 14863–14868 (1998). https://doi.org/10.1073/pnas.95.25.14863
13. Zahiri, J., et al.: Protein complex prediction: a survey. Genomics **112**, 174–183 (2020). https://doi.org/10.1016/j.ygeno.2019.01.011
14. MCODE. https://baderlab.org/Software/MCODE
15. Riolo, G., Cantara, S., Marzocchi, C., Ricci, C.: miRNA targets: from prediction tools to experimental validation. Methods Protoc. **4**, 1 (2020). https://doi.org/10.3390/mps4010001
16. Oliveira, A.C., Bovolenta, L.A., Nachtigall, P.G., Herkenhoff, M.E., Lemke, N., Pinhal, D.: Combining results from distinct microRNA target prediction tools enhances the performance of analyses. Front. Genet. **8**, 59 (2017). https://doi.org/10.3389/fgene.2017.00059

17. Huang, H.-Y., et al.: miRTarBase update 2022: an informative resource for experimentally validated miRNA–target interactions. Nucleic Acids Res. **50**, D222–D230 (2022). https://doi.org/10.1093/nar/gkab1079

18. Kutmon, M., Kelder, T., Mandaviya, P., Evelo, C.T.A., Coort, S.L.: CyTargetLinker: a cytoscape app to integrate regulatory interactions in network analysis. PLoS ONE **8**, e82160 (2013). https://doi.org/10.1371/journal.pone.0082160

19. Kutmon, M., Ehrhart, F., Willighagen, E.L., Evelo, C.T., Coort, S.L.: CyTargetLinker app update: a flexible solution for network extension in cytoscape. F1000Research **7**, 743 (2019). https://doi.org/10.12688/f1000research.14613.2

20. Mullick, A.E.: Modulation of atherosclerosis in mice by Toll-like receptor 2. J. Clin. Invest. **115**, 3149–3156 (2005). https://doi.org/10.1172/JCI25482

21. Liu, X., et al.: Toll-like receptor 2 plays a critical role in the progression of atherosclerosis that is independent of dietary lipids. Atherosclerosis **196**, 146–154 (2008). https://doi.org/10.1016/j.atherosclerosis.2007.03.025

22. Monaco, C., Gregan, S.M., Navin, T.J., Foxwell, B.M.J., Davies, A.H., Feldmann, M.: Toll-like receptor-2 mediates inflammation and matrix degradation in human atherosclerosis. Circulation **120**, 2462–2469 (2009). https://doi.org/10.1161/CIRCULATIONAHA.109.851881

23. Gulliver, C., Hoffmann, R., Baillie, G.S.: The enigmatic helicase DHX9 and its association with the hallmarks of cancer. Future Sci. OA **7**, FSO650 (2021). https://doi.org/10.2144/fsoa-2020-0140

24. Ishikawa, Y., Satoh, M., Itoh, T., Minami, Y., Takahashi, Y., Akamura, M.: Local expression of Toll-like receptor 4 at the site of ruptured plaques in patients with acute myocardial infarction. Clin. Sci. **115**, 133–140 (2008). https://doi.org/10.1042/CS20070379

25. Ha, T., et al.: TLR2 ligands induce cardioprotection against ischaemia/reperfusion injury through a PI3K/Akt-dependent mechanism. Cardiovasc. Res. **87**, 694–703 (2010). https://doi.org/10.1093/cvr/cvq116

26. Shishido, T., et al.: Toll-like receptor-2 modulates ventricular remodeling after myocardial infarction. Circulation **108**, 2905–2910 (2003). https://doi.org/10.1161/01.CIR.0000101921.93016.1C

27. Wang, N., Liang, H., Zen, K.: Molecular mechanisms that influence the macrophage M1â€"M2 polarization balance. Front. Immunol. **5** (2014). https://doi.org/10.3389/fimmu.2014.00614

28. McKellar, G.E., McCarey, D.W., Sattar, N., McInnes, I.B.: Role for TNF in atherosclerosis? Lessons from autoimmune disease. Nat. Rev. Cardiol. **6**, 410–417 (2009). https://doi.org/10.1038/nrcardio.2009.57

29. Rolski, F., Błyszczuk, P.: Complexity of TNF-α signaling in heart disease. J. Clin. Med. **9**, 3267 (2020). https://doi.org/10.3390/jcm9103267

30. Liu, T., Zhang, L., Joo, D., Sun, S.-C.: NF-κB signaling in inflammation. Signal Transduct. Target. Ther. **2**, 17023 (2017). https://doi.org/10.1038/sigtrans.2017.23

31. Li, W., et al.: NF-κB and its crosstalk with endoplasmic reticulum stress in atherosclerosis. Front. Cardiovasc. Med. **9**, 988266 (2022). https://doi.org/10.3389/fcvm.2022.988266

32. Zhang, B., Tian, M., Zhu, J., Zhu, A.: Global research trends in atherosclerosis-related NF-κB: a bibliometric analysis from 2000 to 2021 and suggestions for future research. Ann. Transl. Med. **11**, 57–57 (2023). https://doi.org/10.21037/atm-22-6145

33. Huang, F.-T., et al.: *MiR-143* targeting TAK1 attenuates pancreatic ductal adenocarcinoma progression via MAPK and NF-κB pathway in vitro. Dig. Dis. Sci. **62**(4), 944–957 (2017). https://doi.org/10.1007/s10620-017-4472-7

34. Wang, L., et al.: Suppression of *miR-143* contributes to overexpression of IL-6, HIF-1α and NF-κB p65 in Cr(VI)-induced human exposure and tumor growth. Toxicol. Appl. Pharmacol. **378**, 114603 (2019). https://doi.org/10.1016/j.taap.2019.114603

35. Park, Y.M., Febbraio, M., Silverstein, R.L.: CD36 modulates migration of mouse and human macrophages in response to oxidized LDL and may contribute to macrophage trapping in the arterial intima. J. Clin. Invest. JCI35535 (2008). https://doi.org/10.1172/JCI35535

36. Zhu, F., et al.: A ten-genes-based diagnostic signature for atherosclerosis. BMC Cardiovasc. Disord. 21, 513 (2021). https://doi.org/10.1186/s12872-021-02323-9

37. Zhang, Y., Zhang, H.: Identification of biomarkers of autophagy-related genes between early and advanced carotid atherosclerosis. Int. J. Gen. Med. 15, 5321–5334 (2022). https://doi.org/10.2147/IJGM.S350232

38. Di Taranto, M.D., et al.: Altered expression of inflammation-related genes in human carotid atherosclerotic plaques. Atherosclerosis 220, 93–101 (2012). https://doi.org/10.1016/j.atherosclerosis.2011.10.022

39. Abraham, S., Kogata, N., Fässler, R., Adams, R.H.: Integrin β1 subunit controls mural cell adhesion, spreading, and blood vessel wall stability. Circ. Res. 102, 562–570 (2008). https://doi.org/10.1161/CIRCRESAHA.107.167908

Comparison of VCFs Generated from Different Software in the Evaluation of Variants in Genes Responsible for Rare Thrombophilic Conditions

R. Vrtel[1,2] ⓘ, P. Vrtel[1,2(✉)] ⓘ, and R. Vodicka[1,2] ⓘ

[1] Institut of Medical Genetics, Faculty of Medicine and Dentistry, Palacky University Olomouc, Olomouc, Czech Republic
vrtel@fnol.cz

[2] Department of Medical Genetics, University Hospital Olomouc, Zdravotniku 248/7, 779 00 Olomouc, Czech Republic

Abstract. As part of the implementation and validation of an optimal diagnostic approach based on high-throughput sequencing by Ion Torrent platform in the diagnosis of the rare thrombophilic conditions of protein S (PS), protein C (PC) and antithrombin (AT) deficiency, we compared data from three different software tools – Torrent Suite, Ion Reporter and NextGene – to compare their performance and accuracy in the analysis of each sequence variant detected. A cohort of 31 patients was selected for PS (7), PC (13) and AT (11) deficiency based on defined indication criteria. Within these patient groups, a mutation detection rate of 67.7% was observed. In a cohort of 10 patients who were sequenced in a single sequencing run the three evaluated software detected 16, 19, and 27 variants in the *PROS1* gene; 17, 17, 19 variants in the *PROC* gene; and 15, 15, 16 variants for the *SERPINC1* gene in their baseline settings.

For data generated from the Ion Torrent platform, software from the same provider seems to be more suitable, mainly because of the quality of the false positive filtering.

For further evaluation of the validity of the software used, it will be necessary to expand the cohort of patients examined.

Keywords: DNA sequencing · high-throughput sequencing · Ion Torrent sequencing · data processing · software evaluation · protein S deficiency · protein C deficiency · antithrombin deficiency · mutation detection rate

1 Introduction

Second-generation sequencing techniques, so-called massively parallel sequencing (MPS), are rapidly gaining ground in medical genetics for the diagnosis of genetic diseases due to their speed and capacity [1]. With the ability to simultaneously screen multiple patients for multiple genetic targets, they are replacing previously used molecular diagnostic methods, which was mainly Sanger sequencing. As part of the implementation and validation of an optimal diagnostic approach based on high-throughput

© The Author(s), under exclusive license to Springer Nature Switzerland AG 2023
I. Rojas et al. (Eds.): IWBBIO 2023, LNBI 13920, pp. 456–462, 2023.
https://doi.org/10.1007/978-3-031-34960-7_32

sequencing in the diagnosis of the rare thrombophilic conditions of protein S (PS), protein C (PC) and antithrombin (AT) deficiency, we compared data from three different software tools – Torrent Suite, Ion Reporter and NextGene – to compare their performance and accuracy in the analysis of each sequence variant detected.

2 Methods

A larger project focused on the search for mutations responsible for rare hemostasis disorders was carried out in 2016–2021 at the University Hospital Olomouc. This project included a total of 31 Czech probands, 21 women and 10 men, who met the indication criteria for genetic testing for congenital deficiency of any of the anticoagulant proteins of interest [2, 3]. In all patients, a previous examination excluded mutations of factor V Leiden and factor II prothrombin G20210A.

Ion Torrent S5 and Ion Chef were used for the preparation of amplicon libraries and targeted sequencing of the patient DNA samples [4–6]. AmpliSeq technology based on sequencing of designed amplicons was used for its main advantages, which are: requirement of a small amount of input DNA (10 ng), high coverage of amplicons and better targeting of the PROS1 gene, which is highly homologous to the PROSP pseudogene.

Design of amplicon libraries: Ion AmpliSeq™ Designer software (Thermo Fisher Scientific) was used to design the primers, where Ion AmpliSeq™ On-Demand panels (Thermo Fisher Scientific) were chosen [7]. In this setup, a custom unique gene panel was designed that included the following genes based on the clinical indication: PS deficiency – PROS1; PC deficiency – PROC and PROCR; AT deficiency – SERPINC1. Individual amplicons were in-silico designed to achieve 100% coverage of coding regions and exon/intron boundaries. The design was based on the reference human genome hg19, the designed amplicons ranged from 125–275 bp. For this application, the designed primers were always 2-pool. A unique BED file for each design can be downloaded from the Ion AmpliSeq™ Designer software. This is essential for the analysis of the sequencing data. The software is available from www.ampliseq.com.

In order to compare the quality of the three different evaluation software, 10 patients were selected from one sequencing run in which patients were indicated for PS, PC and AT deficiency, but also for RASopathy (all signed a consent to use the results anonymously). Patients were anonymized and only variants detected in the *PROS1*, *PROC* and *SERPINC1* genes were evaluated.

The study compared variants in terms of their presence/absence in a given VCF cohort. The VCFs compared were those generated from Torrent Suite software using the integrated variantCaller module; VCFs generated from BAM files in Ion Reporter software and VCFs generated using NextGene software. The variantCaller settings in Torrent Suite and Ion Reporter were left at default, see Table 1 for selected key parameters.

Table 1. Selected parameters of the used software for variant calling – basic settings.

	Torrent Suite		Ion Reporter		NextGene	
	INDEL	SNP	INDEL	SNP	INDEL	SNP
MAF	0,25	0,18	0,1	0,1	0,2	0,2
MQ	20	15	10	10	20	20
MC	10	45	10	5	5	5
MCES	3	0	4	0	0	0
MSB	0,95	0,94	0,95	0,98	0,85	0,85

(MAF – minimum allele frequency; MQ – minimum quality; MC – minimum coverage; MCES – minimum coverage of either (each) strand; MSB – maximum strand bias)

3 Analysis of Sequencing Data

3.1 Torrent Suite Software

Primary and secondary analysis was performed according to the basic settings in the Torrent Suite™ software environment (Thermo Fisher Scientific) [8]. For each sequencing run, the software provides sequencing quality control data and coverage of individual samples (for target amplicons and bases separately). For us, the key metric for coverage analysis is coverage of individual amplicons: minimum average amplicon coverage ≥ 200 reads; amplicon coverage of at least 20 reads ≥ 98%.

The output data formats are uBAM, but also FASTQ – the latter can be downloaded within the module – FileExporter. Read assignment to the specified reference sequence (hg19) and target regions (unique BED file for each On-Demand panel design) is also performed automatically – output formats are BAM and BAI. As part of the data analysis in the Torrent Suite software, the BAM format is automatically analyzed in the integrated TVC module – the parameters for VC have remained basic as recommended. The output formats of the module are VCF and XLS. For selected parameters of the basic VariantCaller settings, see. Table 1.

3.2 Ion Reporter

This software from Thermo Fisher Scientific was used for secondary and tertiary analysis of sequencing data obtained from the Torrent Suite software. Patient sequencing data (BAM, VCF format) were loaded into Ion Reporter using the IonReporterUploader module. Ion Reporter is available online at https://ionreporter.thermofisher.com/ [9].

3.3 NextGene

For NextGene data processing, FASTQ files were used, which were downloaded from Torrent Suite using the module – FileExporter [10].

NextGene settings: In the first step, the instrument type – Ion Torrent, applications – SNP/INDEL discovery and sequence alignment were chosen. Further, the reference sequence – Human_v37p10dbsnp135 was set and the corresponding BED file was uploaded, which is unique for each On-Demand panel design. The selected setup parameters for the evaluated software can be found in Tab. 1.

3.4 VCF File Annotation

The VCF files uploaded from the Torrent Suite, Ion reporter and NextGene software were annotated in Ion Reporter. Parameters were left at their default settings. All patients were assessed in this way.

4 Results

Results of a larger project including 31 venous thromboembolism (VTE) probands meeting the indication criteria for genetic testing of congenital deficiency of one of the anticoagulant proteins:
In total, 21 likely pathogenic/pathogenic variants were detected using MPS on the Ion S5 platform. This corresponds to a mutation detection rate (MDR) of 67.7%. Our results are slightly above the published average. Lee et al. 2017 [11] reported a 60.9% MDR in 64 patients with VTE. They focused on searching for rare thrombophilic variants in 55 genes associated with thrombophilia. Data with the largest patient cohort to date were published by Downes et al. 2019 [12] in a cohort of 284 patients referred for thrombosis – an MDR of 48.9% was found. In our cohort the causal variants were detected in 14 females and 7 males. Of the variants detected, 6 were previously unpublished novel variants. Missense variants were the most frequent type of mutations – 76.2%. No SNVs were detected in 10 probands, with subsequent CNV analysis [13] also negative. The results of a larger project to search for mutations responsible for hemostasis disorder were published in Vrtel et al. 2022 [14].

The results of the project comparing three different evaluation software on samples of ten patients analyzed under identical conditions are as follows:
All software successfully detected the causative variant *SERPINC1*:c.79T>C (p.Trp27Arg). In these patients, assessed software detected 16, 19, and 27 variants in the *PROS1* gene; 17, 17, 19 variants in the *PROC* gene; and 15, 15, 16 variants for the *SERPINC1* gene in their baseline settings (shown in Table 2).

Results of individual software:

The **NextGene software** was the only one to detect INDEL variants present in homopolymeric regions: in the *PROS1* gene c.1870+84delA (in patients 3, 4, 6, 8), c.602-23delT (in patients 4, 5, 6, 8, 10), c.469+25delC (in patients 4 and 8), c.1156-9dupT (in patient 10). In the *PROC* gene, c.1212delG (p.Met406TrpfsTer15) (in patients 3 and 9). In the *SERPINC1* gene, this software detected the variant c.409-13_409-12delTT in patient 9. The frequency of alternative alleles for these variants ranged from 20.6–36.9%.

The presence of INDEL variants in homopolymeric regions in multiple patients in the same sequencing run with relatively low frequency suggests that these are sequencing artefacts.

The analysis in **Ion Reporter software** differed from the other compared software by detection of variants in *PROS1* gene c.347-47delT in patients 1 and 7 (frequency of alternative allele 50.98 and 55.49%) and c.939A>T (p.Leu313Phe) in patients 3 and 5 (frequency of alternative allele 20.91 and 21.97%). According to the Integrative Genomics Viewer IGV, the variant was present in all patients within a given sequencing run, but it is not listed in the VarSome genomic viewer. This INDEL located in the A/T rich region has low quality, which is the lowest ranked for VC by the Ion Reporter, so it is probably a sequencing artefact.

Using the **Torrent Suite software**, a variant in the *PROS1* gene c.1156-9_1156 was detected in patient 9 with an alternative allele frequency of 43.96% compared to the other two software. The same variant (c.1156-9dupT) was detected in patient 10 by NextGene software with an alternative allele frequency of 34.7%. This is an INDEL in the A/T rich region – according to IGV present in all patients in a given sequencing run, not listed in VarSome – it is most likely a sequencing artefact.

Table 2. The number of variants detected by the selected software in each patient.

Patient	Torrent Suite			Ion Reporter			NextGene			Total
	PROS1	*PROC*	*SERPINC1*	*PROS1*	*PROC*	*SERPINC1*	*PROS1*	*PROC*	*SERPINC1*	
1	2	2	0	3	2	0	2	2	0	13
2	1	2	0	1	2	0	1	2	0	9
3	3	2	0	4	2	0	4	3	0	18
4	0	2	1	0	2	1	3	2	1	12
5	2	1	3	3	1	3	3	1	3	20
6	2	2	3	2	2	3	4	2	3	23
7	2	1	3	3	1	3	2	1	3	19
8	2	2	3	2	2	3	5	2	3	24
9	2	2	0	1	2	0	1	3	1	12
10	0	1	2	0	1	2	2	1	2	11
Total	16	17	15	19	17	15	27	19	16	161

5   Conclusion

Successful detection of causative gene variants requires a precisely preselected group of patients with suspected congenital deficiency of one of the anticoagulant proteins. The indication criteria summarized by Colucci and Tsakiris 2020 were used [2].

As part of a wider project, a comparison of variants detected by three software tools – Torrent Suite (TVC plugin); Ion Reporter and NextGene – was carried out. All successfully detected the damaging variant *SERPINC1*:c.79T>C (p.Trp27Arg) and all SNVs present. For INDEL variants, especially in homopolymeric regions, we observed the highest number of misdetected variants in NextGene software. Shin et al. 2017 [15] compared two bioinformatics pipelines for optimal processing of data obtained from the Ion Torrent S5 XL sequencer using Torrent Suite (TVC module) and NextGene. All expected variants (681 SNVs, 15 small INDELs and 3 CNVs) were correctly detected in the study. Only one SNV detected by NextGene was a false positive. The Ion Torrent platform has a higher error rate for INDELs compared to the post-synthesis sequencing technology used by Illumina. The literature reports error rates ranging from 0.46–2.4% [15–19]. These imperfections are eliminated by optimizing the bioinformatics pipeline, increasing read depth and uniformity of coverage [15, 19, 20]. Hence, for optimal VC, it is advisable to use primarily the approaches proposed by the platform manufacturer (TVC and Ion Reporter). Shin et al. 2017 [15] optimized a bioinformatics setup that included TVC. For data generated from the Ion Torrent platform, software from the same provider seems to be more suitable, mainly because of the quality of the false positive filtering.

For further evaluation of the validity of the software used, it will be necessary to expand the cohort of patients examined.

Acknowledgement. Supported by the Ministry of Health of the Czech Republic – RVO (FNOl, 00098892).

References

1. Jamuar, S.S., Tan, E.C.: Clinical application of next-generation sequencing for Mendelian diseases. Hum. Genomics **9**, 1–10 (2015)
2. Colucci, G., Tsakiris, D.A.: Thrombophilia screening revisited: an issue of personalized medicine. J. Thromb. Thrombol. **49**, 618–629 (2020)
3. Wypasek, E., Corral, J., Alhenc-Gelas, M., et al.: Genetic characterization of antithrombin, protein C, and protein S deficiencies in Polish patients. Pol. Arch. Int. Med. **127**, 512–523 (2017)
4. Rothberg, J.M., Hinz, W., Rearick, T.M., et al.: An integrated semiconductor device enabling non-optical genome sequencing. Nature **475**, 348–352 (2011)
5. Golan, D., Medvedev, P.: Using state machines to model the Ion Torrent sequencing process and to improve read error rates. Bioinformatics **29**(13), i344–i351 (2013)
6. Kim, M.J.: Complete Guide to Ion Ampliseq Technology – How Does Ion Ampliseq Technology Work? (2017). https://www.thermofisher.com/blog/behindthebench/complete-guide-to-ion-ampliseqtechnology-how-does-ion-ampliseq-technology-work/. Cited 12 Oct 2022
7. Thermo Fisher Scientific. Ion AmpliSeq Designer (2022). https://ampliseq.com/. Cited 12 Sep 2022
8. Thermo Fisher Scientific. Torrent Suite™ Software 5.12 USER GUIDE (MAN0017972) (2019). https://assets.thermofisher.com/TFSAssets/LSG/manuals/MAN0017972_031419_TorrentSuite_5_12_UG_.pdf. Cited 14 Sep 2022
9. Thermo Fisher Scientific. Ion Reporter (2022). https://ionreporter.thermofisher.com/ir/. Cited 11 Sep 2022
10. SoftGenetics. NextGENe (2022). https://softgenetics.com/products/nextgene/. Cited 11 Oct 2022
11. Lee, E.J., Dykas, D.J., Leavitt, A.D., et al.: Whole-exome sequencing in evaluation of patients with venous thromboembolism. Blood Adv. **1**(16), 1224–1237 (2017)
12. Downes, K., Megy, K., Duarte, D., et al.: Diagnostic high-throughput sequencing of 2396 patients with bleeding, thrombotic, and platelet disorders. Blood **134**(23), 2082–2091 (2019)

13. Thermo Fisher Scientific. CNV Detection by Ion Semiconductor Sequencing (2014). https://tools.thermofisher.com/content/sfs/brochures/CNV-Detection-by-Ion.pdf. Cited 14 Sep 2022

14. Vrtel, P., Slavik, L., Vodicka, R., et al.: Detection of unknown and rare pathogenic variants in antithrombin, protein C and protein S deficiency using high-throughput targeted sequencing. Diagnostics 12(5), 1060 (2022)

15. Shin, S., Kim, Y., Oh, S.C.H., et al.: Validation and optimization of the Ion Torrent S5 XL sequencer and Oncomine workflow for BRCA1 and BRCA2 genetic testing. Oncotarget 8(21), 34858–34866 (2017)

16. Loman, N.J., Misra, R., Dallman, T., et al.: Performance comparison of bench-top high-throughput sequencing platforms. Nat. Biotechnol. 30, 434–439 (2012)

17. Quail, M.A., Smith, M., Coupland, P., et al.: A tale of three next generation sequencing platforms: comparison of Ion Torrent, Pacific Biosciences and Illumina MiSeq sequencers. BMC Genomics 13, 1–13 (2012)

18. Glenn, T.C.: Field guide to next-generation DNA sequencers. Mol. Ecol. Resour. 11, 759–769 (2011)

19. Laehnemann, D., Borkhardt, A., McHardy, A.C.: Denoising DNA deep sequencing data – high-throughput sequencing errors and their correction. Brief Bioinform. 17(1), 154–179 (2015)

20. Tarabeux, J., Zeitouni, B., Moncoutier, V., et al.: Streamlined ion torrent PGM-based diagnostics: BRCA1 and BRCA2 genes as a model. Eur. J. Hum. Genet. 22, 535–541 (2014)

Uterine Cervix and Corpus Cancers Characterization Through Gene Expression Analysis Using the KnowSeq Tool

Lucía Almorox$^{(\boxtimes)}$, Luis Javier Herrera, Francisco Ortuño, and Ignacio Rojas

Department of Computer Engineering, Automatics and Robotics,
University of Granada, C.I.T.I.C., Periodista Rafael Gómez Montero,
2, 18014 Granada, Spain
luciaalmorox@correo.ugr.es

Abstract. The characterization of cancer through gene expression quantification data analysis is a powerful and widely used approach in cancer research. This paper describes two experiments that demonstrate its potential in identifying differentially expressed genes (DEGs) and accurately predicting cancer subtypes. To achieve this, RNA-seq data was obtained from TCGA database and subsequently preprocessed and analyzed using the KnowSeq package from Bioconductor. In the first experiment, the study focuses on identifying DEGs in healthy, cervical cancerous, and uterine corpus cancerous tissues. The kNN classifier was employed to evaluate the utility of these genes in predicting a sample belonging to one of these three classes. A gene signature consisting of only three genes produced remarkable results on a 5-fold cross-validation assessment process, with overall test accuracy and F1 values of 99.33% and 96.73%, respectively. The paper provides ontological enrichment, associated diseases, and pathways of the gene signature to shed light on the molecular mechanisms involved in both cancers. The second experiment extends the work by classifying cervical cancer samples into their two most common histological types: adenocarcinoma and squamous cell carcinoma. By using a single gene, the study was able to achieve 100% of test accuracy in a 5-fold cross-validation process. Additionally, the classification of an adenosquamous sample into one of these two categories based on the number of genes used was also examined. Overall, these experiments demonstrate the potential of these techniques to improve cancer diagnosis and treatment. Moreover, the study provides valuable insights into the underlying molecular mechanisms of cervix and uterine corpus cancers, laying the groundwork for further research in this field.

Keywords: Uterine corpus cancer · Cervical cancer · Cervical adenocarcinoma · Cervical squamous cell carcinoma · KnowSeq · RNA-Seq · Differentially expressed genes · Gene signature

© The Author(s), under exclusive license to Springer Nature Switzerland AG 2023
I. Rojas et al. (Eds.): IWBBIO 2023, LNBI 13920, pp. 463–477, 2023.
https://doi.org/10.1007/978-3-031-34960-7_33

1   Introduction and Background

Gynecological cancers present a significant global health concern. Cervical cancer, in particular, ranks as the fourth most common cancer among women worldwide and is the second leading cause of cancer-related deaths in women from developing countries, following breast cancer [2,12]. Conversely, in developed countries, uterine corpus (body) cancer stands as the most prevalent gynecological malignancy, surpassing cervical cancer [9]. Despite these cancers' proximity within the female reproductive system, they differ significantly in their causes, risk factors, and disease characteristics. Consequently, research and treatment approaches for both of them are fundamentally distinct [14].

For example, cervical cancer can be effectively reduced through screening tests (that enable early detection) and HPV vaccination programs, which are designed specifically for this type of cancer. The limited access to these interventions in developing countries is what has led to significant disparities in the disease incidence rates among nations. Conversely, the higher incidence of uterine corpus cancer in developed countries is primarily linked to a specific risk factor for this cancer: obesity [14].

The main histological types that each of these cancers is composed of are also different. Cervical cancer is divided into squamous cell carcinoma (70–75%) and adenocarcinoma (10–25%) [7]. Uterine corpus cancer is divided into adenocarcinoma (more than 90%) and sarcoma (very rare, 2–4%) [1]. For both cancers, each histological type has different prognoses, response to treatment, and risk factors. For instance, compared to cervical squamous cell carcinoma, cervical adenocarcinoma is more aggressive, with a higher rate of metastasis, poorer prognosis, and lower survival rate. Therefore, it is crucial to deepen our understanding of the molecular pathogenesis of these cancers and propose new therapeutic targets and methods for precise diagnosis and personalized management strategies [10].

Differential expression analysis using high-throughput techniques applied to biological samples allows to determine the physiological state of normal cells and the changes that occur during cancer development. Within this discipline, studies have been carried out that identify differentially expressed genes (DEGs) in cervical cancer tissue compared to healthy tissue [15], as well as in uterine corpus cancer tissue compared to healthy tissue [13]. However, identifying DEGs in both types of cancer and healthy samples simultaneously has not yet been achieved (at least not reported). Therefore, this work addresses the determination of genes with expression values that are specific to each cancer type (and not a general characteristic of cancer). Hence, genes with differential expression among the three classes (two tumor types and one normal) were identified. The usefulness of these genes as biomarkers was then evaluated using the kNN algorithm. Furthermore, the aim was to determine a gene signature capable of achieving the highest classification accuracy with as few genes as possible.

After confirming that the two main histological types of cervical cancer were represented in the downloaded samples (which was not the case for uterine corpus cancer samples), the study identified DEGs between these types and evaluated their effectiveness as biomarkers, once again, using the kNN algorithm.

Regarding this classification, another study [11] did something similar using microarray-based data, although the main DEGs identified differ from those we found.

The obtained results showed high accuracy rates. Specifically, the classification of healthy, cervical cancer, and uterine corpus cancer samples using only three genes (CLDN15, VWCE and SERTM1), achieved overall test accuracy and F1 values of 99.33% and 96.73%, respectively, during a 5-fold cross-validation. Moreover, by utilizing a single gene, such as SPRR4, ICA1L, or SMCP, the classification of cervical cancer samples into adenocarcinoma and squamous cell carcinoma yielded a perfect test accuracy of 100%.

The input data for this study was obtained from the TCGA database and consists of clinical and RNA-seq-based gene expression quantification data related to patients with cervical or uterine corpus cancer. The preprocessing and analysis of these data, the feature selection process and the implementation of the kNN classifications were carried out using the bioinformatics package KnowSeq.

2   Materials and Methods

2.1   TCGA Database

The Cancer Genome Atlas (TCGA) is a publicly available database that contains a large collection of genomic and clinical data related to various types of cancer. It was established in 2005 as a collaboration between the National Cancer Institute (NCI) and the National Human Genome Research Institute (NHGRI) in the United States. Since then, the TCGA database has been extensively utilized in cancer research, yielding significant insights into the molecular mechanisms of cancer. While the TCGA program has officially ended, its outcomes remain available for researchers through various sources, including the GDC (Genomic Data Commons) portal, which was used for downloading the data of interest for this work. Specifically, corpus uteri and cervix uteri were selected as primary sites, as part of CESC and SARC TCGA-projects. From that, all STAR-Counts files included as Gene Expression Quantification data types were downloaded. The associated sample sheet and clinical table were also taken.

2.2   KnowSeq

KnowSeq is a freely available R/Bioconductor package for RNA-seq data analysis [4]. It is designed to provide comprehensive and reliable results for transcriptome analysis, including gene expression quantification, differential expression analysis, and functional annotation. Additionally, KnowSeq incorporates machine learning approaches (feature selection and classification processes) that enable the development of predictive models for gene expression. All of this make it a valuable tool for identifying potential therapeutic targets and diagnostic biomarkers, facilitating effective disease management [8].

2.3   Classification of Healthy, Cervical Cancer and Uterine Corpus Cancer Samples

TCGA Data Preprocessing. After downloading the data from GDC, data preprocessing was performed to enable subsequent analysis. First, undersampling of the cancerous classes was carried out to diminish the unbalance of the original dataset (see Table 1). Then, using KnowSeq package, the following steps were performed: count data was transformed into gene expression values for each sample; outliers (samples of questionable quality due to having an expression distribution significantly different from the rest) were detected and removed and, finally, surrogate variable analysis (SVA) model was applied to address batch effects. Table 1 shows the number of samples of each class downloaded from GDC, as well as the number remaining after undersampling and after the elimination of outliers.

Table 1. Downloaded, randomly selected (unders.) and filtered samples of each class.

Class	Description	Project	Downloaded	Rand. selected	Quality samples
CERVIX_TUMOR	Cervix cancer	TCGA-CESC	304	300	295
CORPUS_TUMOR	Uterine corpus cancer	TCGA-SARC	552	300	286
HEALTHY	Non-cancerous	TCGA-CESC/SARC	25	25	21

Identification of the Best Feature Selection Method. To select the most appropriate feature selection method for our classification task, the three methods available in KnowSeq were evaluated: **MRMR** (Maximum Relevance, Minimum Redundancy), **RF** (Random Forest, as a feature selector rather than a classifier), and **DA** (Dissease Association, which selects genes based on their biological relationship with the disease of interest, in this case, uterine disease). For this purpose, the high-quality samples were randomly split (using a 80%-20% scheme) into a training set and a test set. The composition of each of these sets is summarized in Table 2.

The training set was subjected to KnowSeq *DEGsExtraction* function, which performs an analysis to extract DEGs in the three classes of interest. This function was configured with a p-value of 0.001 and a value of 2 for the parameters *lfc* (minimum log fold change) and *cov* (minimum coverage).

The resulting DEGs expression matrix was processed by KnowSeq *FeatureSelection* function configured with three different mode options (*mrmr*, *rf* or *da*) to rank the genes based on their relevance for predicting the output variable. For each of the three rankings obtained, the KnowSeq kNN model was trained 10 times, using 1 to 10 top genes of the ranking as features (in all cases the training set was the same as the one described above). Subsequently, the test set was employed to assess the effectiveness of each classifier.

Table 2. Number of samples of each class in train and test sets.

Class	Train Samples	Test Samples
CERVIX_TUMOR	229	66
CORPUS_TUMOR	234	52
HEALTHY	18	3

5-Fold Cross-Validation Assessment Using MRMR as Feature Selection Method. Based on the outcomes of the previous experiment, MRMR was considered the most effective method for feature selection, and thus it was employed for the remainder of this work.

In order to reduce the impact of sampling variability, the next step was to assess the model's performance on multiple subsets of the data. In particular, in this case a 5-fold cross-validation assessment process was carried out. For each fold, a single MRMR ranking of 10 genes was obtained and used to train the classifier 10 times, with each iteration using a different number of the 10 MRMR-selected genes. For each number of genes, mean training and test accuracy among the five folds was calculated.

Then, a reduced gene signature chosen according to the genes present in the rankings obtained in the 5-fold assessment process, was finally proposed.

Gene Signature Enrichment. The KnowSeq function *geneOntologyEnrichment* was used to construct the gene ontology (GO) categories by biological processes (BP), cellular components (CC) and molecular functions (MF) for the selected genes. Moreover, the related diseases and pathways of these genes were obtained using *DEGsToDiseases* and *DEGsToPathways* KnowSeq functions. The information gathered was cross-referenced with web-based platforms that were useful to condense the most pertinent data for inclusion in this report.

2.4 Classification of Cervical Adenocarcinoma and Cervical Squamous Cell Carcinoma Samples

After ensuring that the main histological types of cervical cancer (adenocarcinoma and squamous cell carcinoma) were sufficiently represented in the downloaded GDC samples, the decision was made to train the kNN model to classify cervical cancer samples into these two types. For this experiment, the samples related to uterine corpus cancer or healthy tissue were omitted.

A preprocessing similar to the one used for the previous classification was carried out. The main difference was the re-labelling of the samples (*ADENO* or *SQUAMOUS*) using the information from the *primary_diagnosis* field in the clinical table (see Table 3). As in the previous experiment, classes are very unbalanced: after removing the outliers, there were 29 samples of adenocarcinoma and 244 of squamous cell carcinoma. For this problem, again, a 5-fold cross-validation assessment process was performed by obtaining a different feature ranking for

Table 3. Number of downloaded cervical cancer samples that share the same primary diagnosis. All samples belonging to the first three types of cervical cancer in the table were labeled as "ADENO"; those belonging to the next 5 types as "SQUAMOUS" and those belonging to the last type were eliminated.

Histological cervical cancer type	Downloaded samples
Adenocarcinoma, endocervical type	42
Adenocarcinoma, NOS	14
Endometrioid adenocarcinoma, NOS	6
Squamous cell carcinoma, NOS	336
Papillary Squamous cell carcinoma	2
Squamous cell carcinoma, keratinizing, NOS	60
Basaloid Squamous cell carcinoma	2
Squamous cell carcinoma, large cell, nonkeratinizing	100
Adenosquamous carcinoma	2

each fold (this time, only MRMR was applied), followed by training and testing the KnowSeq kNN model using 1 to 10 selected genes.

Classifying a Cervical Adenosquamous Cancer Sample as Adenocarcinoma or Squamous Cell Carcinoma. As a final research, we wanted to investigate how a classifier from the previous problem would label the mixed (adenosquamous) samples that were eliminated from the beginning of the experiment (see Table 3). For this purpose it was necessary to repeat the preprocessing including these samples (since their expression values must be normalized with respect to those of the rest of the samples). However, in this preprocessing, one of the two mixed samples was detected as outlier and, therefore, eliminated. To classify the remaining mixed sample based on the number of genes, the kNN model that had been trained with the last fold of the previous cross-validation was used.

Note: All of the R Markdown files developed for the creation of this article are available in this GitHub repository[1].

3   Results

3.1   Classification of Healthy, Cervical Cancer and Uterine Corpus Cancer Samples

Identification of the Best Feature Selection Method. Based on the training set (Table 2) and the defined statistical parameters, 21 genes were identified

---

[1] https://github.com/Almorox/Uterine-Cervix-and-Corpus-cancers-characterization-through-gene-expression-analysis-using-Knowseq.

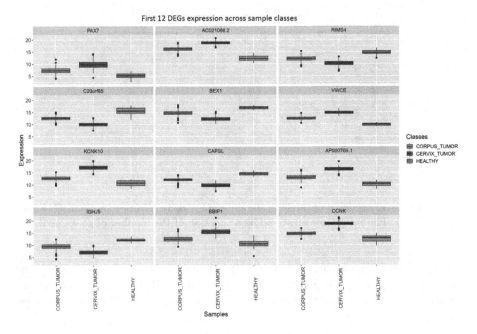

Fig. 1. Boxplots showing the expression of the 12 first genes detected as DEGs in each sample class using the train set.

as DEGs among the 3 classes. Figure 1 shows, for each of the first 12 extracted genes, a boxplot of their expression in each sample class. It can be seen how, for all of these genes, the mean expression value in uterine corpus cancer samples falls between the means of the other two sample classes, indicating that the classifier is likely to effectively separate cervical cancerous and healthy classes, but may face greater difficulty in distinguishing uterine corpus cancer from either of the other two classes.

Figure 2 shows the test accuracy of the three classifiers trained using the same set of samples but with different feature selection methods (MRMR, RF, or DA) as a function of the number of genes used.

The RF method achieves the highest accuracy when using a single gene, but the MRMR method appears to be more effective with a larger number of genes. Therefore, based on the results depicted in the graph, MRMR was selected as the feature selection method for this work (despite the potential impact of the random train-test partition on these results).

5-Fold Cross-Validation Using MRMR as Feature Selection Method.
Figure 3 shows the mean training and test accuracy achieved by the 5-fold cross-validation using MRMR as the feature selection method. The values are presented in relation to the number of genes used. The graph clearly shows that utilizing more than one gene results in high accuracy values surpassing 0.98, indi-

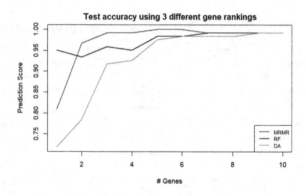

Fig. 2. kNN test accuracy obtained using different feature selection methods (MRMR, RF and DA). The values are presented as a function of the number of genes used.

cating a promising outcome. Interestingly, when exactly two genes are used, the mean test accuracy exceeds the mean train accuracy. However, with an increasing number of genes, the training accuracy values surpass the test accuracy ones, as expected.

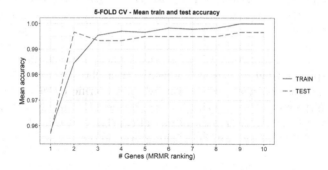

Fig. 3. kNN mean train and test accuracy of 5-fold cross-validation using MRMR. The values are presented as a function of the number of genes used.

It is important to note that the healthy class is very underrepresented, so high error rates in its classification have little impact on the whole classification accuracy value. This is why another interesting graphical representation of the classification results is the sum of the test confusion matrices from each fold. This matrix provides a comprehensive overview of the hits and misses of the classification process for all quality samples. Figure 4 presents the aggregated test confusion matrix obtained when using three genes, which indicates that, as predicted, there is no confusion between cervical cancerous and healthy samples or vice versa. The most frequent error is the misclassification of healthy samples as samples of uterine body cancer, with an error rate of 4.76%. The fact that the

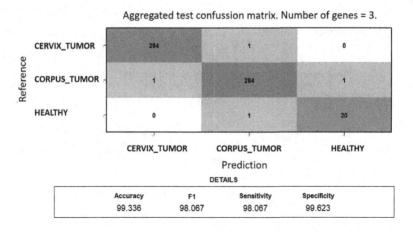

Fig. 4. Sum of the test confusion matrices of each fold of the 5-fold cross-validation when using the first 3 MRMR selected genes.

highest error rate occurs in the classification of the least represented class results in a decrease in the overall test F1 value (98.06%) compared to that of precision (99.33%), which is important to consider. Table 4 displays the MRMR ranking of each fold from the 5-fold cross-validation. Interestingly, CLDN15 and VWCE were ranked as the top two genes in four out of five rankings. Furthermore, in three out of these four rankings, the third gene coincides: SERTM1. Therefore, the genes CLDN15, VWCE, and SERTM1 were selected to compose a gene signature, with which the 5-fold cross-validation was repeated.

Table 4. Top 10 MRMR selected genes for each fold (train set) of the cross-validation.

	Gene1	Gene2	Gene3	Gen4	Gene5	Gene6	Gene7	Gene8	Gene9	Gene10
Fold1	CLDN15	VWCE	SERTM1	CCNK	AP000769.1	SHQ1	CD34	CAPSL	SERPINB5	AC021066.2
Fold2	CLDN15	VWCE	TMEM53	SHQ1	KCNK10	SERTM1	SERPINB5	CD34	AC021066.2	AP000769.1
Fold3	SERPINB5	AC021066.2	VWCE	FAM216B	CLDN15	AP000769.1	KCNK10	CCNK	SERTM1	CD34
Fold4	CLDN15	VWCE	SERTM1	CCNK	KCNK10	AP000769.1	AC021066.2	SERPINB5	CD34	AL079303.1
Fold5	CLDN15	VWCE	SERTM1	SHQ1	AP000769.1	CCNK	AC021066.2	CAPSL	KCNK10	SERPINB5

5-Fold Cross-Validation Using the Gene Signature. Figures 5 and 6 allow to gain insight into the ability of CLDN15, VWCE and SERTM1 genes to effectively distinguish between the three classes of interest. Specifically, Fig. 5 displays boxplots of the gene expression levels in each sample class, while Fig. 6 is a heatmap that visualizes the expression patterns of these genes across the different classes. Note that only 21 randomly selected samples from each class were included in the heatmap, as undersampling of the cancerous classes was necessary to enhance visualization. Both figures clearly demonstrate the significant differential expression of the three genes in the classes considered (and once again, it can be observed that the greatest differences occur between the healthy and cervical cancerous classes).

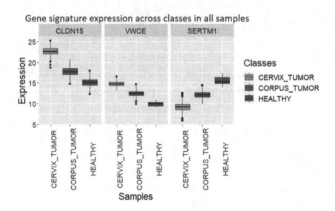

Fig. 5. Boxplots showing the expression of CLDN15, VWCE and SERTM1 (gene signature) in each sample class using all quality samples.

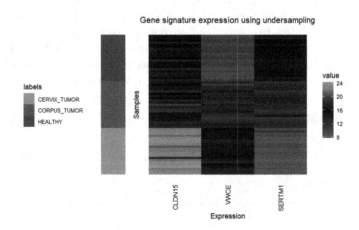

Fig. 6. Heatmap showing the expression of CLDN15, VWCE and SERTM1 (gene signature) in 21 randomly selected samples of each class (undersampling of cancerous classes was carried out using all quality samples).

The mean training and test accuracies achieved through 5-fold cross-validation using CLDN15, VWCE, and SERTM1 as selected features were both above 0.995, which demonstrate the effectiveness of the signature in accurately characterizing the three sample classes. In this case, the aggregated test confusion matrix (Fig. 7) shows clear separation not only between the healthy and cervical cancerous classes but also between the two cancerous classes. On the other hand, the error of classifying uterine corpus cancer samples as healthy samples

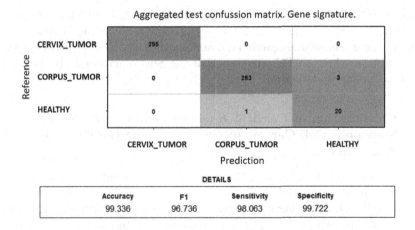

Fig. 7. Sum of the test confusion matrices of each fold of the 5-fold cross-validation using the complete gene signature as feature selection.

has increased (from 0.35% to 1.04%), and the misclassification of healthy samples as samples of uterine corpus cancer maintains the same error rate. The overall test accuracy and F1 values are 99.33% and 96,73%, respectively. These results are still remarkable, especially considering the small size of the gene signature.

CLDN15 Gene Enrichment. This gene encodes a member of the claudin family. Claudins are components of tight junction strands, which serve as a physical barrier to prevent solutes and water from passing freely through the paracellular space between epithelial or endothelial cell sheets. This is why CLDN15 GO annotations include: identical protein binding (GO:0042802); cell adhesion (GO:0007155) and it is active in the plasmatic membrane (GO:0005886). CLDN15 related pathways are Blood-Brain Barrier and Immune Cell Transmigration and its associated diseases include Collagenous Colitis [3]. This gene has been linked to endometrial cancer among other types of cancer [5].

VWCE Gene Enrichment. Biological products of VWCE (also known as URG11) are predicted to: enable calcium ion binding activity (GO:0005509); be involved in cellular response to virus (GO:0098586) and be located in cytoplasm (GO:0005737). They may be a regulatory element in the beta-catenin signaling pathway. Diseases associated with VWCE include Tarsal-Carpal Coalition Syndrome [3]. There are articles linking this gene to various types of neoplasms, including prostate, stomach, and lung cancer, among others [6].

SERTM1 Gene Enrichment. Biological products of SERTM1 (Serine-Rich And Transmembrane Domain-Containing Protein 1) are predicted to enable protein binding (GO:0005515) and to be located in intracellular membrane-bounded organelle (GO:0043227). This gene lacks ontology related to biological processes [3]. Furthermore, although certain types of cancer, including central nervous system cancer and ovarian carcinoma, are associated with it, the scores of these associations are low [5]. This suggests that the gene remains poorly studied and warrants further investigation.

3.2 Classification of Cervical Adenocarcinoma and Cervical Squamous Cell Carcinoma Samples Using the KnowSeq kNN Algorithm

The 5-fold cross-validation results for classifying cervical cancer samples into their two main histological types (labeled as *ADENO* or *SQUAMOUS*), using MRMR as the feature selection method, were optimal. Regardless of the number of genes used (ranging from 1 to 10), the training and test accuracy values for all folds were consistently 100%. This indicates that the MRMR selected genes are highly informative in distinguishing between these two types of cancer. Moreover, depending on the fold, the number of extracted DEGs varied between 1148 and 1150.

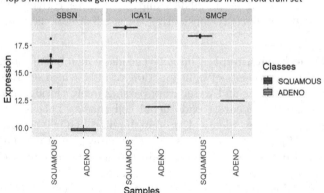

Fig. 8. Boxplots showing the expression of the top 3 MRMR genes in each sample class using the fifth fold (train set) of the cross-validation.

While the precision graphs and confusion matrices were not included in this article due to their trivial nature, expression boxplots of some of the selected genes are of interest. Specifically, Fig. 8 shows boxplots of the expression of the top 3 MRMR selected DEGs across classes, using the train set from the fifth fold. The distinctiveness between the two sample classes is evident from the significant separation between their mean expression values, with no overlap between the outlier values of each class. This striking differential expression is responsible for the excellent kNN separation between the two classes, even with a single gene. The gene in the first position showed variations in MRMR rankings across folds (SPRR4, SPRR3, TASR5, LCEP3 and SBSN, for each fold) while genes in positions 2, 3, and 4 remained consistent (ICA1L, SMCP and CALML3, respectively).

Classifying an Adenosquamous Sample in the ADENO or SQUA-MOUS Class. By using the latest classifier from the previous cross-validation to label the adenosquamous sample as either *ADENO* or *SQUAMOUS*, the probabilities of belonging to each class were obtained based on the number of genes used. In the majority of cases, the probability of belonging to the *ADENO* class was 1 (p($SQUAMOUS$) = 0). However, when exclusively using the first gene, the probability of belonging to the *ADENO* class was 0 (p($SQUAMOUS$) = 1), and when using the first 6 genes, the probability was 0.16 (p($SQUAMOUS$) = 0.84). While these results suggests that the expression value of the sixth gene in the mixed sample is likely *much* closer to the mean expression value in the *SQUA-MOUS* class than that of the *ADENO* class, the same may not be true for the first gene, as it is not being influenced by the additive effect of the rest of the genes.

To confirm this, it is interesting to compare the expression values of the 10 genes used by the classifier in the adenosquamous sample (unique expression value) and in the *ADENO* and *SQUAMOUS* classes (in the form of boxplots), as shown in Fig. 9. It can be seen that for gene 1, the expression value in the mixed sample is slightly closer to the *SQUAMOUS* class than the *ADENO* class. However, for gene 6, the expression value in the mixed sample almost perfectly matches that of the *SQUAMOUS* class, which, added to the effect of the first five genes, causes the classifier to change its decision and assign the sample to the *SQUAMOUS* class with a probability of 0.84.

4   Conclusions

Real gene expression quantification data obtained from TCGA was analyzed using KnowSeq, allowing identification of differentially expressed genes among healthy, cervical cancerous and uterine corpus cancerous tissue. By utilizing the MRMR feature selection method, a gene signature composed of only three genes (CLDN15, VWCE and SERTM1), was identified. The 5-fold cross-validation results showed a high level of performance, with overall test accuracy and F1 values of 99.33% and 96.73%, respectively. Furthermore, KnowSeq provided gene ontologies, diseases and pathways related to the gene signature. The three genes are protein coding genes linked to various neoplasms, though only CLDN15 has been previously associated with one of the cancers studied in this work (endometrial cancer, which is the most common uterine corpus cancer).

In another experiment, cervical cancer samples were classified into its two principal histological types: squamous cell carcinoma and adenocarcinoma. A 5-fold cross-validation was performed using 1 to 10 MRMR selected genes, and remarkably, the training and test accuracy remained consistently perfect across all folds and irrespective of the number of selected genes. Finally, in the presence of an adenosquamous sample in the downloaded samples, it was tested how it would be classified based on the number of genes, and in 8 out of 10 cases, the sample was classified as adenocarcinoma.

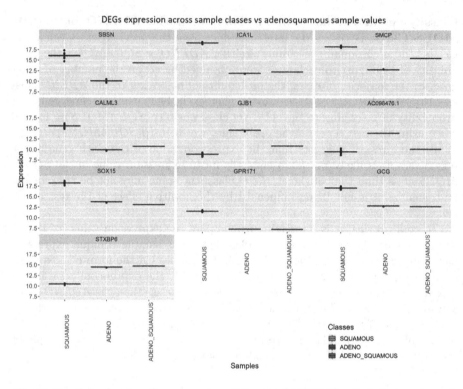

Fig. 9. Boxplots showing the expression of the top 10 MRMR selected genes in each sample class using the last fold (train set) of the 5-fold cross-validation. The third column of each subgraph shows the expression value in the mixed sample.

Overall, this study provides valuable insights into the molecular basis of cervical and uterine corpus cancer and lays the foundation for future studies aimed at improving diagnosis and treatment outcomes.

Acknowledgements. This work was funded by the Spanish Ministry of Sciences, Innovation and Universities under Project PID2021-128317OB-I00 and the projects from Junta de Andalucia P20-00163.

References

1. American society of clinical oncology. Uterine Cancer - Introduction – cancer.net. https://www.cancer.net/cancer-types/uterine-cancer/introduction. Approved by the Cancer. Net Editorial Board, 02/2022. Accessed 3 Feb 2023

2. Cervical cancer – who.int. https://www.who.int/news-room/fact-sheets/detail/cervical-cancer. Accessed 25 Mar 2023

3. Genecards human gene database. https://www.genecards.org/. Accessed 4 Mar 2023

4. KnowSeq – bioconductor.org. https://www.bioconductor.org/packages/release/bioc/html/KnowSeq.html. Accessed 20 Feb 2023

5. Open Targets Platform – platform.opentargets.org. https://platform.opentargets.org/. Accessed 4 Mar 2023

6. VWCE von Willebrand factor C and EGF domains [Homo sapiens (human)] - Gene - NCBI – ncbi.nlm.nih.gov. https://www.ncbi.nlm.nih.gov/gene/220001. Accessed 3 Mar 2023

7. Balcacer, P., Shergill, A., Litkouhi, B.: MRI of cervical cancer with a surgical perspective: staging, prognostic implications and pitfalls. Abdom. Radiol. **44**(7), 2557–2571 (2019). https://doi.org/10.1007/s00261-019-01984-7

8. Castillo-Secilla, D., et al.: KnowSeq R-Bioc package: the automatic smart gene expression tool for retrieving relevant biological knowledge. Comput. Biol. Med. **133**, 104387 (2021). https://doi.org/10.1016/j.compbiomed.2021.104387

9. Crosbie, E.J., Zwahlen, M., Kitchener, H.C., Egger, M., Renehan, A.G.: Body mass index, hormone replacement therapy, and endometrial cancer risk: a meta-analysis. Cancer Epidemiol. Biomarkers Prev. **19**(12), 3119–3130 (2010). https://doi.org/10.1158/1055-9965.epi-10-0832

10. He, Z., et al.: The value of HPV genotypes combined with clinical indicators in the classification of cervical squamous cell carcinoma and adenocarcinoma. BMC Cancer **22**(1) (2022). https://doi.org/10.1186/s12885-022-09826-4

11. Kim, Y.W., et al.: Target-based molecular signature characteristics of cervical adenocarcinoma and squamous cell carcinoma. Int. J. Oncol. **43**(2), 539–547 (2013). https://doi.org/10.3892/ijo.2013.1961

12. Martínez-Rodríguez, F., et al.: Understanding cervical cancer through proteomics. Cells **10**(8), 1854 (2021). https://doi.org/10.3390/cells10081854

13. Okuda, T., et al.: Genetics of endometrial cancers. Obstet. Gynecol. Int. **2010**, 1–8 (2010). https://doi.org/10.1155/2010/984013

14. Sung, H., et al.: Global cancer statistics 2020: GLOBOCAN estimates of incidence and mortality worldwide for 36 cancers in 185 countries. CA: A Cancer J. Clin. **71**(3), 209–249 (2021). https://doi.org/10.3322/caac.21660

15. Wu, X., et al.: Identification of key genes and pathways in cervical cancer by bioinformatics analysis. Int. J. Med. Sci. **16**(6), 800–812 (2019). https://doi.org/10.7150/ijms.34172

Sensor-Based Ambient Assisted Living Systems and Medical Applications

Smart Wearables Data Collection and Analysis for Medical Applications: A Preliminary Approach for Functional Reach Test

João Duarte[1], Luís Francisco[1], Ivan Miguel Pires[2,3], and Paulo Jorge Coelho[1,4(✉)]

[1] Polytechnic of Leiria, Leiria, Portugal
{2200782,2203882}@my.ipleiria.pt, paulo.coelho@ipleiria.pt
[2] Instituto de Telecomunicações, Universidade da Beira Interior, Covilhã, Portugal
impires@it.ubi.pt
[3] Polytechnic of Santarém, Santarém, Portugal
[4] Institute for Systems Engineering and Computers at Coimbra (INESC Coimbra), Coimbra, Portugal

Abstract. The Functional Reach Test (FRT) is a commonly used clinical tool to evaluate the dynamic balance and fall risk in older adults and individuals with specific neurological conditions. Several studies have highlighted the importance of using FRT as a reliable and valid measure for assessing functional balance and fall risk in diverse populations. Additionally, FRT is sensitive to changes in balance function over time and can provide critical information for designing rehabilitation programs to improve balance and reduce the risk of falls. The FRT has also been used as a screening tool for identifying individuals who may benefit from further assessment or intervention. Thus, the FRT is a valuable clinical instrument for assessing functional balance and fall risk and should be incorporated into routine clinical practice. This paper intends to describe the preliminary results and future directions for implementing the FRT with various sensors gathered from smartphones or smart wearables to provide valuable indicators to aid professional healthcare practitioners in evaluating and following up on the elderly but possibly extending to other age groups.

Keywords: Functional Reach Test · Smart Wearables · Monitoring Apps

1 Introduction

Balance issues are a common concern in older adults [1, 2], and people with neurological conditions, such as Parkinson's disease or multiple sclerosis [3–5]. Functional Reach Test (FRT) is a commonly used clinical assessment tool to evaluate an individual's balance and fall risk. The test measures the maximum distance a person can reach forward while maintaining a standing position without taking a step [6, 7]. Other similar tests used to evaluate balance include the Timed Up and Go (TUG) test [8, 9] and the Berg Balance Scale (BBS) [10]. These tests assess different aspects of balance, with the FRT focusing on the individual's ability to maintain stability while reaching forward.

I. Rojas et al. (Eds.): IWBBIO 2023, LNBI 13920, pp. 481–491, 2023.
https://doi.org/10.1007/978-3-031-34960-7_34

These tests help healthcare professionals to assess an individual's balance and fall risk, which allows them to develop an appropriate treatment plan to improve balance and prevent falls. Improving balance can involve a combination of exercise, physical therapy, and lifestyle modifications [11, 12]. Therefore, these tests can help identify balance issues early and initiate appropriate interventions to prevent falls and improve the overall quality of life [13].

Sensors and mobile devices are closely related, as sensors are often integrated into mobile devices to provide a variety of functions and features. Mobile devices such as smartphones and tablets typically have a variety of built-in sensors, including accelerometers, gyroscopes, proximity sensors, ambient light sensors, GPS sensors, magnetometers, and barometers [14].

After researching the main applied test evaluation methods and respective equipment used, these were considered in terms of applicability, namely regarding the evaluation of balance in the elderly population. In the last few years, some authors have proposed the use of different technologies to determine the FRT. These include motion capture systems or console controls that integrate the necessary sensors to perform the test (accelerometer, gyroscope, and magnetometer), video capture systems, and recently, smart wearables [5, 15–18]. The research also included identifying mobile devices and data from their sensors to evaluate, monitor, and follow people's clinical status and predict adequate therapy [18].

We propose a tool to assist in evaluating individuals' dynamic balance, which is applied in this case study to the FRT. This tool consists of a mobile application that allows the collection and analysis of data from sensors present in a mobile device in which it is installed or in devices paired with it. Consequently, the data to be collected should allow for obtaining coordinates of position, velocities, inclination, etc. Thus, it is proposed to collect inertial data (accelerometer, gyroscope, and magnetometer) to estimate the FRT result.

The main contributions of this study are:

- Provide a preliminary approach for a low-cost solution using devices massively used by everyone, which makes it easily replicable worldwide. No additional devices other than a conventional smartphone to use it, although with the possibility of adding other wearable devices;
- Promote the possibility of carrying out tests, even in underprivileged areas, since the data can be stored locally on the device and/or in a cloud service;
- Allow in the future to prepare a report to be integrated into the individual file of each patient. The continuity of these reports will allow for tracking the evolution of possible diseases, such as Parkinson's disease, post-stroke, etc.

2   Methods

2.1   Description of the Test

The FRT is a clinical trial used to evaluate the dynamic balance of people. The most used method for determining the FRT is a scale, usually a graduated ruler, fixed horizontally on a wall at shoulder height. The subject should stretch the dominant arm to the body at a 90° angle. The fingertips should be positioned at the reference point of the scale, i.e., the starting point of the ruler. The individual should tilt the body in the anterior direction, only rotating the hip as far as possible without raising the ankles. Figure 1 depicts the initial and final positions for the previous test. The test should be performed more than once (two or three depending on the authors/studies), and the FRT value corresponds to the average of the values obtained. The evolution of an individual's dynamic balance can be evaluated by comparing two or more consecutive FRT results obtained at significantly spaced moments in time (months). A reach of less than 15cm clinically indicates a high fall risk [6, 19].

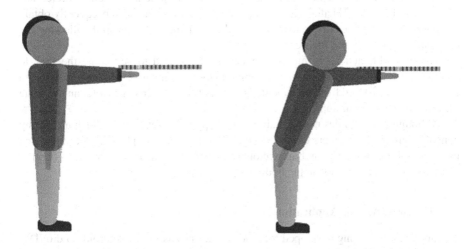

Fig. 1. Example of the conventional FRT balance test. On the left is the starting position for the procedure, and on the right is the final position without losing balance.

The script for informed consent was validated by the Universidade da Beira Interior Ethics Committee, which approved the study with code number CE-UBI-Pj-2020-035. The development of this guide, the processing and availability of data collected throughout the project comply with the European Commission's guide for the ethical processing of personal data.

2.2 Proposed Approach

As stated previously, a variety of technologies, including different types of sensors, have been reported, including, more recently, smart wearables to determine the FRT [5, 15–18].

We propose to evolve and integrate into the conventional FRT test using a traditional smartphone and other smart wearables. To that objective, an application to be run on a common device such as a smartphone, tablet, or similar is proposed, which will allow the data capture in an easy, non-invasive way, being able to aid healthcare professionals in a fast, secure, and reliable manner.

The proposed methodology, as depicted in Fig. 2, consists in recording the result of the FRT through the conventional manual method and recording the data from smart wearables obtained during the same test, using for this purpose an Android-developed application (Android App), described in detail in the Subsect. 2.3.

The data collected through the application results from placing a mobile device on the extremity of one of the persons in test upper limbs when performing the manual method. The instrument will be a smartphone on the individual's hand or a smart band on the wrist. Note that this method can be applied to other devices, such as smartwatches or smart rings. The collected data will be processed to obtain the position of the inertial sensors of the mobile devices, which are usually composed of an inertial measurement unit (IMU). In this study, we propose the implementation of an Attitude and Heading Reference System (AHRS) [20, 21] algorithm to estimate the orientation and subsequently obtain the position coordinates of the extremity of the participant's upper limb, allowing the determination of the FRT metric.

The data obtained using mobile devices are then used to determine the FRT to identify individuals with a high risk of falling. In the preliminary tests, the data will data acquired and registered from both methods to be able to in later stages infer and validate the effectiveness of our approach.

Although other studies use wearables, most apply external sensors such as gaming console commands or others custom designed for this purpose [19, 22]. Our approach uses smartphones and/or wearables without needing additional devices, not vests or other elements, to fix the sensors on the body.

2.3 Proposed Mobile Application

The application is being developed for Android devices using the Android Studio IDE, version Dolphin (2021.3.1), with the following requirements:

- Programming language: Java;
- Minimum SDK: Android 5.0 (Lollipop);
- Communication with other mobile devices: Bluetooth;
- Sensors to be collected: accelerometer, gyroscope, and magnetometer;
- Data storage through Google LLC's Firebase services.

The development of the application considers three distinct phases:

- Collection and storage of data from inertial sensors of smartphones and devices paired with them;
- Elaboration of the data structure, namely regarding the differentiation of collection centers (data collection sites), individuals/patients, and evaluations/tests;
- Implementation and integration of methods that make it possible to obtain test results immediately, with registration in the application.

On first access to the application, the user must register (Sign Up) to be able to connect (Log in) and disconnect (Logout), as shown in Fig. 3. Figure 4 shows the menu of the mobile application. Once in the application, the user can add a collection center (which may, e.g., correspond to a Nursing Home). Each collection center can be associated with several individuals, who, in turn, can have several collections/assessments related to them, as demonstrated in Fig. 5. In addition, the application allows the registration of

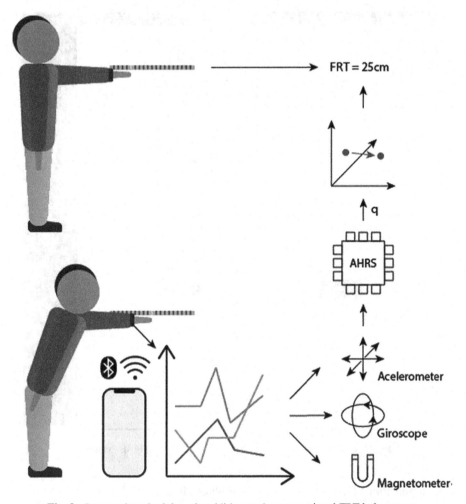

Fig. 2. Proposed methodology in addition to the conventional FRT balance test.

the critical aspects of the individuals in the analysis and evaluation of the data obtained (age, clinical condition, among others).

The evaluation begins with selecting the test to be performed, which is only possible by choosing the FRT. Even so, this test selection page allows the quick expandability of the application to other tests. The evaluation page includes the start, stop, and validation buttons, which would enable, respectively, to start and stop the test and validate its performance so that the data can be stored, as presented in Fig. 6.

The processing of the data collected for the acquisition of the appropriate metrics to analyze then goes through the following stages:

- Obtaining the position of the sensors relative to a spatial referential (XYZ) by implementing an algorithm for orientation estimation, namely an Attitude and Heading Reference System (AHRS) algorithm [20, 21];

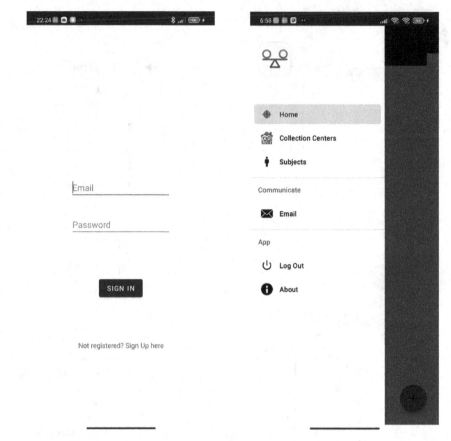

Fig. 3. Login Screen.   **Fig. 4.** Menu.

- Determination of distance metrics as a result of FRT;
- Developing an algorithm for binary classification of a high risk of falling in the elderly population.

The initial tests for capturing motion from the IMU are at the final developing stage, and data capture in a real environment is planned.

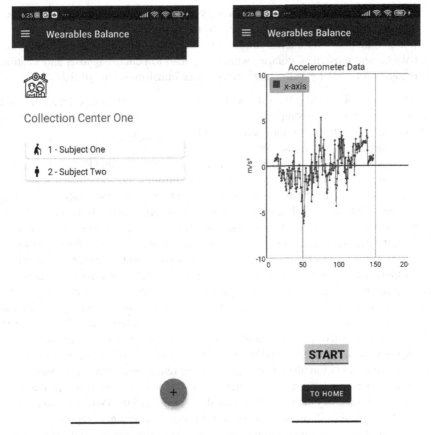

Fig. 5. Collection Center Screen.   **Fig. 6.** Assessment Screen.

3 Expected Results

The Functional Reach Test (FRT) is a joint clinical assessment to evaluate balance and fall risk in older adults or individuals with balance impairments. Wearable technology, such as accelerometers and gyroscopes, can measure body movements during the FRT, providing objective and detailed information about the test performance.

Some of the expected results of implementing the FRT with wearables may include the following:

- Improved accuracy: Wearable technology can provide accurate measurements of body movements, reducing the potential for human error in manually recording the results of the test;
- Objective data: Wearables can provide accurate data on body movements during the FRT, allowing for a more comprehensive analysis of balance and movement control;
- Detailed analysis: Wearable technology can provide detailed information about specific movements during the FRT, such as the speed and range of motion, which may not be detectable by the naked eye;

- Increased efficiency: Wearable technology can help to streamline the FRT process, allowing for quicker and more efficient testing;
- Better diagnosis: Wearables can provide more accurate and detailed data about balance and movement control, which may lead to better diagnosis and treatment recommendations for individuals with balance impairments or fall risk.

Overall, implementing the FRT with wearable technology can enhance the test's accuracy, efficiency, and comprehensiveness, leading to better diagnosis, treatment, and fall prevention strategies for individuals with balance impairments.

In future work, several actions are proposed. First, regarding data collection, two collections are planned from elderly individuals (users of a Nursing Home). In the initial phase, user data will be collected (demographic information and clinical status), and the first assessment of the FRT will be carried out (manually and with the aid of the application). The second data collection will occur approximately 60 days after the initial collection. The aim is thus to compare the two evaluations spaced out in time to analyze the evolution of the functional balance of the users and the effectiveness of the exercises performed by the users of the Nursing Home with the help of the therapists.

Although the initial developments efforts for data acquisition, future developments involve implementing methods for determining the position of the sensors and extracting the necessary metrics. In the data analysis, the development of a binary classification method is foreseen for identifying a high risk of falls in elderly individuals. After this process, it is intended to make available these data to the scientific community.

Algorithms for orientation estimation have been developed in the field of air navigation, positioning of industrial robotic arms, and real-time monitoring of human body movement. In recent years, the use of inertial and magnetic sensors in orientation estimation has been driven by the reduced costs and dimensions of the IMU. Current proposals for AHRS algorithms have also applied quaternions to represent orientation rather than Euler angles and associated rotation matrices. The advantage of using quaternions is that they allow the representation of rotations in $\mathfrak{R}^4$ instead of $\mathfrak{R}^3$, eliminating the problem of the existence of singularities. The AHRS algorithms that have been proposed consider the use of complementary filters (FC) and gradient descent (GD) algorithms. In both cases, gyroscope data are used to determine orientation, while accelerometer and magnetometer data allow for the reduction of the error associated with gyroscope drift. AHRS algorithms based on the Extended Kalman Filter (EKF) have been proposed using quaternions derived from FC and GD algorithms to reduce system errors and the impact of noise associated with sensors. In this work, it is proposed to implement an AHRS algorithm based on an EKF that uses quaternions derived from a DG algorithm to determine the orientation of the sensors. The position of the extremity of the participant's upper limb results from the rotation of the initial position by mapping the EKF quaternions.

Additionally, all the data gathered from the various collections will be made available to the scientific community to aid further developments in this expertise field.

4   Discussion and Conclusions

This study proposes a low-cost method for acquiring movement data associated with balance tests to help caregivers or health professionals identify people at increased risk of falling. Although it is at a preliminary stage, it already presents encouraging developments based on the state-of-the-art studies that led to the implementation of laboratory tests.

The results of the FRT with wearables can provide valuable insights into an individual's balance and postural stability. For example, wearable devices can measure various parameters during the test, such as the range of motion, joint angles, and muscle activity, which can identify potential areas of weakness or imbalance. These insights can be used to develop personalized interventions or exercises to improve balance and reduce the risk of falls.

One potential benefit of using wearables during the FRT is obtaining objective balance and postural stability measurements. Traditional FRT measurements rely on the subjective judgment of the tester, which can be influenced by factors such as fatigue or bias. Wearables provide an objective measure of balance and postural stability, which can help track progress over time and determine the effectiveness of interventions.

Another benefit of using wearables during the FRT is the ability to collect data in real-time. This can be particularly useful for identifying potential risk factors for falls or balance issues during the test, such as excessive body sway or inadequate weight shifting. Real-time data can also provide immediate feedback to the individual, which can help improve performance and reduce the risk of falls.

In conclusion, wearing wearables during the FRT can provide valuable insights into an individual's balance and postural stability. In addition, objective measurements and real-time data from wearables can be used to identify potential areas of weakness or imbalance, develop personalized interventions, and track progress over time. Overall, the combination of wearables and the FRT can provide a comprehensive assessment of balance and postural stability in individuals.

Future work will collect and process data from at least the set of data acquisitions described. After this phase, it will be necessary to extract metrics and analyze the collected data to infer relevant characteristics that aid healthcare. Finally, the goal will be to make it available, adequately treated, and anonymized to the scientific community for further studies.

Acknowledgments. This work is funded by FCT/MEC through national funds and co-funded by FEDER – PT2020 partnership agreement under the project **UIDB/00308/2020**.

This work is funded by FCT/MEC through national funds and co-funded by FEDER–PT2020 partnership agreement under the project **UIDB/50008/2020**.

This article is based upon work from COST Action CA18119 (Who Cares in Europe?), COST Action CA21118 Platform Work Inclusion Living Lab (P-WILL), and COST Action CA19101 Determinants of Physical Activities in Settings (DE-PASS). More information on www.cost.eu.

References

1. Gaspar, A.G.M., Lapão, L.V.: eHealth for addressing balance disorders in the elderly: systematic review. J. Med. Internet Res. **23**(4), e22215 (2021). https://doi.org/10.2196/22215
2. Cuevas-Trisan, R.: Balance problems and fall risks in the elderly. Clin. Geriatr. Med. **35**(2), 173–183 (2019). https://doi.org/10.1016/j.cger.2019.01.008
3. Park, J.-H., Kang, Y.-J., Horak, F.B.: What is wrong with balance in Parkinson's disease? JMD **8**(3), 109–114 (2015). https://doi.org/10.14802/jmd.15018
4. Norbye, A.D., Midgard, R., Thrane, G.: Spasticity, gait, and balance in patients with multiple sclerosis: a cross-sectional study. Physiother. Res. Int. **25**(1), e1799 (2020). https://doi.org/10.1002/pri.1799
5. Gheitasi, M., Bayattork, M., Andersen, L.L., Imani, S., Daneshfar, A.: Effect of twelve weeks pilates training on functional balance of male patients with multiple sclerosis: randomized controlled trial. J. Bodyw. Mov. Ther. **25**, 41–45 (2021). https://doi.org/10.1016/j.jbmt.2020.11.003
6. Duncan, P.W., Weiner, D.K., Chandler, J., Studenski, S.: Functional reach: a new clinical measure of balance. J. Gerontol. **45**(6), M192–M197 (1990). https://doi.org/10.1093/geronj/45.6.M192
7. Hannan, A., Shafiq, M.Z., Hussain, F., Pires, I.M.: A portable smart fitness suite for real-time exercise monitoring and posture correction. Sensors **21**(19), 6692 (2021). https://doi.org/10.3390/s21196692
8. Soto-Varela, A., et al.: Modified timed up and go test for tendency to fall and balance assessment in elderly patients with gait instability. Front. Neurol. **11**, 543 (2020). https://doi.org/10.3389/fneur.2020.00543
9. Kear, B.M., Guck, T.P., McGaha, A.L.: Timed up and go (TUG) test: normative reference values for ages 20 to 59 years and relationships with physical and mental health risk factors. J. Prim Care Commun. Health **8**(1), 9–13 (2017). https://doi.org/10.1177/2150131916659282
10. Blum, L., Korner-Bitensky, N.: Usefulness of the berg balance scale in stroke rehabilitation: a systematic review. Phys. Ther. **88**(5), 559–566 (2008). https://doi.org/10.2522/ptj.20070205
11. Halvarsson, A., Dohrn, I.-M., Ståhle, A.: Taking balance training for older adults one step further: the rationale for and a description of a proven balance training programme. Clin. Rehabil. **29**(5), 417–425 (2015). https://doi.org/10.1177/0269215514546770
12. Papalia, G.F., et al.: The effects of physical exercise on balance and prevention of falls in older people: a systematic review and meta-analysis. JCM **9**(8), 2595 (2020). https://doi.org/10.3390/jcm9082595
13. Bjerk, M., Brovold, T., Skelton, D.A., Bergland, A.: A falls prevention programme to improve quality of life, physical function and falls efficacy in older people receiving home help services: study protocol for a randomised controlled trial. BMC Health Serv. Res. **17**(1), 559 (2017). https://doi.org/10.1186/s12913-017-2516-5
14. Oniani, S., Pires, I.M., Garcia, N.M., Mosashvili, I., Pombo, N.: A review of frameworks on continuous data acquisition for e-Health and m-Health. In: Proceedings of the 5th EAI International Conference on Smart Objects and Technologies for Social Good, pp. 231–234. ACM. Valencia Spain (2019). https://doi.org/10.1145/3342428.3342702
15. Springer, S., Friedman, I., Ohry, A.: Thoracopelvic assisted movement training to improve gait and balance in elderly at risk of falling: a case series. CIA **13**, 1143–1149 (2018). https://doi.org/10.2147/CIA.S166956
16. Adıguzel, H., Elbasan, B.: Effects of modified pilates on trunk, postural control, gait and balance in children with cerebral palsy: a single-blinded randomized controlled study. Acta Neurol. Belg, **122**(4), 903–914 (2022). https://doi.org/10.1007/s13760-021-01845-5

17. de Paula Gomes, C.A.F., Dibai-Filho, A.V., Biasotto-Gonzalez, D.A., Politti, F., de Carvalho, P.D.T.C.: Association of pain catastrophizing with static balance, mobility, or functional capacity in patients with knee osteoarthritis: a blind cross-sectional study.J. Manipulative Physiol. Ther. **41**(1), 42–46 (2018). https://doi.org/10.1016/j.jmpt.2017.08.002

18. Pires, I.M., Garcia, N.M., Zdravevski, E.: Measurement of results of functional reach test with sensors: a systematic review. Electronics **9**(7), 1078 (2020). https://doi.org/10.3390/ele ctronics9071078

19. Williams, B., et al.: Real-time fall risk assessment using functional reach test. Int. J. Telemed. Appl. **2017**, 1–8 (2017). https://doi.org/10.1155/2017/2042974

20. Munguía, R., Grau, A.: A practical method for implementing an attitude and heading reference system. Int. J. Adv. Rob. Syst. **11**(4), 62 (2014). https://doi.org/10.5772/58463

21. Wang, L., Zhang, Z., Sun, P.: Quaternion-based Kalman filter for AHRS using an adaptive-step gradient descent algorithm. Int. J. Adv. Rob. Syst. **12**(9), 131 (2015). https://doi.org/10.5772/61313

22. Allen, B., et al.: Evaluation of fall risk for post-stroke patients using bluetooth low-energy wireless sensor. In: 2013 IEEE Global Communications Conference (GLOBECOM). IEEE, pp. 2598–2603 Atlanta (2013). https://doi.org/10.1109/GLOCOM.2013.6831466

Radar Sensing in Healthcare: Challenges and Achievements in Human Activity Classification & Vital Signs Monitoring

Francesco Fioranelli[✉], Ronny G. Guendel, Nicolas C. Kruse, and Alexander Yarovoy

Microwave Sensing Signals & Systems (MS3) Group,
Department of Microelectronics, TU Delft, Delft, The Netherlands
f.fioranelli@tudelft.nl
https://radar.tudelft.nl/

Abstract. Driven by its contactless sensing capabilities and the lack of optical images being recorded, radar technology has been recently investigated in the context of human healthcare. This includes a broad range of applications, such as human activity classification, fall detection, gait and mobility analysis, and monitoring of vital signs such as respiration and heartbeat. In this paper, a review of notable achievements in these areas and open research challenges is provided, showing the potential of radar sensing for human healthcare and assisted living.

Keywords: Radar sensing · radar signal processing · machine learning · human activity classification · vital signs monitoring

1 Introduction

Radar is a technology conventionally associated to applications in the context of defence and security to monitor objects located at rather long distances. More recently, radar sensors are broadly used in autonomous vehicles for perception of multiple objects at distances of a few tens of meters, leveraging on their robustness to low-visibility weather and light conditions, if compared to cameras and lidar [1,2].

Besides these more conventional usages of radar, there is an increasing body of studies in the literature demonstrating the potential of radar sensing technologies in the context of human healthcare [3–6]. More specifically, these studies can be broadly categorized into three macro-areas:

- **Monitoring of vital signs**, typically including respiration rate and heartbeat. The principle behind this application is the capability of radar to detect the small physiological movements of the thorax and abdomen due to the respiration cycle and the beating of the heart [7–10]. More recently, research also investigates the usage of radar for other vital signs, such as the arterial pulse wave due to circulation in blood vessels [11], and the combination/fusion of multiple vital indicators to assess higher-level physiological functions, such as sleep stages and quality [12,13].

© The Author(s), under exclusive license to Springer Nature Switzerland AG 2023
I. Rojas et al. (Eds.): IWBBIO 2023, LNBI 13920, pp. 492–504, 2023.
https://doi.org/10.1007/978-3-031-34960-7_35

- **Human activity recognition** (HAR), which can include the prompt detection of critical events such as falls, but also the more general pattern of usual activities an individual performs in their home environment. This can help assess the general fitness level and everyday physical activities, as well as the presence of outliers or anomalies that, if repeated, may be signs of physical and/or cognitive decline. In literature, HAR includes the monitoring and classification of movements and activities such as walking, sitting, standing, bending, crouching, carrying objects, and (quasi-)static postures while standing, sitting, or sleeping [14–21]. These basic motions can also be considered in combination to form higher-level activities, such as food preparation, cleaning, or personal hygiene, which constitute the expected daily activities of healthy subjects living independently at home.
- **Gait analysis and gait parameter extraction**, which can include the identification of impaired vs normal gait with radar sensing, such as episodes of frozen gait, limping, dragging feet, as well as the quantification from the radar signatures of gait-related parameters that would otherwise require a visit to a specialized gait laboratory [22–25]. The emphasis on gait analysis is related to the fact that more irregular and slower gait patterns may indicate a worsening in the individual's physical and cognitive health.

The aforementioned applications share the approach to use radar sensors and radar data to monitor physiological quantities and processes that can help form an overall 'health picture' of subjects, and where needed spot critical events and/or anomalies. The assumption is that radar has the potential to be well-embedded in home environments, with the objective to unlock forms of proactive care and diagnostics that alert medical professional of possible problems before the subjects themselves succumb to illness, become bedridden, and may then require costly and invasive hospitalization.

One may ask: why radar as a sensor for healthcare applications and not others? The potential advantage of radar sensing comes from its non-intrusive, contactless capability to measure small and large movements and postures of subjects [3,26,27]. Hence, individuals do not need to carry, wear, or interact with electronic devices, which can provide an advantage for users' compliance, especially for those affected by cognitive impairments. Additionally, it is important to highlight that radar sensing does not generate optical images or videos of subjects and their private environments at home; this can provide an advantage in terms of perceived privacy infringement and help users' acceptance.

It is important to highlight that the word 'radar' in the context of the aforementioned healthcare applications includes in reality many different types of hardware architectures and operational processing parameters. An in-depth discussion on these aspects goes beyond the scope of this paper and the readers are referred to review papers such as [3,5,15], amongst others.

In the remainder of this article, a brief overview of the basic principles of radar signal processing in healthcare applications is provided in Sect. 2, with representative results and outstanding research challenges discussed in Sect. 3. While the discussion in this section remains at a general high level, additional technical details for each challenge are presented in the provided references. Finally, a brief conclusion is drawn in Sect. 4.

2 Principles of Radar Signal Processing in Healthcare

The working principle of any radar system is based on transmitting and receiving sequences of electromagnetic waves, whose amplitude and/or frequency is modulated according to suitable waveform patterns. These waveforms are transmitted into the environment, propagate in open-air, and are reflected back to the radar by objects present in the area under test. By digitizing these received backscattered waveforms, one can extract information on the objects of interest. For a parallel in terms of working principles, one can make a comparison with echolocation in animals, such as bats or dolphins navigating in their living environments and hunting for prey. However, they utilize acoustic waves (i.e., ultrasound) rather than electromagnetic waves as it is the case for man-made radar systems.

The basic architecture of a radar system comprises of a transmitter and receiver block to generate, condition, and receive electromagnetic waves, antennas that act as the transducers to/from open-air propagation, and a digital block to process and store the radar data with suitable signal processing techniques, briefly outlined later in this section. While a detailed discussion on radar architectures goes beyond the scope of this paper, two main families of radar architectures have been mainly used in the literature for healthcare applications [28]. One includes the so-called CW (Continuous Wave) radars, with simpler architectures and processing capable of measuring the velocity of moving objects (e.g., the bulk walking velocity of a person) but not their location. The other family includes radars whose waveform is modulated in frequency to occupy a certain band (either pulsed Ultra Wide Band, UWB, radars, or Frequency Modulated Continuous Wave, FMCW, radars); the notable implication is that with these radars the distance of objects can also be measured together with their velocity. Recently, radars with multiple transmitters/receivers (MIMO, multiple input multiple output radars) are used to also estimate the angular position of objects of interest in azimuth and/or elevation.

For all the applications mentioned in Sect. 1, the detection in space and the characterisation over time of the movements of body parts of the subjects are crucial. This includes the very minute movements of chest and internal organs for vital signs monitoring, as well as larger-scale movements of limbs and the whole body for activity classification and fall detection. Hence, the typical signal processing operations on radar data for healthcare aim to characterise the subject's posture and movements in three domains: *range*, as the physical distance at which the person and their body parts are located with respect to the radar; *time*, intended as the evolution of the position of the person or their body parts; and *velocity*, as the speed and regularity at which such changes over time happen. As mentioned before, with modern MIMO radars the *angular* position can also be estimated, which can be very important to separate different subjects present together in the field of view of the radar, or even differentiate body parts of a single subject.

It is important to highlight that radar systems measure velocities via the Doppler effect, i.e. the change/shift in the frequency of the received radar waveforms induced by the movement of objects. As a simple example, we can think of a person sitting and facing a radar while breathing, hence with chest moving back and forth during the respiration cycle. This will induce a positive Doppler shift when the chest is moving towards the radar as more electromagnetic wavefronts will be scattered back to the radar in a given time unit; on the contrary, the Doppler shift will be negative when the chest is moving away from the radar. Therefore, measuring velocity with radar implies measuring the frequency components/modulations of the received radar waveforms. This can be accomplished using Fourier analysis, as done in many other fields of engineering (e.g., acoustics, mechanical vibrations), and specifically using the Fast Fourier Transform algorithm (FFT).

Figure 1 shows a simplified signal processing chain for an example of data where a person was walking back and forth in front of a radar system.

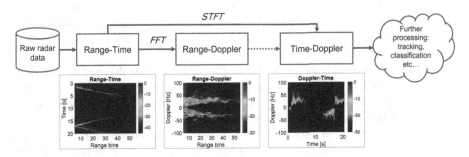

Fig. 1. Simplified signal processing chain for radar data, with examples of Range-Time, Range-Doppler, and Doppler-Time patterns for a person walking back and forth in front of a radar operating in the X-frequency band.

The processing chain starts from raw radar data, i.e., the temporal sequence of the digitized samples of the received radar waveforms. Such data are normally structured into a matrix, in which each radar waveform will include range bins, i.e., digitized samples along one dimension of the matrix that are related to the physical distance of possible targets; the sequence of waveforms along the other dimension of the matrix can be associated to the concept of physical time. This matrix is often referred to as *Range-Time-Intensity (RTI)* matrix. As the person was walking back and forth in front of the radar, a diagonal zig-zag pattern is visible in the RTI image of Fig. 1, with the 'signature' of the person moving first away from the radar (i.e., the range bin index increases over time), and then back towards the radar (with the range bin indexes decreasing over time). As discussed, an FFT operation is used to estimate the velocity of objects of interest via the Doppler effect. By applying this FFT across the sequence of radar waveforms, i.e., the time dimension of the RTI matrix, a new matrix is generated, the *Range-Doppler (RD)* matrix. An example of RD is the middle picture of Fig. 1, with both positive and negative Doppler contributions due to the person walking towards the radar (positive Doppler) and away (negative Doppler) in the considered unit of time. The RD matrix characterizes the overall velocity components present across the (relatively) long observation time of the whole measurement, but it is difficult to map the precise time when each movement happened. For this, a different processing approach is needed, the Short Time Fourier Transform (STFT). This approach applies multiple FFT operations on the data across shorter, possibly overlapped time windows. Each FFT produces one vector with the dimensions of Doppler/velocity, and by placing such vectors one next to each other in their temporal order, a Doppler vs time 2D matrix is generated, the so-called 'spectrogram'. In Fig. 1, positive and negative Doppler contributions are visible in the pattern of the spectrogram. Each contribution has a central, more intense signature denoted by red and yellow colour due to the movement of the torso and main body, with additional less intense streaks around the main signature denoted by light blue colour and physically attributable to the limbs. This is a typical pattern for a person walking, with the bulk movement contribution due to the torso and main body moving, and additional oscillating movements of the arms and legs. As these movements from the limbs produce smaller, additional Doppler frequencies around the main bulk Doppler, they are called 'micro-Doppler' in the radar literature [29, 30].

The importance of micro-Doppler in the context of healthcare applications comes from the fact that each specific activity, movement, or type of gait will exhibit its specific micro-Doppler pattern or signature. Thus, to achieve human activity classification it is possible to use machine learning techniques to 'teach' algorithms to recognize these patterns as well as to utilize this information to support tracking multiple people in the scene. A selection of representative results is reported in the following section.

3   Representative Results and Open Challenges

Representative results and open challenges from the literature in the context of healthcare applications of radar-based sensing are discussed in this section.

1. *What is the most suitable representation and format of the radar data to characterize or infer healthcare-related information?* Conventionally, most studies use Doppler-time representations generated by the STFT, as explained in the previous section. This allows to study the temporal patterns of Doppler modulations due to the different body parts moving [14, 29]. As Doppler modulations are directly related to the velocity of the body parts, algorithms operating on this data representation can characterize human movements, from the small ones related to vital signs, to the larger movements due to gait and/or complex activities. These micro-Doppler patterns can then be exploited as input to classification pipelines based on machine learning to

learn the relevant information, including neural networks with various architectures, or more conventional classifiers such as Support Vector Machines (SVM) [5,31].

A couple of notable examples of micro-Doppler signatures are shown in this section. Figure 2 shows the spectrograms of six human activities recorded with a radar operating at 5.8 GHz, the same frequency as Wi-Fi networks commonly used indoors. The six activities included sitting on a chair, standing up from a chair, bending to tie shoelaces, bending to pick up a pen, crouching to the floor and standing back up, and finally a simulated frontal fall. One can see that each activity has a distinct pattern of positive/negative Doppler components over time, that can be potentially learnt by algorithms for automatic classification.

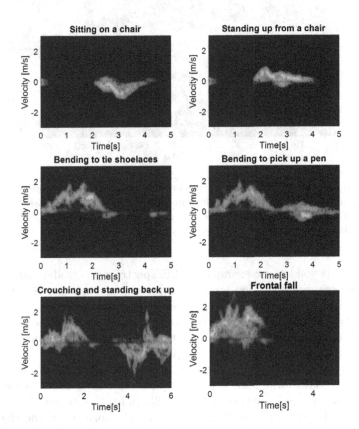

Fig. 2. Example of six Velocity-Time (spectrogram) patterns for six human activities performed by the same volunteer and recorded by a 5.8 GHz radar. It should be noted that these activities were performed in a simplified manner in isolation, with the volunteer starting and ending the movement in a stationary posture.

Figure 3 shows another example of a spectrogram for vital signs monitoring, specifically respiration. In this case, a volunteer was sitting in front of a UWB radar operating in the X-frequency band and performing different respiration patterns on purpose, namely normal breathing, holding breath, deep exhalation, and simulated fast breathing. The different breathing patterns can be easily recognized 'by eye' in the spectrogram, and its envelope can be used to estimate the breathing rate in normal, semi-stationary conditions, as well as the presence of anomalous or irregular patterns.

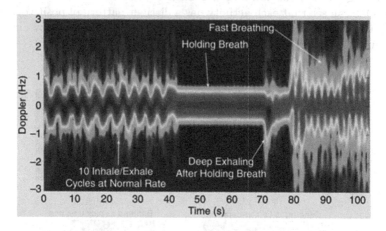

Fig. 3. Spectrogram for a person sitting on a chair at about 60 cm from the radar and simulating different respiration patterns. Specifically, the subject was asked to perform 10 cycles of normal respiration followed by a long period of holding breath and then fast breathing. The radar used for this test worked in the X-frequency band.

However, beyond these examples of 2D spectrograms, radar data can take very diverse representation formats combining into a tensor the dimensions of range or distance, Doppler or velocity, angular information in azimuth and/or elevation, and time. Furthermore, radar data are typically complex-valued, unlike optical images. Hence, they can be represented with their real & imaginary values or magnitude & phase [32–35].

An outstanding research question remains the identification of the most suitable radar data format for a given application and a desired classification algorithm. This also includes the possibility of fusing different data representations and their salient features to maximize performances, and the robustness of any proposed approach and choice to the diversity of environments and individual movement patterns.

2. *How to analyze radar data for HAR that present continuous and diverse sequences of activities?* Studies in radar-based HAR began with the analysis of separated motions and actions performed and collected in isolation, as if they were sort of 'snapshots', whereas in reality human activities are performed continuously. They appear as a seamless sequence, with duration and

transitions between the different motions not being predefined. Furthermore, such sequences are diverse, in the sense that they may include a mixture of full-body actions (e.g., walking, vacuum cleaning the room), actions involving movements of limbs while remaining on the same location (e.g., sitting, picking up an object from the floor), and broadly stationary intervals with small or no movements (e.g., reading a book while sitting, watching TV). These sequences are also highly diverse depending on the subject's gender, age, physical condition, and environmental constraints from surrounding furniture and objects [36–38]. This undoubtedly adds additional complexity in the task of learning the salient information related to the activities being performed.

As a visual example of such diversity, Fig. 4 shows six velocity/time patterns (spectrograms) for the same continuous sequence of five activities performed by six different subjects. Notable aspects are: i) how diverse the signatures appear for the different subjects, despite the sequences being nominally labelled as exactly the same activity-wise; this would set an interesting classification challenge to train automatic algorithms with these diverse data with the same nominal set of labels; ii) how challenging it can be to clearly identify the transitions in order to segment the different activities, i.e., where one stops and a new one starts [39].

For this reason, the formulation of approaches for segmenting, interpreting, and classifying these continuous sequences of human activities that may also include data portions for vital signs extraction and gait parameters analysis is an open research challenge.

3. *How to effectively and wisely use deep learning techniques for radar-based healthcare applications?* Similar to plenty of other disciplines in science and engineering, deep learning approaches are increasingly used for radar data, including for healthcare applications discussed in this paper [5,31].

As a typical example of this trend, neural network architectures have been proven as an effective tool for the classification of patterns in radar signatures for HAR. These architectures include Convolutional Neural Networks (CNNs) and recurrent networks such as Long-Short Term Memory (LSTM) and Gated Recurrent Units (GRUs), as well as their combinations, to classify bi-dimensional and temporal patterns in the radar data. Additional architectural choices include Auto-Encoders (AEs) to perform unsupervised feature extraction; Generative Adversarial Networks (GANs) to generate synthetic radar data to complement experimental datasets; and attention-based Transformer architectures [5,40–45]. However, while deep learning techniques have clear advantages in their ability to learn patterns and features that cannot be easily captured with conventional approaches, their usage also poses practical challenges to address. First, the need to have datasets that are large (in order to train deep architectures with many hyperparameters), well labelled (in order to use supervised learning approaches), and representative (in order to capture enough diverse subjects and environments with their specific multipath phenomena and clutter signatures, so that the classification networks can generalise well to new people and situations).

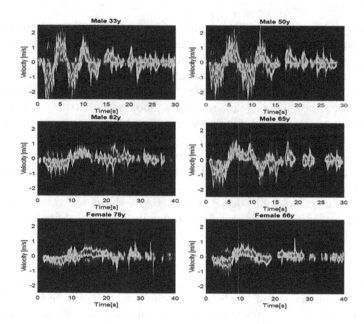

Fig. 4. Radar spectrograms of the same sequence of five daily activities (namely walking back and forth, sitting on a chair, standing up, bending down and coming back up, and drinking from a cup) performed by 6 subjects of different age, gender, physical conditions. The radar operated at 5.8 GHz carrier frequency.

To the best of our knowledge, a dataset with the aforementioned desired characteristics that is publicly shared and accepted by the radar research community as 'the benchmark dataset' does not exist at the moment. However, some steps have been taken to address this challenge of data scarcity by increasingly sharing datasets, as in the several examples reported in [46], or the relatively large dataset our research groups shared in [47,48]. While these activities are still fragmented at the level of the initiative of each individual research group, they still represent a positive sign to help further develop deep learning techniques in this context. With established datasets, one would be able to perform a deep, comprehensive statistical study to compare the different relevant techniques on the same data, scenario, and radar system.

Another important challenge related to the utilization of deep learning techniques in the context of radar-based healthcare comes from the often difficult interpretation and explanation of the decisions made by the neural networks and their rationale. The importance of explainable approaches combined with the aforementioned issue of data scarcity can discourage the usage of deep data-driven end-to-end architectures, even if they have proved very effective in other research domains such as image and audio processing. On the other hand, this challenge can be turned into an opportunity to incorporate physics-based information into the networks. This can leverage both aspects of electromagnetic scattering & propagation, as well as human kinematics, since both are phenomena that can be well characterized a priori with models and equations, which do not need to be learnt from the data since they are well understood from physics and human physiology.

4. *How should a classification algorithm manage the situation of receiving unforeseen data?* This is the so called "open-set" problem [49,50], for which a typical example in the context of radar-based HAR is the occurrence of a movement or activity pattern that was not present in the training data. A closely related aspect to account for, is the fact that many critical, potentially life-threatening activities that a classifier should always recognize (e.g., fall instances) are in a sense part of these unexpected, unforeseen data. This is because they cannot be instigated on purpose to generate representative training data, thus leading to datasets that can be potentially unsuitable to train for robust performances and to deal with the actual critical cases appearing in the test set. The development of robust techniques to deal with these cases remains an open research question.

5. *How can radar be integrated into a wider suite of sensors for healthcare?* For healthcare applications such as those presented in this work, it is expected that the combination of data from multiple sensors located in future smart home environments will yield better performances than using each sensor in isolation. This is because each technology has its own distinctive advantages and disadvantages [26,27], thus a proper synergy in a multimodal fashion can be beneficial. Radar is no exception to this idea, and its acceptance in the perhaps not immediately obvious context of healthcare can be facilitated by its usage together with other sensors that are more familiar to the end-users. An open research challenge for this is the formulation of algorithms for data processing and sensor management to combine information from different, heterogeneous sensors. Incidentally, these approaches can also include networks of multiple radar sensors [24,47], either with similar radars looking at the scene under test from different aspect angles, or with radars operating at different carrier frequencies to 'perceive' complementary characteristics of the objects of interest due to the different probing waveforms.

4 Conclusions

This paper provides a detailed overview of recent developments in the field of radar sensing for healthcare, along with the associated research challenges. Notably, radar technology has allowed contactless monitoring of vital signs such as respiration and heartbeat, as well as the classification of human activity patterns, including fall detection and gait monitoring/analysis through parameter extraction. Nevertheless, research challenges remain in radar signal processing and machine learning including the most modern deep learning techniques for these applications. Specifically, researchers are investigating the most promising approaches for extracting essential information from radar data and formulating effective, robust, and scalable classification algorithms for diverse subjects and activities in indoor environments.

Acknowledgments. The authors thank the many students and collaborators for their contributions to the data collection and processing, especially Dr J. Le Kernec at the University of Glasgow. Financial support from UK EPSRC (grant EP/R041679/1 *INSHEP*), Dutch Research Council NWO (grant NWO KLEIN *RAD-ART*), and Dutch Government Sector Plan is also acknowledged.

References

1. Waldschmidt, C., Hasch, J., Menzel, W.: Automotive radar - from first efforts to future systems. IEEE J. Microwaves **1**(1), 135–148 (2021)
2. Sun, S., Petropulu, A.P., Poor, H.V.: MIMO radar for advanced driver-assistance systems and autonomous driving: advantages and challenges. IEEE Signal Process. Mag. **37**(4), 98–117 (2020)
3. Fioranelli, F., Le Kernec, J., Shah, S.A.: Radar for health care: recognizing human activities and monitoring vital signs. IEEE Potentials **38**(4), 16–23 (2019)
4. Le Kernec, J., et al.: Radar signal processing for sensing in assisted living: the challenges associated with real-time implementation of emerging algorithms. IEEE Signal Process. Mag. **36**(4), 29–41 (2019)
5. Gurbuz, S.Z., Amin, M.G.: Radar-based human-motion recognition with deep learning: promising applications for indoor monitoring. IEEE Signal Process. Mag. **36**(4), 16–28 (2019)
6. Cippitelli, E., Fioranelli, F., Gambi, E., Spinsante, S.: Radar and RGB-depth sensors for fall detection: a review. IEEE Sens. J. **17**(12), 3585–3604 (2017)
7. Li, C., Mak, P.-I., Gómez-García, R., Chen, Y.: Guest editorial wireless sensing circuits and systems for healthcare and biomedical applications. IEEE J. Emerg. Select. Top. Circuits Syst. **8**(2), 161–164 (2018)
8. Iwata, S., Koda, T., Sakamoto, T.: Multiradar data fusion for respiratory measurement of multiple people. IEEE Sens. J. **21**(22), 25870–25879 (2021)
9. Koda, T., Sakamoto, T., Okumura, S., Taki, H.: Noncontact respiratory measurement for multiple people at arbitrary locations using array radar and respiratory-space clustering. IEEE Access **9**, 106895–106906 (2021)
10. Paterniani, G., et al.: Radar-based monitoring of vital signs: a tutorial overview. In: Proceedings of the IEEE, pp. 1–41 (2023)
11. Oyamada, Y., Koshisaka, T., Sakamoto, T.: Experimental demonstration of accurate noncontact measurement of arterial pulse wave displacements using 79-GHZ array radar. IEEE Sens. J. **21**(7), 9128–9137 (2021)
12. Hong, H., et al.: Microwave sensing and sleep: noncontact sleep-monitoring technology with microwave biomedical radar. IEEE Microw. Mag. **20**(8), 18–29 (2019)
13. Kang, S., et al.: Non-contact diagnosis of obstructive sleep apnea using impulse-radio ultra-wideband radar. Sci. Rep. **10**, 5261 (2020)
14. Kim, Y., Ling, H.: Human activity classification based on micro-doppler signatures using a support vector machine. IEEE Trans. Geosci. Remote Sens. **47**(5), 1328–1337 (2009)
15. Amin, M.G., Zhang, Y.D., Ahmad, F., Ho, K.D.: Radar signal processing for elderly fall detection: the future for in-home monitoring. IEEE Signal Process. Mag. **33**(2), 71–80 (2016)
16. Li, X., He, Y., Fioranelli, F., Jing, X.: Semisupervised human activity recognition with radar micro-doppler signatures. IEEE Trans. Geosci. Remote Sens. **60**, 1–12 (2022)

17. Gorji, A., Bourdoux, A., Sahli, H.: On the generalization and reliability of single radar-based human activity recognition. IEEE Access **9**, 85334–85349 (2021)
18. Piriyajitakonkij, M., et al.: SleepPoseNet: multi-view learning for sleep postural transition recognition using UWB. IEEE J. Biomed. Health Inform. **25**(4), 1305–1314 (2021)
19. Zhu, S., Guendel, R.G., Yarovoy, A., Fioranelli, F.: Continuous human activity recognition with distributed radar sensor networks and CNN-RNN architectures. IEEE Trans. Geosci. Remote Sens. **60**, 1–15 (2022)
20. Zhao, Y., Yarovoy, A., Fioranelli, F.: Angle-insensitive human motion and posture recognition based on 4D imaging radar and deep learning classifiers. IEEE Sens. J. **22**(12), 12173–12182 (2022)
21. Kim, Y., Alnujaim, I., Oh, D.: Human activity classification based on point clouds measured by millimeter wave MIMO radar with deep recurrent neural networks. IEEE Sens. J. **21**(12), 13522–13529 (2021)
22. Wang, F., Skubic, M., Rantz, M., Cuddihy, P.E.: Quantitative gait measurement with pulse-doppler radar for passive in-home gait assessment. IEEE Trans. Biomed. Eng. **61**(9), 2434–2443 (2014)
23. Seifert, A.-K., Amin, M.G., Zoubir, A.M.: Toward unobtrusive in-home gait analysis based on radar micro-doppler signatures. IEEE Trans. Biomed. Eng. **66**(9), 2629–2640 (2019)
24. Li, H., Mehul, A., Le Kernec, J., Gurbuz, S.Z., Fioranelli, F.: Sequential human gait classification with distributed radar sensor fusion. IEEE Sens. J. **21**(6), 7590–7603 (2021)
25. Gurbuz, S.Z., Rahman, M.M., Kurtoglu, E., Martelli, D.: Continuous human activity recognition and step-time variability analysis with FMCW radar. In: 2022 IEEE-EMBS International Conference on Biomedical and Health Informatics (BHI), pp. 01–04 (2022)
26. Chaccour, K., Darazi, R., El Hassani, A.H., Andrès, E.: From fall detection to fall prevention: a generic classification of fall-related systems. IEEE Sens. J. **17**(3), 812–822 (2017)
27. Debes, C., Merentitis, A., Sukhanov, S., Niessen, M., Frangiadakis, N., Bauer, A.: Monitoring activities of daily living in smart homes: understanding human behavior. IEEE Signal Process. Mag. **33**(2), 81–94 (2016)
28. Li, C., et al.: A review on recent progress of portable short-range noncontact microwave radar systems. IEEE Trans. Microw. Theory Tech. **65**(5), 1692–1706 (2017)
29. Fioranelli, F., Griffiths, H., Ritchie, M., Balleri, A.: Micro-Doppler Radar and Its Applications. Institution of Engineering and Technology (2020). https://digital-library.theiet.org/content/books/ra/sbra531e
30. Chen, V.: The Micro-Doppler Effect in Radar, 2nd edn. Artech, New Jersey (2019)
31. Gurbuz, S.Z.: Deep Neural Network Design for Radar Applications. Institution of Engineering and Technology (2020). https://digital-library.theiet.org/content/books/ra/sbra529e
32. Guendel, R.G., Fioranelli, F., Yarovoy, A.: Phase-based classification for arm gesture and gross-motor activities using histogram of oriented gradients. IEEE Sens. J. **21**(6), 7918–7927 (2021)
33. Scarnati, T., Lewis, B.: Complex-valued neural networks for synthetic aperture radar image classification. In: 2021 IEEE Radar Conference (RadarConf21), pp. 1–6 (2021)
34. Brooks, D.A., Schwander, O., Barbaresco, F., Schneider, J.Y., Cord, M.: Complex-valued neural networks for fully-temporal micro-doppler classification. In: 2019 20th International Radar Symposium (IRS), pp. 1–10 (2019)

35. Yang, X., Guendel, R.G., Yarovoy, A., Fioranelli, F.: Radar-based human activities classification with complex-valued neural networks. In: 2022 IEEE Radar Conference (RadarConf22), pp. 1–6 (2022)
36. Shrestha, A., Li, H., Le Kernec, J., Fioranelli, F.: Continuous human activity classification from FMCW radar with Bi-LSTM networks. IEEE Sens. J. **20**(22), 13607–13619 (2020)
37. Chen, P., et al.: Multi-view real-time human motion recognition based on ensemble learning. IEEE Sens. J. **21**(18), 20335–20347 (2021)
38. Li, H., Shrestha, A., Heidari, H., Le Kernec, J., Fioranelli, F.: Bi-LSTM network for multimodal continuous human activity recognition and fall detection. IEEE Sens. J. **20**(3), 1191–1201 (2020)
39. Kruse, N., Guendel, R., Fioranelli, F., Yarovoy, A.: Segmentation of micro-doppler signatures of human sequential activities using rényi entropy. In: International Conference on Radar Systems (RADAR 2022), vol. 2022, pp. 435–440 (2022)
40. Seyfioğlu, M.S., Özbayoğlu, A.M., Gürbüz, S.Z.: Deep convolutional autoencoder for radar-based classification of similar aided and unaided human activities. IEEE Trans. Aerosp. Electron. Syst. **54**(4), 1709–1723 (2018)
41. Erol, B., Gurbuz, S.Z., Amin, M.G.: Motion classification using kinematically sifted ACGAN-synthesized radar micro-doppler signatures. IEEE Trans. Aerosp. Electron. Syst. **56**(4), 3197–3213 (2020)
42. Rahman, M.M., Gurbuz, S.Z., Amin, M.G.: Physics-aware generative adversarial networks for radar-based human activity recognition. IEEE Transactions on Aerospace and Electronic Systems, pp. 1–15 (2022)
43. Zheng, L., et al.: Dynamic hand gesture recognition in in-vehicle environment based on FMCW radar and transformer. Sensors, **21**(19), 6368 (2021). https://www.mdpi.com/1424-8220/21/19/6368
44. Chen, S., He, W., Ren, J., Jiang, X.: Attention-based dual-stream vision transformer for radar gait recognition. In: ICASSP 2022–2022 IEEE International Conference on Acoustics, Speech and Signal Processing (ICASSP), pp. 3668–3672 (2022)
45. Guo, Z., Guendel, R., Yarovoy, A., Fioranelli, F.: Point transformer based human activity recognition using high dimensional radar point clouds. In: Accepted for IEEE Radar Conference 2023, San Antonio, TX, USA (2023)
46. Fioranelli, F., Zhu, S., Roldan, I.: Benchmarking classification algorithms for radar-based human activity recognition. IEEE Aerosp. Electron. Syst. Mag. **37**(12), 37–40 (2022)
47. Guendel, R.G., Unterhorst, M., Fioranelli, F., Yarovoy, A.: Dataset of continuous human activities performed in arbitrary directions collected with a distributed radar network of five nodes (2021). https://data.4tu.nl/articles/dataset/Dataset_of_continuous_human_activities_performed_in_arbitrary_directions_collected_with_a_distributed_radar_network_of_five_nodes/16691500
48. Yang, S., et al.: The human activity radar challenge: benchmarking based on the 'radar signatures of human activities' dataset from Glasgow university. IEEE J. Biomed. Health Inform. **27**, 1–13 (2023)
49. Yang, Y., Hou, C., Lang, Y., Guan, D., Huang, D., Xu, J.: Open-set human activity recognition based on micro-doppler signatures. Pattern Recogn. **85**, 60–69 (2019)
50. Bhavanasi, G., Werthen-Brabants, L., Dhaene, T., Couckuyt, I.: Open-set patient activity recognition with radar sensors and deep learning. IEEE Geosci. Remote Sens. Lett. **20**, 1–5 (2023)

Author Index

Printed in the United States
by Baker & Taylor Publisher Services